NUTRITIONAL SUPPLEMENTS IN SPORTS AND EXERCISE

NUTRITIONAL SUPPLEMENTS IN SPORTS AND EXERCISE

Edited by

MIKE GREENWOOD, PhD
Department of Health, Human Performance, and Recreation
Baylor University
Waco, Texas

DOUGLAS S. KALMAN, PhD, RD
Division of Nutrition and Endocrinology
Miami Research Associates
Miami, Florida

and

JOSE ANTONIO, PhD
NOVA Southeastern University
Fort Landerdale, Florida

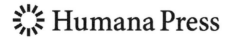 Humana Press

Editors
Mike Greenwood
Department of Health, Human Performance,
and Recreation
Baylor University
Waco, Texas

Douglas S. Kalman
Division of Nutrition and Endocrinology
Miami Research Associates
Miami, Florida

Jose Antonio
NOVA Southeastern University
Fort Landerdale, Florida

ISBN: 978-1-58829-900-0   e-ISBN: 978-1-59745-231-1
DOI: 10.1007/978-1-59745-231-1

Library of Congress Control Number: 2008932055

Cover illustration: Designed by Shirley Karina.

Printed on acid-free paper

9 8 7 6 5 4 3 2 1

springer.com

Contents

Preface .... *vii*

About the Editors and Authors .... *ix*

Part I. Industrial Nature of the Supplement Game

1 Effect of Government Regulation on the Evolution
   of Sports Nutrition *3*
   Rick Collins and Douglas Kalman

2 Psychology of Supplementation in Sport and Exercise:
   Motivational Antecedents and Biobehavioral Outcomes ... *33*
   Rafer Lutz and Shawn Arent

Part II. Nutritional Basics First

3 Role of Nutritional Supplements Complementing Nutrient-
   Dense Diets: *General Versus Sport/Exercise-Specific
   Dietary Guidelines Related to Energy Expenditure* *75*
   Susan Kleiner and Mike Greenwood

4 Macronutrient Intake for Physical Activity *95*
   Thomas Buford

5 Essential and Nonessential Micronutrients and Sport .. *121*
   Kristen M. Beavers and Monica C. Serra

6 Fluid Regulation for Life and Human Performance ... *167*
   Allyn Byars

Part III. Specialized Nutritional Supplements and Strategies

7 Muscle Mass and Weight Gain Nutritional Supplements *189*
   Bill Campbell

8 Weight Loss Nutritional Supplements *225*
   Joan M. Eckerson

9 Effective Nutritional Supplement Combinations *259*
   Matt Cooke and Paul J. Cribb

10 Nutritional Supplements for Strength Power Athletes .   *321*
       Colin Wilborn

11 Nutritional Supplements for Endurance Athletes.....   *369*
       Christopher J. Rasmussen

12 Nutritional Supplements to Enhance Recovery *409*
       Tim N. Ziegenfuss, Jamie Landis, and Mike Greenwood

13 Nutritional Supplementation and Meal Timing *451*
       Jim Farris

Part IV. Present and Future Directions of Nutritional Supplements

14 Future Trends: *Nutritional Supplements in Sports
       and Exercise*...................................   *491*
       Marie Spano and Jose Antonio

Index *509*

Preface

Over the past two decades, the area of sport nutrition and nutritional supplementation has escalated in monumental proportions. An enormous number of qualified professionals—sport nutritionists, athletic coaches, athletic trainers, sports medicine personal, strength and conditioning coaches, personal trainers, medical representatives, and health practitioners, among others—as well as a variety of athletic and exercise participants have searched for viable dietary ergogenic aids to attain optimal training and performance levels. A multitude of universities have infused relevant sport nutrition courses, academic majors, and critical research agendas into their curriculums to further investigate this popular and dynamic aspect of our society. Professional organizations such as The International Society of Sport Nutrition have evolved to place a scientifically based approach to help further understand this billion dollar a year industry. What an extraordinary challenge this undertaking has been due to the multitude of nutritional supplements that are currently on the market and the plethora of these products that surface on a regular basis.

The editors and the authors noted in this textbook firmly believe that the public has the right to know the truth regarding nutritional supplement ingestion—health, safety, efficacy—and quality based scientific research is the accepted approach supported by those contributing to this published endeavor. However, it should be noted that even highly acclaimed researchers are not always in agreement regarding specific scientific findings, but such is the nature of research in any realm. Based on this reality, there is a critical need for professionals to bridge the gap between scientific results and common sense approaches related to practical sports and exercise nutritional supplement strategies.

A major purpose of this book is to provide detailed analysis of nutritional supplementation supported by, whenever possible, replicated scientific research regarding sports and exercise performance. The book is divided into four sections to accomplish this goal. Section 1 delves into the industrial component as well as the psychological nature of the consumer-based nutritional supplement game. In Section 2, strong emphasis is placed on nutrient-dense food/fluid ingestion basics. Section 3 provides information regarding specialized nutritional supplements and strategies, and Section 4 addresses the present and future status of nutritional supplements in sports and exercise environments. This book provides the readers with a viable up-to-date reference guide, keeping in mind that publication time frames limit the inclusion of current research outcomes. We have attempted to include revelant nutritional supplement information that is timely and useful in an ever-evolving industry.

Mike Greenwood
Douglas Kalman
Jose Antonio

About the Editors and Contributors

EDITORS

MIKE GREENWOOD PhD, FISSN, FNSCA, FACSM, CSCS*D
Mike Greenwood, PhD, is currently a Professor in the Department of Health, Human Performance, and Recreation at Baylor University. At Baylor he serves as the HHPR Graduate Director for Exercise Physiology/Strength and Conditioning as well the Research Coordinator primarily involved with the Center for Exercise, Nutrition, and Preventive Health and the Exercise and Sport Nutrition Laboratory [http://www3.baylor.edu/HHPR/]. Dr. Greewood is a Fellow of the International Society of Sport Nutrition (ISSN), National Strength and Conditioning Association (NSCA), and American College of Sport Medicine (ACSM). He is certified as a strength and conditioning specialist recognized with distinction by the NSCA and has previously served as a NCAA collegiate baseball/basketball coach as well as a strength training/conditioning

professional. He is currently serving on the Advisory Board of the ISSN and the NSCA Certification Commission Executive Council. His primary lines of research are in the area of sport/exercise nutrition and strength and conditioning. Dr. Greenwood serves on the Editorial Boards of the *Strength and Conditioning Journal* and *Journal of the International Society of Sports Nutrition* and has reviewed manuscripts for *Medicine and Science in Sports and Exercise*. In 2003, he received the Educator of The Year Award from the National Strength and Conditioning Association.
Mike_Greenwood@baylor.edu

DOUGLAS S. KALMAN PhD, RD, FISSN, FACN
Dr. Kalman is currently a Director in the Nutrition and Endocrinology Division of Miami Research Associates. He has been active in the sports nutrition community for more than a decade. Dr. Kalman has worked with professional, collegiate, and high-level amateur athletes inside and outside the United States. His publications have been announced in numerous scientific journals and popular media outlets. Dr. Kalman is a co-founder of The International Society of Sports Nutrition (ISSN) and an associate editor for targeted journals. In addition, he is on the faculty of New York Chiropractic College and Florida International University.
dkalman@miamiresearch.com

JOSE ANTONIO, PhD, FISSN, FACSM, CSCS
Jose Antonio, Ph.D. is the Chief Executive Officer of the International Society of Sports Nutrition (ISSN) and one of its co-founders. Dr. Antonio earned his Ph.D. from the *University of Texas Southwestern Medical Center* (UTSWMC) in Dallas and completed a postdoctoral fellowship in endocrinology and metabolism at the UTSWMC. His latest book project is the *Essentials of Sports Nutrition and Exercise* (2008 Humana Press). In addition to heading the ISSN, he is a scientific consultant to VPX and Javalution, and a popular consultant to the sports supplement industry. For more information, go to www.joseantoniophd.com

CONTRIBUTORS

SHAWN M. ARENT, PhD, CSCS
Dr. Arent is an Assistant Professor in the Department of Exercise
Science and Sport Studies at Rutgers University, where he is also the
Director of the Human Performance Laboratory. He completed
both his MS and PhD in Exercise Science at Arizona State University. He received his BA from the University of Virginia and is a
Certified Strength and Conditioning Specialist with the NSCA. He
is on the national staff for the U.S. Soccer Federation and provides
performance enhancement advice for a variety of athletes. His
research focuses on the relations between the stress response, health,
and performance and has been funded by various sources including
the NIH, RWJF, and nutritional biotechnology companies.
shawn.arent@rutgers.edu

KRISTEN M. BEAVERS, MS, RD, CPT
Mrs. Beavers is a PhD student in Exercise, Nutrition, and Preventive
Health at Baylor University in the Department of Health, Human
Performance, and Recreation. She received her BS degree from the
Division of Nutritional Sciences at Cornell University and her MPH
degree from the School of Public Health at the University of North
Carolina at Chapel Hill. Mrs. Beavers is a Registered Dietitian and
is certified through the American College of Sports Medicine as a
Personal Trainer.
Kristen_Beavers@baylor.edu

THOMAS BUFORD, MS, CSCS
Mr. Buford is currently pursuing a PhD in Exercise, Nutrition, and
Preventative Health at Baylor University and is a research assistant

in the Exercise and Sport/Biochemical Nutrition Laboratories. He has a Master's degree in Applied Exercise Science from Oklahoma State University, where he was awarded the Graduate Research Excellence Award. Mr. Buford also holds a Bachelor's degree from Oklahoma Baptist University in Secondary Education. He holds CSCS and NSCA-CPT certifications from the National Strength and Conditioning Association as well as the Certified Sports Performance Coach certification through United States Weightlifting. He is a member of several professional organizations. Thomas_Buford@baylor.edu

ALLYN BYARS, PhD, CSCS
Dr. Byars received a BA in Physical Education from Henderson State University, an MSEd in Exercise Physiology from Baylor University, and a PhD in Exercise Science from the University of Mississippi. Dr. Byars was employed at Arkansas State University from 1992 to 1999, where he served as a faculty member teaching graduate and undergraduate courses in exercise science. He later served at Hardin-Simmons University in the same capacity before joining the Angelo State University faculty in the spring of 2005. Dr. Byars teaches courses in exercise physiology, cardiopulmonary assessment, research design, and statistics. His areas of research interests include cardiopulmonary assessment, sport supplements, blood lipids, and measurement studies.
allyn.byars@angelo.edu

BILL CAMPBELL, PhD, FISSN, CSCS
Dr. Campbell is an assistant professor of exercise physiology at the University of South Florida, where he conducts research focusing on nutritional supplements and their effects on human performance, body composition, and metabolism. During the past few years, Dr. Campbell has coordinated clinical research on the nation's top selling nutritional supplements. He has authored more than 50 scientific papers and abstracts in relation to exercise and nutrition. In addition to his research agenda, he is an elected officer of the International Society of Sports Nutrition and assists the National Strength and Conditioning Association in their marketing efforts targeting students.
Campbell@coedu.usf.edu

RICK COLLINS, ESQ., JD

Mr. Collins is a principal in the law firm of Collins, McDonald & Gann, P.C. (www.cmgesq.com). The author of *Legal Muscle*, he is nationally recognized as a legal authority on sports performance-enhancing substances and is counsel to the International Society of Sports Nutrition and the International Federation of BodyBuilders. He received his law degree from Hofstra School of Law, where he attended on a full academic scholarship and served on the *Law Review*. He is admitted to practice in New York, Massachusetts, Pennsylvania, Texas, District of Columbia, and various federal courts.

www.cmgesq.com

MATTHEW COOKE, PhD
Dr. Cooke received his Bachelor of Science (Biomedical Sciences) with honors and his PhD in Exercise Science from Victoria University in Australia. Dr. Cooke then completed his Post-Doctoral Fellowship at Baylor University during 2006–2007. He has recently been appointed as an Assistant Professor in Exercise Physiology and Nutrition at Baylor University, where he will serve as a faculty member teaching graduate and undergraduate courses in exercise physiology and nutrition. His research interests include skeletal muscle adaptations to exercise, aging and sarcopenia, exercise performance, obesity, and sport supplements.
Matt_Cooke@baylor.edu

PAUL J. CRIBB, PhD, CSCS
Dr. Cribb completed his PhD in Exercise and Nutritional Biochemistry at Victoria University, Victoria, Australia. His innovative research earned him a number of awards, including Investigator of the Year by The Australian Association for Exercise and Sports Science and Research Fellow of The Australian Academy of Technological Sciences. Dr. Cribb's technical/engineering experience and his worldwide product manufacture and marketing expertise make him a leading consultant in the health food industry.
pcribb@bigpond.net.au

JOAN ECKERSON, PhD, CSCS
Dr. Eckerson is an Associate Professor in the Department of Exercise
Science and Athletic Training at Creighton University. She has more
than 30 publications in the area of body composition and dietary
supplementation and received the 2004 NSCA Research Achievement Award in recognition of her research in sports nutrition, primarily in the area of creatine supplementation.
JOANECKERSON@creighton.edu

JAMES (JIM) W. FARRIS, PT, PHD
Dr. Farris is an associate professor of physical therapy at A.T. Still
University Arizona School of Health Sciences in Mesa, Arizona. He
is a member of the American Physical Therapy Association and is
active in the Education, Orthopedic, and Cardiovascular sections of
the Association. His current primary research focus is on cardiovascular disease prevention in overweight children and their families.
Dr. Farris's interest in sports nutrition began with his master's work
at California State University Fresno and was further defined during his doctoral work at The Ohio State University, where he
focused on carbohydrate supplementation during prolonged exercise. He actively maintains his interest in nutrition and athletic
performance through an online sports nutrition continuing education course for health professionals, and privately working with
select clientele from team sport athletes to long-distance runners
and cyclists desiring to optimize their nutrition for training and
competition. Dr. Farris's interest in nutrition goes beyond the optimization of sports performance and into the area of injury repair
and rehabilitation.
jfarris@atsu.edu

SUSAN M. KLEINER, PHD, RD, FACN, CNS, FISSN
Dr. Kleiner is the owner and president of High Performance Nutrition, LLC™ (Mercer Island, Washington), a consulting firm specializing in media communications, industry consulting, and personal and team counseling. She is the author of seven books, including *Power Eating*®, 3rd edition (Human Kinetics, 2007), *The Powerfood Nutrition Plan* (Rodale, 2006), and *The Good Mood Diet* (Springboard Press, 2007).
susan@powereating.com

Jamie Landis, PhD, MD, CSCS
Dr. Landis received a BS in Biology from Ferris State University, followed by an MS in Endocrine Physiology and a PhD in Neuroscience, both from Bowling Green State University. His MD was earned at the University of Toledo (formerly MCO), followed by an internship in medicine at Michigan State University and a residency appointment in Physical Medicine and Rehabilitation at the Mayo Clinic. He also holds the CSCS. He volunteers his time as a youth weightlifting, conditioning, and football coach. He has presented his workd at national meetings since 1989 and has co-authored several textbook chapters and research articles. Currently an Associate Professor of Biology, Dr. Landis was selected for the 2006 Lakeland College Excellence in Teaching Award and was recognized as one of the Best Educators in the State by Ohio Magazine.
jlandis@lakelandcc.edu

RAFER LUTZ, PhD

Dr. Lutz is an associate professor of sport and exercise psychology in the Department of Health, Human Performance, and Recreation at Baylor University. A former winner (2001) of the Dissertation Award by the American Alliance for Health, Physical Education, Recreation, and Dance's Sport Psychology Academy, Dr. Lutz's research focuses primarily on the psychology of exercise participation and related health psychology topics, although he has secondary interests in the study of sport performance enhancement. In his brief career, Dr. Lutz has published 22 papers, 13 as first or second author, in well respected, peer-reviewed journals such as *Behavioural Brain Research*, *Journal of Sport and Exercise Psychology*, *American Journal of Health Behavior*, and *Psychology of Sport and Exercise*. Additionally, Dr. Lutz has been the primary presenter or co-presenter of more than 45 offerings to national and international scientific societies such as the North American Society for the Psychology of Sport and Physical Activity, American College of

Sports Medicine, American Psychological Association, and American Psychological Society. A former Academic All-American golfer, Dr. Lutz has served as a performance enhancement consultant for a variety of business organizations, individual athletes, and athletic teams.
Rafer_Lutz@baylor.edu

CHRISTOPHER J. RASMUSSEN, MS, CSCS
Christopher J. Rasmussen MS, CSCS, serves as the Research Coordinator for the Exercise & Sport Nutrition Laboratory within the Department of Health & Kinesiology at Texas A & M University in College Station, TX. Christopher received his Bachelor's degree from Kansas University and Master's degree from the University of Memphis. Contact information: crasmussen@hlkn.tamu.edu

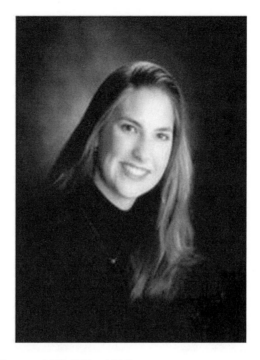

Monica C. Serra, MS, RD, ATC
Ms. Serra is a PhD student in Exercise, Nutrition, and Preventative Health at Baylor University in the Department of Health, Human Performance, and Recreation. She is an adjunct lecturer in the Department of Health, Human Performance, and Recreation at Baylor. Ms. Serra received her BS degree from Duquesne University in Athletic Training and her MS in Nutrition from Case Western Reserve University. She is a Registered Dietitian and Certified Athletic Trainer.
Monica_Serra@baylor.edu

MARIE SPANO, MS, RD, FISSN

Ms. Spano is a nutrition communications expert and consultant. She has written several freelance articles for trade publications and popular press magazines as well as two book chapters for sports nutrition textbooks. She has counseled hundreds of individuals on weight loss and sports nutrition, is the Vice President and Fellow of the International Society of Sports Nutrition, and is a registered dietitian with an MS in Nutrition from the University of Georgia and BS in Exercise and Sports Science from the University of North Carolina, Greensboro.

mariespano@comcast.net

COLIN WILBORN, PhD, CSCS, NSCS-CPT, ATC
Dr. Wilborn is an Assistant Professor of Exercise and Sport Science at The University of Mary Hardin Baylor. He has a PhD in Exercise, Nutrition, and Preventive Health and is a Certified Strength and Conditioning Specialist and Certified Athletic Trainer. He served as the Weight Training Studies Coordinator for the Exercise and Sport Nutrition Laboratory at Baylor University and has been involved in numerous clinical studies investigating the effects of various sports supplements on health and performance. In addition, Dr. Wilborn has designed both nutrition and strength and conditioning programs for collegiate football, soccer, basketball, and softball athletes. He has published research on the effects of sport supplements and exercise on body composition, metabolism, and performance.
cwilborn@umhb.edu

TIM N. ZIEGENFUSS, PHD, FISSN, CSCS
Dr. Ziegenfuss is the Chief Executive Officer of The Ohio Research
Group of Exercise Science and Sports Nutrition. He is a well known
author, speaker, and researcher with expertise in exercise training,
nutrition, dietary supplements, and sports performance. He is a
Fellow of the International Society of Sports Nutrition, a Certified
Strength and Conditioning Specialist, and has recently been
appointed Chair of the Exercise Physiology and Sports Nutrition
program at Huntington College of Health Sciences. Dr. Ziegenfuss
earned a Bachelor of Science from Lock Haven University, a Master
of Science from Purdue University, and a Doctorate from Kent
State University. During his 5-year tenure as a professor, he taught
undergraduate and graduate level courses in exercise physiology,
metabolism and body composition, anatomy, physiology, patho-
physiology, and sports nutrition at Lock Haven University and
Eastern Michigan University. In addition to authoring and co-
authoring book chapters on nutritional ergogenic aids, antioxidants, protein, and over-the-counter hormones, he has written
numerous articles for *Physical Magazine*, *Muscular Development*,

and *Flex*. Because of his unique ability to "connect the dots" between diet, supplements, and exercise science, Dr. Ziegenfuss is routinely sought as a consultant by top supplement companies and elite athletes.

tziegenfuss@wadsnet.com

Part I
The Industrial Nature of the
Supplement Game

1 Effect of Government Regulation on the Evolution of Sports Nutrition

Rick Collins and Douglas Kalman

Abstract

The sports nutrition segment of the dietary supplement industry enjoyed nearly a decade of unfettered growth under federal legislation passed in 1994. A series of breakthroughs in the dietary supplement field led to the development and marketing of innovative products designed to enhance performance, build muscle, or lose excess fat. As the popularity of these products soared and evolved into a multi-billion dollar industry, the sports nutrition supplement market drew the attention of federal and state regulatory bodies and sports antidoping authorities. Growing concerns over potential health risks and unfair athletic advantages have spurred government regulators and legislators to heighten the scrutiny of this market, leading to recent legislative amendments and increased government enforcement action.

Key words

Dietary Supplement Health and Education Act of 1994 · Food and Drug Administration · Dietary supplements · FTC · Nutrition

1. DIETARY SUPPLEMENT HEALTH AND EDUCATION ACT OF 1994

The Dietary Supplement Health and Education Act of 1994 (DSHEA) was passed with the unanimous consent of Congress. This statute was enacted amid claims that the Food and Drug Administration (FDA) was distorting the then-existing provisions of the Food, Drug, and Cosmetic Act (FDCA) *(1)* to try improperly to deprive the public of safe and popular dietary supplement products. The FDA was perceived as engaging in antisupplement

From: *Nutritional Supplements in Sports and Exercise*
Edited by: M. Greenwood, D. Kalman, J. Antonio,
DOI: 10.1007/978-1-59745-231-1_1 © Humana Press Inc., Totowa, NJ

policy and enforcement tactics, provoking a groundswell of legislative criticism. In its official report about the need for DSHEA to curtail excessive regulation of dietary supplements by the FDA, the Senate Committee on Labor and Human Resources charged that the "FDA has been distorting the law in its actions to try to prevent the marketing of safe dietary supplement substances" *(2)*. The Senate Committee also concluded, the "FDA has attempted to twist the statute (i.e., the provisions of the FDCA, as it then existed) in what the Committee sees as a result-oriented effort to impede the manufacture and sale of dietary supplements" *(3)*.

DSHEA represented a sharp rebuke to the FDA's regulatory tactics. However, although DSHEA defined "dietary supplements" and "dietary ingredients," set certain criteria for "new dietary ingredients," and prevented FDA from overreaching, it did not, as some critics have charged, leave the entire industry unregulated. The dietary supplement industry is in fact regulated by the FDA as a result of DSHEA. The Center for Food Safety and Applied Nutrition (CFSAN), a branch of the FDA, along with the Office of Dietary Supplements (ODS) assist the FDA in the regulation of dietary supplements. This power ensures the FDA's authority to provide legitimate protections for the public health. The Federal Trade Commission (FTC) also continues to have jurisdiction over the claims (marketing, from implied to direct claims) that manufacturers make about their products. The FDA and FTC work together to regulate the dietary supplement industry. There is increased sharing of information and overlapping of jurisdiction with regard to marketing and advertising of dietary supplements.

2. GOVERNMENT PROTECTIONS FROM DIETARY SUPPLEMENT HAZARDS AND RISKS

How is the FDA authorized to protect the public in the realm of dietary supplements? What if evidence showed that a particular supplement product was causing an acute epidemic of illnesses and fatalities? What could the FDA do about it? The FDA is an Operating Division of the U.S. Department of Health and Human Services (HHS), which is headed by the Secretary of Health and Human Services. The Secretary has the power to declare a dangerous supplement to be an "imminent hazard" to public health or safety and immediately suspend sales of the product *(4)*.

The FDA also has the authority to protect consumers from dietary supplements that *do not* present an imminent hazard to the public but *do* present certain *risks* of illness or injury to consumers. The FDCA prohibits introducing *adulterated* products into interstate commerce *(5)*. Several grounds exist by which unsafe dietary supplements can be deemed to be adulterated *(6)*. Two provisions are relevant to our examination.

The first provision, which applies to all dietary supplements, states that a supplement shall be deemed adulterated if it presents "a significant or unreasonable risk of illness or injury under... conditions of use recommended or suggested in labeling, or ... if no conditions of use are suggested or recommended in the labeling, under ordinary conditions of use *(7)*." The standard does not require proof that consumers have *actually* been harmed or even that a product will harm anyone. It was under this provision that the FDA—after 7 years, numerous criticisms including a negative report from the General Accounting Office, and a storm of public debate—concluded that dietary supplements containing ephedra presented an unreasonable risk. However, the conclusion the FDA drew and its reasoning to declare products containing ephedra adulterated utilized a new and novel approach when it ruled, in 2004, that dietary supplements containing ephedrine alkaloids present an "unreasonable risk of illness or injury" and that there is no acceptable dose of the ingredient. Although there is still a debate as to whether this standard is appropriate under DSHEA, the fact remains that the FDA's rule has been established. The criminal penalties for a first conviction of introducing adulterated supplement products into interstate commerce can include a fine of up to $1000, imprisonment for up to 1 year, or both *(8)*. Subsequent convictions, or convictions for offenses committed with the intent to defraud or mislead, can include fines of up to $10,000, imprisonment of up to 3 years, or both *(9)*.

3. NEW DIETARY INGREDIENTS

The second provision by which supplements may be deemed adulterated addresses only dietary supplements containing "new dietary ingredients" for which the FDA believes there may be inadequate information to provide a reasonable assurance that the ingredient does not present a significant risk of illness or injury. Recognizing that new and untested dietary supplement products may pose unknown health issues, DSHEA distinguishes between products containing dietary ingredients that were already on the market and products containing new dietary ingredients that were not marketed prior to the enactment of the law *(10)*. A "new dietary ingredient" (NDI) is defined as a dietary ingredient that was not marketed in the United States before October 15, 1994 *(11)*. DSHEA grants the FDA greater control over supplements containing new dietary ingredients. A new dietary ingredient is deemed adulterated and subject to FDA enforcement sanctions unless it meets one of two exemption criteria: either 1) the supplement in question contains "only dietary ingredients which have been present in the food supply as an article used for food in a form in which the food has not been chemically altered"; or 2) there is a "history of use or other evidence of safety" provided by the manufacturer or distributor to the FDA at least 75 days before introducing the product into interstate commerce *(12)*. The first criterion is silent as to how and by whom presence in the food supply as food articles without chemical alteration is to be established. The second criterion—applicable only to new dietary ingredients that have not been present in the food supply—requires manufacturers and distributors of the product to take certain actions. Those actions include submitting, at least 75 days before the product is introduced into interstate commerce, information that is the basis on which a product containing the new dietary ingredient is "reasonably be expected to be safe *(13)*." That information would include: 1) the name of the new dietary ingredient and, if it is an herb or botanical, the Latin binomial name; 2) a description of the dietary supplement that contains the new dietary ingredient, including the a) the level of the new dietary ingredient in the product, b) conditions of use of the product stated in the labeling or if no conditions of use are stated the ordinary conditions of use, and c) a history of use or other evidence of safety establishing that the dietary ingredient, when used under the conditions recommended or suggested in the labeling of the dietary supplement, is reasonably expected to be safe. There is no guidance as to what evidence is required to establish a reasonable expectation of safety *(14)*. In fact, the FDA specifically states that the person submitting the application is responsible for determining what information provides the basis for the conclusion that the product is reasonably expected to be safe. By not providing guidance, the FDA could arguably be claimed to be giving itself a wide berth to decide arbitrarily what ingredients to approve or disapprove. The only hint given is that the FDA expects the applicant to "consider the evidence of safety found in the scientific literature, including an examination of adverse effects associated with the use of the substance *(14)*." Thus, it appears that the question should be one of safety alone, rather than a safety and efficacy analysis, which in turn naturally progresses to a risk/benefit analysis. This is a much different and more difficult, if not impossible, standard for an NDI to meet.

4. NEW DIETARY INGREDIENT REVIEW: APPLICATION PROCESS

If a supplement manufacturer seeks to market an ingredient to the public that was not previously sold on the United States market (prior to October 15, 1994), a dossier of either animal and human safety data and/or proof of historical use as a food must be compiled. Unless the ingredient has been present in the food supply as an article used for food in a form in which the food has not been chemically altered, the history of use or other evidence of safety must be presented to the FDA/CFSAN at least 75 days before introducing the product into interstate commerce. What sort of information should be presented? By what process would the FDA evaluate the data to determine if the ingredient should be allowed on the market? Let us take an example.

Excluding discussions of whether animal safety studies have utility as related to efficacy outcome-oriented research in humans, let us assume that the product of interest was found to be of acceptable safety margins when used in human doses. This means that the animal studies found the ingredient to be noncarcinogenic, to have a high-LD_{50} (median lethal dose) and not to be organotoxic. In addition, let us assume that the human studies also found the product not to have an effect on blood pressure or heart rate or on markers of safety as denoted by specific blood tests (e.g., liver, kidney). Based on the compilation of animal and brief human studies, we can believe that the product is nontoxic, although further and more invasive safety data are warranted.

For example, the dietary supplement popularly known as 7-Keto®, also known as 3-acetyl-7-oxo dehydroepiandrosterone, was approved by the FDA via the NDI premarket notification process. Prior to marketing 7-Keto as a dietary supplement it was submitted for a review of safety to the FDA in the form of an NDI premarket notification. This document, which can be viewed at the FDA website, received no comments or concerns expressed from the FDA. Subsequent to this initial filing, another NDI premarket notification has been filed specific to the use of 7-Keto for weight loss in adults at the prescribed dosage. This notification also has received no comments or concerns expressed from the FDA (June 24, 1997).

The FDA has a searchable database of submitted NDIs that the public can view, located online at http://www.fda.gov/ohrms/dockets/dockets/95s0316/95s0316.htm/. In addition to 7-Keto, other popular dietary supplements (e.g., creatine ethyl ester, vinpocetine, Diosmin 95) have successfully undergone the NDI process. Some applications to the FDA fail to meet the bar of demonstrating relative safety, usefulness, and other criteria requested by the FDA in the application, and thus the rejection or failure rate for NDI applications is thought to be about 65% to 70% of the applications submitted.

5. FDA REGULATORY ACTION—EPHEDRA SUPPLEMENTS

On February 6, 2004, the FDA issued a final rule prohibiting the sale of dietary supplements containing ephedrine alkaloids, reasoning that this category of supplements presented an unreasonable risk of illness or injury (based on a risk-to-benefit evaluation). The rule took effect on April 12, 2004, sixty days from the date of publication, at which time companies that continued to sell supplements containing ephedra alkaloids found themselves subject to a variety of enforcement possibilities, including seizure of the product, injunction against the manufacturers and distributors of such products, and criminal prosecution of violators. Three months after the rule was announced, Utah-based Nutraceutical Corp. filed suit challenging the ban, specifically the FDA's risk-to-benefit analysis, arguing that the FDA had not shown ephedra supplements to present an undue risk at low doses.

On April 13, 2005, a federal court in Salt Lake City (U.S. District Court for the District of Utah, Central Division) issued its decision on a legal challenge to the FDA's 2004 Final Rule banning all ephedrine-alkaloid dietary supplements. Judge Tena Campbell's decision made two key points.

- It held that the analysis used by the FDA was incorrect and improper. The FDA's analysis weighed risks against benefits. DSHEA, however, requires a straightforward *risk* assessment. The court held that requiring supplement companies to demonstrate a benefit as a precondition to marketing violated DSHEA by shifting the burden from the FDA to industry.
- It held that the FDA did not have adequate scientific evidence to find that a daily dose of 10 mg or less of ephedrine alkaloids presented a "significant or unreasonable risk of illness or injury" [under 21 U.S.C. § 342(f)(1)(A)]. The court effectively held that it is improper to ban all ephedra supplements because the FDA lacks data to determine what dosage might be safe.

The FDA filed a notice of appeal in the 10th Circuit Court of Appeals.

On August 17, 2006, the U.S. Circuit Court of Appeals in Denver reversed and remanded the Utah ruling that challenged the FDA ban on products containing ephedra. The Federal Appeals court overturned Judge Campbell's decision, ruling that the FDA was correct in its 2004 analysis of ephedrine products, concluding that the FDA had properly examined the facts when it ruled, in 2004, that dietary supplements containing ephedrine alkaloids present an "unreasonable risk of illness or injury," and that there is no acceptable dose of the ingredient. Pursuant to this the government has since seized numerous products containing the herbal ingredient.

Following the Federal Appeals court decision, Nutraceutical filed a petition for Rehearing on September 28, 2006. In its appeal, Nutraceutical requested that the entire 10th Circuit court rehear the case, rather than the three judges in the Circuit Court of Appeals case. Nutraceutical contended that the three-judge panel violated the Supreme Court's canons of statutory construction and raised questions of exceptional importance by reversing Judge Campbell's April 2005 ruling against the FDA's risk–benefit analysis method. The petition was denied, and Nutraceutical filed a petition for a writ of certiorari to the U.S. Supreme Court. Subsequently, Nutraceutical filed a motion for summary judgment with the Federal District Court in Utah, and the FDA responded with a cross-motion for summary judgment.

On March 16, 2007, the Federal District Court in Utah ruled in favor of the FDA, finding that FDA's rulemaking with regard to dietary supplements containing ephedrine alkaloids was procedurally and substantively proper. They thus granted the FDA's cross-motion for summary judgment.

The U.S. Supreme Court denied Nutraceutical's petition for certiorari on May 14, 2007, refusing to consider their appeal. This is likely the final action in this case, establishing a precedent for the applicable legal standards and confirming the FDA's regulatory authority over the issues.

6. FDA REGULATORY ACTION—ANDROSTENEDIONE

The ban on androstenedione is another example of the authority of the FDA to prohibit the marketing of ingredients the agency believes are adulterated. On March 11, 2004, the FDA pronounced that dietary supplement products containing androstenedione were adulterated new dietary ingredients under DSHEA. For the second time in as many months the FDA took regulatory action against the sports nutrition industry. There was no evidence of an imminent health hazard posed by androstenedione. However, instead of the formal administrative procedure of issuing a proposed rule and inviting public comment, the FDA took unilateral action, issued a press release, held a news conference, and sent warning letters to 23 companies that had manufactured, marketed, or distributed the products containing androstenedione. In its warning letters, the FDA threatened possible enforcement actions for noncompliance. The effect was to cause retailers, manufacturers, and distributors alike to cease selling products containing androstenedione. No meaningful dialogue between the FDA and industry occurred prior to the FDA taking this action.

Supplements containing androstenedione were introduced during the mid-1990s and were promoted as a natural way to help increase strength and muscle mass as well as to combat the effects of the aging process in older men, much of which is attributed to declining testosterone levels. Like dehydroepiandrosterone (DHEA), androstenedione is a naturally derived precursor to testosterone. Androstenedione converts directly to testosterone in the metabolic pathway. The fact that it is naturally derived and, as described below, present in the food supply is important in relation to the action taken by the FDA. In its press release *(15)* and warning letters, the FDA declared androstenedione to be an adulterated new dietary ingredient based on its position that no evidence demonstrates "that androstenedione was lawfully marketed as a dietary ingredient in the United States before October 15, 1994 *(16)*." It seems to be correct that androstenedione was not marketed before 1994, given that the first commercial marketing of products containing androstenedione appears to have been in 1996. Furthermore, a review of the FDA's electronic database indicates no submission of an application for a new dietary ingredient involving androstenedione *(17)*. Interestingly, however, the FDA went beyond the explicit words of the statute and used the term "lawfully marketed" in their letters instead of simply "marketed." The implication was that to receive "grandfathered" status into DSHEA as a pre-1994 supplement ingredient the product must not only have been marketed but must have met the additional requirement of having been *lawfully* marketed. At least one commentator has interpreted this language to impose a burden on industry to prove the product was generally regarded as safe (GRAS) pre-1994—an impossible standard for any product that was not explicitly affirmed as such by the FDA prior to the enactment of DSHEA *(18)*. Assuming that androstenedione is indeed a new dietary ingredient, the FDA could determine that products containing androstenedione are adulterated under DSHEA unless they meet either of the two exemption criteria stated above.

Accordingly, it appears that the question of exemption turns on 1) whether androstenedione is present in the food supply as an article used for food without chemical alteration and 2) if not, could the product satisfy the requirement of reasonable expectation of safety. With respect to the first exemption, according to scientific journals androstenedione is indeed present in the food supply without chemical alteration *(19)*. Had there been open communication between the FDA and industry, the scientific evidence that androstenedione is present in the foods we eat could have been presented and discussed. Moreover, until 1998, which is the date for the most recent information, there were no reports of adverse events reported on the FDA's database *(20)*. Adverse events comprise one of the few specific pieces of information that the FDA sets forth in their "information" about what safety data they require *(21)*. The FDA's requirements to show safety have never been fully articulated; arguably, the FDA's policy creates a nearly impossible procedure to demonstrate safety. Unlike the situation regarding ephedra supplements, Industry did not formally challenge the FDA's regulatory action regarding androstenedione; and sales of androstenedione ceased.

The FDA's action on androstenedione suggests a heightened enforcement policy against what the agency deems to be adulterated new dietary ingredients. Bringing products to market that do not meet either of the two exemption criteria of 21 U.S.C. 350(b) may not be overlooked in the future. If a new dietary ingredient is exempted from adulterated status because it is present in the food supply as an article used for food in a form in which the food has not been chemically altered, it is prudent to document that information prior to marketing the product or even to communicate that information to the FDA. If a new dietary ingredient is not exempted from adulterated status based on the food supply exemption, premarket notification of history of use or other evidence of safety establishing that the dietary ingredient, when used under the conditions recommended or suggested in the labeling of the dietary supplement, is reasonably expected to be safe must be provided to the FDA at least 75 days before the product is introduced into interstate commerce.

7. DOPING—CONTAMINATED SUPPLEMENTS AND BANNED INGREDIENTS

Research conducted by the United States Olympic Committee in 2004 found that 90% of athletes use some form of dietary supplements. To ensure a "level playing field" and protect the health of athletes, sports bodies are free to create their own lists of banned ingredients. Athletes are generally held under a standard of "strict liability" by their respective sport bodies and are therefore held responsible for everything they put into their bodies.

Highly publicized cases, however, have shown that when athletes fail drug tests tainted dietary supplements may be blamed, and expensive litigation may follow. Such lawsuits may seek not only compensation for the athlete's lost potential income during the ever-lengthening suspensions from athletic bodies and the tarnishing of the athlete's name in the publicity that follows a positive test but also punitive damages in the tens of millions. One of the most well known lawsuits against a dietary supplement company involved world-class swimmer Kicker Vencill, who tested positive for the anabolic steroid nandrolone. He blamed it on contaminated dietary supplements—vitamin capsules. When he received a 2-year suspension from competition, he sued the sports nutrition company Ultimate Nutrition and received a nearly $600,000 verdict, which was later appealed. The case was later settled for an undisclosed amount with nodirect admission of guilt by the manufacturer.

National Football League (NFL) running back Mike Cloud, a Boston College graduate playing for the Patriots, tested positive for norandrostenedione and androstenediol (two nandrolone metabolite). He claimed that a tainted whey protein powder (Nitro-Tech)^{TM} caused his positive test and consequent four-game suspension. The supplement's Canadian manufacturer, MuscleTech, countersued, maintaining that the allegations were false and amounted to trade libel. Pavle Jovanovic, a U.S. bobsledder, tested positive for 19-norandrosterone (also a nandrolone metabolite) in 2001. He was suspended for 2 years and disqualified from the Salt Lake City Olympics in 2002. He blamed his troubles on MuscleTech's whey protein, saying it was cross-contaminated by one of its prohormone products. Jovanovic also filed suit, and MuscleTech countersued. Both Cloud and Jovanovic submitted tests showing containers of the product contained nandrolone metabolites not listed on the label (these results were also televised nationally on Bryant Gumbel's HBO Inside Sports show).

Graydon Oliver, a tennis player, tested positive in 2003 for the prohibited substance hydrocholorothiazide, a diuretic used as a masking agent for other banned substances (used medically typically for hypertension). He blamed a purportedly homeopathic Chinese herbal sleeping aid called Relax-Aid. He filed suit on October 1, 2004 against Keimke Inc. (a.k.a. Barry's Vitamin and Herbs), a Boca Raton purveyor of food and health supplements. The Association of Tennis Professionals panel found that Oliver was aware of the ATP warnings regarding using supplements and that he failed to investigate the product as thoroughly as possible. ATP suspended Oliver for 2 months and directed him to forfeit $5000 in prize money and championship points. Oliver had retained a sample of Relax-Aid for testing (*note*: his sample of Relax-Aid also tested positive for chlordiazepoxide, the active ingredient in Librium). Allegedly the store owner, even after being informed that the user was a professional athlete subject to mandatory testing, told Oliver's mother who purchased the product that the dietary supplement was safe for all sports organizations as it contained no banned ingredients. Graydon brought suit alleging $15 million in damages (economic and noneconomic losses).

A 2002 International Olympic Committee (IOC) study titled, "Analysis of Non-Hormonal Nutritional Supplements for Anabolic-Androgenic Steroids" done by an IOC-accredited drug testing laboratory found that 94 of the 634 (14.8%) dietary supplement samples it studied contained substances not listed on the label that would trigger positive drug tests. The dietary supplements were from 12 countries. The dietary supplements from the United States tested positive in 45 of the 240 products tested, at a fail rate of 18.8%. During the 2002 Salt Lake City Winter Games, athletes from The Netherlands submitted 55 supplements to be confidentially analyzed for banned substances. In total, 25% of the supplements tested positive for prohibited substances.

8. ANABOLIC STEROID CONTROL ACT

During the past 4 years, the use of performance-enhancing substances in sports has been in the media spotlight like never before, with publicized positive doping tests in major and minor league professional baseball, professional football, track and field, cycling, weightlifting, tennis, inline skating, boxing, soccer, swimming, softball, Paralympics, and even horse racing. Chemically induced advantages can undermine the traditional principle of a level playing field, and the abuse of these substances can lead to health risks. The war against the use of performance-enhancing substances in sports has been waged mostly on two fronts: 1) prohibition of the substances by athletic bodies that have implemented drug testing of players, and 2) federal and state legislation of the substances as dangerous drugs with criminal penalties imposed on violators. Federal legislators responded to the reports of extensive use of anabolic steroids and steroid precursors among sports competitors by subjecting possessors of steroid precursor products—openly sold in U.S. health food stores until January 2005—to arrest and prosecution. The law, passed by Congress in 2004, was an expansion of antisteroid legislation passed in 1990 and demonstrates the evolution of government regulation in this area.

8.1. Anabolic Steroid Control Act of 1990

During the mid-1980s, reports of the increasing use of anabolic steroids in organized sports, including a purported "silent epidemic" of high school steroid use, came to the attention of Congress. When Canadian sprinter Ben Johnson tested positive for the steroid stanozolol (popularly known as Winstrol) at the 1988 Seoul Olympics and was stripped of his gold medal, the ensuing media frenzy galvanized the U.S. Congress into action. Between 1988 and 1990, Congressional hearings were held to determine whether the Controlled Substances Act should be amended to include anabolic steroids *(22)*. Significantly, medical professionals and representatives of regulatory agencies (including the FDA, DEA, and National Institute on Drug Abuse) testified *against* the proposed amendment to the law. Even the American Medical Association opposed it, maintaining that steroid abuse does *not* lead to the physical or psychological dependence required for scheduling under the Controlled Substances Act *(23)*. However, any "psychologically addictive" properties of steroids or public health dangers seemed to be secondary considerations to Congress. Most of the witnesses at the hearings were representatives from competitive athletics whose testimony, and apparently Congress' main concern, focused on the purported need for legislative action to solve an athletic "cheating" problem *(24)*.

Congress passed the Anabolic Steroid Control Act of 1990 *(25)*, criminalizing the possession of anabolic steroids without a valid prescription. This was accomplished by amending Title 21 of the United States Code (U.S.C.) § 812(c), which contains the initial schedules of controlled substances *(26)*. Anabolic steroids were listed under subsection (e) of Schedule III. The law placed steroids in the same legal class as barbiturates, ketamine, LSD precursors, and narcotic painkillers such as Vicodin. To this day, anabolic steroids remain the only hormones in the schedules.

Once the law became effective, in 1991, mere unlawful possession of any amount of anabolic steroids, even without any intent to sell or distribute, became a federal crime *(27)*. A conviction is punishable by a term of imprisonment of up to 1 year and/or a minimum fine of $1000; and prior state or federal drug convictions increase the possible sentence. Unlawful steroid distribution or possession with intent to distribute is punishable by up to 5 years in prison for a first offender or 10 years for a prior drug offender *(28)*.

The 1990 law listed only 27 compounds, along with their salts, esters, and isomers *(29)*. In theory, however, there are hundreds or even thousands of anabolic steroidal compounds—many of which might enhance athletic performance—that could be created in laboratories and offered for human use. By the early part of the current decade, some of these substances were being openly marketed as performance-enhancing dietary supplements. Called "prohormones" or in some cases "prosteroids," these products were frequently metabolic precursors to testosterone or other listed anabolic steroids. Prosecution of those responsible for selling these compounds—including androstenedione, norandrostenedione, norandrostenediol, 1-testosterone, and 4-hydroxytestosterone—was hampered by the absence of these compounds from the list *(30)*.

8.2. Anabolic Steroid Control Act of 2004

Although dietary supplements have never been embraced by "anti-doping" agencies, prohormone supplements were of particular concern, presenting at least four problems beyond any perceived health issues. First, by their very nature and design they defied traditional sports values: They were little pills that might give the player who swallowed them a chemically induced advantage over the player who did not. Second, some of the steroid precursor products shared metabolites with banned anabolic steroids, raising the specter of false-positive tests *(31)*. Third, traditional drug screening might fail to detect some of the newer "designer" steroid configurations. Lastly, poor quality control at the manufacturing level presented the possibility that some dietary supplement products might inadvertently contain steroid precursors by "cross-contamination," resulting in false positive tests for anabolic steroids *(32)*. Amid the searing media attention to the issue, legislators publicly cried out for broader and stiffer steroid laws. Congress drafted bills and held hearings *(33)*. On October 22, 2004, President Bush signed into law the Anabolic Steroid Control Act of 2004, and it took effect in January 2005 *(34)*. The new law expanded the original steroid law that had been passed in 1990, also providing $15 million for educational programs for children regarding the dangers of anabolic steroids and directing the U.S. Sentencing Commission to consider revising federal guidelines to increase the penalties for steroid possession and distribution.

The law added 26 new steroid compounds to the previous list of substances that are legally defined as "anabolic steroids" and classified them as Schedule III controlled substances *(35)*. An exhaustive analysis of all the new compounds is beyond the scope of this chapter, but a few observations are in order. Some of these compounds were being marketed as dietary supplements, whereas others, such as bolasterone, calusterone, furazabol, and stenbolone, are actually early pharmaceutical steroids that were missed in the original federal law (note, however, that some states, among them California, did include some of these compounds in their own steroid laws). These dusty old compounds were likely added to the list after the highly publicized reemergence of norbolethone (also added to the list) in an Olympic urine sample. Listed also is tetrahydrogestrinone, or THG, the so-called "designer steroid" that precipitated the BALCO scandal. Mere possession of any of these products is now a basis for a person's arrest and prosecution as a federal drug criminal.

The new law also changes the general requisite elements of an anabolic steroid. Ironically, no longer is there any requirement for evidence that an anabolic steroid is "anabolic" (i.e., that it promotes muscle growth). It simply needs to be chemically and pharmacologically related to testosterone and either on the new list of substances or be any salt, ester, or ether of a substance on the list. The omission of the criterion of promoting muscle growth profoundly affects the process by which a newly created "designer" steroidal compound may be added to the list. Even under the 1990 law, the U.S. Attorney General had the authority *(36)* to schedule additional or newly discovered steroidal compounds without going back to Congress for approval. However, under the old law, for a compound to qualify as an anabolic steroid the Attorney General was required to prove that the compound had anabolic properties. Now, for administrative scheduling, the Attorney General must only establish that the compound is chemically and pharmacologically related to testosterone *(37)*.

After a protracted battle on the issue among members of Congress, the law permits the continued sale of DHEA *(38)* as a dietary supplement by adding it to the other hormonal substances explicitly excluded from scheduling (estrogens, progestins, and corticosteroids). The law also fixes some of the mistakes and poor draftsmanship of the 1990 law *(39)*. Compounds that had been erroneously listed twice, under alternate names, are now listed only once (reducing by correction the original list of 27 compounds to 23). Gone are methandrostenolone *(40)*, stanolone *(41)*, chlorotestosterone *(42)*, and methandranone *(43)*. Also, the misspelling of formebolone as "formebulone" has been corrected. For further clarity, the new law provides the chemical names of all the compounds.

The new law retains the "catch-all" provision of the 1990 law concerning certain variations of the listed compounds and includes specific isomers of a compound under that compound's heading *(44)*. Not all prohormone products fall under the new law, nor do all conceivable anabolic steroids. Chemical nomenclature is the only way that some of the newly listed substances are represented. It remains to be seen how regulatory authorities and in-the-field law enforcement officers will make determinations as to which products remain legal and which do not.

9. ADVERSE EVENTS REGULATION & LEGISLATION

In response to growing criticism of the dietary supplement industry, which is often inaccurately characterized by mass media and sometimes the U.S. government as "unregulated," the 109th Congress passed the first mandatory Adverse Event Reporting (AER) legislation for the dietary supplement industry. On December 22, 2006, President Bush signed into law the Dietary Supplement and Nonprescription Drug Consumer Protection Act. This Act, which took effect on December 22, 2007, was sponsored by Senator Orrin Hatch (Utah) and co-sponsored by Senators John Cornyn (Texas), Michael Enzi (Wyoming), Edward Kennedy (Massachusetts), Richard Durbin (Illinois), and Tom Harkin (Iowa). After much debate in Congress and input from the FDA, the American Medical Association (AMA), many of the major supplement trade associations, and a host of others, the group finally agreed that the legislation was necessary and the final version was approved by all. In short, the Act requires that all "serious adverse events" regarding dietary supplements be reported to the Secretary of Health and Human Services.

An adverse event, as defined in section (a)(1) of this new law, is any health-related event associated with the use of a dietary supplement that is adverse. A *serious* adverse event, as defined in section (a)(2)(A), is an adverse event that results in: (i) death; (ii) a life-threatening experience; (iii) inpatient hospitalization; (iv) a persistent or significant disability or incapacity; (v) or a congenital anomaly or birth defect or (B) requires, based on reasonable medical judgment, a medical or surgical intervention to prevent an outcome described under subparagraph (A).

Once it is determined that a serious adverse event has occurred, the manufacturer, packer, or distributor of a dietary supplement whose name appears on the label of the supplement shall submit to the Secretary of Health and Human Services any report received of the serious adverse event accompanied by a copy of the label on or within the retail packaging of the dietary supplement.

This law strengthens the regulatory structure for dietary supplements and builds greater consumer confidence in this category of FDA-regulated products, thus ensuring and protecting Americans' continued access to safe, beneficial dietary supplements. Consumers have a right to expect that if they report a serious adverse event to a dietary supplement manufacturer the FDA will be advised about it. The Council for Responsible Nutrition, Natural Products Association, American Herbal Products Association and the Consumer Healthcare Products Association all support the AER legislation and have structured educational presentations for the dietary supplement industry in order to educate and implement programs for all companies to comply with this useful law.

10. CONTAMINATION OR ADULTERATION: A NEED FOR BETTER CONSUMER CONFIDENCE

We must ask ourselves: Can the dietary supplement/nutrition industry do anything more to enhance the image and/or quality of the products being sold? Most health professionals would emphatically state that conducting clinical trials that examine the products as they are intended to be used or as they are marketed to ensure that they deliver on their promise is well worthwhile.

The research industry intersects with many other industries in more than one way. Research is used to plan marketing, create a product, and learn more about the product. Research can be used to delineate consumer demographics, within the dietary supplement industry to define safety and efficacy, and for intellectual property means. However, a point to consider is whether there exists a responsibility on the part of a company that markets a product to research and learn the unknowns about their products. To paraphrase former U.S. Secretary of Defense Donald Rumsfeld: Are you responsible for known-unknowns? Before we explore the potential answers, one must also wonder just who in the chain of product retailing is really responsible for the product dossier? Many companies purchase their finished products from other companies and simply relabel the products for their own marketing purposes (private labeling). Yet other companies source raw materials or sometimes branded ingredients from a supplier and then retail it as their own in the finished product. This is all perfectly legal and quite common in many industries.

To utilize a popular ingredient for the purposes of this chapter, the example of blue-green algae is examined. Blue-green algae are often skimmed or collected from surface waters. Among the most popular site in the United States for cultivating blue-green algae is the Upper Klamath Lake in southern Oregon. In 1996, the state of Oregon noted that the Upper Klamath Lake was experiencing an extensive growth of *Microcystis aeruginosa* (a type of blue-green alga) that is known to produce hepatotoxins (microcystins). A local public uproar occurred, and the local health departments decided to test the waters and blue-green algae dietary supplements (61 to be exact) for the presence of microcystins. Among the dietary supplements tested was spirulina (15 samples), which is also considered a blue-green alga and is not from the Upper Klamanth Lake. The researchers established a "no-observed adverse effect level" for the presence of microcystins via animal data and guidance from Health Canada along with the World Health Organization. The "tolerable human dose" was determined to be $0.04\,\mu g/kg/day$ or 2.4 total μg for a 60-kg person. Most people who use these types of supplement ingest $2\,g/day$; thus the safe dose of the hepatotoxin from blue-green algae was determined to be $1.0\,\mu g/g$ of product.

The results of the study were surprising (and perhaps not well distributed among the companies that sell these types of dietary supplement). In general, the average microcytin level of the blue-green algae from this lake was found to be $> 2.15\,\mu g/g$ of product (more than double the "safe" limit). Some samples tested were from the same lot (meaning that three bottles of product were purchased from the same lot), and the variation within the same lot from bottle to bottle ranged from $< 30\%$ to 99%. This indicates that a wide variation and potential for this particular hepatotoxin exists within this class of dietary supplements harvested from this region. The spirulina dietary supplements did not contain any serious amounts of hepatotoxins.

The exposure to high levels of microcystins is known to disrupt liver function and can result in intrahepatic hemorrhage as well as hypovolemic shock; less is known about the risks of exposure to low levels of this hepatotoxin over time. In animals, chronic exposure to low doses of this hepatotoxin is correlated with tumor progression (liver cancer). The blue-green algae often harvested for dietary supplements are harvested during the bloom when microcystins in the surface water are at their maximum. Because spirulina is grown under controlled conditions, the contamination risk is less likely. Blue-green algae are harvested from the surface waters; therefore, the microcystins contained therein are known also to contain neurotoxins that are produced from cyanotoxins.

Under DSHEA it is the marketer's responsibility to ensure the safety of the dietary supplement being retailed to the public. It appears that the most common "adverse effect" associated with blue-green algae supplements is gastrointestinal disturbance and that this side effect is sometimes interpreted by the industry to be "detoxification." There are no known cases in the FDA database of serious AERs regarding blue-green algae. Once the appearance of microcystin hepatotoxin in surface water and in blue-green algae dietary supplements became known, some states have enacted public health measures (Vermont and Oregon).

It may be wholly possible that every company in the product distribution chain from the original raw material supplier through the wholesalers and distributors (with the possible exception of the retail store outlet selling other companies' brands) should conduct safety studies *(45,46)*. These safety studies might run the gamut from animal toxicity surveillance all the way to human safety studies. In addition, the safety dossier may include laboratory analysis for the presence of known carcinogens, adulterated medications, and other standard safety parameters. Perhaps it is the responsibility of the raw material company and the finished goods retailer to know directly with first-hand evidence that their product is safe from known agents that can negatively affect the heart, kidneys, liver, and so on. This is where it is important for both types of company to employ the services of a firm or in-house individual to create a dossier ("Product Master File") that has direct and/or third party peer-reviewed published science denoting the safety of the ingredient and to contract with a laboratory—either private or at a university—to do first-hand product safety studies. Companies that do all of this can feel comfortable that they have directly satisfied the safety aspects of DSHEA for their products; consumers should feel more confident as well. It is true that unknown safety issues may still be lurking; however, a company that is responsible in doing its due diligence by having a complementary file on safety compiled from third-party science and direct "owned" science is taking the steps that the public should expect members of the sports nutrition supplement industry to take. This type of dilemma has led to the implementation of Good Manufacturing Practices (GCP's). GCPs have been enacted by the FDA in order to increase quality controls and product reliability.

11. SUBSTANTIATION FROM THE PERSPECTIVE OF RESEARCH

If you were set up on a blind date or better yet get involved in online dating and were told that the person who you will be meeting for drinks was six-foot two, in decent physical shape, and had brown hair and hazel eyes, but when you got to the club you saw no such person, would you wonder what was going on? You find the club just fine; it is located where he said it would be; the music is just as cool as you thought it would be; and then a man taps you on the shoulder, says your name ("Hi Staci, nice to meet you!). You stare at him wondering, "Who are you"? Finally he says his name and thanks you for meeting him. He compliments your attire and perfume, and notes that your drink of choice is one of his favorites as well. Still, you remember the online picture you saw of him, the details of his height, build, even the type of work that he said he did. So why does he not appear as well as he should? Why is not the real thing (the guy) *substantiating* the claims he made about himself in his online dating profile? Do you think he was fudging to hedge his bet? To perhaps increase the likelihood he could *make the sale*?

The scenario occurs not only socially but in business as well. For example, years ago you could go to a used car lot and test-drive any car there. The car would start, roar even; however, after you plunked down your money for the vehicle of your temporary dreams, the car would start to stutter and often not even start. After an almost epidemic of bad cars being sold, many states enacted laws to protect the car buyers—known as "lemon laws." The typical scenario encompasses a car being test-driven and enjoyed for its supposedly superior strength and looks that just does not pan out to have the "muscle" you expected. Simply put, the advertising was not *substantiated*. Can you think of other industries that may also need a "lemon law"?

12. ADVERTISING

Imagine you read an advertisement for a dietary supplement that stated "other natural supplements appear to treat only 15%, or one type of pain." Would you then believe that the supplement ad is implying that the particular product being promoted treats or is useful for *all* types of pain? Because a product is being directed or sold to consumers, should it not have direct research demonstrating support for its marketing claims? In the case of comparing one product versus others for a specific effect, once a specific effect is mentioned there is a further implication that substantiation must exist. A lawyer or even a regulatory agency might say that the advertisement is playing fast and loose with the scientific record.

Another example of a common advertisement that we all see in the major periodicals and trade journals is for a product purported to "reduce stress, improve sleep quality, diminish PMS, enhance mental sharpness, and reduce negative side effects of caffeine." This product is popular; in fact, it is a branded ingredient in many products. The studies on the branded ingredient have been carried out mostly in Japan, with few having been conducted in the United States. One may wonder if studies carried out in a foreign land, on people who may have genotypes and phenotypes different from those in the typical U.S. population, would yield the same results here. In other words, does the research *substantiate* the advertising claims that are made in the U.S. market? The current Draft Guidance for Substantiation released by the FDA in November 2004 addresses this very issue. In fact, the FDA indicates that foreign studies may not have equivalence for U.S. substantiation but will be considered as part of the portfolio. This is surely something that must be a consideration for any dietary supplement company and thus some motivation to organize a clinical trial platform is there. It is also a factor that consumers should consider when reading supplement marketing materials *(47–50)*.

13. IS THERE LEGAL PRECEDENCE?

In January 2002 the FDA issued guidance regarding claims and compliance guidelines for dietary supplements. According to the Guidance for Industry Regarding Structure/Function Claims, claims can be made on or for dietary supplements if you have substantiation that the claims are truthful and not misleading. The substantiation must be in place prior to the claim being made; and, in fact, the FDA is to be notified within 30 days of first marketing the product. So although there is clear guidance regarding what constitutes a structure/function claim (for more information see the set of 10 criteria in section 101.93(g) of Title 21 Code of Federal Regulations), it appears from the above two product advertisement examples that these laws are not being followed. Section 101.93(g) of Title 21 contains guidance regarding claims, and this section keys in on disease or symptom claims, implied claims, and much more. This is of utmost importance because the document contains clear guidance regarding what a company could say or how a company could structure the label, advertisements, and other product-supportive literature when marketing the product.

However, the lack of clarity as to what constitutes substantiation for a claim is a concern. For example, if you had one small-scale pilot open-label study on what you considered the key ingredient (the "active") in your product and the small study found efficacy of that ingredient, is this enough substantiation on which to base advertising claims? Or think about a situation in which you have a single-ingredient product, and studies on that ingredient have already been published in decent scientific journals and in the United States. Would this be considered substantiation? In this case, the answer appears to be that the substantiation of prior "third-party" (borrowed science, if you will) is valid if the product that you sell has the same exact dosage and quality of the studies on which you are basing your claims. In other words, if there are five studies on "product X" and in those studies the ingredient is dosed at 250 mg three times per day but you decide to sell it at 100 mg for twice per day usage, the published science does not support your claims.

The Federal Trade Commission (FTC) has announced its intention to be more active in policing the advertising of weight loss products. In fact, the FTC's publication "A Reference Guide for Media on Bogus Weight Loss Claim Detection" detailed the types of claims that the Agency believes to be almost impossible to substantiate. The FTC was granted this power in the Federal Trade Commission Act by the simple words within the act that note the prohibition of "unfair or deceptive acts or practices." It is clear that deceptive claims are those that are misleading or false in some way because facts are misstated or omitted or important information was not disclosed (47–53). Even if "puffery" is used in an advertisement, it can be considered deceptive if substantiation for the basis of the claim is not real or valid. There are many other areas that the FTC and FDA consider when evaluating if an advertisement (of any form) is valid, and these parameters should not be discarded. However, one should be cognizant of the FTC advertising and substantiation policy (known as the "substantiation doctrine") that was first enacted in 1972 and then further articulated in

1984. Within the FTC's actions, use of the "Pfizer factors" in noting if a claim is substantiated is typically used *(54)*. The factors evaluating substantiation include 1) type of product; 2) type of claim; 3) the benefits of a truthful claim; 4) the cost/feasibility of developing substantiation; 5) consequences of a false claim; and 6) the amount of substantiation that experts in the field believe is reasonable. Did you know that the FTC's experts have stated that weight loss beyond one pound per week without dieting or exercise should be considered scientifically not feasible?

14. THE COST OF NONTRUTH

Although the FTC has not clearly defined what constitutes substantiation, it has provided a global overview of how the Agency analyzes marketing claims. In addition, the FDA along with the FTC point to a 1994 ruling as related to weight loss claims that states that at least two well designed randomized clinical trials are needed to support weight loss and appetite suppressant claims. Because both the FDA and FTC point to this 1994 ruling, we now have some specific guidance as to what constitutes substantiation from the perspective of the amount of clinical trials needed for claims support *(55)*.

Companies that have run afoul of either the FTC or FDA guidelines for advertising and marketing have been pursued in courts, private actions, and have paid financially. Fines appear to have ranged from the cost of consumer redress to outright fines payable to the Agency. In a recent case, the manufacturer and subsequent retailer of one popular weight loss supplement paid $100,000 to the FTC and consented to not advertise any weight loss supplements that did not have substantiation (not all parties in this suit have settled with the FTC). In another FTC action, one company touting an oral growth hormone product paid the FTC $485,000 for consumer redress with a balloon clause of $5.9 million if the individual violates the consent order. Two other companies who also marketed oral growth hormone products have consented to pay the FTC up to $20 million dollars for their unsubstantiated marketing claims. In addition to the $20 million dollar notation, the companies and officers named in the FTC action may have to pay up to an additional $80 million dollars if the FTC finds that they misrepresented their personal and corporate finances. The consent order notes that substantiation is needed for claims and that the defendants have agreed to acquire the proper substantiation for future products they wish to retail. Do you think that with the possibility of losing $100 million dollars this company will spend the money on studies to support their marketing? The FTC has been very active over the past few years, and it appears that they are more active than ever in enforcing the laws regarding misleading advertising and substantiation *(56–60)*. This should be motivation enough for any dietary supplement company to learn and to conduct themselves in the right manner. In the big picture, because research and development is tax-credible (IRS Codes 41 and 174) why not spend the money now rather than paying fines, facing possible disbarment from the industry, and, heaven forbid, jail time later for being guilty of unsubstantiated marketing claims?

15. CONCLUSION

As we have seen, after a decade of quietude, recent Congressional legislation and FDA regulatory actions have targeted sports and fitness supplements, banning some products and criminalizing others. The future may yet see additional regulatory efforts, or even legislative initiatives, as the market continues to evolve. Although the rules and regulations set forth by the Federal Trade Commission and the Food and Drug Administration were designed with the best intentions, the role that political pressure and potential lobbying by other industries plays in the enforcement of these rules and regulations cannot be ignored. Part of the problem lies with the industry itself, which could assist in uplifting the perception of sports supplements by creating a structured "self-policing" policy to ensure product purity and quality. Meanwhile, despite certain governmentally imposed limitations, sports nutrition supplements will likely continue to grow in popularity. The Internet has provided consumers with access to a wealth of information to research dietary supplements from a multitude of perspectives conveniently, helping individuals to make more informed decisions. Millions of members of the consuming public continue to enjoy the right to decide for themselves whether to take a wide variety of dietary supplements.

REFERENCES

1. Pub. L. No. 75-717, 52 Stat. 1040 (1938) [codified as amended 21 U.S.C. 301 et seq. (1994)].
2. *Ibid*, citing Senate Report No. 103-410, 103d Congress, 2d Session, Committee on Labor and Human Resources (October 8, 1994), at page 16.
3. *Ibid*, citing Senate Report No. 103-410, at page 22.
4. Provided that the Secretary then initiates on-the-record rulemaking to affirm or withdraw the declaration. *See* 21 U.S.C. § 342(f)(1).
5. 21 U.S.C. § 331(a) and (v).
6. 21 U.S.C. § 342(f)(1) and 350b(a).
7. 21 U.S.C. § 342(f)(1)(A).
8. 21 U.S.C. § 333(a)(1).
9. 21 U.S.C. § 333(a)(2).
10. A "dietary ingredient" may be a vitamin, a mineral, an herb or other botanical, an amino acid, a dietary substance for use by man to supplement the diet by increasing the total dietary intake, or a concentrate, metabolite, constituent, extract, or combination of any of these. *See* 21 U.S.C. 321 (ff)(1).
11. 21 U.S.C. § 350b(c).
12. 21 U.S.C. § 350b(a).
13. See, FDA, Center for Food Safety & Applied Nutrition (CFSAN), "New Dietary Ingredients in Dietary Supplements," available at http://www.cfsan. fda.gov/~dms/ds-ingrd.html#whatis. See also, 21 CFR § 190.6.
14. "To date, we have not published guidance defining the specific information that the submission must contain. Thus, you are responsible for determining what information provides the basis for your conclusion." FDA, CFSAN, available at http://www.cfsan.fda.gov/~dms/ds-ingrd.html/.
15. FDA, CFSAN, Androstenedione Press Release, available at http://www.fda. gov/bbs/topics/news/2004/hhs_031104.html.
16. FDA, CFSAN, Androstenedione Warning Letters, available at http://www. cfsan.fda.gov/~dms/andrlist.html#letter/.
17. FDA, 9/10/01 UPDATE to FDA's Table of New Dietary Ingredient Notifica-tions, available at http://www.cfsan.fda.gov/~dms/ds-ingrd.html#whatis/.
18. Siegner W. "FDA's Actions on Ephedra and Androstenedione: Understanding How They Erode the Protections of DSHEA." Paper presented at FDLI's 47th Annual Conference, April 2004.
19. Johnson SK, Lewis PE, Inskeep EK. Steroids and cAMP in follicles of post-partum beef cows treated with norgestomet. J Anim Sci 1991;69:3747–3753. Braden TD, King ME, Odde KG, Niswender GD. Development of preovula-tory follicles expected to form short-lived corpora lutea in beef cows. J Reprod Fertil 1989;85:97–104. Wise TH, Caton D, Thatcher WW, Lehrer AR, Fields MJ. Androstenedione, dehydroepiandrosterone and testosterone in ovarian vein plasma and androstenedione in peripheral arterial plasma during the bovine oestrous cycle. J Reprod Fertil 1982;66:513–518.
20. Margolis M. Dietary supplements: caveat athlete. Available at http://www.law. uh.edu/healthlawperspectives/Food/980910Dietary.html/.

21. See footnote 14. (FDA, CFSAN, available at http://www.cfsan.fda.gov/~dms/ds-ingrd.html/.)

22. See generally, Legislation to Amend the Controlled Substances Act (Anabolic Steroids): Hearings on H.R. 3216 Before the Subcomm. on Crime of the House of Representatives Comm. on the Judiciary, 100th Cong., 2d Sess. 99, July 27, 1988; Steroids in Amateur and Professional Sports—The Medical and Social Costs of Steroid Abuse: Hearings Before the Senate Comm. on the Judiciary, 101st Cong. 1st Sess 736, April 3 and May 9, 1989; Abuse of Steroids in Amateur and Professional Athletics: Hearings Before the Subcomm. on Crime of the House Comm. on the Judiciary, 101st Cong., 2d Sess. 92, March 22, 1990; Hearings on H.R. 4658 Before the Subcomm. on Crime of the House Comm. on the Judiciary, 101st Cong., 2 nd Sess. 90, May 17, 1990.

23. Pursuant to 21 U.S.C. 812(b), a substance in Schedule III is to be placed there if: A) the drug or other substance has a potential for abuse less than the drugs or other substances in schedules I and II; (B) the drug or other substance has a currently accepted medical use in treatment in the United States; and (C) abuse of the drug or other substance may lead to moderate or low physical dependence or high psychological dependence.

24. John Burge, Legalize and Regulate: A Prescription for Reforming Anabolic Steroid Legislation, 15 Loy. L.A. Ent. L.J., 33, 45 (1994).

25. Pub. L. No. 101-647, Sec. 1902, 104 Stat. 4851 (1990).

26. Revised schedules are published in the Code of Federal Regulations, Part 1308 of Title 21, Food and Drugs.

27. 21 U.S.C. § 844.

28. 21 U.S.C. § 841(b)(1)(D).

29. For a comprehensive examination of the 1990 steroid legislation, see this author's treatise, Legal Muscle: Anabolics in America (2002) (www.legalmusclebooks.com).

30. Androstenedione or "andro," a precursor to the sex hormone testosterone, achieved national notoriety in 1998 when a bottle of the pills was spotted in the locker of St. Louis Cardinals slugger Mark McGwire.

31. Norsteroid prohormones, for example, can result in a positive test for nandrolone because both nandrolone and the norsteroid prohormones give rise to the urinary excretion of the metabolites norandrostenedione and norandrostenediol.

32. The problem of positive doping results from cross-contamination made national headlines when U.S. bobsledder Pavle Jovanovic was disqualified from the Salt Lake City Olympics for a test result he blamed on a cross-contaminated protein powder supplement and when research conducted at an IOC-accredited drug testing laboratory in 2002 found that 94 of the 634 samples contained substances not listed on the label that would trigger positive drug tests.

33. Examples: Anabolic Steroid Penalties: Hearings on H.R. 3866 Before the Subcomm. on Crime of the House of Representatives Comm. on the Judiciary, 108th Cong., 2 nd Sess., March 16, 2004; Consolidating Terrorist Watchlists: Hearings on H.R. 3866 Before the Subcomm. on Crime, Terrorism, and

Homeland Security, House of Representatives Comm. On the Judiciary, 108th Cong., 2 nd Sess., March 25, 2004; Anabolic Steroid Control Act of 2004: Hearings on S. 2195 Before the Senate Comm. on the Judiciary, 108th Cong., 2 nd Sess., October 6th, 2004.

34. Public L. No. 108-358; 118 Stat. 1661 (2004).

35. The new compounds are androstanediol; androstanedione; androstenediol; androstenedione; bolasterone; calusterone; *1-dihydrotestosterone (a.k.a. "1-testosterone"); furazabol; 13β-ethyl-17α-hydroxygon-4-en-3-one; 4-hydroxy-testosterone; 4-hydroxy-19-nortestosterone; mestanolone; 17α-methyl-3β, 17β-dihydroxy-5α-androstane; 17α-methyl-3α,17β-dihydroxy-5α-androstane; 17α-methyl-3β,17β-dihydroxyandrost-4-ene; 17α-methyl-4-hydroxynandrolone; methyldienolone; methyltrienolone; 17α-methyl-*1-dihydrotestosterone (a.k.a. "17α-methyl-1-testosterone"); norandrostenediol; norandrostenedione; norbolethone; norclostebol; normethandrolone; stenbolone; and tetrahydrogestrinone.

36. 21 U.S.C. § 811.

37. Note that litigation may be required to explore what "pharmacologically related" means with respect to steroidal compounds, including whether a steroid's anabolic capacity is inherent to its pharmacologic effect.

38. DHEA, or dehydroepiandrosterone, is a steroid hormone that serves as a metabolic precursor to the sex hormones estrogen and testosterone. It has become a popular dietary supplement among middle-aged and elderly consumers seeking improved vigor.

39. The errors in the 1990 law were exposed by coauthor Collins in *Legal Muscle* (see footnote 8). However, at least one typographic error appears in the new law: 13α-ethyl-17α-hydroxygon-4-en-3-one (a demethylated version of norbolethone) should read as 13β-ethyl-17β-hydroxygon-4-en-3-one. A new bill [S. 893] was introduced on April 25, 2005 to fix the problem (it also amends the chemical nomenclature for stanozolol).

40. Now listed only as methandienone.

41. Now listed only as 4-dihydrotestosterone.

42. Now listed only as clostebol.

43. A substance that does not appear in the Merck Index and appears to have resulted from a typographic error.

44. The definition of an anabolic steroid under the old law included, under subparagraph (xxviii), is "any salt, ester, or isomer of a drug or substance described or listed in this paragraph, if that salt ester, or isomer promotes muscle growth." The new law says, "any salt, ester, or ether of a drug or substance described in this paragraph." Note that the reference to muscle growth in the old law has been omitted [from what is now, in the longer list, subparagraph (xlx) of the new law] and that the word "isomer" has been replaced by "ether."

45. Gilroy DJ, et al. Assessing potential health risks from microcystin toxins in blue-green algae dietary supplements. Environ Health Perspect 2000; 108:435–439.

46. www.cdc.gov/hab/cyanobacteria/pdfs/facts.pdf/. Accessed March 26, 2007.

47. http://www.cfsan.fda.gov/~dms/sclmguid.html/.

48. http://www.cfsan.fda.gov/~lrd/fr000106.html/.

49. Edward W Correia. The Federal Trade Commission's Regulation of Weight Loss Advertising Claims. Food and Drug Law Journal 2004;59(4).
50. www.ftc.gov/bcp/conline/edcams/redflag/index.html/.
51. http://www.ftc.gov/bcp/conline/edcams/redflag/falseclaims.html/.
52. 15 USC 45(a)(1)(2000).
53. www.ftc/gov/bcp/guides/ad3subst.htm [104 FTC 648, 839 (1984)].
54. Pfizer Inc. 81 FTC at 91-93.
55. 118 FTC 1030, 1123, 1127 1994 Consent Order. FTC v. Schering.
56. http://www.ftc.gov/opa/2005/07/xenadrine.htm/.
57. http://www.ftc.gov/opa/2005/06/creaghanharry.htm/.
58. http://www.ftc.gov/opa/2005/06/greatamerican.htm/.
59. http://www.ftc.gov/opa/2005/06/fiberthin.htm/.
60. http://www.ftc.gov/opa/2005/06/avsmarketing.htm/.

2

Psychology of Supplementation in Sport and Exercise
Motivational Antecedents and Biobehavioral Outcomes

Rafer Lutz and Shawn Arent

Abstract

Research concerning the physiological and biobehavioral effects of supplements commonly used in sport or exercise settings has multiplied rapidly over the last decade. However, less attention has been directed to understanding the motivational pathways leading to sport and exercise supplement use. This chapter summarizes known usage rates for sport/fitness supplements and describes motivational theories and constructs that may be of use for understanding individuals' use of these substances. In this respect, we contend that researchers should consider behavioral approaches, the theory of planned behavior, balance theory, achievement goal theory, social physique anxiety, and muscle dysmorphia as useful for developing an understanding of the psychological influences on supplement use. For some of the latter theories/constructs, research has already shown support for their explanatory abilities, whereas research is scant and the utility for understanding sport/exercise supplement use is yet to be determined for many of the theories. In addition to describing the motivation behind supplement use, this chapter summarizes the biobehavioral effects of a select group of supplements commonly used to improve performance, fitness, or health. Specifically, we consider psychobiological effects of caffeine, creatine, *Ginkgo biloba*, and St. John's wort related to enhanced arousal, improved memory and cognition, enhanced brain function and protection, and reduced depression. There is promising initial evidence for the efficacy of these compounds in producing favorable psychological outcomes, although certain shortcomings of many studies on these compounds must be taken into account before reaching definitive conclusions.

Key words

Attitudes · Norms · Persuasion · Goal orientations · Motivational climate · Body image · Creatine · Caffeine · *Ginkgo* · St. John's wort

From: *Nutritional Supplements in Sports and Exercise*
Edited by: M. Greenwood, D. Kalman, J. Antonio,
DOI: 10.1007/978-1-59745-231-1_2, © Humana Press Inc., Totowa, NJ

1. INTRODUCTION

Of the more amazing news stories in the sports world over the years was shocking news that Terrell Owens, all-pro receiver for the Dallas Cowboys had attempted suicide and had been rushed to the hospital. Later, reports indicated that he was hospitalized owing to a reaction caused by mixing painkillers prescribed for his broken hand and his daily supplements. Terrell denied that he had attempted suicide, and Dallas police closed their investigation, labeling it an "accidental overdose." Although there is a lot of drama and attention surrounding such high profile cases, it does show some of the reason for concern in a largely unregulated and sometimes poorly researched industry. As supplement use has grown in sport and exercise settings, it becomes increasingly important to understand the reasons for using supplements. Additionally, many supplements target physiological systems, and it is important to understand that these systems have an impact on behavior (and vice versa). Such interactions of biological processes and behavior are termed biobehavioral effects and represent the intersection of the fields of physiology and psychology.

The growth of the overall supplement industry over the past decade is startling. As long ago as the 1994 congressional Dietary Supplement Health and Education Act (DSHEA) *(1)*, it was reported that there were an estimated 600 dietary supplement manufacturers in the United States producing in the neighborhood of 4000 products. The total annual sales of these products were estimated to be at least $4 billion *(1)*. In 1998, this figure had grown to $13.9 billion and by 2006 equaled $21.3 billion *(2)*. Anecdotally, it seems there are several motivations for this explosion in the use of supplements in sport and exercise: The increase in overweight and obesity and the pursuit of an ideal body shape has likely spurred the growth of supplements purporting to aid weight loss; media portrayal of "ideal" body images for males may be causing adolescent and adult males to increase supplement use to increase muscle size; and the ever-increasing stakes in the sporting world seems to be causing athletes to continue to strive for new ways to gain an edge on the competition.

As an initial foray into the psychology of supplementation, this chapter proposes to look at three questions: 1) What do we know about the prevalence of supplement use, particularly considering supplementation among athletic populations or among those trying to gain muscle mass or lose weight? 2) What accounts for the motivations causing the explosive growth in the sport and exercise/ weight loss supplement industry? 3) What do we know about psychobiological outcomes related to sport performance, fitness, and health for a select group of supplements with purported biobehavioral effects? Reviewing the literature considering each of the latter three questions reveals some interesting research findings as well as significant gaps in our body of knowledge. It is our hope that an initial review concerning supplement use in sport and exercise from a psychological perspective will serve as an impetus to further research, allowing a better understanding of the reasons for sport and exercise supplement use, their effects, and areas of potential concern.

2. USE RATES AND MOTIVATION

2.1. Definition of a Dietary Supplement

In the United States, a summary definition of dietary supplements as defined by the DSHEA of 1994 (1) can be stated as "a product (other than tobacco) intended to supplement the diet that bears or contains one or more of the following dietary ingredients: a) A vitamin, b) a mineral, c) an herb or other botanical, d) an amino acid, e) a dietary substance for use by man to supplement the diet by increasing the total dietary intake, or f) a concentrate, metabolite, constituent, extract, or combination of any ingredient described in clauses a, b, c, d, or e...." It should be noted, additionally, that any claims a manufacturer or individual makes about a supplement might change its classification. Dietary supplements claimed to cure, mitigate, or treat disease would be considered to be an unauthorized new drug rather than a supplement. Researchers have also differentiated "nonvitamain, nonmineral supplements" (NVNM) as those primarily consisting of herbal, botanical, protein/amino acid, brewer's yeast, and shark cartilage and a variety of other plant-based and nonplant dietery supplements such as enzymes and fish oil (3,4). Finally, in the arena of competitive sport specifically, it should be noted that there are both "accepted" and "illegal/banned" substances, including some supplements. Making this distinction somewhat difficult, various sports' governing bodies do not necessarily agree about which supplements should constitute banned substances. Among substances on the National Collegiate Athletic Association (NCAA) Banned Drug list are supplements such as ephedra and bitter orange (5).

In an interesting quandary for the field of performance enhancement, many supplements marketed to athletes contain banned substances—either overtly or because of impurities in these supplements. Geyer and colleagues' (6) International Olympic Committee (IOC)-commissioned study examined nonhormonal supplements to determine the prevalence of anabolic-androgenic steroids (AASs) in these products that were not listed on the label. Researchers bought supplements from 215 suppliers in 13 countries testing 634 nonhormonal supplements. A meaningful percentage of the supplements (14.8%) contained substances that would lead to a positive drug test. These results indicate that a proportion of supplements would be considered drugs; furthermore, they would be considered drugs with potentially deleterious side effects. The difficulty for many athletes and the various sports' governing bodies is that it is difficult to determine which supplements may truly be supplements and which contain substances that would be considered drugs.

Despite the possibility for failed doping tests, athletes typically take supplements because they want an advantage over their competition. Supplement use should not be surprising considering that some athletes are willing to take illegal/banned drugs to improve performance. Thus, it appears that the desire for money, fame, and feelings of achievement associated with athletic success are driving forces for the use of sport supplements. Yet not all athletes use supplements, and some are extremely cautious about the substances they ingest. Perhaps there are more subtle psychological factors at work that should be considered.

Problems also abound for individuals who use supplements to achieve added weight loss and/or muscle gain (or even improved recovery after workouts) from their exercise programs. Products of dubious efficacy are plentiful, and little is known about drug–supplement or supplement–supplement interactions that may be hazardous to the user's health *(7,8)*. Substances such as brindleberry (*Garcinia cambogia*/indica), capsaicin, caffeine, L-carnitine, chromium picolinate, and *Ginkgo biloba* have purported weight loss benefits; however, not all of these substances have research support in the published literature *(9)*. Considering a worldwide ongoing obesity epidemic *(10)*, and problems particularly in the United States, it is not surprising that many individuals are seeking new ways to lose weight. Supplements promise, though probably seldom deliver, a magic bullet of sorts: easy, hassle-free weight loss with little in the way of dietary sacrifice. Motivation for users of weight loss supplements, however, is likely not simple or straightforward. A range of issues ranging from body image concerns to obsessive tendencies may be important to consider.

2.2. Prevalence of Supplement Use

There have been several large-scale surveys of supplement use among United States citizens. The Slone Survey *(11)* used random digit dialing to survey 2590 U.S. citizens regarding commonly used herbals/supplements (NVNM) and vitamins and minerals. The ten most commonly used substances in these categories are reported in Tables 1 and 2, respectively. Additionally, 14.0% of individuals reported use of a herbal/supplement over the previous 7-day period. Popular reasons given for using herbal/supplements included health (16%), arthritis (7%), memory improvement (6%), energy (5%), and immune booster (5%).

Other large surveys have been conducted to examine supplement use and find, generally, greater usage rates among older individuals, the nonobese, Caucasians, females, nonsmokers, physically active individuals, those with higher levels of educational attainment, and those with high fruit and fiber intake *(3,12,13)*. In a study of 1000 university students, Perkin and colleagues *(4)* found that 26.3% indicated use of an NVNM supplement and 16% had used in the past. *Ginseng*, *Echinacea*, protein powder/amino acids, and *Gingko biloba* were the most frequently used supplements. Reasons for use included improve energy (61.2%), promote weight loss (38%), burn fat (36.1%), supplement inadequate diet (35%), build muscle (27.8%), and relieve stress/improve mood (24.7%). It appears the reasons for use in the university population are more performance- and appearance-driven than

Table 1
Ten Most Commonly Used Vitamins/Minerals in the United States: 1-Week Percentage Prevalence by Sex and Age[a]

Rank	Herbal/supplement	Men			Women			Total
		18–44 Years	45–64 Years	≥ 65 Years	18–44 Years	45–64 Years	≥ 65 Years	
1	Multivitamin	19	29	31	25	29	33	26.0
2	Vitamin E	3	18	14	4	19	19	10.0
3	Vitamin C	4	13	12	5	16	14	9.1
4	Calcium	3	4	7	6	19	23	8.7
5	Magnesium	1	2	5	2	6	5	3.0
6	Zinc	2	2	3	1	5	3	2.2
7	Folic acid	<1	3	4	2	3	4	2.2
8	Vitamin B_{12}	<1	2	2	2	3	3	2.1
9	Vitamin D	<1	<1	2	<1	5	7	1.9
10	Vitamin A	<1	3	2	<1	4	3	1.8
	Any use	24	46	47	35	51	59	40.0

Adapted from Kaufman et al. (11).
[a]Percentages are weighted according to household size. Numbers (excluding multivitamins) indicate prevalence of use in nonmultivitamin products.

Table 2
Ten Most Commonly Used Herbals/Supplements in the United States: 1-Week Percentage Prevalence by Sex and Age[a]

Rank	Herbal/supplement	Men			Women			Total
		18–44 Years	45–64 Years	≥ 65 Years	18–44 Years	45–64 Years	≥ 65 Years	
1	Ginseng	4	4	<1	2	5	2	3.3
2	Ginko biloba extract	<1	4	1	1	4	5	2.2
3	Allium sativum	<1	4	4	1	3	3	1.9
4	Glucosamine	<1	2	4	<1	5	4	1.9
5	St. John's wort	<1	2	0	2	3	<1	1.3
6	Echinacea augustifolia	1	1	0	1	3	<1	1.3
7	Lecithin	<1	<1	1	1	3	1	1.1
8	Chondroitin	<1	1	1	0	3	2	1.0
9	Creatine	4	0	0	0	0	0	0.9
10	Serenoa repens	1	1	4	0	<1	0	0.9
	Any use	12	17	11	10	23	14	14.0

Adapted from Kaufman et al. (11).
[a]Percentages are weighted according to household size.

the reasons in the general U.S. population, which are more health-focused *(11)*.

2.2.1. SPORT-SPECIFIC USE

Athletes undoubtedly account for a large portion of those who use dietary supplements, and there are a variety of products that are marketed directly at competitive athletes. Sobal and Marquart's *(14)* meta-analytic review of vitamin/mineral supplement use among athletes reported an overall use rate of 46%. They also found that elite athletes tended to take supplements more commonly than college or high school athletes, and women used supplements more often than men. Another conclusion was that some athletes take doses high enough to lead to nutritional problems. Regarding herbal or other agents (e.g., AASs), use rates widely vary depending on the sport population investigated or the definition of supplement. Some studies report lower use rates—often when investigating younger or less competitive athletes and when using a definition of supplement that excludes sports drinks and vitamins. Scofield and Unruh *(15)* found that 22.3% of adolescent athletes reported supplement use in a small sample in Nebraska. In their study, athletes defined supplements on their own terms, and most did not consider sport drinks to be a supplement. Considering elite Canadian athletes participating at the Atlanta and Sydney Olympics, respectively, prevalence rates of 69% and 74% were reported *(16)*. Vitamin use was most common (58%–66%), whereas nutritional supplements were used commonly (Atlanta: 35% of men, 43% of women; Sydney: 43% of men, 51% of women), often consisting of creatine and/or amino acid supplementation. Nutritional supplement use occurred most often in cycling (100%) and swimming (56%) populations. Based on results overall, it appears that supplementation increases with the competitive level of the sport and is somewhat higher for female athletes.

2.3. Motivational Theories Applied to Supplement Use

Examining the literature as a whole, it is well established that supplement use is high among athletic populations and those who want to either build muscle or lose weight. What existing theoretical paradigms, however, might inform our future study of this area concerning reasons for use and potential abuse? To answer this question, it is perhaps best to consider three specific categories: supplement use to produce athletic performance benefits; supplement use to build muscle for aesthetic purposes or body image concerns; and supplement use to lose weight for aesthetic purposes, body image concerns, or health. Each of the three reasons for use likely has different motivational underpinnings. Therefore, in our description of theoretical paradigms that may be applied to understand supplement use, we have tried to identify the areas where each theory may be particularly effective for understanding supplementation. Overall, it should be considered that there are likely to be multifaceted, overlapping motivations for supplement use. Table 3 gives an overview of psychological/motivational theories and constructs that may be related to supplement use. As is obvious upon examination of this table, little direct inquiry has been conducted to explain motivation to use supplements in sport and exercise settings.

Certainly, there is a behavioristic explanation possible for the use of supplements in that athletes' use may lead to reward contingencies (e.g., more prize money), thereby driving future behavior. Similarly, supplements that build muscle or promote weight loss could produce rewarding results. Also, there are undoubtedly social influences at work considering that coaches, parents, athletic trainers, and peers have been reported to be influential regarding the decision to take supplements *(15,17,18)*. Finally, other extant motivational theories may be useful for predicting supplement use and abuse. Most of these potential explanations have undergone limited research, if any, in the context of supplement use, so our primary purpose is to describe how theory or previous research would predict the stated constructs' explanatory ability and direct future inquiry to understand motivation for supplementation.

2.3.1. OPERANT CONDITIONING

If supplements work quickly and effectively to produce performance changes or body shape changes resulting in reward or praise, it is possible to use operant conditioning as a means of explaining the choice to take supplements. Operant conditioning focuses on the manner in which our behavior and action are influenced by the outcomes that follow them *(19)*. Derived from

Table 3
Understanding the Motives for Supplement Use in Sport and Exercise Settings

Theory and pertinent construct(s)	Expected relation to supplement use	Research support
Behaviorism/operant conditioning		
Positive reinforcement	Supplement use that leads to reward/praise should promote future use	No known direct support
Punishment	Supplement use leading to punishment/sanction should reduce future use	Indirect support (success of doping sanctions)
Theory of planned behavior		
Attitude	Attitudes that supplements are good/healthy should promote use	Supported with limited research
Subjective norm	Beliefs that others think you ought (injunctive norm) to use, or beliefs that others commonly do use (descriptive norm) should promote use	Supported with limited research
Perceived behavioral control	Perceptions that supplements are easy to use, available, or inexpensive should promote use	Supported with limited research

Balance theory	If the subject likes a celebrity spokesperson promoting a supplement and perceives that the celebrity approves of the supplement, the subject should be more willing to try the supplement as he/she should like it more (to achieve psychological balance)	Well supported generally No known direct support
Achievement goal theory		
Goal orientation	Individuals who have a high ego orientation for their sport would be expected to be more willing to use supplements even if these supplements are potentially harmful	No known direct support
Motivational climate	Individuals within ego-promoting climates would be expected to be more willing to use supplements even if these supplements are potentially harmful	No known direct support
Other constructs		
Social physique anxiety	High social physique anxiety promotes supplement use in some circumstances	No known direct support
Muscle dysmorphia	High levels of muscle dysmorphia may promote supplement use	Supported with limited research

the behavioristic research tradition *(20)*, the sum of findings in this area dictate that some outcomes/stimuli strengthen the behavior that preceded them, and others weaken the likelihood of the behavior that preceded them. Outcomes or consequences that increase the likelihood of behavior are known as reinforcers, and those that decrease the likelihood of behavior are known as punishment. In the present context, prize money, praise from others, or rewards due to improved performance are reinforcers of the behavior to take supplements. However, it should be apparent that this theory demands that the reward is contingent upon taking the supplement—in other words, the supplement must work effectively and ostensibly. Otherwise, we must use other motivational explanations to understand supplement use. Because most legal supplements likely would not produce dramatic sport performance gains, muscle mass gains, or weight loss results, perhaps the best explanation for use is found in other theories. Behavioristic explanations, however, might be highly applicable considering the use of illegal substances such as steroid use.

2.3.2. PERSUASION AND CONFORMITY

A set of ideas/principles that might best explain supplement use across areas (performance enhancement, weight loss, muscle building) can be found by examining research on persuasion and conformity to norms. Edwin Moses, one of the greatest hurdlers of all time, once estimated the use rate of illegal drugs in track and field to be about 50% at the elite level *(21)*. Such a statement is reflective of a descriptive norm indicating what you believe others actually do. In an exercise setting, this would manifest in one's belief that others commonly take weight loss supplements or muscle-building supplements. The power of such norms to influence behavior is well documented *(22–24)*; and if supplement use is perceived as the norm, there will be social pressure to conform—even in the face of negative outcomes that might be due to use.

Among theories adopting the concept of normative influence on behavior, the Theory of Planned Behavior (TPB) may be useful for understanding motivation to use supplements. This theory proposes the existence of three psychological constructs that are believed to influence behavior through the mediator of intent: normative influences, attitude, and perceived behavioral control *(25)* (Fig. 1).

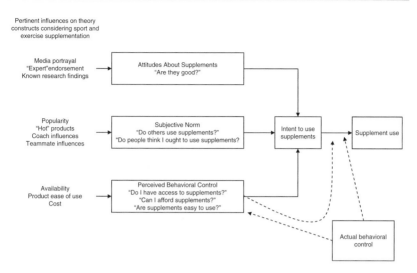

Pertinent influences on theory
constructs considering sport and
exercise supplementation

Media portrayal
"Expert"endorsement
Known research findings

Attitudes About Supplements
"Are they good?"

Popularity
"Hot" products
Coach influences
Teammate influences

Subjective Norm
"Do others use supplements?"
"Do people think I ought to use supplements?"

Intent to use
supplements

Supplement use

Availability
Product ease of use
Cost

Perceived Behavioral Control
"Do I have access to supplements?"
"Can I afford supplements?"
"Are supplements easy to use?"

Actual behavioral
control

Fig. 1. Theory of planned behavior applied to sport/exercise supplement use. Adapted from Ajzen and Fishbein *(25)*.

Within the category normative influences, it is important to consider what one believes concerning what others think you ought to do (known as the *injunctive norm*) in addition to the actual *descriptive norm* relating to what others do themselves *(25,26)*. Attitudes pertain to the degree that a behavior is positively or negatively valued. Specifically, in this context, would taking supplements be considered good? Perceived behavioral control pertains to one's perceptions about factors that might facilitate/impede performance of a behavior. Factors such as cost and availability of supplements would be expected to influence perceptions of behavioral control.

No known study has focused on supplement use among athletes/exercisers using this theory, although it has shown good predictive ability examining supplement use in other populations. In a study of 400 randomly selected women from the UK Women's Cohort Study including 15,000 participants who completed variables pertinent to TPB constructs and supplement use, Conner and colleagues *(26)* found that intent accurately predicted supplement use (82.9% accuracy) and that attitude was the strongest predictor of intent, although the subjective norm and perceived behavioral control were also significant predictors of intent. The latter three variables predicted an astonishing 70% of variance in intent to use supplements. Certainly, these findings suggest that attitudes, normative beliefs, and perceptions of control are important correlates of supplement use among women. What is not known, however, are the ultimate reasons for use in this sample (e.g., weight loss, health benefits). Future research examining the TPB's predictive ability given different reasons for supplement use may be informative as these reasons may moderate the relative association of these predictors with intent to use and actual supplement use.

When trying to change attitudes about whether supplements are good or bad, it is likely that some individuals are more persuasive than others. Individuals are more persuasive if they are seen as trustworthy or having pertinent expertise *(27)*. The supplement industry often uses exactly such a strategy to help market their products. University research and "expert" sport and exercise nutritionists are increasingly being used to support the efficacy of performance enhancing, muscle building, or weight loss supplements. This is a wise strategy on the part of the supplement industry from an economic standpoint, but consumers should also consider that a company may contract with three universities to test their product and report only the results of the positive outcomes in their advertisements. Such a practice should be viewed as unethical; yet this is certainly a possibility. Even in the published literature there is likely to be a bias to the benefits of supplements in contrast to studies documenting no effects due to the file drawer problem in current scientific practice where significant results are published more frequently than nonsignificant results *(28,29)*. Scientists conducting investigations in the area of supplements should attempt to publish nonsignificant and significant findings alike because of these issues.

Another theory that might inform persuasive efforts to effect attitude change is Heider's *(30)* balance theory. According to this theory, people want to view the world in a consistent manner. In other words, if your favorite athlete is Mark McGwire (who has admitted using androstenedione, a controversial supplement) but you do not generally approve of the use of performance enhancing substances, you are likely to change either your opinion of McGwire or of the acceptability of substance use to achieve harmony of thought and action. In this manner, we might expect that overt or covert messages from well liked or well respected athletes that indicate their use of supplements will cause others to view use as acceptable even if they initially resist. This would also support supplement companies' use of well liked spokespersons to tout the use of their products. Similarly, using well liked spokespersons to advertise supplements with purported weight loss benefits should have similar effects. As a recent example of this strategy, actress/ model Carmen Electra recently signed on to help promote a "weight-loss beauty pill" known as NV™.

2.3.3. ACHIEVEMENT GOAL THEORY

Although the latter theoretical explanations allow some insight into the use of supplements among athletic populations, perhaps no theory addresses supplement use in this population as precisely as the achievement goal theory (31,32). Within this theoretical paradigm, it is assumed that there are differences in the manners by which athletes judge their competence or success. Such differences of goal involvement may be influenced by environmental/situational influences, termed the climate (31), and individual differences, termed goal orientations (33). Individuals who are task-oriented tend to judge their success on the basis of personal improvement (e.g., I am successful if I learn or improve), whereas those who are ego-oriented tend to judge their success on the basis of social comparison with others (e.g., I am successful if I win). Task-oriented individuals typically view personal ability as changeable and exhibit strong motivation regardless of their perceptions of competence (34). Those who are ego-oriented, in contrast, tend to view ability as more static and are thus more likely to engage in questionable strategies to ensure winning (35) and would be expected to engage in more frequent doping activities and perhaps a greater willingness to use supplementation strategies. Interestingly, there is little direct evidence to link ego goal orientations or ego climates with supplement use. However, this is an important topic of inquiry in future studies. Certainly, it is an important topic of study in youth sport. If young athletes who are ego involved begin using supplements, we must consider that they may not be as likely to comprehend the potential for drug–supplement interactions.

2.3.4. BODY IMAGE AND EATING DISORDERS

The desire to win in athletics couched within the achievement goal framework appears to be a good model for understanding sport supplement use. In exercise and fitness settings, however, theories directed to body image may be more useful. Although obesity rates have dramatically increased in the United States over the past few decades *(36)*, a similar increase in the ideal body size has not occurred in the female population. In fact, the "ideal" waist size for females may have become unhealthily small *(37)*. Because of these two contradictory trends, it is no surprise that the use of supplements targeted at weight loss has increased dramatically during this same time period. The nation is getting heavier and feeling worse about it, especially the female segment. Although we know that a large portion of supplement use is driven by a desire to lose weight *(4)*, less is known about more severe body image disturbance in females and how it may affect unhealthy supplement practices. Risky supplementation involving taking high doses or mixing supplements may occur among individuals who have more severe body image disturbances. More research is needed to investigate the relations between social physique anxiety, the degree to which people are anxious or nervous when others are observing or evaluating their physique *(38)*, and supplementation practices. Social physique anxiety has been shown to be positively related to percent body fat and body dissatisfaction *(38,39)*, and it is logical that such dissatisfaction would be an impetus for the use of weight loss supplements that may be used in exercise settings.

In a similar vein, it may be informative to look at disordered eating practices and supplement use. One of the diagnostic criteria for bulimia *(40)* is "recurrent inappropriate compensatory behavior in order to prevent weight gain, such as self-induced vomiting; misuse of laxatives, diuretics, enemas, or other medications; fasting; or excessive exercise" (p. 549). It might be expected that individuals suffering from bulimia, and perhaps anorexia nervosa, might have excessive or unhealthy supplementation practices. In one survey, research showed that among women at risk for eating disorders approximately 65% engage in frequent use of "diet pills" *(41)*, of which it is reasonable to assume a large portion would be considered supplements. Again, as the supplement industry has surged ahead, promoting an ever-increasing range of products, researchers must try to determine how this may pose yet another risk for those who have eating disorders. It may also be that a high level of supplementation is a warning sign of an eating disorder. Simply, more research is needed in this area. We do know that a warning sign of eating disorders is excessive exercise *(40)*, and those working in exercise settings should be alert to this warning sign.

2.3.5. ADONIS COMPLEX

Typically, when one considers issues related to body image concerns in the United States one focuses on female populations. However, Pope and colleagues *(42)* have identified a similar but opposing set of preoccupations afflicting males termed the Adonis complex, which seems to be afflicting boys and men more specifically during the last few decades. Those afflicted with the Adonis complex may compulsively lift weights or exercise, engage in steroid abuse, elect to undergo plastic surgery, or suffer from eating disorders or body dysmorphic disorders, all in attempts to gain muscle mass, change fat distribution, or otherwise alter their appearance to some ideal.

In one of the seminal works in this area, Pope and colleagues *(43)* interviewed 108 bodybuilders (55 steroid users and 53 nonuser controls) and found a higher than normal incidence of anorexia nervosa (2.8%) and a surprising incidence of "reverse anorexia" (8.3%), with some of the respondents believing that they appeared small and weak despite their large, muscular appearance. The latter finding indicated that some of these bodybuilders exhibited unusual preoccupations with their appearance. Such pathological preoccupation with muscularity has been termed *muscle dysmorphia*, and research concerning this issue has found it to be relatively common among adolescents and young males *(42)*. As an important link to potential supplement use or abuse, in Pope and colleagues' *(43)* research all of the bodybuilders indicating muscle dysmorphia (then termed "reverse anorexia") were in the sample of steroid users, and many reported that the symptoms of muscle dysmorphia were a factor that led to steroid use. If these individuals are driven to use illegal substances, one might expect higher than normal use of any substance (i.e., muscle building supplements) purported to promote strength or muscle mass gains. As an indication of the degree of this obsession, individuals with this affliction have reported lifting weights for hours a day while sacrificing other areas of their lives. For example, some of these individuals reported earning degrees in business, law, or medicine but did not pursue a career or gave up a career in these areas because they needed more time to lift weights *(44)*.

Recent research indicates that bodybuilders suffering from higher levels of muscle dysmorphia are more likely to experience body dissatisfaction, social physique anxiety, and use muscle-building- or fat-reducing-targeted supplements *(45)*. Research has questioned the association between "muscle dissatisfaction" and frequency of exercise, indicating that the incidence of muscular dissatisfaction is similar among frequent and infrequent exercisers *(46)*, the latter study used perhaps an overly simplistic measure of muscular dissatisfaction likely not sensitive to indications of muscle dysmorphia. Nonetheless, further research is necessary to determine the incidence of muscular dissatisfaction and muscle dysmorphia and their impact on a variety of important outcomes such as supplement use. At present, there is some evidence that supplement use is greater among individuals with muscle dissatisfaction *(46)* or muscle dysmorphia *(45)*. It also appears that illegal supplement use may accompany muscle dysmorphia as data indicate that 1 million or more U.S. males have used these substances (e.g., anabolic steroids) primarily to promote muscle growth as opposed to performance enhancement purposes *(42)*. Finally, it should also be noted that research finds that some men have become preoccupied with fat, as opposed to muscle, and, in contrast to attempting to gain weight, may develop eating disorders *(42)*. This suggests that body image concerns among males may drive some to attempt obsessively to build muscle mass whereas others may obsessively work to lose fat. In both cases it is likely that legal or illegal supplementation is a common means to achieve such goals.

3. BIOBEHAVIORAL EFFECTS OF SELECTED SUPPLEMENTS COMMONLY EMPLOYED FOR PERFORMANCE, FITNESS, AND HEALTH

In addition to the psychological reasons underlying supplement use, there are also a number of psychological and biobehavioral effects that are associated with certain supplements or ergogenic aids. In some illegal ergogenic aids [i.e., anabolic androgenic steroids (AASs)], many of these effects are considered negative side effects associated with use of the drug or substance. For example, AAS abuse has been reported to result in mood changes, irritability, aggression, and psychotic or manic behavior *(47–51)*. There has been mixed support for these effects in experimental studies using supraphysiological doses of testosterone *(52–54)*. However, it appears that this may be due to considerable individual variability in the psychological responses to AAS use, with those individuals predisposed to aggression or hostility most likely to respond with increased aggressive behaviors or anger *(52,53)*. It is worth noting, though, that owing to ethical restrictions in human subjects research, even the "supraphysiological" doses of testosterone used in the above studies (200–600 mg·wk^{-1}) fall well below what some individuals typically use when on an AAS cycle. The negative psychological responses likely become more apparent as the dosage increases.

In contrast to the negative psychological outcomes associated with AAS administration, there are many positive psychobiological responses for some of the legal ergogenic aids and nutritional supplements used in sport or exercise settings, such as caffeine, creatine, *Ginkgo biloba*, and St. John's wort. Some of these effects include enhanced arousal, improved memory and cognition, enhanced brain function and protection, and reduced depression. Unfortunately, definitive conclusions are difficult for each of these compounds owing to the typically small-scale studies conducted on each of them and the likely publication bias in some cases *(55)*. In light of this, emphasis is placed herein on meta-analytic results and findings of well controlled, randomized trials where available when reviewing the evidence for each supplement.

3.1. Caffeine

Caffeine, particularly in the form of coffee, is one of the most commonly used ergogenic aids and central nervous system (CNS) stimulants *(56)*. In fact, according to Sinclair and Geiger *(57)*, caffeine is the worlds' most abused substance. Its effects on physical performance (particularly endurance) are well established, but there are also many studies that have examined the role of caffeine as a psychological ergogenic. Caffeine would be used by athletes as a performance aid in athletic situations, yet it is likely that this is a widely used substance among those populations in exercise settings as well. Research has generally supported a beneficial effect of caffeine intake on sensorimotor performance as well as vigilance and attention. Some of these findings include improvements in simple and choice reaction time *(56,58–60)*, faster response rate for a tapping task *(60)*, sustained attention and vigilance *(60–62)*, and improved decision-making *(63)*. Furthermore, these effects appear to be even more pronounced under the situation of sleep deprivation or fatigue *(63–65)*. It has been suggested that many of these findings could be due to reversal of withdrawal effects after caffeine abstinence rather than actual performance benefits per se, but there is recent evidence that this is not the case *(66)*.

Lieberman et al. *(67)* examined the impact of three doses of caffeine (100, 200, or 300 mg) and a placebo on cognitive performance in 68 Navy SEAL trainees under extreme conditions during their Hell Week. At the time of testing, the SEAL trainees had been sleep-deprived for approximately 72 hours. Furthermore, owing to standard SEAL training policy, they had refrained from caffeine intake for the entire week, thus controlling for possible "reversal of withdrawal" effects. Results indicated significant improvements in visual vigilance, choice reaction time, fatigue, repeated acquisition, and alertness in both the 200- and 300-mg groups, with the greatest effects at 1 hour after ingestion but persisting at 8 hours. There were no differences between the 200-mg and 300-mg groups. Additionally, although marksmanship was not improved, caffeine also did not adversely affect it, which had previously been expressed as a concern regarding fine-motor steadiness tasks due to muscle tremors associated with caffeine *(68,69)*. It could be that even if caffeine disrupts "steadiness" it improves target acquisition and detection *(70,71)*. For tasks requiring less fine-motor control (i.e., a tennis serve), accuracy has been found to improve with caffeine administration *(72)*. In general, the arousal effects induced by caffeine appeared to benefit most areas of cognitive and psychological performance tested.

Caffeine can be linked to a state of arousal through specific physiological mechanisms. Research has shown that caffeine intake leads to more positive frontal P2 and parietal P3 components of event-related potentials (ERPs) *(73)*. This increase in CNS activity indicates elevated levels of arousal, heightened information processing, and control-oriented mechanisms *(73)*. The arousal-inducing properties of caffeine appear to be tied primarily to its binding to adenosine A_1 and A_{2A} receptors, although recent research has suggested that the A_{2A} receptor is primarily responsible for mediating these effects *(74)*. This may help explain the positive effects that caffeine ingestion appears to have on the risk of Parkinson's disease, which is characterized by central dopamine deficiency *(75)*. Because of an interaction between dopamine D_2 receptors and adenosine A_{2A} receptors *(76)*, caffeine appears to have dopamine agonist-like effects *(77)*. Although two small-scale clinical trials have failed to find an effect of caffeine on Parkinson's symptoms, the dosage used was extremely high (>1000 mg), which may have eradicated the motor benefits seen at lower caffeine doses *(78)*.

According to Ferrauti and colleagues *(72)*, caffeine can also produce improvements in psychomotor coordination and neuromuscular function through its effect on contractile status. Other effects include increased heart rate and the release of epinephrine, norepinephrine, and cortisol *(73,79)*. Because of caffeine's stimulatory effects on the hypothalamic-pituitary-adrenal (HPA) axis *(80)*, it has the potential to induce an endocrine response similar to that seen under mental or physical stress. Individuals classified as "at risk" for hypertension appear to be especially sensitive to this HPA activation *(81,82)*, particularly when also exposed to a psychosocial stressor *(82)*. In light of this, it appears prudent to caution those at risk for hypertension on the use of caffeine, particularly during periods of high stress.

When evaluating the psychological or biobehavioral effects of caffeine, it is important to keep in mind that plasma concentrations typically reach their peak within 30 to 60 minutes after ingestion *(83)*. Therefore, test administration must be timed appropriately. Additionally, some studies have evaluated the psychological and cognitive effects using absolute doses whereas others have standardized dose by body weight. Depending on the heterogeneity of the study sample, these two approaches could provide distinctly different interpretations of a dose-response effect. This can become even more problematic if pretreatment caffeine intake is not accounted for or controlled *(84)*. This may be particularly important for the study of skills requiring concentration, reaction speed, and accuracy. Finally, Mikalsen et al. *(85)* found mixed evidence for potential expectancy and placebo effects for caffeine and arousal. Further research in this area appears warranted to determine the extent of these effects and the impact on psychological performance.

3.2. Creatine

Creatine use is common in athletic and exercise populations. Much like the research on caffeine, there is abundant evidence that creatine supplementation improves physical performance, particularly for tasks requiring strength and power. Therefore and because of the role of creatine in brain function and neuroprotection *(86)*, interest has risen in investigating the role of creatine in the treatment of certain clinical conditions such as traumatic brain injury (TBI) and Huntington's disease as well as general cognitive functioning.

Overall, TBI affects approximately 7 million people each year in North America alone *(87)*. It is also estimated that approximately 300,000 individuals suffer TBI due to sport participation annually *(88)*, which can result in subdural hematomas, deterioration of cognitive function, or even death *(89)*. These injuries are occurring despite improvements in safety equipment, rule changes, and training methods. Furthermore, there is currently little in the way of "therapeutic intervention" to help prevent or even treat this type of injury *(90)*. It has been hypothesized that creatine might exert neuroprotective effects through provision of sufficient ATP immediately after trauma or by inhibiting the mitochondrial permeability transition pore (mPTP) *(91,92)*.

In an effort to examine the potential neuroprotective effects of creatine on this type of injury as well as underlying mechanisms, Sullivan et al. *(91)* employed an animal model of experimentally induced TBI. They found that chronic creatine administration reduced cortical damage due to experimentally induced TBI by as much as 36% in mice and 50% in rats. They also found that chronic creatine supplementation resulted in maintenance of cellular ATP levels, reduction of free radical formation, decreased intramitochondrial Ca^{2+}, and maintenance of mitochondrial membrane potential *(91)*. There appeared as well to be inhibition of mPTP, although recent research using ubiquitous mitochondrial creatine kinase (UbMi-CK)-deficient mice, suggests that the neuroprotective effects of creatine are most likely due to the maintenance of ATP/ADP and phosphocreatine/creatine (PCr/Cr) levels rather than mPTP inhibition *(93)*. Nonetheless, the overall findings of Sullivan et al. *(91)* are encouraging and warrant further investigation into the potential use of creatine supplementation in the prevention and treatment of TBI.

The use of a human model in a true-experimental design of TBI and creatine supplementation would likely not be ethically acceptable or logistically feasible at this point. However, there is evidence supporting the cognitive benefits of creatine in humans. Using a double-blind, placebo-controlled, crossover design, Rae et al. *(93)* found that 6 weeks of creatine supplementation (5 g·day^{-1}) in 45 young adult vegetarians resulted in improved brain function as evidenced by significant improvements on speed of processing tasks, including working memory and intelligence. Similarly, Watanabe et al. *(94)* found that supplementing with 8 g of creatine per day for 5 days reduced mental fatigue resulting from a repeated serial calculation task. Perhaps more importantly, they found that creatine ingestion was associated with increased oxygen utilization in the brain during the task as measured with near-infrared spectroscopy *(94)*. This is consistent with in vitro results demonstrating that mitochondrial oxygen consumption is greater in the presence of higher creatine concentrations *(95)*.

Another potential area of clinical utilization for creatine supplementation has been in the treatment of Huntington's disease (HD). HD is a progressive neurodegenerative disease for which there really is no effective therapy. One of the side effects to HD that might contribute to neuronal death is impaired energy metabolism, particularly in the brain *(96)*. Given creatine's demonstrated neuroprotective effects and its role in bioenergetics, researchers have begun to examine the utility of the supplement in delaying or alleviating the symptoms associated with HD. Fortunately, a transgenic mouse model of HD was developed *(97)* that has allowed examination of mechanistic effects of HD and potential treatments. Using this model, Ferrante et al. *(96)* found that creatine supplementation enhanced brain creatine concentrations, delayed brain atrophy, delayed atrophy of striatal neurons, and significantly improved survival. Body weight and motor performance were also improved with creatine supplementation. Matthews et al. *(98)* used a slightly different animal model of HD by inducing HD-like striatal lesions with malonate and 3-nitropropionic acid (3-NP). They found that both creatine and cyclocreatine provided protective effects against malonate and 3-NP lesions in a dose-response manner. There was a reduction in the neuroprotection afforded by creatine supplementation at its highest dose, suggesting a curvilinear (U-shaped) dose-response curve for treatment efficacy. It is also worth noting that creatine appeared to function as either a free-radical scavenger or an inhibitor of hydroxyl radical formation *(98)*. The authors suggested that by preventing ATP depletion creatine and cyclocreatine serve to prevent the cascade of events leading to free radical generation or accumulation and cell death *(98)*. Unfortunately, these effects have not always translated into symptom alleviation in studies utilizing creatine supplementation in human HD patients *(99)*. However, it is possible that the supplementation method was less than ideal for attenuating declines associated with this disease in humans or that differential effects may be seen depending on the stage of the disease. Given the promising results from animal studies, further research is clearly warranted to determine whether the effects seen in an animal model of HD translate to the human model.

3.3. Ginkgo biloba

The use of *Ginkgo biloba* as a therapeutic aid can be traced back over 5000 years to early Chinese medicine *(100)*. It was not until the 1960s that *Ginko biloba* extract (EGb) was introduced into Western medicine *(100)*. EGb is commonly found in energy drinks that would be commonly used by sport and exercise populations, although the doses in these drinks are usually quite low. Athletes may be persuaded to take EGb because of its professed effects on mental alertness or focus. Purported clinical uses now include treatment for memory impairment, dementia, Alzheimer's disease, and intermittent claudication *(55,101)*. There is even evidence that EGb might help prevent symptoms of seasonal affective disorder (SAD) during the winter months *(102)*. There have been well over 40 controlled trials examining these effects, but much of the early research was relatively small scale and of questionable methodological quality *(101)*. Meta-analyses that have included the more rigorous of these trials have generally found positive results for the effects of EGb on memory impairment, Alzheimer's disease, and dementia *(101,103–105)*. Despite the generally positive findings, however, Birks and Grimly Evans *(105)* concluded that the evidence for EGb having significant benefit for treatment of dementia or cognitive impairment was both inconsistent and unconvincing. This conclusion appears to be partly based on concern for publication bias in this area *(101,105)*. Ernst *(55)* argued that the evidence for EGb's effects on normal cognitive function is not compelling but that the average effect associated with EGb and Alzheimer's-related dementia is likely to be clinically relevant. There is some evidence that acute EGb ingestion can provide enhanced cognitive effects in even healthy, young adults, although this effect was significantly more pronounced when EGb was combined with *Panax ginseng (106)*. Future research should consider interactive effects of EGb with other common supplements.

Most EGb used for medical and research purposes (EGb 761) contains 24% flavone and 6% terpenoids *(107)*. It is believed that the flavonoids contribute the antioxidant effects of EGb, thereby making it an effective free radical scavenger *(107)*. This might be one of the primary mechanisms through which EGb exerts its effects on dementia and cognition. Oxidative damage to polyunsaturated fatty acids (PUFAa) in brain cells results in impaired neurotransmitter uptake, and cellular damage resulting from free radical production appears to be a key contributor to the pathology and neurotoxicity associated with Alzheimer's disease *(108)*. Research using an animal model for Alzheimer's disease found that both in vivo and in vitro treatment with EGb 761 resulted in significant attenuations in reactive oxygen species (ROS) *(109)*. This is consistent with epidemiological findings in humans demonstrating an association between flavonoid *(110)* and vitamin E *(111)* intake and cognitive functioning in older adults.

Results of a few, large-scale randomized clinical trials (RCT) have provided somewhat mixed results regarding the effects of EGb administration on symptoms of dementia, although most of the evidence supports the efficacy of EGb treatment. In a 24-week study of 156 subjects with Alzheimer's induced dementia or multi-infarct dementia, Kanowski et al. *(112)* found that EGb 761 administration resulted in improved cognitive performance, behavior (i.e., coping skills, independence), and electroencephalographic (EEG) mapping characteristics compared to placebo. LeBars et al. *(113)* randomly assigned 202 subjects with dementia of mild to moderate severity to receive either EGb 761 (120 mg·day^{-1}) or a placebo for 52 weeks. They found that EGb successfully attenuated continued declines in cognitive functioning and the signs and symptoms of the disease compared to placebo, although the effects were relatively modest. There were also some patients who improved while taking EGb. A reanalysis of the data *(114)* revealed that the changes in the experimental group were dependent on the severity of the disease at baseline. In those individuals classified as having "mild" cognitive impairment, EGb resulted in improvements. In those individuals having "severe" cognitive impairments, progression of the disease was only stabilized or slowed by EGb but was still better than placebo *(114)*. A more recent 24-week placebo-controlled, double-blind EGb trial with 123 patients *(115)* failed to support the efficacy of two different doses of EGb (240 or $160 \, mg \cdot d^{-1}$) for subjects with age-associated memory impairment or dementia. However, these results have been criticized as difficult to interpret owing to the mixture of subjects suffering from dementia with subjects who, in some cases, had little cognitive impairment *(107)*. In light of this criticism, interpretation becomes even more challenging owing to the fact that subjects in each of the original EGb dose groups were randomly assigned to continue with EGb or a placebo after 12 weeks. Overall, it appears that EGb has potentially beneficial effects on Alzheimer's disease and dementia symptoms, but further well designed RCTs are necessary to examine long-term efficacy and dose-response effects considering the mixed findings that have been presented in recent years.

3.4. St. John's Wort

The use of St. John's wort may be less common for sport- or exercise-specific reasons compared to caffeine, creatine, or EGb. However, this supplement is widely used in the general population *(11)* and has interesting purported benefits and concerning possible side effects. Also, this supplement has the purported benefit of alleviating depression similar to that attributed to physical exercise, and its study in conjunction with exercise settings is warranted. Several meta-analytic reviews *(116–121)*, most of which have included only RCTs in their analyses, have supported the efficacy of St. John's wort (SJW) (*Hypericum perforatum* L.) for the treatment of depression of mild to moderate severity compared to placebo. Many of these meta-analyses *(116,118,120)* have also concluded that SJW is as effective as some of the conventional antidepressants, including tricyclic antidepressants (TCAs) and fluoxetine, typically with fewer side effects. However, Whiskey et al. *(120)* urged caution when interpreting the apparent equivalence due to the low statistical power inherent in most of the RCTs that have compared SJW to antidepressants. Nonetheless, most RCTs continue to support the meta-analytic conclusions [see reviews *(122,123)*], although evidence for efficacy in treating major depression appears mixed at best *(124–126)*.

The two primary substances contained in SJW that are believed to be responsible for the antidepressant effect are hyperforin and hypericin *(122)*. Evidence now points to hyperforin being the main contributor to this effect *(127,128)*. Various mechanisms have been identified to explain how SJW might exert its antidepressant effects, although the precise mechanism remains unclear *(122)*. SJW, or its isolated components, has been shown to inhibit serotonin (5-hydroxytryptamine, 5-HT), norepinephrine, and dopamine reuptake *(129)*; inhibit γ-aminobutyric acid (GABA) and L-glutamate uptake *(127,130)*; and inhibit substance-P (SP) production of proinflammatory cytokines [e.g., interleukin-6 (IL-6)] *(131)* that have been implicated in depressive etiology. Additionally, SJW has been shown to inhibit monoamine oxidase (MAO), although this inhibition appears to be of insufficient magnitude to account for significant antidepressant effects *(132)*. It is entirely possible that SJW functions to reduce depression through a combined effect of each of these mechanisms.

Despite the established benefits of SJW in the treatment of mild to moderate depression, there is a major concern with its use that focuses on apparent drug-interaction effects. There have been various reports of SJW having negative interactions with drugs such as warfarin, cyclosporine, human immunodeficiency virus (HIV) protease inhibitors, digoxin, theophylline, and oral contraceptives by decreasing their circulating concentrations *(133,134)*. The mechanism behind these interactions appears to be the combined result of induction of the cytochrome P450 isoenzymes CYP3A4, CYP2C9, and CYP1A2 and the transport protein P-glycoprotein by SJW *(133,134)*. It also appears that SJW can compound the effects of other selective serotonin reuptake inhibitors (SSRIs) and potentially lead to serotonin excess if SJW and SSRIs are combined because of their common pharmacological mechanisms *(133,135, 136)*. Considering the commonality of the mechanisms identified above to explain SJW's effects on depression and those that have been advanced to explain the effects of exercise on depression [see review *(137)*], it is imperative that future studies begin to explore the potential interactions between SJW (and many other popular herbal supplements) and exercise. Although there might be a synergistic effect, we must also consider the physiological responses induced by exercise and how they might affect the metabolism of these supplements. Dose-response models derived from primarily sedentary populations may not apply to more active groups.

4. CONCLUSION

Although supplement use continues to grow in the United States, little attention has been given to developing a theoretical understanding of the motivational forces at work in this area. Researchers should consider that most individuals in sport and exercise environments are looking for quick results, reflected in performance gains, body shape change, or health. If supplements provide such changes, their use could be said to be reinforced through reward contingencies. Thus, in one sense, the motivation to take supplements is quite simple—those who take supplements are (sometimes) rewarded. This might help explain why many muscle-building supplements have been shown to contain illegal constituents that would lead to a failed drug test *(6)*. Supplement companies want their products to work effectively in order to capitalize on the motivating properties of reward. On the other hand, we know that the effectiveness of many supplements is not well established and that it is likely that few supplements can legitimately offer immediate improvement. Therefore, researchers must consider other motivational explanations for supplement use in sport and exercise settings.

Among theories or constructs that may explain supplement use in sport and exercise, the theory of planned behavior, the balance theory, the achievement goal theory, social physique anxiety, and muscle dysmorphia would be important to consider. The theory of planned behavior considers the importance of norms, attitude, and perceived behavioral control as predictors of intent to engage in behavior that is partially under our volitional control. Understanding how supplements are considered good/bad (attitude), commonly used or have received approved (subjective norm), and available and easy to use (perceived behavioral control) may allow researchers to understand the pathways directing supplement use. Closely related to the idea of normative influence on behavior, we should consider the persuasive effects of well-liked spokespersons who tout the efficacy of certain supplements. According to balance theory, individuals may change their action (begin to use supplements) or attitudes (condone use) when faced with a situation that causes imbalance of thought and action (i.e., one of their favorite athletes uses supplements). Achievement goal theory proposes that individuals who are motivated to beat others or win (i.e., judge success based on social comparison) are more likely to use legal or even illegal supplementation to achieve such goals. Individuals driven by personal improvement (i.e., task-involved motivation) should be less likely to use supplements. Finally, upon considering the forces driving supplementation in sport and exercise, it is important to consider how body image and social physique anxiety relate to supplement use. This may be one of the more important psychological considerations considering the dramatic findings of Pope and colleagues *(42)*. If individuals are willing to risk their career to lift weights to change their body shape, it suggests an alarming obsession with body image that has been already linked to supplement use. What seems to be lacking in the literature is an understanding of motivational factors and their relation to supplement use generally but more importantly to dangerous (i.e., oversupplementation or mixed supplementation) supplementation practices.

Consideration of the psychological outcomes associated with some of the best-selling supplements (e.g., caffeine, creatine, *Ginkgo biloba*, St. John's wort) has revealed encouraging findings for efficacy in improving such things as arousal and attention, neuroprotection, cognitive function, neurological diseases, and mild to moderate depression. Furthermore, there is reason to be encouraged by the RCTs and meta-analytic reviews that have been conducted on many of these compounds. Some of these studies, particularly those employing animal models, have provided important evidence for the physiological mechanisms potentially underlying these psychological outcomes. Although considerable work remains to be done to continue to clarify these mechanisms and examine their applicability to the human model, biological plausibility for these supplements' psychological utility appears to have been established. Dose-response issues, on the other hand, remain unresolved for many of the substances, and this is an area that warrants further consideration in order to establish optimal use.

Along these same lines, effects that might be affected by exercise need to be evaluated for *Ginkgo biloba* and St. John's wort. This is particularly important for St. John's wort considering the established interaction effects that it has been shown to have with many drugs *(133–136)*. Some of the effects induced by these drugs have also been found to be induced by exercise *(137)*, and this point needs to be considered in future research.

Finally, before definitive efficacy conclusions can be reached for creatine and neuroprotection, *Ginkgo biloba* and dementia/ Alzheimer's disease and St. John's wort and depression, larger-scale, long-term RCTs are needed. To this point, trials of these supplements have been too small and too short *(55)*. Generally speaking, though, there appears to be a positive risk–benefit profile for each of the reviewed supplements, thus warranting dedication of resources to further examination of their application to enhancing performance as well as to various diseases and conditions.

4.1. Practical Applications

- The use of supplements having obvious efficacy to the user is likely to cause future use.
- Information leading to favorable attitudes about supplements is likely to promote their use. Promotion or data from "experts" can be persuasive in developing (or degrading) favorable attitudes regarding supplements.
- Individuals who perceive that supplement use is common among their peers are more likely to use supplements. Individuals who believe that they "should" be taking supplements are more likely to take supplements. One should be aware of the messages sent by coaches, trainings, and professionals regarding the frequency of use of supplements and the need for supplement use.
- Individuals are more likely to take supplements if they believe that they are easy to use or cheaper.
- Well liked individuals (e.g., professional athletes, celebrities) who state that they use supplements may lead their fans to use supplements.
- Athletes motivated primarily to win may engage in greater supplementation. Care may be taken to ensure that win-focused athletes do not engage in illegal or dangerous supplementation practices.

- There appears to be a portion of the population compulsive about their appearance, which may cause an increase in supplementation (either to gain muscle or lose weight). Appearance of these compulsions should be taken seriously, as we are just beginning to learn about the nature of these issues. Professionals working with such individuals should be on the lookout for illegal or dangerous supplementation practices.

- We know little about the overlap of eating disorders and supplementation use, but individuals who work with those attempting to lose weight should consider preoccupation with weight an important risk factor. Studies show that intense, prompt treatment is the best course of action to help individuals with eating disorders. Professionals working in this area should monitor traditional warning signs and supplementation practices alike.

- Caffeine doses in the range of 200 mg appear to produce positive changes in attention, vigilance, and arousal, particularly if the individual is fatigued or sleep-deprived. This may be particularly useful to athletes who require sustained vigilance or individuals exposed to extreme conditions and physical and psychological demands (i.e., military, firefighters).

- Creatine not only enhances physical performance but also has apparent neuroprotective effects. Its use as a prophylactic might be beneficial to athletes involved in sports where the incidence of TBI is high (i.e., mixed martial arts, football, soccer, boxing). Additionally, creatine use might be beneficial for individuals suffering from Huntington's disease to help attenuate neurodegeneration. Moderate doses appear most effective.

- *Ginkgo biloba* supplementation might help attenuate or even reverse some of the cognitive symptoms of Alzheimer's-related dementia. It appears to do this through a reduction in oxidative stress at the level of the brain cells.

- St. John's wort can be used to help reduce symptoms of mild to moderate depression. However, caution must be observed if the individual is taking other drugs, such as warfarin, cyclosporine, HIV protease inhibitors, digoxin, theophylline, or oral contraceptives because it may interact with these drugs and lower their circulating levels. Combining it with SSRIs can exacerbate their effects. Interactions with exercise also must be considered because of many overlapping mechanisms.

REFERENCES

1. Dietary Supplement Health and Education Act of 1994. Public Law 103-417. 103rd Congress. Downloaded from http://www.fda.gov/opacom/laws/dshea. html on May 7, 2007.
2. Nutrition Business Journal's Supplement Business Report. Downloaded from http://www.nutritionbusiness.com on May 7, 2007.
3. Radimer KL, Subar AF, Thompson FE. Nonvitamin, nonmineral dietary supplements: issues and findings from NHANES III. J Am Diet Assoc 2000; 100:447–414.
4. Perkin JE, Wilson WJ, Schuster K, Rodriguez J, Allen-Chabot A. Prevalence of non-vitamin, non-mineral supplement usage among university students. J Am Diet Assoc 2002;102:412–414.
5. NCAA Banned-Drug Classes 2006-2007. Downloaded from http://www1.ncaa. org/membership/ed_outreach/health-safety/drug_testing/banned_drug_classes. pdf on May 7, 2007.
6. Geyer H, Parr MK, Mareck U, Reinhart U, Schrader Y, Schanzer W. Analysis of non-hormonal nutritional supplements for anabolic-androgenic steroids: results of an international study. Int J Sports Med 2004;25:124–129.
7. Almeida JC, Grimsley EW. Coma from the health food store: interaction between kava and alprazolam. Ann Intern Med 1996;125:940–941.
8. Fennell D. Determinants of supplement usage. Prev Med 2004;39:932–939.
9. Egger G, Cameron-Smith D, Stanton R. The effectiveness of popular, non-prescription weight loss supplements. Med J Aust 1999;171:604–608.
10. Popkin BM, Doak C. The obesity epidemic is a worldwide phenomenon. Nutr Rev 1998;56:106–114.
11. Kaufman DW, Kelly JP, Rosenberg L, Anderson TE, Mitchell AA. Recent patterns of medication use in the ambulatory adult population of the United States: the Slone survey. JAMA 2002;287:337–344.
12. Foote JA, Murphy SP, Wilkens LR, Hankin JH, Henderson BE, Kolonel LN. Factors associated with dietary supplement use among healthy adults of five ethnicities: the Multiethnic Cohort Study. Am J Epidemiol 2003;157: 888–897.
13. Harrison RA, Holt D, Pattison D, Elton PJ. Who and how many people are taking herbal supplements? A survey of 21,923 adults. Int J Vit Nutr Res 2004;74:183–186.
14. Sobal J, Marquart LF. Vitamin/mineral supplement use among athletes: a review of the literature. Int J Sport Nutr 1994;4:320–334.
15. Scofield DE, Unruh S. Dietary supplement use among adolescent athletes in central Nebraska and their sources of information. J Strength Cond Res 2006;20:452–455.
16. Huang SH, Johnson K, Pipe, AL. The use of dietary supplements and medications by Canadian athletes at the Atlanta and Sydney Olympic games. Clin J Sport Med 2006;16:27–33.
17. Nieper A. Nutritional supplement practices in UK junior national track and field athletes. Br J Sports Med 2005;39:645–649.

18. Perko MA, Eddy JM, Bartee RT, Dunn MS. Giving new meaning to the term "taking one for the team": influences on use/non-use of dietary supplements among adolescent athletes. J Am Health Stud 2000;16:99–106.
19. Martin G, Pear J. Behavior Modification: What It Is and How to Do It. 6th ed. Prentice-Hall, Englewood Cliffs, NJ, 1998.
20. Watson JB. Psychology as the behaviorist views it. Psychol Rev 1913; 20:158–177.
21. Moses E. An athlete's Rx for the drug problem. Newsweek p 57, October 10, 1988.
22. Asch SE. Opinions and social pressures. Sci Am 1955;193:31–35.
23. Okun MA, Karoly P, Lutz R. Clarifying the contribution of subjective norm to predicting leisure-time exercise. Am J Health Behav 2002;26:296–305.
24. Reno RR, Cialdini RB, Kallgren CA. The transsituational influence of norms. J Pers Soc Psychol 1993;64:104–112.
25. Ajzen I, Fishbein M. The influence of attitudes on behavior. In: Albarracin D, Johnson BT, Zanna MP (eds) The Handbook of Attitudes (pp 173–221). Erlbaum, Mahwah, NJ, 2005.
26. Conner M, Kirk SF, Cade JE, Barrett JH. Why do women use dietary supplements? The use of the theory of planned behaviour to explore beliefs about their use. Soc Sci Med 52:621–33.
27. Perloff RM. The Dynamics of Persuasion. Lawrence Erlbaum Associates, Hillsdale, NJ, 1993.
28. Rosenthal R. The "file drawer problem" and tolerance for null results. Psychol Bull 1979;86:638–641.
29. Scargle JD. Publication bias: the "file-drawer" problem in scientific inference. J Sci Exp 2000;14:91–106.
30. Heider F. The Psychology of Interpersonal Relations. Wiley, New York, 1958.
31. Ames C. Classrooms: goals, structures, and student motivation. J Educ Psychol 1993;84:261–271.
32. Dweck C. Motivational processes affecting learning. Am Psychol 1986;41:1040–1048.
33. Nicholls JG. The Competitive Ethos and Democratic Education. Harvard University Press, Cambridge, MA, 1989.
34. Sarrazin P, Biddle S, Famose JP, Cury F, Fox K, Durand M. Goal orientations and conceptions of the nature of sport ability in children: a social cognitive approach. Br J Soc Psychol 1996;35:399–414.
35. Duda JL, Olson LK, Templin TJ. The relationship of task and ego orientation to sportsmanship attitudes and the perceived legitimacy of injurious acts. Res Q Exerc Sport 1991;62:79–87.
36. Flegal KM, Carroll MD, Ogden CL, Johnson CL. Prevalence and trends in obesity among US adults, 1999-2000. JAMA 2002;288:1723–1727.
37. Owen PR, Laurel-Seller E. Weight and shape ideals: thin is dangerously in. J Appl Soc Psychol 2000;30:979–990.
38. Hart EH, Leary MR, Rejeski WJ. The measurement of social physique anxiety. J Sport Exerc Psychol 1989;11:94–104.
39. Hausenblas HA, Fallon EA. Relationship among body image, exercise behavior, and exercise dependence symptoms. Int J Eat Disord 2002;32:179–185.

40. American Psychiatric Association. Diagnostic and Statistical Manual of Mental Disorders. 4th ed. APA, Washington, DC, 1994.
41. Celio CI, Luce KH, Bryson SW, et al. Use of diet pills and other dieting aids in a college population with high weight and shape concerns Int J Eat Disord 39:492–497.
42. Pope HG Jr, Phillips KA, Olivardia R. The Adonis Complex: The Secret Crisis of Male Body Obsession. Free Press, New York, 2000.
43. Pope HG Jr, Katz DL, Hudson JI. Anorexia nervosa and "reverse anorexia" among 108 male bodybuilders. Compr Psychiatry 1993;34:406–409.
44. Pope HG Jr, Gruber AJ, Choi P, Olivardia R, Phillips KA. Muscle dysmorphia: an underrecognized form of body dysmorphic disorder. Psychosomatics 1997;38:548–557.
45. Hildebrandt T, Langenbucher J, Schlundt DG. Muscularity concerns among men: development of attitudinal and perceptual measures. Body Image 2004;1:169–181.
46. Raevuori A, Keski-Rahkonen A, Bulik CM, Rose RJ, Rissanen A, Kaprio J. Muscle dissatisfaction in young adult men. Clin Pract Epidemiol Ment Health 2006;2:6.
47. Bond A, Choi PYL, Pope HG. Assessment of attentional bias and mood in users and non-users of anabolic-androgenic steroids. Drug Alcohol Depend 1995;37:241–45.
48. Choi PY. Alarming effects of anabolic steroids. Psychologist 1993;6:258–260.
49. Dalby JT. Brief anabolic steroid use and sustained behavioral reaction. Am J Psychiatry 1992;149:271–272.
50. Pope HG, Katz DL. Psychiatric and medical effects of anabolic-androgenic steroid use. Arch Gen Psychiatry 1994;51:375–382.
51. Su T, Pagliaro M, Schmidt PJ, Pickar D, Wolkowitz O, Rubinow DR. Neuropsychiatric effects of anabolic steroid in male normal volunteers. JAMA 1993;269:2760–2764.
52. O'Connor DB, Archer J, Hair WM, Wu FCW. Exogenous testosterone, aggression, and mood in eugonadal and hypogonadal men. Physiol Behav 2002;75:557–566.
53. Pope HG, Kouri EM, Hudson JI. Effects of supraphysiological doses of testosterone on mood and aggression in normal men. Arch Gen Psychiatry 2000;57:133–140.
54. Tricker R, Casaburi R, Storer TW, et al. The effects of supraphysiological doses of testosterone on angry behavior in healthy eugonadal men: a clinical research center study. J Clin Endocrinol Metab 1996;81:3754–3758.
55. Ernst E. The risk-benefit profile of commonly used herbal therapies: Ginkgo, St. John's wort, ginseng, echinacea, saw palmetto, and kava. Ann Intern Med 2002;136:42–53.
56. Kruk B, Chmura J, Krzeminski K, et al. Influence of caffeine, cold and exercise on multiple choice reaction time. Psychopharmacology 2001;157:197–201.
57. Sinclair CJ, Geiger JD. Caffeine use in sports: a pharmacological review. J Sports Med Phys Fitness 2000;40:71–79.
58. Doyle T, Arent SM, Lutz RS. Dose-response effects of caffeine on performance in college fencers. J Strength Cond Res 2006;20:e41.

59. Kenemans JL, Lorist MM. Caffeine and selective visual processing. Pharmacol Biochem Behav 1995;52:461–471.
60. Rees K, Allen D, Lader M. The influences of age and caffeine on psychomotor and cognitive function. Psychopharmacology 1999;145:181–188.
61. Fine BJ, Kobrick L, Lieberman HR, Marlowe B, Riley RH, Tharion WJ. Effects of caffeine or diphenhydramine on visual vigilance. Psychopharmacology 1994;114:233–238.
62. Koelega HS. Stimulant drugs and vigilance performance: a review. Psychopharmacology 1993;111:1–16.
63. Reyner LA, Horne JA. Early morning driver sleepiness: effectiveness of 200 mg caffeine. Psychophysiology 2000;37:251–256.
64. Lorist MM, Snel J, Kok A, Mulder G. Influence of caffeine on selective attention in well-rested and fatigued subjects. Psychophysiology 1994;31:525–534.
65. Patat A, Rosenzwieg P, Enslen M, et al. Effects of a new slow release formulation of caffeine on EEG, psychomotor and cognitive functions in sleep deprived subjects. Hum Psychopharmacol Clin Exp 2000;15:153–170.
66. Smith A. Effects of repeated doses of caffeine on mood and performance of alert and fatigued volunteers. J Psychopharmacol 2005;19:620–626.
67. Lieberman HR, Tharion WJ, Shukitt-Hale B, Speckman KL, Tulley R. Effects of caffeine, sleep loss, and stress on cognitive performance and mood during U. S. Navy SEAL training. Psychopharmacology 2002;164:250–261.
68. Bovim G, Naess P, Helle J, Sand T. Caffeine influence on the motor steadiness battery in neuropsychlogical tests. J Clin Exp Neuropsychol 1995;17:472–476.
69. Loke WH, Hinrichs JV, Ghoneim MM. Caffeine and diazepam: separate and combined effects on mood, memory, and psychomotor performance. Psychopharmacology 1985;87:344–350.
70. Tikuisis P, Keefe AA, McLellan TM, Kamimori G. Caffeine restores engagement speed but not shooting precision following 22 h of active wakefulness. Aviat Space Environ Med 2004;75:771–776.
71. Gillingham RL, Keefe AA, Tikuisis P. Acute caffeine intake before and after fatiguing exercise improves target shooting engagement time. Aviat Space Environ Med 2004;75:865–871.
72. Ferrauti A, Weber K, Struder HK (1997). Metabolic and ergogenic effects of carbohydrate and caffeine beverages in tennis. J Sports Med Phys Fitness 1997;37:258–266.
73. Ruijter J, Lorist LM, Snel J, De Ruiter M. (2000) The influence of caffeine on sustained attention: an ERP study. Pharmacol Biochem Behav 2000;66:29–37.
74. Huang Z-L, Qu W-M, Eguchi N, et al. Adenosine A2A, but not A1, receptors mediate the arousal effect of caffeine. Nat Neurosci 2005;8:858–859.
75. Ross GW, Abbott RD, Petrovich H, et al. Relationship between caffeine intake and Parkinson's disease. JAMA 2000;283:2674–2679.
76. Ferré S, Fredholm BB, Morelli M, Popoli P, Fuxe K. Adenosine-dopamine receptor-receptor interactions as an integrative mechanism in the basal ganglia. Trends Neurosci 1997;20:482–487.
77. Garrett BE, Griffiths RR. The role of dopamine in the behavioral effects of caffeine in animals and humans. Pharmacol Biochem Behav 1997;57:533–541.

78. Schwarzschild MA, Chen J-F, Ascherio A. Caffeinated clues and the promise of adenosine A2A antagonists in PD. Neurology 2002;58:1154–1160.
79. Gilbert DG, Dibb WD, Plath LC, Hiyane SG. Effects of nicotine and caffeine, separately and in combination, on EEG topography, mood, heart rate, cortisol, and vigilance. Psychophysiology 2000;37:583–595.
80. Lovallo WR, al'Absi M, Blick K, et al. Stress-like adrenocorticotropin responses to caffeine in young healthy men. Pharmacol Biochem Behav 1996;55:365–369.
81. Al'Absi M, Lovallo WR, Sung BH, et al. Persistent adrenocortical sensitivity to caffeine in borderline hypertensive men. FASEB J 1993;7:A552.
82. Al'Absi M., Lovallo WR, McKey B, Sung BH, Whitsett TL, Wilson MF. Hypothalamic-pituitary-adrenocortical responses to psychological stress and caffeine in men at high and low risk for hypertension. Psychosom Med 1998;60: 521–527.
83. Blanchard J, Sawers SJA. Comparative pharmacokinetics of caffeine in young and elderly men. J Pharmacokinet Biopharm 1983;11:109–126.
84. Jacobson BH, Thurman-Lacey SR. Effect of caffeine on motor performance by caffeine-naïve and -familiar subjects. Percept Mot Skills 1992; 74:151–157.
85. Mikalsen A, Bertelsen B, Flaten MA. Effects of caffeine, caffeine-associated stimuli, and caffeine-related information on physiological and psychological arousal. Psychopharmacology 2001;157:373–380.
86. Wyss M, Schulze A. Health implications of creatine: can oral creatine supplementation protect against neurological and atherosclerotic disease? Neuroscience 2002;112:243–260.
87. McNair ND. Traumatic brain injury. Nurs Clin North Am 1999;34:637–659.
88. Thurman DJ, Branche CM, Sniezek JE. The epidemiology of sports-related traumatic brain injuries in the United States: recent developments. J Head Trauma Rehabil 1998;13:1–8.
89. Clark K. Epidemiology of athletic head injury. Clin Sports Med 1998; 17:1–12.
90. Klivenyi P, Calingasan NY, Starkov A, et al. Neuroprotective mechanisms of creatine occur in the absence of mitochondrial creatine kinase. Neurobiol Dis 2004;15;610–617.
91. Sullivan PG, Geiger JD, Mattson MP, Scheff SW. Dietary supplement creatine protects against traumatic brain injury. Ann Neurol 2000;48:723–729.
92. Klivenyi P, Ferrante RJ, Matthews RT, et al. Neuroprotective effects of creatine in a transgenic animal model of amyotrophic lateral sclerosis. Nat Med 1999;5:347–350.
93. Rae C, Digney AL, McEwan SR, Bates TC. Oral creatine monohydrate supplementation improves brain performance: a double-blind, placebo-controlled, cross-over trial. Proc R Soc Lond Biol 2003;270:2147–50.
94. Watanabe A, Kato N, Kato T. Effects of creatine on mental fatigue and cerebral hemoglobin oxygenation. Neurosci Res 2002;279–285.
95. Jacobus WE, Diffley DM. Creatine kinase of heart mitochondria: control of oxidative phosphorylation by the extramitochondrial concentrations of creatine and phosphocreatine. J Biol Chem 1986;261:16579–16583.

96. Ferrante RJ, Andreassen OA, Jenkins BG, et al. Neuroprotective effects of creatine in a transgenic mouse model of Huntington's disease. J Neurosci 2000;20:4389–4397.

97. Mangiarini L, Sathasivam K, Seller M, et al. Exon 1 of the HD gene with an expanded CAG repeat is sufficient to cause a progressive neurological phenotype in transgenic mice. Cell 1996;87:493–506.

98. Matthews RT, Yang L, Jenkins BG, et al. Neuroprotective effects of creatine and cyclocreatine in animal models of Huntington's disease. J Neurosci 1998;18:156–163.

99. Verbessem P, Lemiere J, Eijnde BO, et al. Creatine supplementation in Huntington's disease: a placebo-controlled pilot trial. Neurology 2003;61: 925–930.

100. Mahady GB. Ginkgo biloba: a review of quality, safety, and efficacy. Nutr Clin Care 2001;4:140–147.

101. Kleijnen J, Knipschild P. Ginkgo biloba for cerebral insufficiency. Br J Clin Pharmacol 1992;34:352–358.

102. Lingaerde O, Foreland AR, Magnusson A. Can winter depression be prevented by Ginkgo biloba extract? A placebo-controlled trial. Acta Psychiatr Scand 1999;100:62–66.

103. Hopfenmüller W. Evidence for a therapeutic effect of Ginkgo biloba special extract: meta-analysis of 11 clinical studies in patients with cerebrovascular insufficiency in old age. Arzneimittelforschung 1994;44;1005–1013.

104. Oken BS, Storzbach DM, Kaye JA. The efficacy of Ginkgo biloba on cognitive function in Alzheimer's disease. Arch Neurol 1998;55:1409–1415.

105. Birks J, Grimly Evans J. Ginkgo biloba for cognitive impairment and dementia. Cochrane Database Syst Rev 2007;2.

106. Scholey AB, Kennedy DO. Acute, dose-dependent cognitive effects of Ginkgo biloba, Panax ginseng and their combination in healthy young volunteers: differential interactions with cognitive demand. Hum Psychopharmacol Clin Exp 2002;17:35–44.

107. Gertz H-J, Kiefer M. Review about Ginkgo biloba special extract EGb 761 (Ginkgo). Curr Pharm Des 2004;10:261–264.

108. Butterfield DA, Howard B, Yatin S, et al. Elevated oxidative stress in models of normal brain aging and Alzheimer's disease. Life Sci 1999;65: 1883–1892.

109. Vining Smith J, Luo Y. Elevation of oxidative free radicals in Alzheimer's disease models can be attenuated by Ginkgo biloba extract EGb 761. J Alzheimers Dis 2003;5:287–300.

110. Letenneur L, Proust-Lima C, Le Gouge A, Dartigues JF, Barberger-Gateau P. Flavonoid intake and cognitive decline over a 10-year period. Am J Epidemiol 2007;165:1364–1371.

111. Grodstein F, Chen J, Willett WC. High-dose antioxidant supplements and cognitive function in community-dwelling elderly women. Am J Clin Nutr 2003;77:975–984.

112. Kanowski S, Herrmann WM, Stephan K, Wierich W, Horr R. Proof of efficacy of the Ginkgo biloba special extract EGb 761 in outpatients suffering

from mild to moderate primary degenerative dementia of Alzheimer type or multi-infarct dementia. Pharmacopsychiatria 1996;29:47–56.

113. LeBars PL, Katz MM, Berman N, Itil TM, Freedman AM, Schatzberg A. A placebo-controlled, double-blind, randomized trial of an extract of Ginkgo biloba for dementia. JAMA 1997;278:1327–1332.

114. LeBars PL, Velasco FM, Ferguson JM, Dessain EC, Kieser M, Hoerr R. Influence of the severity of cognitive impairment on the effect of the Ginkgo biloba extract EGB 761 in Alzheimer's disease. Neuropsychobiology 2002; 45:19–26.

115. Van Dongen M, van Rossum E, Kessels A, Sielhorst H, Knipschild P. Ginkgo for elderly people with dementia and age-associated memory impairment: a randomized clinical trial. J Clin Epidemiol 2003;56:367–376.

116. Kim HL, Streltzer J, Goebert D. St. John's wort for depression: a meta-analysis of well-defined clinical trials. J Nerv Ment Dis 1999;187:532–539.

117. Linde K, Ramirez G, Mulrow CD, Pauls A, Weidenhammer W, Melchart D. St. John's wort for depression: an overview and meta-analysis of randomized clinical trials. BMJ 1996;313:253–258.

118. Linde K, Mulrow CD. St. John's wort for depression. Cochrane Database Syst Rev 2005;Apr 18(2):CD000448

119. Gaster B, Holroyd. St. John's wort for depression: a systematic review. Arch Intern Med 2000;160:152–156.

120. Whiskey E, Werneke U, Taylor D. A systematic review and meta-analysis of Hypericum perforatum in depression: a comprehensive clinical review. Int Clin Psychopharmacol 2001;16:239–252.

121. Williams JW, Mulrow CD, Chiquette E. A systematic review of new pharmacotherapies for depression in adults. Ann Intern Med 2000;132:743–756.

122. Barnes J, Anderson LA, Phillipson JD. St. John's wort (Hypericum perforatum L.): a review of its chemistry, pharmacology and clinical properties. J Pharm Pharmacol 2001;53:583–600.

123. Bilia AR, Gallori S, Vincieri FF. St. John's wort and depression: efficacy, safety, and tolerability—an update. Life Sci 2002;3077–3096.

124. Shelton RC, Keller MB, Gelenberg A, et al. Effectiveness of St. John's wort in major depression: a randomized controlled trial. JAMA 2001;285: 1978–1986.

125. Lecrubier Y, Clerc G, Didi R, Kieser M. Efficacy of St. John's wort extract WS 5570 in major depression: a double-blind, placebo-controlled trial. Am J Psychiatry 2002;159:1361–1366.

126. Fava M, Alpert J, Nierenberg AA, et al. A double-blind, randomized trial of St. John's wort, fluoxetine, and placebo in major depressive disorder. J Clin Psychopharmacol 2005;25:441–447.

127. Chatterjee SS, Nolder M, Koch E, Erdelmeier C. Antidepressant activity of hypericum perforatum and hyperforin: the neglected possibility. Pharmacopsychiatry 1998;31:7–15.

128. Müller WE, Singer A, Wonnemann M, Hafner U, Rolli M, Schäfer C. Hyperforin represents the neurotransmitter reuptake inhibiting constituent of hypericum extract. Pharmacopsychiatry 1998;31:16–21.

129. Müller WE, Rolli M, Schäfer C, Hafner U. Effects of Hypericum extract in biochemical models of anti-depressant activity. Pharmacopsychiatry 1997;30: 102–107.

130. Wonnemann M, Singer A, Müller WE. Inhibition of synaptosomal uptake of ^{3}H-L-glutamate and ^{3}H-GABA by hyperforin, a major constituent of St. John's wort: the role of amiloride sensitive sodium conductive pathways. Neuropsychopharmacology 2000;23:188–197.

131. Fiebich BL, Höllig A, Lieb K. Inhibition of substance P-induced cytokine synthesis by St. John's wort extracts. Pharmacopsychiatry 2001; 34:S26–S28.

132. Cott JM. In vitro receptor binding and enzyme inhibition by Hypericum perforatum extract. Pharmacopsychiatry 1997;30:108–112.

133. Henderson L, Yue QY, Bergquist C, Gerden B, Arlett P. St. John's wort (Hypericum perforatum): drug interactions and clinical outcomes. Br J Clin Pharmacol 2002;54:349–356.

134. Hennessy M, Kelleher D, Spiers JP, et al. St. John's wort increases expression of P-glycoprotein: implications for drug interactions. Br J Clin Pharmacol 2002;53:75–82.

135. Müller WE, Rolli M, Schäfer C, Hafner U. Effects of Hypericum extract on the expression of serotonin receptors. Geriatr Psychiatry Neurol 1994;7:63–64.

136. Demmott K. St. John's wort tied to serotonin syndrome. Clin Psychiatry News 1998;26:28.

137. Landers DM, Arent SM. Exercise and mental health. In: Eklund RC, Tenenbaum G (eds) The Handbook of Sport Psychology (pp 469–491). 3rd ed. Wiley, New York, 2007.

Part II
Nutritional Basics First

3

Role of Nutritional Supplements Complementing Nutrient-Dense Diets

General Versus Sport/Exercise-Specific Dietary Guidelines Related to Energy Expenditure

Susan Kleiner and Mike Greenwood

Abstract

A nutrient-dense diet is a critical aspect in attaining optimal exercise training and athletic performance outcomes. Although including safe and effective nutritional supplements in the dietary design can be extremely helpful in promoting adequate caloric ingestion, they are not sufficient for promoting adequate caloric ingestion based on individualized caloric expenditure needs without the proper diet. Specifically, a strategic and scientifically based nutrient-dense dietary profile should be created by qualified professionals to meet the sport/exercise-specific energy demands of any individual involved in select training intensity protocols. Finally, ingesting the right quantity and quality of nutrient dense calories at precise windows of opportunity becomes vital in attaining desired training and/or competitive performance outcomes.

Key words

Nutrient dense · Nutritional supplements · Caloric intake · Caloric expenditure · Nutrient timing · Restoration · Macronutrient profiles

1. ESTABLISHING ADEQUATE DIETARY FOUNDATIONS

One of the important differences between any athlete and a champion is the keen ability to pay attention to details regarding a variety of training categories. Although one might think that

From: *Nutritional Supplements in Sports and Exercise*
Edited by: M. Greenwood, D. Kalman, J. Antonio,
DOI: 10.1007/978-1-59745-231-1_3, © Humana Press Inc., Totowa, NJ

the important details of sports nutrition center on targeting the right supplements, that assumption could not be farther from the truth. Nothing is more important than a good nutritional foundation. This chapter focuses on the details of nutrient density—how to pack in the most nutrition for every calorie you eat from food and putting your food to work for you. By maximizing nutrient density your body can be primed to respond to the supplements you may choose to add to enhance your nutrition program. When one builds a strong nutritional foundation, the muscular fitness, strength, and thus optimal performance outcomes will follow.

2. NUTRIENT DENSITY DEFINED

The most important factor in an athlete's diet is the amount of energy available. Calories (or joules) are the key to activity, feeling energized, building muscle, increasing power, and fueling endurance. Hand-in-hand with calories are the nutrients one consumes per calorie. If one just focuses on getting enough calories to fuel activity, foods high in simple sugars and fats would be the ideal choices; there is no doubt that this would fuel exercise. However, a diet robust in calories, along with the right composition of proteins, carbohydrates, healthy fats, vitamins, and minerals, catalyzes metabolism and tissue building, getting the athlete farther in training at a faster rate, with a superior outcome.

A common example of nutrient density compares a sports drink to orange juice and water. Even if the amount of sports drink is doubled to make the calories equivalent to those in orange juice, the nutrient density of the two liquids is not even close. The macronutrients and micronutrients noted in Table 1 help clarify this issue in greater detail.

The sports drink is the performance beverage of choice during exercise to promote rehydration and glucose/electrolyte replenishment. If a sports drink is the beverage of choice all day long, however, many calories are consumed without further nutrition. If orange juice is chosen as a beverage during the day, there is greater density of nutrients per calorie, building the nutritional foundation throughout the day. Finally, water adds no calories and a minor amount of nutrients, but the ongoing fluid replacement is essential for an athlete. By eliminating the "empty calories" from sports

Table 1
Macronutrient and Micronutrient Contents of Select Fluids

	8 oz. Sports drink	8 oz. Orange juice	8 oz. Water
Calories	50	112	0
Protein (g)	0	2	0
Carbohydrate(g)	14	27	0
Fat (g)	0	0	0
Vitamin A (IU)	0	266	0
Vitamin C (mg)	0	97	0
Folate (mg)	0	110	0
Ca (mg)	0	0	5
Mg (mg)	0	0	2
K (mg)	30	473	0
Na (mg)	110	2	5

drinks during nonexercise times, the athlete can eat more nutrient-dense foods, allowing a greater amount and variety of nutrients in the diet and enhancing the nutritional foundation for performance.

3. NUTRIENT-DENSE DIET

It is often assumed that a nutrient-dense diet is also a diet full of variety, but this is not always so. One could easily have a diet that is high in certain nutrients but missing others due to the elimination of an entire food group. For instance, if dairy is eliminated from the diet, one could easily consume enough of most nutrients, but it would be difficult to consume enough calcium and vitamin D. Along with choosing foods that are dense in nutrients, a nutrient-dense diet must also be rich in a variety of foods from all the food groups to be both nutritionally dense and complete (Table 2).

In addition to choosing foods from all of the food groups, a variety of foods from within each food group should be selected. For example, one could select only wheat bread, apples, celery, yogurt, hamburger, and butter to create a diet that includes foods from all the food groups. However, if this is what was eaten each day it would not provide a nutritionally complete diet despite

Table 2
Foundational Food Group Categories

Grains, breads and starches
Fruit
Nonstarchy vegetables
Milk/dairy
Meat and meat substitutes
    Very lean
    Lean
    Medium fat
    High fat
Fats

representation from all of the food groups. Selecting apples, peaches, and grapefruit from the fruit group; celery, broccoli, and
onions from the vegetable group; wheat, buckwheat, and winter
squash from the grains and starches group; and so on represents
variety among the food groups as well as from within each group.

4. NUTRITIONAL SUPPLEMENTS DEFINED

Within the area of sports, performance-enhancing aids have been
defined and classified by Williams *(1)*. Describing them as sports
ergogenics, Williams categorizes them as nutritional aids, pharmacological aids, and physiological aids. Nutritional supplements fall
under the category of nutritional aids and serve to "increase muscle
tissue, muscle energy supplies, and the rate of energy production in
the muscle." Some nutritional supplements are also used to enhance
mental focus and energy *(1)*.

Most nutritional supplements are obvious, but some are not as
clear. Supplements included in the broad nutritional aids category
include fluid replacements, carbohydrates, fats, protein/amino
acids and their metabolites, vitamins, minerals, plant extracts, mis-
cellaneous food factors, phytochemicals, and engineered dietary
supplements.

5. ROLE OF SUPPLEMENTS IN A NUTRIENT-DENSE DIET

There are many circumstances where nutritional supplementation plays an important role in both health and performance. The most common instance is when food groups or key foods in the diet must be eliminated intentionally owing to food allergies or intolerance. Supplementation with the minerals and vitamins associated with bone health is essential when individuals are allergic to cows' milk protein. Supplementation may come in the form of food fortification, as with the addition of calcium and vitamins A and D to soy milk. Supplementation can also be in a daily dose of the nutrients through liquids, pills, or capsules.

Often the diet is unintentionally incomplete owing to lack of nutritional knowledge, a hectic lifestyle, or other reasons. In this case, daily multivitamin/mineral supplementation has been suggested as an "insurance policy" for health promotion and disease prevention *(2)*.

Nutritional supplementation in sports is undertaken with the goal of enhancing performance. As stated above, in this case nothing can substitute for a complete, nutrient-dense diet from foods. However, there are circumstances in which athletes benefit from supplementation owing to increased requirements but an inability to consume increased amounts of nutrients from foods. Such an example is protein. The protein requirements of a lightweight or middleweight strength and power athlete may be difficult to meet with food alone. Because dense sources of protein most often occur with fats and/or carbohydrates (egg whites are the exception here), the total caloric intake for the diet may be too high or the foods too inconvenient to eat throughout the day. A powdered protein supplement can offer pure protein that can be easily mixed with water any time during the athlete's day.

Athletes may also benefit from increased amounts of specific nutrients but do not necessarily benefit from consuming increasing amounts of the whole foods that are the source of those nutrients. Creatine is an excellent example of this notion. Although the research is clear that many athletes benefit from creatine supplementation, consuming enough creatine from food (meat) to achieve dose requirements would be both difficult and unhealthy.

In both of the above examples, nutritional supplementation plays an important role in performance enhancement. However, neither

of the supplements would be as effective if used in place of a
nutrient-dense diet rather than in addition to a nutrient-dense diet.

6. DESIGNING A NUTRIENT-DENSE DIET

6.1. Establishing Viable Energy Requirements

Signs and symptoms associated with decrements in performance
and safety related to health markers have been linked to a number of
aspects including chronic energy consumption deficits (3,4). To set
the stage for reaching optimal training/performance outcomes, one
need not look any further than establishing a properly designed
dietary plan. Although this may seem to be an easy goal to accomplish, the commitment, time, and cost associated with a quality
dietary strategy often make it difficult to attain this option. It is a
widely accepted fact that athletes involved with high intensity training
and competition do not ingest the right type and/or amounts of
macronutrients to offset their energy expenditure. Although nutritional supplementation is a viable alternative for athletes to consume
and meet their dietary needs, it should be noted that this practice
is not a healthy option to replace a quality nutrient-dense diet. This
is exactly why it is referred to as nutritional *supplementation*—it
complements/supplements a properly designed nutrient-dense diet.
Nutritional necessities surrounding an "eat to compete" philosophy
are addressed here in relation to strategies to promote recovery of
athletic and exercise populations.

The most critical aspect of establishing a properly implemented
dietary strategy to accomplish optimal performance outcomes is to
ensure that exercise participants ingest quality caloric needs to
balance specific energy expenditure (3,4). When considering this
nutritional approach, it is always important to include and calculate
individual differences regarding select exercise training intensities.
Although recommended daily allowance (RDA) guidelines have
been established regarding daily dietary consumption for general
populations, these suggested initiatives really do not apply to
athletic or exercise populations involved with intense training protocols because of their greater caloric requirements. For example,
the daily caloric intake needs of untrained individuals are based on
the number of kilocalories per kilogram of body weight per day,
which usually averages 1900 to 3000 kcal daily (5,6). Without

question, when adding various factors of exercise to the equation, the frequency, duration, and intensity demands of the training protocol requires increased nutritional intake to maintain an effective energy balance. Individuals involved in low intensity exercise lasting 30 to 40 minutes a day performed three times per week typically require 1800 to 2400 kcal/day owing to minimal physical exertion and energy expenditure *(4,7)*. Athletes undertaking moderate exercise protocols defined as five or six times a week for 2 to 3 hours a day or intense training 5 or 6 days a week 3 to 6 hours a day obviously require greater dietary needs (2500–8000 kcal/day depending on body weight) compared to individuals involved in light exercise protocols *(4,8)*.

When accurately evaluating the amount of caloric values needed for individuals involved in the previously mentioned levels of exercise training, it becomes increasingly evident that athletes have a difficult time maintaining ingestion of enough calories by simply consuming a well balanced diet. Owing to the enormous energy expenditure for high volume intensity training, the combination of nutritional supplementation to a quality nutritional dietary profile makes it much more feasible for athletes to consume enough energy to replace caloric needs. The proper replacement of caloric needs based on energy expenditure not only helps control a person's health status but definitely heightens the recovery process needed for future optimal training/performance bouts.

Although a balanced energy status is vital for all athletes in formalized training, this aspect becomes even more imperative for large athletes who must consume huge amounts of quality calories to offset the energy expenditures acquired from high volume and intensity training. Obviously, the ramifications of inappropriate dietary strategies can lead not only to tremendous weight loss but make the athlete more susceptible to the various signs and symptoms of physiological and psychological decline. Furthermore, there is scientific evidence that the athlete who undertakes intense training has a greater preponderance to display suppressed appetite, which increases the possibility of health risk factors and ultimately promotes performance decrements *(9)*. A successful strategic plan is to develop a multidisciplinary team approach that combines athletic coaches, athletic trainers, sports nutritionists, strength and conditioning coaches, parents, and physicians to closely monitor and evaluate the athlete's nutrient-dense dietary status in an effort to maintain body weight and enhance restoration, thereby promoting optimal performance outcomes.

Although the caloric ingestion concerns of large athletes have been mentioned, additional athletic groups require close monitoring in relation to meeting caloric energy demands. Specifically, female athletes such as gymnasts, figure skaters, and distance runners are highly susceptible to eating disorders and place themselves in jeopardy of not meeting specific energy caloric needs. This may also hold true for athletes who participate in sports such as horse racing (jockeys), boxing, and wrestling and select unsafe dietary strategies to meet the requirements for being in a particular weight class for competition.

7. DETERMINING MACRONUTRIENT PROFILES

The goals of training and competition guide the determination of dietary macronutrient profiles. Although the differences in the dietary recommendations may appear small, research has shown that they lead to significant differences in outcomes in endurance enhancement, muscle growth, strength building, and power output. In all cases, adequate energy consumption is the most critical factor, with macronutrient distribution following close behind. For a detailed discussion of research and nutrient-dense recommendations for energy and macronutrient profiles and diet plans, see *Power Eating*, 3rd edition *(10)*.

7.1. Endurance Enhancement

Carbohydrates and fats are long-distance fuels. A diet low in either can temper performance advancement. Protein is essential for repair and recovery of damaged tissues and to keep the body healthy relative to participating in distance and duration activities. Typical recommendations for carbohydrate are 5 to 7 g/kg body weight per day for general training. With exercise of increased distance and duration, a carbohydrate intake of 8 to 10 g/kg/day has been shown to be a successful dietary strategy for fueling performance. Male athletes are more likely than female athletes to achieve these recommendations. Owing to their lower energy requirements, it is often difficult to consume this much carbohydrate and maintain control of body weight, an essential factor in long-distance sports *(11)*.

The protein needs of endurance athletes are close to twice the Dietary Reference Intakes for the general population. According to research studies, a protein intake of 1.4 to 1.6 g/kg/day is an important target for support of endurance exercise *(12)*.

The difficulty with these recommendations is that once you add up the carbohydrate and protein needs of the diet, there is little room left for fat. However, the critical role that monounsaturated and polyunsaturated fats play in health promotion, disease prevention, hormone production, weight control, mood, and cognitive function requires their adequate inclusion in the diet. The proportion of calories from fat in the diet should not go below 25% for an extensive length of time. The role of fats is so important that if calories are restricted it is advisable to reduce the proportion of carbohydrates in the diet to leave room for adequate protein and fat *(13)*.

7.2. Weight Gain and Muscle Growth

There is little disagreement on the nutritional requirements of building strength and power. Carbohydrate is required to fuel muscle building and sports-specific exercise; protein is essential to tissue recovery, repair, and growth; and fat is essential for production of the hormones that allow the entire anabolic process to move forward.

The protein needs of strength trainers and bodybuilders are higher than that of endurance athletes. When energy intake is adequate, protein needs during a building phase range from 1.8 to 2.0 g/kg/day. Carbohydrate needs are lower than that of endurance athletes and range from 4.5 g/kg/ for women to 7.0 g/kg.day for men, depending on the intensity and frequency of training. Depending on energy intake, 25% to 30% of total calories come from healthy fats *(10)*.

8. NUTRITIONALLY DENSE RESTORATION TIMING CONSIDERATION

The issue of rest and recovery for athletes involved with specific training intensities and competitive situations involves so much more than adhering to adequate sleep patterns. Specifically, sport nutritionists and researchers have placed large emphasis on the value of adequate dietary timing to promote viable recovery ingestion strategies. Understanding the value of select but quality nutritional timing is critical for athletes not only to increase strength and muscle mass but to enhance optimal performance outcomes.

Sport nutrition researchers in the Exercise and Sport Nutrition Laboratory at Baylor University promote the following guidelines for nutritional timing for athletic and exercise populations: To enhance the digestive process, athletes are encouraged to eat a full meal complete with high energy carbohydrates 4 to 6 hours before practice or competition. An example includes ingestion of a high carbohydrate breakfast for afternoon training sessions and a carbohydrate snack for events prior to noon. Then, 30 to 60 minutes before practice/competition athletes should consume a combination carbohydrate (30–50 g) and protein (5–10 g) snack or milkshake to help provide needed energy and to reduce catabolism. Ready-to-drink products and bars are convenient options for pretraining and precompetition events that can help control markers of overtraining. One concern associated with posttraining and competition dietary ingestion is that select athletes are not hungry after the intense event. However, this is one of the most critical times to replenish dietary energy balance to offset huge energy expenditure. A viable recommendation is to ingest a post-workout snack within 30 to 60 minutes after the event comprised of a light carbohydrate/protein (50–100 g of carbohydrates and 30–40 g of protein) snack until the individual is ready to consume a complete dietary meal within the 2-hour supported nutritional recovery window. The postworkout/competition meal should be high in carbohydrates and protein because this is when the body is most receptive to energy replenishment that helps sustain the critical energy repletion and balance. Overall, the general dietary suggested guidelines for athletes during heavy training periods include 55% to 65% calories from carbohydrates, 15% from protein, and less than 30% from fat *(3)*. Sport nutritionists and researchers recommend that athletes involved in heavy, intense training eat as many as four to six meals daily. However, it is recommended that power and/or strength athletes do not need as much carbohydrate as suggested in these guidelines.

In a recent book entitled *Nutrient Timing*, the authors further drive home the importance of proper dietary ingestion based on a finely tuned nutritional schedule *(14)*. This exceptionally written and scientifically based reading promotes the importance of a "nutritional timing system" that is comprised of three vital phases: 1) energy; 2) the anabolic process; and 3) growth. Although most nutritional research has been focused on what to eat, this cutting edge contribution places emphasis on what and when to eat. The primary concepts of this book not only allow the athlete to reach optimal potentials but to prepare for the next training or competitive bout.

There are numerous formulas for determining energy needs. Table 3 is a representation of accumulated data translated into an easy-to-use format (15). Under circumstances of endurance and ultra-endurance training, energy demands may be significantly higher, and the sport-specific energy demand should be added to the energy estimation. Numerous charts of energy expenditure during exercise are available from exercise physiology texts, at online sites, and in nutrient analysis software.

Table 3
Estimation of Daily Energy Needs of Men and Women Based on Activity Intensity

Level of activity[a]	Estimated energy expenditure (Kcal/kg/day)
Very light	
Men	31
Women	30
Light	
Men	38
Women	35
Moderate	
Men	41
Women	37
Heavy	
Men	50
Women	44

From Food and Nutrition Board, Institutes of Medicine of the National Academy (15).
[a] Very light, walking/standing; Light, walking up to 3.0 mph, housecleaning; Moderate, walking 3.5–4.0 mph, heavy housecleaning; Heavy, football, soccer, other serious athletic endeavors, manual labor.

9. TRANSLATING NUTRIENTS INTO FOOD

Once macronutrient numbers are determined, they must be applied to the design of a meal plan and a menu. Meal plans can be based on accepted dietary guidelines but must be broadened and increased based on the needs of athlete. For a sedentary population, dietary guidelines often serve as a complete meal plan as calorie and nutrient needs are relatively low. In an active population, dietary guidelines serve as the scaffolding on which a diet more dense in calories and nutrients can be built. Without the guidelines it becomes more cumbersome to determine a basic outline of foods to include in the diet. The guidelines also ensure food group variety in the diet.

Several dietary guidelines are in use. The U.S. Department of Agriculture (USDA) Food Guide Pyramid is the most widely disseminated food guide in the United States (www.mypyramid.gov) (16). Worldwide, many countries have developed their own dietary guidelines based on cultural preferences and regional foods. The Healthy Eating Pyramid from Harvard School of Public Health (http://www.hsph.harvard.edu/nutritionsource/pyramids.html) (17), the Mediterranean Diet Pyramid from Oldways Preservation and Exchange Trust (www.oldwayspt.org) (18), and the Healing Foods Pyramid from the University of Michigan Integrative Medicine Clinic (http://www.med.umich.edu/umim/clinical/pyramid/index.htm) (19) are also well accepted and scientifically founded food guides.

An adjunct to the food guides is the use of an accepted database of calories and macronutrients in foods. This can be done by hand using a rough equivalency chart based on The Exchange Lists for Meal Planning from the American Dietetic Association and The American Diabetes Association (Table 4) (20). Added to this chart is the notation for a teaspoon of added sugar. In all food guides, sugar is an optionally added food. Sugar can be used by an athlete to gain the greatest fuel advantage. Sugar also needs to be controlled to avoid diluting the nutrient density of the diet.

One can determine the amount of added sugar in foods by first selecting a sugar-free prepared food, such as shredded wheat cereal, and observing the amount of sugar listed per serving on the Nutrition Facts label. This sugar-free cereal contains 0 g of sugar. Therefore, any cereal that contains sugar has it added as an ingredient (unless fruit is added as an ingredient, and even then the fruit

Table 4
Nutrients and Calories per Serving of Food from Each Food Group and from
Teaspoons of Added Sugar

Food groups	Carbohydrates (g)	Protein (g)	Fat (g)	Calories
Grains, breads, starches	15	3	0–1	72–81
Fruit	15	–	–	60
Milk				
Fat free	12	8	0	80
Reduced fat	12	8	1–5	89–120
Whole	12	8	8	152
Nonstarchy vegetables	5	2	–	28
Meat and meat substitutes				
Very lean	–	7	0–1	28–37
Lean	–	7	3	55
Medium fat	–	7	5	73
High fat	–	7	8	100
Fat	–	–	5	45
Added sugars (1 tsp)	4	–	–	16

Adapted from American Dietetic Association *(20)* and Kleiner and
Greenwood-Robinson *(10)*.

adds a minor amount of sugar). You can determine the number of
teaspoons of added sugar in foods by using this strategy and knowing that 1 teaspoon of sugar contains 4 g. Note that certain foods
contain a natural amount of sugar; milk, for instance contains
12 g of milk sugar per 1 cup serving. This should be subtracted
from any amount greater than 12 g per cup in a milk product with
added sugar *(10)*.

Nutrient database software is also available for meal planning
and menu design. There are a number of options available, from
free and subscription online services to personal and professional
software programs. Take the time to investigate the quality of the
software and the nutrient database used for establishing nutrient

values of foods consumed. The database should be research quality from reputable sources. The USDA Nutrient Database for Standard Reference is the gold standard used by nearly all modern software programs. Release 18 is the most recent edition. Other research quality databases may specialize in certain nutrients. For instance, the University of Minnesota Nutrition Coordinating Center Database is well known for its collection of foods important in the research of lipid metabolism, cancers, and other disease states.

If you are going to use the software for meal planning, it must be versatile enough to allow customization of calorie levels and nutrient composition. Many software programs generate menus based on guidelines that you enter; others can use only preset guidelines. Editing should be easy and fairly intuitive. Files should be able to be exported and used in word processing or calendar programs or compressed for e-mail or other online utilities *(21)*.

10. EXAMPLE DIET PLANS

Creating an effective diet plan requires an understanding of the goals of the athlete and his/her sport. When training and competition seasons alter energy and nutrient requirements, one diet plan may not be enough. Many athletes, especially youth athletes, participate in more than one sport and move quickly from a strength and power sport to the next season where speed and agility are primary. They frequently request a weight gain diet prior to the first season and a weight loss diet prior to the second.

For some athletes the diet during the competitive season is lower in calories than during the preseason. During the preseason for football, increases in weight and strength can be desirable. Training activity is high, especially during training camps when twice-a-day practices are common, along with strength training and position-specific drills. Energy needs are high. Once the season begins, activity levels decrease—so calorie intakes must decrease to avoid fat gain.

Calorie and nutrient levels must be established based on physical goals and seasonal changes. Diets can be designed for weight maintenance, muscle building, or fat loss. Macronutrient composition can be set to enhance endurance capacity or strength and power. Refer to other chapters in this text for sport-specific macronutrient breakdowns.

Once the calorie and macronutrient compositions are set, choose a food guide to start the distribution of nutrients into foods, portions, and servings. Use the values in Table 4 to calculate the amount of calories and macronutrients gained from the number of portions from each food group. In most cases, the number of servings suggested in the food guide do not add up to the calorie and macronutrient levels set for your diet plan. At this point begin to fill out the plan by adding more servings from the various food groups to yield the set values and still allow for variety among all the food groups. An example follows.

10.1. Maintenance Diet for a Male Athlete Training ≥ 5 Days per Week

Refer to Kleiner and Greenwood-Robinson (10) for more information. This diet is for a man whose body weight is 180 pounds (81.5 kg), height is 6 feet 2 inches, and age is 25 years.

Calories (42 calories/kg/day)	3423 calories
Protein (1.4 g/kg/day)	115 g
Carbohydrate (6.0 g/kg/day)	486 g
Fat (~1.4 g/kg/day)	113 g

If you now use the USDA Food Guide Pyramid at www. mypyramid.gov and enter this personal data, the results are quite close in calories: 3200 (16). The distribution of food groups is as follows.

Grains, breads, starches	10 servings
Fruit	5 servings
Milk	3 servings
Nonstarchy vegetables	4 servings
Meat and meat substitutes	7 servings (ounces)
Fats	11 servings

Now the actual distribution of macronutrients must be calculated and food group servings adjusted to meet the guidelines set for the maintenance diet. More detail must be added to the meat group and the servings adjusted to control for unhealthy saturated fats, leaving room for healthy and essential monounsaturated and polyunsaturated fats. More carbohydrate in the form of added sugar must be added to allow for glucose nutriture before, during, and after exercise. Although the final numbers are not exactly the same as the goal, they are close. It is often difficult to create a diet that exactly meets the goal values. An example of a final dietary meal plan can be see in Table 5.

The next step is to distribute these foods throughout the day, creating a meal plan that follows the principles of food combinations and timing of eating that enhance sports performance and that is also practical and palatable for the athlete. The food group categories are then translated into actual food selections to create a day's menu. An example of a daily menu is provided in Table 6.

Table 5
Final Meal Plan for Maintenance Diet

Food groups	No. of servings	Carbohydrate (g)	Protein (g)	Fat(g)	Calories
Grains, breads, starches	11	165	33	0	792
Fruit	9	135	–	–	540
Nonfat milk	3	36	24	0	240
Nonstarchy vegetables	6	30	12	–	168
Meat, meat substitutes					
Very lean	4	–	28	0	112
Lean	3	–	21	9	165
Medium fat	1	–	7	5	73
Fat	19	–	–	95	855
Teaspoon of added sugar	29	116	–	–	464
Totals (goal)		482 (486)	125 (115)	109 (113)	3409 (3423)

Table 6
Food Group Categories: One Day's Menu

Food group servings	Menu
Preworkout snack	
	Water
1 Milk	1 c Plain yogurt
1 Fruit	$^3/_4$ c Fresh blueberries
Added sugar 3 tsp	1 tbsp Honey
Workout	
	Water
Added sugar 16 tsp	32 oz Sports drink
Breakfast	
	Water
2 Grain/bread/starch	2 slices Whole-grain bread
1 Milk	1 c Fat-free milk
3 Fruit	1 c Orange juice
	1 c Melon cubes
Added sugar 6 tsp	2 tbsp 100% Fruit spread for bread
1 Medium-fat meat/ substitute	1 Whole egg, scrambled
5 Fat	½ Avocado cooked with eggs
	1 tsp Heart healthy spread for cooking eggs
Snack	
2 Grains/bread/starch	8 Whole wheat crackers
1 Vegetable	1 c Celery sticks
6 Fat	3 tbsp Natural peanut butter
Lunch	
5 Grains/bread/starch	Foot-long Subway sandwich (choose from "6 grams of fat or less" list)
2 Vegetable	Fill sandwich with vegetable choices
1 Fruit	1 Banana
4 Very lean meat/ substitute	4 oz Meat included in sandwich
2 Fat	2 tsp Olive oil or 2 tbsp salad dressing

(*Continued*)

Table 6 (Continued)

Food group servings	Menu
Snack	
2 Fruit	8 Dried apricots
1 Milk	1 Tall nonfat latte
Added sugar 2 tsp.	2 tsp Sugar in latte
3 Fat	18 Almonds
Dinner	
	Green tea or other tea
2 Grains/bread/starch	1 Baked sweet potato
2 Fruit	6 oz or ~30 Red grapes
Added sugar 2 tsp	2 tsp Sugar or honey for tea
3 Vegetable	½ c Steamed asparagus
	2 c Mixed green salad
3 Lean mean/	3 oz. Grilled salmon
substitute	
3 Fat	2 tbsp Salad dressing
	1 tsp Olive oil for grilling salmon

11. CONCLUSION

- A nutrient-dense diet is the foundation to athletic performance and optimal results.
- The most important factor in an athlete's diet is the amount of energy available.
- Nutrient density is defined as a rich amount of nutrients per calorie consumed.
- To attain optimal performance outcomes via quality nutritional intake, exercise and athletic populations must match caloric ingestion with caloric expenditure.
- Along with choosing foods that are dense in nutrients, a nutrient-dense diet must also be rich in a variety of foods from all the food groups to be both nutritionally dense and complete.
- Nutritional supplements are not a complete substitute for a well balanced nutrient-dense diet. However, nutritional supplementation strategies, in addition to a nutrient-dense diet, are vital in

assisting the athlete in replacing the necessary caloric requirements lost through high intensity energy expenditure.

- Because matching caloric intake to energy expenditure is critical, athletes engaged in intense training (2–3 hr/day) should ingest 60 to 80 kcal/kg/day. The quality caloric requirement should be based on the intensity of training and the total energy expenditure.
- Because it is difficult to consume large quantities of food in one setting and difficult to maintain a quality energy balance, athletes are encouraged to eat four to six meals per day. Ingesting carbohydrate/protein snacks between meals helps offset energy expenditure.
- Because athletes are susceptible to a negative energy balance during intense training periods, dietary options should be comprised of the following combinations: carbohydrate (8–10 g/kg/day), high quality protein (1.5–2.0 g/kg/day) and low to moderate fat intake (< 30% of diet). This recommendation is suggested 4 to 6 hours before training whenever possible. Recommended fat intake for athletes attempting to lose weight is 0.5 to 1.0 g/kg/day.
- Nutritional timing is imperative as a dietary strategy. To help maintain energy balance and reduce catabolic states, athletes are encouraged to consume the following 30 to 60 minutes prior to exercise: 50 to 100 g of carbohydrate and 30 to 40 g of protein.

REFERENCES

1. Williams MH. The Ergogenics Edge (p 12). Human Kinetics, Champaign, IL, 1998.
2. Fletcher RH, Fairfield KM. Vitamins for chronic disease prevention in adults. JAMA 2002;287:3127–3129.
3. Kreider RB, Leutoholtz B. Nutritional considerations for preventing overtraining. In: Antonio J, Stout JR (eds) Sports Supplements (pp 199–208). Lippincott Williams & Wilkins, Philadelphia, 2001.
4. Kreider RB, Almada AL, Antonio J, et al. ISSN exercise and sport nutrition review: research recommendations. Sports Nutr Rev J 1:1–44, 2004.
5. Berning JR. Energy intake, diet, and muscle wasting. In: R.B. Kreider RB, A. C. Fry AC, O'Toole ML (eds) Overtraining in Sport (pp 275–288). Champaign, IL, Human Kinetics, 1998.
6. American College of Sports Medicine. Encyclopedia of Sports Sciences and Medicine (pp 1128–1129). Macmillan, New York, 1999.
7. Leutholtz B, Kreider RB. Exercise and sport nutrition. In: Temple N, Wilson T (eds) Nutritional Health (pp 207–239). Humana, Totowa, NJ, 2001.

8. Bloomstrand E, Hassmen P, Newsholme E. Effect of branch-chain amino acid supplementation on mental performance. Acta Physiol Scand 1991;143: 225–226.

9. Blomstrand E, Celsing F, Newshome EA. Changes in plasma concentrations of aromatic and branch-chain amino acids during sustained exercise in man and their possible role in fatigue. Acta Physiol Scand 1988;133:115–121.

10. Kleiner SM, Greenwood-Robinson M. Power Eating. 3rd ed. Human Kinetics, Champaign, IL, 2007.

11. Burke LM, Cox GR, Culmmings, NK, Desbrow B. Guidelines for daily carbohydrate intake: do athletes achieve them? Sports Med 2001;31:267–299.

12. Lemon PW. Beyond the zone: protein needs of active individuals. J Am Coll Nutr 2000;19(Suppl):513S–521S.

13. Wells A, Read NW, Macdonald IA. Effects of carbohydrate and lipid on resting energy expenditure, heart rate, sleepiness and mood. Phys Behav 1998;63:621–628.

14. Ivy J. Portman P. The Future of Sports Nutrition: Nutrient Timing (pp 7–14). Basic Health Publications, North Bergen, NJ, 2004.

15. Food and Nutrition Board, Institute of Medicine of the National Academies. Dietary Reference Intakes for Energy, Carbohydrate, Fiber, Fat, Fatty Acids, Cholesterol, Protein, and Amino Acids (Macronutrients). The National Academies Press, Washington, DC, 2005.

16. United States Department of Agriculture, Center for Nutrition Policy and Promotion. The Food Guide Pyramid. www.mypyramid.gov/, 2005.

17. Willett WC. Eat, Drink, and Be Healthy. Free Press/Simon & Schuster, New York, 2001, 2005 (http://www.hsph.harvard.edu/nutritionsource/pyramids. html)

18. Oldways Preservation and Exchange Trust. The Mediterranean Diet Pyramid. www.oldwayspt.org/, 1999.

19. The University of Michigan Integrative Medicine Clinic. The Healing Foods Pyramid. http://www.med.umich.edu/umim/clinical/pyramid/index.htm/, 2004.

20. American Dietetic Association, American Diabetes Association. Exchange Lists for Meal Planning. ADA, Alexandria, VA, 2003.

21. Prestwood E. Nutrition software: 101 questions to ask before you buy. Todays Dietitian 2000;2(2).

4 Macronutrient Intake for Physical Activity

Thomas Buford

Abstract

Proper nutrition is an essential element of athletic performance, body composition goals, and general health. Although natural variability among persons makes it impossible to create a single diet that can be recommended to all; examining scientific principles makes it easier for athletes and other physically active persons to eat a diet that prepares them for successful training and/or athletic competition. A proper nutritional design incorporates these principles and is tailored to the individual. It is important for the sports nutritionist, coach, and athlete to understand the role that each of the macronutrients plays in an active lifestyle. In addition, keys to success include knowing how to determine how many calories to consume, the macronutrient breakdown of those calories, and proper timing to maximize the benefits needed for the individual's body type and activity schedule.

Key words

Sport-specific eating · Carbohydrate · Fat · Protein · Nutrient timing · Energy expenditure · URTI

1. INTRODUCTION

Proper nutrition is an essential element of athletic performance, body composition goals, and general health. Although the natural variability among persons makes it impossible to create a single diet that can be recommended to all; examining scientific principles makes it easier for athletes and other physically active persons to eat a diet that prepares them for successful training or athletic competition. The purpose of this chapter is to discuss the three major nutrients that make up the bulk of energy intake and how

From: *Nutritional Supplements in Sports and Exercise*
Edited by: M. Greenwood, D. Kalman, J. Antonio,
DOI: 10.1007/978-1-59745-231-1_4, © Humana Press Inc., Totowa, NJ

they fit into the diet of those with a physically active lifestyle. Throughout the chapter the term athlete is used often, but it is meant to refer to anyone with a physically active lifestyle.

Designing a nutritional program for an athlete can be viewed much like the process of exercise prescription. The "nutritional design" should be individualized, taking into account factors such as age, size, sex, training regimen, and bioenergetic demands of the activity or sport. Much like the concept of sport-specific training, to maximize the benefits of training and/or event performance one needs to implement a system of "sport-specific eating." To do this, it is necessary to understand the primary energy systems used during a particular activity. Proper calorie needs, macronutrient ratios, and nutrient timing issues can then be addressed.

The aim of this chapter is to provide a framework that allows athletes, coaches, and sports nutritionists to make successful food and supplement choices. Hopefully, these choices will enable the athlete to train at maximal capacity, compete at maximal ability, and/or reach fitness goals while maintaining proper health. However, this framework must not be confused with a recipe. Each individual responds differently to a given diet, so an athlete must take these recommendations and adjust if her/his body does not respond as wished.

2. MACRONUTRIENTS

Macronutrients consist of the three nutrients that are required in large quantities in the diet: carbohydrates, proteins, and fats. These nutrients provide the energy required to maintain the body's functions as well as uphold cellular structure and homeostasis. Whether in energy production or cellular structure, these nutrients play a vital role in athletic performance as well as the overall health of an individual.

2.1. Carbohydrates

Carbohydrates are naturally occurring compounds that are composed of carbon, hydrogen, and oxygen. It was once thought that carbohydrates adhered to a chemical structure of $C_x(H_2O)_y$. However, this structure does not encompass all carbohydrates, but it does include other noncarbohydrate compounds such as acetic acid. A newer definition defines carbohydrates as polyhydroxy aldehydes or ketones and their derivatives.

There are three major classes of carbohydrates: monosaccharides, oligosaccharides, and polysaccharides. Monosaccharides are single sugar molecules and include glucose (also known as dextrose in the diet), fructose, and galactose. Oligosaccharides are chains of sugars that contain two to ten monosaccharides, the most common being disaccharides such as lactose, maltose, and sucrose. Polysaccharides are complex carbohydrates that contain possibly thousands of monosaccharides. Starches and fibers are the primary types of dietary polysaccharide. Glycogen is a polysaccharide that is also the storage form of glucose in the body and is found primarily in the liver and skeletal muscle.

Simple carbohydrates are primarily more calorie-dense, yet less nutrient-dense, than complex carbohydrates. Monosaccharides are a major problem in the sedentary population because of the low metabolic cost and ease in which they can convert to fat when in calorie excess. Polysaccharides are generally promoted because of their ability to be absorbed more slowly and provide greater nutrient value. Starch is the storage form of carbohydrate in plants, and it can be found in grains, nuts, legumes, and vegetables. It is a viable energy source because it digests slowly and provides energy for longer periods than does simple carbohydrate. Fiber, on the other hand, is indigestible and is useful in slowing the digestive rate of food, removing toxins, and adding bulk to the feces. It is found in foods such as vegetables, fruits, nuts, and legumes. The National Cancer Institute recommends 20 to 30 g of fiber per day for proper health. Two types of fiber exist: soluble and insoluble. Soluble fibers dissolve in water and slow the rate at which food travels through the small intestine, thereby maximizing nutrient uptake time. Soluble fibers can be found in foods such as wheat, rye, rice, and bran. Insoluble fiber, or cellulose, on the other hand does not dissolve in water and serves to remove toxins from and add bulk to the fecal matter. Cellulose can be found in oats, legumes, beans, and many fruits and vegetables.

It may sound as if simple carbohydrates are vastly inferior to complex carbohydrates, but a mix of carbohydrate types is beneficial for supplying athletes with energy. In fact, isolated dextrose and fructose can be useful in sports drinks or carbohydrate gels for athletes, and complex carbohydrates may increase glucose levels greatly. Although first developed for use with diabetics, the glycemic index (GI) provides a useful tool for helping athletes with food choices (Table 1). The GI value is a measure of how much and how long a particular food raises blood glucose levels. The values are based on a standard of 100, which is the value for glucose or white bread. Often, people mistake the GI as a function of whether foods are simple or complex. However, some complex carbohydrates (e.g., baked potatoes)

Table 1
Glycemic Index of Common Foods[a]

Low-GI foods	GI	Medium GI foods	GI	High GI foods	GI
Roasted and salted peanuts	14	Boiled potatoes	56	Mashed potato	70
Low-fat yogurt with sweetener	14	Sultanas	56	White bread	70
Cherries	22	Pita bread	57	Watermelon	72
Grapefruit	25	Basmati rice	58	Swede	72
Pearl barley	25	Honey	58	Bagel	72
Red lentils	26	Digestive biscuit	59	Bran flakes	74
Whole milk	27	Cheese and tomato pizza	60	Cheerios	74
Dried apricots	31	Ice cream	61	French fries	75
Butter beans	31	New potatoes	62	Coco Pops	77
Fettuccine pasta	32	Coca cola	63	Jelly beans	80
Skimmed milk	32	Apricots, canned in syrup	64	Rice cakes	82
Low-fat fruit yogurt	33	Raisins	64	Rice Krispies	82
Whole grain spaghetti	37	Shortbread biscuit	64	Cornflakes	84

Table 1
(continued)

Low-GI foods	GI	Medium GI foods	GI	High GI foods	GI
Apples	38	Couscous	65	Jacket potato	85
Pears	38	Rye bread	65	Puffed wheat	89
Tomato soup, canned	38	Pineapple, fresh	66	Baguette	95
Apple juice, unsweetened	40	Cantaloupe melon	67	Parsnips, boiled	97
Noodles	40	Croissant	67	White rice, steamed	98
White spaghetti	41	Shredded wheat	67		
All Bran	42	Mars bar	68		
Chick peas, canned	42	Ryvita	69		
Peaches	42	Crumpet, toasted	69		
Oranges	44	Wholemeal bread	69		
Macaroni	45				
Green grapes	46				
Orange juice	46				
Peas	48				
Baked beans in tomato sauce	48				
Carrots, boiled	49				
Milk chocolate	49				
Kiwi fruit	52				
Crisps	54				
Banana	55				
Raw oat bran	55				
Sweetcorn	55				

GI, glycemic index.
[a]Based on glucose (GI = 100) as a standard.
See Foster-Powell (1) for a more complete list.

increase glucose levels similarly to glucose. These values can be helpful for athletes to determine food choices when they want to increase glucose/glycogen levels quickly.

2.2. Protein

Proteins are nitrogen-containing compounds composed of dozens, hundreds, or thousands of amino acids. Amino acids are joined by peptide bonds, and several amino acids joined together become a polypeptide. Polypeptide chains then bond together and form various proteins. Chemically, proteins can be divided into groups: simple or conjugated. Simple proteins contain only amino acids or their derivatives. More recognizable to the nutritionist are conjugated proteins, which contain some nonprotein substance such as sugar molecules (glycoproteins), lipids (lipoproteins), or phosphate groups (phosphoproteins).

The human body is composed of 18% protein on average. Proteins provide structure to bodily tissues such as skeletal muscle, connective tissue, bone, and organs. In addition, nonstructural proteins act as hormones, catalysts (enzymes), buffer systems, cellular water balance regulators, lubricants, and immunoregulatory cells. Noting these numerous functions, the value of proteins in the body cannot be underestimated. In contrast to carbohydrates and fat, the body has no physiological reserve of protein stores. Therefore, if the body is not sufficiently supplied with protein, it catabolizes tissue proteins and cellular function is lost.

For nutritional purposes, amino acids can be divided into two general groups: essential and nonessential. Determining whether an amino acid is essential or nonessential hinges on whether the body synthesizes sufficient amounts to meet its own needs (nonessential) or if the diet must provide them (essential). The essential and nonessential amino acids are listed in Table 2. The most important of the essential amino acids are the branched-chain amino acids (BCAAs). The BCAAs are made up of leucine, isoleucine, and valine. These amino acids are available for uptake directly by the skeletal muscle without having to be metabolized by the liver. Many new protein supplements are either supplementing whole protein with the BCAAs or are simply marketed as the BCAAs themselves.

In addition, a helpful stratification of dietary proteins is to determine if the protein is complete or incomplete. Primarily in animal

Table 2
Essential and Nonessential Amino Acids

Essential	Nonessential
Histidine[a]	Alanine
Isoleucine	Arginine
Leucine	Asparagine
Lysine	Aspartic acid
Methionine	Cysteine
Phenylalanine	Glutamic acid
Threonine	Glutamine
Tryptophan	Glycine
Valine	Proline

[a]Some adults may be able to synthesize histidine on their own

products, proteins that contain the proper quantity and balance of essential amino acids are known as complete proteins. Meat, fish, eggs, milk, and cheese are all good sources of complete proteins. Incomplete proteins, on the other hand, lack one or more essential amino acids or are imbalanced in regard to essential amino acids. A few animal products may be incomplete, but they are primarily located in plant protein sources such as grains, beans, or vegetables. It can often be a challenge for vegetarian athletes to obtain proper amounts of all essential amino acids when forced to eat many plant sources in combination. (Further discussion of the vegetarian diet can be found in the sidebar at the end of the chapter.)

2.3. Fat

Lipids comprise a broad group of water-insoluble, energy-dense compounds made up of carbon, hydrogen, and oxygen. Often the terms *fat* and *lipid* are used interchangeably, although they are indeed different. Lipids include fats and oils in the body as well as fatty compounds such as sterols and phospholipids. Fats are specifically esters formed when fatty acids react with glycerol. Many other lipids exist and are significant in the diet. The lipids of greatest importance in the body and the diet are triglycerides, fatty acids, phospholipids, and cholesterol *(2)*. For the purposes of this chapter, the terms fat and fats are used to refer to dietary lipids.

Weight gain is often attributed to fats because they contain far more energy per gram than does carbohydrate or protein (9 kcal/g vs. 4 kcal/g, respectively). Because of this tendency (weight gain with overconsumption of fats) and their role in disease development, fats are often viewed in a highly negative light. However, fats serve many functions in the body: They provide energy for tissues and organs, membrane makeup, nerve signal transmission, and vitamin transport as well as cushioning and insulation for internal organs. In addition, in endurance athletes they are a vital fuel source for skeletal muscle.

The primary fats found in large quantities in foods are triglycerides. Triglycerides are composed of three fatty acids and one glycerol molecule. Fatty acids can be grouped by the amount of hydrogen they contain, otherwise known as saturation. Saturated fatty acid chains contain no double bonds; monounsaturated fatty acids contain one double bond; and polyunsaturated fats have multiple double bonds. Triglycerides typically contain a mix of the three fatty acid types. The ratio of unsaturated to saturated fatty acids is known as the P/S ratio. Animal fats usually have a low P/S ratio, whereas most vegetable oils (except tropical plant oils) have a high P/S ratio. The fatty acid makeup of the triglycerides is important to its metabolism in the body. For example, saturated fats may increase cholesterol in the body, whereas unsaturated fats may have no effect or lower cholesterol.

Cholesterol is found in small amounts in food and is generated by the body. High density lipoproteins (HDLs) are a type of cholesterol composed of a high protein to fat ratio. HDL is typically known as "good" cholesterol because of its protective nature against heart disease, whereas low-density lipoproteins (LDLs) are negative risk factors for heart disease. LDL is primarily fat with low amounts of protein.

There are two types of fatty acid in the diet that require pay special attention. First, essential fatty acids are not synthesized in the body and therefore must be taken in through the diet. The essential fatty acids are linoeic (omega-6) and linolineic (omega-3), both 18-carbon fatty acids. Linoeic acid is found in oils of plant origin, whereas marine oils are a good source of linolenic acid. The other type of fatty acid that needs attention is the trans fatty acid. "Trans fats" as they are commonly known, are often oils that are solidified through a process known as dehydrogenation, although some amounts are found naturally. Trans fats are in foods such as margarine, shortening, and some dairy products. The reduction of trans fats has become a point of public scrutiny in places such as fast food and packaged foods because although they are unsaturated fats they behave like saturated fats in the body. Trans fats appear to promote myriad diseases, including heart disease, diabetes, and obesity *(3)*.

Generally, fruits and vegetables contain little fat. Animal products such as meat, milk, cheese, and eggs, as well as baked goods, generally contain high amounts of saturated fats. Nuts and peanut/canola oils can be good sources of monounsaturated fats. Polyunsaturated fats, including the essential fatty acids, can be found in fish, nuts, and corn, soy, and sunflower oils. Lastly, margarine, shortening, cookies, pastries, and fried foods have high levels of trans fats.

2.4. Metabolic Usage

Paramount to understanding "sport-specific" eating and proper food decisions is a basic knowledge of the metabolic usage of the macronutrients in the body. This chapter provides only a cursory overview of the bioenergetics of activity and exercise, yet these principles are essential to a proper nutritional design. Gluconeogenesis, or energy production from protein sources, is a minor source of energy production, but its major functions include structural and enzymatic functions. Therefore, the discussion here is limited to energy production from carbohydrates and fats.

The primary energy systems used depend on the intensity and duration of the exercise. Short, quick bursts (e.g., vertical jumping or throwing events) are supplied by the ATP system. Sprints (quick events up to 10 seconds) are supplied via the ATP + phosphocreatine (PCr) system. Anaerobic power endurance events (e.g., 200 or 400 meter dashes) are supplied by ATP, PCr, and anaerobic glycolysis. Endurance events (> 800 meters) are supplied successively by glycolysis, the tricarboxylic acid (TCA) cycle, the electron transport chain (all carbohydrate metabolism), and eventually fat oxidation. In terms of carbohydrate and fat oxidation, a simple rule is the higher the intensity the more carbohydrate that is burned. In addition, the longer the duration, the more fat is utilized. This means that even activity that begins at high or moderate levels tapers off due to glycogen depletion, and fat becomes the primary fuel after 20 to 30 minutes of continual exercise. These basic concepts are essential to "nutrient load" adequately for the activity or event.

When comparing energy reserves, the human body has far more fat than glycogen stores. Skeletal muscle glycogen stores number around 400 g for an 80 kg individual, with an additional 100 g stored in the liver. In comparison, that same 80 kg individual may have more than 12,000 g of fat stored in adipose tissue. When factoring in the fact that fat is more than twice as calorie dense as carbohydrate, the energy stores are quite unbalanced. However, although each fatty acid provides 147 ATP and each triglyceride provides 460 ATP, glucose metabolism (36 ATP/molecule) is more efficient per unit of oxygen at providing energy. In addition, it takes roughly 20 minutes for free fatty acids to be liberated for use through lipolyis. Per unit of time, therefore, glucose is more efficient and thus the preferred fuel for high intensity exercise. These data help detail the merits of both fats and carbohydrates as fuel sources. The importance of these fuels should not be underestimated by the athlete when considering intake or expenditure.

3. DETERMINING INTAKE NEEDS

Possibly the most important piece of the nutritional puzzle for athletes is the understanding of how to determine the right amount of calories and macronutrients to consume. After all, what good is an understanding of what the nutrients are for athletes if they do not know how much they need to take in?

The initial consideration when determining macronutrient needs is to determine the goal of the nutritional design. Is the goal to maintain, lose, or gain weight? Or is there a body composition goal such as gaining muscle mass or maximizing strength while maintaining a certain body weight? Second, one must consider the bioenergetics of the event or training required to meet the goal. For example, high intensity, high frequency resistance training requires higher levels of protein intake than other forms of exercise. Finally, physiological factors such size, age, and sex play a role in caloric needs as well as the macronutrient distribution of those calories. For example, older athletes may need higher protein intake to prevent muscle loss and/or bone resorption, and highly active women at risk for amenorrhea may need to increase caloric intake and fat consumption.

3.1. Determining Caloric Needs

To determine macronutrient intake needs, one must begin by determining the caloric needs of the individual. Total energy expenditure (TEE) is composed of four factors: resting metabolic rate (RMR), exercise energy expenditure, thermogenesis, and activities of daily living.

The RMR accounts for the greatest percentage of calorie expenditure. RMR is positively correlated with the size and the amount of lean body mass a person has. The percentage of TEE from the RMR depends on the fitness and activity level of the individual, but it generally accounts for 60% to 70% of daily energy expenditure. As a person becomes more active, the RMR begins to account for a slightly lower percentage of energy expenditure. RMR values below 50% of the TEE have even been reported in male endurance athletes *(4)*.

The simplest and most accessible method for determining RMR is to use the Harris and Benedict equations *(5)*. These formulas require the height (in centimeters), weight in kilograms, and age (in years) to predict the daily RMR. The formulas are as follows.

$$\text{Males: RMR (kcal/day)} = 66.47 + 13.75(\text{weight}) + 5(\text{height}) - 6.76(\text{age})$$

$$\text{Females: RMR (kcal/day)} = 655.1 + 9.56(\text{weight}) + 1.85(\text{height}) - 4.68(\text{age})$$

Even though these equations do not take into account the fat free mass, they have been reported to predict the RMR within 200 kcal/day in endurance athletes of both sexes *(6)*. The most accurate ways to measure RMR include chamber indirect calorimetry and the use of metabolic carts equipped with RMR software. These methods are quite pricey, however, and not practical for most sports nutritionists, coaches, or athletes. Newer methods are becoming increasingly available to provide more individualized data than the formulas in a less expensive, portable form. Although several brand models exist (not discussed here), most of these portable RMR devices are fast and relatively accurate compared to the gold standard methods.

Once the RMR has been established, the next step is to determine the daily energy expenditure from physical activity. Exercise energy expenditure and activities of daily living are often combined into a physical activity. If this is done, it is important to remember to factor in daily activities such as walking up/down stairs, yardwork, or even shopping. These activities take small amounts of energy to complete but energy nonetheless. The athlete's training routine must also be taken into account. Caloric expenditure tables provide estimates of the energy requirements of many activities per minute (Table 3). To find the energy expenditure from physical activity per day, simply multiply the expenditure per minute by the number of minutes participating in the activity per day.

The final component of TEE is thermogenesis, which is energy expenditure not accounted for by RMR or activity. The most important form of thermogenesis is the thermic effect of food. The metabolic rate increases during the digestive processes, which increases energy expenditure. Fats and simple sugars have the lowest metabolic cost, whereas proteins and complex carbohydrates take more energy to digest. The thermic effect of food generally accounts for 5% to 10% of the TEE.

The TEE determines the number of calories the athlete needs per day to stabilize his or her body weight. If, however, a change in body weight is desired, as in the case of weight-dependent or muscle building sports; the daily caloric intake must be adjusted as such. One pound is equivalent to 3500 calories, therefore a 3500 calorie deficit/excess over a given period of time results in a loss/gain of 1 pound. Safe weight loss/gain guidelines generally recommend weight change of 1 to 2 pounds per week. Therefore, to achieve weight loss of 1 pound per week, for example, the athlete would need to consume 500 fewer kilocalories per day than the TEE.

3.2. Determining Macronutrient Intake

Once the daily caloric intake is determined, the question becomes how to determine what foods to eat. Although each athlete has individualized needs based on the sport's requirements, a good starting point

Table 3
Metabolic Expenditure of Given Activities

Activity	Metabolic expenditure (kcal/min), by body weight (kg/lb)																								
	45.0/ 100.0	48.0/ 105.0	50.0/ 110.0	52.0/ 115.0	55.0/ 120.0	57.0/ 125.0	59.0/ 130.0	61.0/ 135.0	64.0/ 140.0	66.0/ 145.0	68.0/ 150.0	70.0/ 155.0	73.0/ 160.0	75.0/ 165.0	77.0/ 170.0	80.0/ 175.0	82.0/ 180.0	84.0/ 185.0	86.0/ 190.0	89.0/ 195.0	91.0/ 200.0	93.0/ 205.0	95.0/ 210.0	98.0/ 215.0	100.0/ 220.0
Light work, cleaning	2.7	2.9	3.0	3.1	3.3	3.4	3.5	3.7	3.8	3.9	4.1	4.2	4.4	4.5	4.6	4.8	4.9	5.0	5.2	5.3	5.4	5.6	5.7	5.9	6.0
Basketball, vigorous	6.5	6.8	7.2	7.5	7.8	8.2	8.5	8.8	9.2	9.5	9.9	10.2	10.5	10.9	11.2	11.5	11.9	12.2	12.5	12.9	13.2	13.5	13.8	14.2	14.5
Bicycling																									
10 mph	4.2	4.4	4.6	4.8	5.1	5.3	5.5	5.7	5.9	6.1	6.4	6.6	6.8	7.0	7.2	7.4	7.6	7.9	8.1	8.3	8.5	8.7	8.9	9.1	9.4
15 mph	7.3	7.6	8.0	8.4	8.7	9.1	9.5	9.8	10.0	10.5	10.9	11.3	11.6	12.0	12.4	12.7	13.1	13.4	13.8	14.2	14.5	14.9	15.3	15.6	16.0
Golf, twosome (no cart)	3.6	3.8	4.0	4.2	4.4	4.6	4.7	4.9	5.1	5.3	5.4	5.6	5.8	6.0	6.2	6.4	6.6	6.7	6.9	7.1	7.3	7.4	7.6	7.9	8.0
Running																									
5 mph	6.0	6.3	6.6	7.0	7.3	7.6	7.9	8.2	8.5	8.8	9.1	9.4	9.7	10.0	10.3	10.6	10.9	11.2	11.6	11.9	12.2	12.5	12.8	13.1	13.4
7 mph	8.5	8.9	9.3	9.8	10.2	10.6	11.0	11.5	11.9	12.3	12.8	13.2	13.6	14.1	14.5	14.9	15.4	15.8	16.2	16.6	17.1	17.5	17.9	18.4	18.8
9 mph	10.8	11.3	11.9	12.4	12.9	13.5	14.0	14.6	15.1	15.7	16.2	16.8	17.3	17.9	18.4	19.0	19.5	20.1	20.6	21.2	21.7	22.2	22.8	23.3	23.9
11 mph	13.3	14.0	14.6	15.3	16.0	16.7	17.3	18.0	18.7	19.4	20.0	20.7	21.4	22.1	22.7	23.4	24.1	24.8	25.4	26.1	26.8	27.5	28.1	28.8	29.5
Tennis, competitive	6.4	6.7	7.1	7.4	7.7	8.1	8.4	8.7	9.1	9.4	9.8	10.1	10.4	10.8	11.1	11.4	11.8	12.1	12.4	12.8	13.1	13.4	13.7	14.1	14.4
Walking																									
2 mph	2.1	2.2	2.3	2.4	2.5	2.6	2.8	2.9	3.0	3.1	3.2	3.3	3.4	3.5	3.6	3.7	3.9	4.0	4.1	4.2	4.3	4.4	4.5	4.6	4.7
3 mph	2.7	2.9	3.0	3.1	3.3	3.4	3.5	3.7	3.8	3.9	4.1	4.2	4.4	4.5	4.6	4.8	4.9	5.0	5.2	5.3	5.4	5.6	5.7	5.9	6.0
4 mph	4.2	4.4	4.6	4.8	5.1	5.3	5.5	5.7	5.9	6.1	6.4	6.6	6.8	7.0	7.2	7.4	7.6	7.9	8.1	8.3	8.5	8.7	8.9	9.1	9.4
Weight training	5.2	5.4	5.7	6.0	6.2	6.5	6.8	7.0	7.3	7.6	7.8	8.1	8.3	8.6	8.9	9.1	9.4	9.7	9.9	10.2	10.5	10.7	11.0	11.2	11.5
Wrestling	8.5	8.9	9.3	9.8	10.2	10.6	11.0	11.5	11.9	12.3	12.8	13.2	13.6	14.1	14.5	14.9	15.4	15.8	16.2	16.6	17.1	17.5	17.9	18.4	18.8

Adapted from *Nutrition for Sport and Exercise* (pp 262–265), ed 2. Aspen, New York, 1998.

for prescribing macronutrient intake is to try to have a diet that is near 60% carbohydrates (with most of them being complex), 15% protein, and 25% fat. These numbers can be slightly adjusted based on the needs of the individual, as diets with various macronutrient mixtures have been proven to be effective for training and performance *(7–9)*. For example, an athlete in a weight-dependent sport who is trying to lose weight may want to reduce fat intake because of the calorie density of the substrate. On the other hand, athletes participating in resistance training may benefit from an increased percentage of protein compared to the other two fuel sources. Vegetarian athletes also require greater protein requirements owing to a lack of high quality meat proteins in the diet.

A second factor to consider is the daily recommendations for carbohydrate and protein. Beginning with protein, the general guideline for sedentary individuals is to intake ~0.8 g/kg body weight. Endurance athletes have been recommended to consume up to 1.4 g of protein/kg/day *(10)*. These athletes may require larger amounts of protein owing to repetitive motion breakdown of contractile proteins. In addition, BCAAs may be important for endurance athletes in delaying fatigue in respect to the central fatigue hypothesis. New research reported that dietary protein intake as high as 1.8 g/kg stimulated protein synthesis following endurance exercise, but further research is needed *(11)*. For athletes participating in regular resistance training, larger amounts of protein are needed to maintain an anabolic environment and increase muscle mass. Protein at 1.7 to 1.8 g/kg is generally recommended for strength training *(12)*, although 2.0 g/kg may ensure adequate intake. There is no evidence indicating the usefulness of > 2.0 g/kg, and it has been reported that 2.4 g/kg provided no greater benefit in increasing protein synthesis than a moderate protein diet *(13)*. On the other end of the spectrum, the athletes at greatest risk for deficient protein intake are those in weight-restrictive sports (e.g., wrestling, gymnastics) who are restricting calories.

Carbohydrate intake is of great importance to endurance athletes who train for durations of longer than 90 minutes per day to replenish muscle and liver glycogen levels. However, each gram of glycogen requires extra water to be stored and may inhibit performance in training or events shorter than 90 minutes. generally, individuals need carbohydrate intake of 6 to 10 g/kg daily to restore muscle and liver glycogen levels, but athletes training for periods longer than 90 minutes may require 8 to 10 g/kg/day *(14)*. In fact, Fallowfield and Williams *(15)* determined that even when isocaloric diets were consumed a high carbohydrate (8.8 g/kg) diet was significantly better at maintaining running time than a low carbohydrate (5.8 g/kg) diet. Benefits vary among individuals, however, and some may experience gastrointestinal problems on a high carbohydrate diet; therefore, it is important for athletes to determine what works best for them.

Once the total grams of protein and carbohydrate have been determined, multiply each factor by 4 to find the total number of calories from each of the respective substrates. Once this step is completed, subtract the number from the total energy needs determined earlier. The remainder of the necessary calories then come from fat. Divide the fat calories by 9 to determine the total number of fat grams to be consumed per day.

Although fats are the last macronutrient to be prescribed in the diet, they are not simply throw-away calories. It is not the unimportance of fats but, rather, the importance of carbohydrates and proteins that makes fat the final consideration. In fact, high fat diets (~35% kcal) have been reported to enhance endurance performance in some athletes *(7)*. However, although fat is a significant fuel source for many endurance or ultra-endurance athletes due to the "carbohydrate sparing effect," fats are not in short supply and a high-fat diet inhibits high intensity training and reduces endurance because glycogen stores are limited. In addition, per unit of oxygen, fats are less efficient than glucose at providing energy. The effects of a high fat diet also vary among individuals, so caution should be used in recommending one. Caution must be taken, however, not to restrict fat to a point where lipolysis and fat oxidation are inhibited as well *(16)*.

Once the macronutrient intake is determined based on body weight and activity recommendations, the percentage of each of the fuels should be compared to the original percentage goals. Because the daily recommendations are ranges, they should be reconciled with the percentages. For example, if the levels of 9 g/kg for carbohydrate and 1.4 g/kg for protein break down to 80% carbohydrate, 12% protein, and 8% fat, they need to be adjusted to lower the carbohydrate intake and increase the fat (and possibly protein slightly).

3.3. Case Study

Ellen is an endurance athlete who comes to you for diet counseling. She is 27 years old, 5 ft 4 in. (163 cm) tall, and weighs 121 lb (55 kg). During her training, she performs well at the beginning of her sessions, but she feels fatigued sooner than she would like or expect. She trains 5 or 6 days per week for 90 minutes per session at a 7 mph pace. She informs you that she consumes 2197 calories per day with 302 g of carbohydrate, 83 g of protein, and 73 g of fat.

You inform Ellen that she first needs to consume more calories. Based on her RMR (1356 kcal), exercise (918 kcal), other physical activity ($\sim$200 kcal), and thermogenesis (124 kcal), you inform her that her calorie intake during regular training periods should be 2498 kcal. The added calories should help her to improve her endurance.

In addition, you inform her that her current carbohydrate intake is only 54% of her diet and should be closer to 65% to 70%. You recommend that she begin to consume daily 440 g of carbohydrate, 77 g of protein, and 48 g of fat. This diet will provide her with 8.0 and 1.4 g/kg of carbohydrate and protein, respectively, while maintaining a 70.5/12.3/17.2 carbohydrate/protein/fat ratio. However, you inform her that some research has shown the beneficial effect of a high fat diet in some individuals. Therefore, if the current high carbohydrate diet does not work for her, she may consider increasing dietary fat and reducing carbohydrate intake.

You inform her that this diet should help maximize muscle glycogen stores while still providing enough fat for oxidation and protein for contractile protein repair. After beginning this new diet, Ellen discovers that her ability to maintain intensity during the ends of her workouts improves, and she feels more ready to train properly for her competitions.

4. NUTRIENT TIMING

One of the hottest topics in the sports nutrition field is the concept of "nutrient timing." Nutrient timing suggests that it is not merely what you eat and how much but also when. To build lean mass properly, replace glycogen stores, or simply maximize athletic performance, one needs to be conscious of the proper time to ingest food or macronutrient supplements. Not only is the timing an issue in terms of proper metabolic usage, but improper timing can also cause gastrointestinal or psychological discomfort. In general, pre- and postexercise supplements are preferred to be in liquid form to limit discomfort to the gastrointestinal tract.

Aside from improving performance, the primary nutritional goal for endurance athletes is to maximize (pre) and replenish (post) glycogen stores. For many years, endurance athletes have used "carbohydrate loading" as a tool for increasing muscle glycogen stores. Variations of carbohydrate loading regimens exist, but loading requires several days of high carbohydrate intake combined with tapered exercise the week before competition. Yet endurance athletes should not forget about the importance of fats as a fuel either. For muscle glycogen to be spared during long-duration endurance training, fats must be metabolized preferentially to glycogen. Depending on the intensity of exercise, carbohydrate supplementation immediately before or during exercise may inhibit lipolysis due to insulin increases. Fat oxidation seems to be inhibited by carbohydrate intake during lower intensity exercise (~45% VO$_2$max) but not during moderate intensity exercise *(17,18)*. In contrast, increased dietary fat appears to increase lipolysis and fat oxidation during exercise *(19)*. Therefore, it may be advisable to consume carbohydrate prior to moderate to high intensity exercise so muscle glycogen is maximized but to consume small amounts of fat prior to low intensity exercise to increase fat oxidation.

To maximize glycogen replenishment after exercise, it is necessary to ingest a carbohydrate supplement immediately after and every 2 hours (up to 6 hours) following exercise *(20)*. In addition, adding protein to the carbohydrate supplement appears to increase glycogen storage by acting synergistically on insulin secretion *(21)*. Restoring the glycogen levels properly after exercise allows proper recovery and supports the next day's training or competition.

For resistance training athletes, the goals are to increase amino acid uptake and anabolic hormone release to enhance protein synthesis as well as to replenish glycogen stores. At present, it appears that providing protein and/or carbohydrates immediately before and after resistance exercise may provide the optimal environment for enhanced muscle growth *(22)*. Whereas consumption of protein *(23)* or carbohydrate *(24)* following exercise has been shown to increase protein synthesis, the combination of the two has shown even greater success before and after exercise *(25,26)*. The protein–carbohydrate combination consumed prior to and after a workout has also been shown to increase growth hormone levels significantly *(27,28)*. In addition, it currently appears that consuming the postexercise supplement as soon as possible is extremely important and more effective than waiting for extended periods *(22,29)*.

Research in the area of nutrient timing and its connection with athletic performance is greatly expanding. As more research is completed, further information on the proper timing of macronutrient intake will most certainly come to light.

5. MAINTAINING OPTIMAL HEALTH DURING TRAINING AND COMPETITION

One of the most overlooked aspects of nutrition is the role of macronutrients in maintaining proper health during training and/or competition. Poor nutritional status plays a major role in the development of upper respiratory tract infections (URTIs), the most common health risk limiting physical activity. Several other factors including physical and mental stress, training intensity, injury, and environmental status (e.g., exposure to damp locker rooms) have also been linked to URTI development (Fig. 1) *(30,31)*. Athletes with extremely rigorous training schedules put themselves at risk for

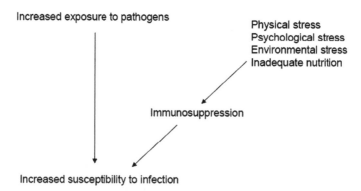

Fig. 1. Factors contributing to the incidence of infection in athletes. Adapted from Gleeson *(30)*.

infection, and the risk is greatly increased with improper macronutrient nutrition *(32)*. URTIs can greatly disrupt the quality of training or competition as well decreasing the quality of life for the individual. Specifically, infections with certain pathogens can cause appetite suppression, malabsorption of and increased need for nutrients, and increased loss of endogenous nutrients *(33)*.

In the public media, much attention is focused on micronutrients such as vitamin C and the B vitamins. However, deficiencies in protein and carbohydrate play a major role in immune dysfunction. Less is known about the contribution of fat to the immune response with exercise. Protein is essential for rapid cell replication and production of immunoregulatory proteins such as immunoglobulins, acute-phase proteins, and cytokines *(33)*. Carbohydrate feeding is also necessary for reducing the inflammatory response to exercise, including increased cortisol, catecholamines, and cytokine production. In fact, there is significant evidence that carbohydrate feeding during exercise is the best supplement for preventing immune dysfunction *(34,35)*. Although often overlooked, the role of proper macronutrient nutrition in immune function is critical to proper performance and should be examined when infections are occurring.

6. CONCLUSION

Proper macronutrient nutrition is part of the basis of any successful training and competition program. Prior to using nutritional supplements, athletes must take care to design their diet properly for their health in addition to making it "sport-specific." A proper nutritional design is as essential to success as a well tailored exercise program. The athlete must not only meet the body's need for calories but also meet specific needs for carbohydrates, protein, and fat. A proper, macronutrient-balanced diet can help manage body weight and provide the energy necessary for training and competition while at the same time promote a healthy immune system. In addition, it is now known that not only are the amounts of calories and individual macronutrients important but the timing of nutrient intake is as well. Each of these factors plays a critical role in "sport-specific eating".

6.1. Nutritional Considerations for the Vegetarian Athlete

6.1.1. JEAN JITOMIR

Although little long-term experimental research has been conducted on vegetarian athletes, the available evidence suggests that these athletes can carefully plan their diets to obtain adequate energy and nutrient consumption (36). Ovo (only eggs included), lacto (only dairy products included), or ovo-lacto (both eggs and dairy are included) vegetarians are at minimal risk for macronutrient deficiencies. On the hand, vegan athletes, who avoid all animal products, must carefully plan their diets to maintain optimal athletic performance and general health (36).

6.1.2. ENERGY

The three macronutrients—protein, carbohydrate, fat—provide energy in the diet and have countless essential functions. Additionally, the combined macronutrient energy contribution determines whether an athlete maintains, loses, or gains weight. Athletes should consult with dietitians and coaches to determine an intake that satisfies energy needs and prepares the athlete for competition. An athlete's energy requirements vary widely based on athletic event, training level, sex, and individual metabolism, among many others. Hence, a broad energy intake recommendation is not appropriate.

6.1.3. PROTEIN

The amount of protein required for an athlete is somewhat controversial. The American recommended daily allowance (RDA) value indicates that a protein intake of 0.8 g/kg is sufficient for most of the population; however, sports nutrition experts advocate a higher intake for athletes. Based on the RDA, a 50 kg (110 lb) athlete requires only 40 g protein/day. The American College of Sports Medicine (ACSM) recommends 1.2 to 1.4 g/kg/day for endurance athletes and 1.6 to 1.7 g/kg/day for power athletes (37). Because vegans eat only plant proteins, which are of lower quality than animal proteins, the RDA for the general population may not provide adequate levels of the essential amino acids. A vegan athlete would likely benefit from consuming protein at the top of the recommended range to ensure that essential amino acids are consumed in adequate quantities. For example, a 50 kg vegan athlete consuming protein at 1.7 g/kg should consume 85 g/day. Good vegan sources of protein include soy products, beans, nuts, and nut flours. These protein sources, in combination with the grains eaten on a vegan diet, provide complete and adequate protein for the vegan athlete.

Lacto-ovo vegetarians should also strive to meet the ACSM guidelines. However, high quality protein is easily accessible with the inclusion of dairy and egg products.

6.1.4. CARBOHYDRATE

According to the ACSM, athletes require 6 to 10 g of carbohydrate/kg/day (37). Glycogen is an important source of fuel for all athletes but especially endurance athletes; carbohydrate consumption in the higher range of the recommendation ensures maximal glycogen storage. As such, an endurance athlete may benefit from 8 to 10 g of carbohydrate/day, depending on the level of training. The 50 kg athlete consuming 8 g/kg/day should consume 400 g. A nonendurance athlete requires less energy and may choose to consume 6 g/kg/day.

It is particularly important to include carbohydrates before exercise to maintain glucose levels throughout training. Either a high-carbohydrate meal 3 to 4 hours before training or a low-carbohydrate meal about 1 hour before training is appropriate *(37)*. Additionally, a source of carbohydrate should be consumed within 30 minutes of exercise completion to replace lost glycogen. The postworkout carbohydrate meal should contain at least 1.5 g/kg, or 75 g, for the 50 kg athlete. Clearly, an athlete must consume a substantial amount of total daily carbohydrates soon before and after exercise *(36)*. Carbohydrate sources are abundant in the vegetarian diet and include grains, beans, vegetables, and fruit. Fast-absorbing carbohydrates (e.g., ripe bananas) are good choices before and immediately after workouts.

6.1.5. FAT

The ACSM recommends a moderate dietary fat intake of 20% to 25% of the total daily energy intake *(37)*. Furthermore, a fat intake lower than 15% of total energy from fat is not recommended by the ACSM, as some fats are essential and adequate dietary fat is needed to digest and transport the fat-soluble vitamins. Considering the 50 kg female athlete on a 2500 calorie diet, fat intake of 62 g/day provides the remainder of her energy requirements when she also consumes 85 g of protein and 400 g of carbohydrate. A fat intake of 62 g/day provides about 22% of total energy as fat.

Additionally, athletes should consume dietary fat sources rich in essential fats; nuts, fish, and oils are good choices. Saturated and trans fats from cheese, meats, and processed foods harm long-term health and do not provide essential fats.

6.1.6. Creatine Supplementation

Creatine supplementation has ergogenic effects for power athletes who depend on quick bursts of energy or sports performance; this category of athletes includes sprinters, weightlifters, and many others. Athletes who eat meat may obtain an additional gram of creatine each day from the diet in addition to endogenous production. Vegetarians do not obtain creatine from the diet and have lower intramuscular creatine levels than nonvegetarians. Research has provided evidence that creatine supplementation is particularly efficacious for vegetarian power athletes *(38,39)*. In fact, one study that compared creatine supplementation in vegetarians and nonvegetarians showed a greater increase in muscle phosphocreatine, total creatine, type II fiber area, lean tissue mass, and total work performed in the vegetarian group *(39)*. Even so, it is still recommended that athletes consult with coaches to determine the appropriateness, safety, and dosage before taking a dietary supplement.

As detailed above, vegetarian and vegan diets can, with careful planning, fuel an athletic body for optimal performance.

REFERENCES

1. Foster-Powell K, Holt SH, Brand-Miller JC. International table of glycemic index and glycemic load values: 2002. Am J Clin Nutr 2002;76:5–56.
2. Remiers K, Ruud J. Nutritional factors in health and performance. In: Essentials of Strength and Conditioning. 2 nd ed. Human Kinetics, Champagne, IL, 2000.
3. Duyff RL, American Dietetic Association. American Dietetic Association Complete Food and Nutrition guide (pp xii, 658). 2 nd ed. Wiley, Hoboken, NJ, 2002.
4. Thompson JL, Manore MM, Skinner JS. Resting metabolic rate and thermic effect of a meal in low- and adequate-energy intake male endurance athletes. Int J Sport Nutr 1993;3:194–206.

5. Harris JA, Benedict FG. A Biometric Study of Basal Metabolism in Man. Vol 279. Carnegie Institute, Washington, DC, 1919.

6. Thompson JL, Manore MM. Predicted and measured resting metabolic rate of male and female endurance athletes. J Am Diet Assoc 1996;96:30–34.

7. Muoio DM, Leddy JJ, Horvath PJ, Awad AB, Pendergast DR. Effect of dietary fat on metabolic adjustments to maximal VO_2 and endurance in runners. Med Sci Sports Exerc 1994;26:81–88.

8. Pendergast DR, Horvath PJ, Leddy JJ, Venkatraman JT. The role of dietary fat on performance, metabolism, and health. Am J Sports Med 1996; 24:S53–S58.

9. Phinney SD, Bistrian BR, Evans WJ, gervino E, Blackburn gL. The human metabolic response to chronic ketosis without caloric restriction: preservation of submaximal exercise capability with reduced carbohydrate oxidation. Metabolism 1983;32:769–776.

10. Moore DR, Phillips SM, Babraj JA, Smith K, Rennie MJ. Myofibrillar and collagen protein synthesis in human skeletal muscle in young men after maximal shortening and lengthening contractions. Am J Physiol Endocrinol Metab 2005;288:E1153–E1159.

11. Bolster DR, Pikosky MA, gaine PC, Martin W, Wolfe RR, Tipton KD, et al. Dietary protein intake impacts human skeletal muscle protein fractional synthetic rates after endurance exercise. Am J Physiol Endocrinol Metab 2005;289:E678–E683.

12. Lemon PWR. Effects of exercise on dietary protein requirements. Int J Sports Nutr 1998;8:426–447.

13. Tarnopolsky MA, Atkinson SA, MacDougall JD, Chesley A, Phillips S, Schwarcz HP. Evaluation of protein requirements for strength trained athletes. J Appl Physiol 1992;73:1986–1995.

14. Costill DL, Bowers, R, Branam G, Sparks K. The role of dietary carbohydrate in muscle glycogen resynthesis after strenuous running. Am J Clin Nutr 1981;34:1831–1836.

15. Fallowfield JL, Williams C. Carbohydrate intake and recovery from prolonged exercise. Int J Sports Nutr 1993;3:150–164.

16. Coyle EF, Jeukendrup AE, Oseto MC, Hodgkinson BJ, Zderic TW. Low-fat diet alters intramuscular substrates and reduces lipolysis and fat oxidation during exercise. Am J Physiol Endocrinol Metab 2001;280:E391–E398.

17. Horowitz JF, Mora-Rodriguez R, Byerly LO, Coyle EF. Lipolytic suppression following carbohydrate ingestion limits fat oxidation during exercise. Am J Physiol Endocrinol Metab 1997;273:E768–E775.

18. Horowitz JF, Mora-Rodriguez R, Byerly LO, Coyle EF. Substrate metabolism when subjects are fed carbohydrate during exercise. Am J Physiol Endocrinol Metab 1999;276:828–835.

19. Zderic TW, Davidson CJ, Schenk S, Byerly LO, Coyle EF. High-fat diet elevates resting intramuscular triglyceride concentration and whole body lipolysis during exercise. Am J Physiol Endocrinol Metab 2004;286:E217–E225.

20. Ivy JL. Glycogen resynthesis after exercise: effect of carbohydrate intake. Int J Sports Med 1998;19:S142–S145.

21. Williams MB, Raven PB, Fogt DL, Ivy JL. Effects of recovery beverages on glycogen restoration and endurance exercise performance. J Strength Cond Res 2003;17:12–19.

22. Volek J. Influence of nutrition on responses to resistance training. Med Sci Sports Exerc 2004;36:689–696.

23. Tipton KD, Ferrando AA, Philips SM, Wolfe RR. Postexercise net protein synthesis in human muscle from orally administered amino acids. Am J Physiol 1999;276:E628–E634.

24. Roy BD, Tarnopolsky MA, MacDougall JD, Fowles J, Yarasheski KE. Effect of glucose supplement timing on protein metabolism after resistance training. J Appl Physiol 1997;82:1882–1888.

25. Rasmussen BB, Tipton KD, Miller SL, Wolf SE, Wolfe RR. An oral amino acid-carbohydrate supplement enhances muscle protein anabolism after resistance exercise. J Appl Physiol 2000;88:386–392.

26. Tipton KD, Rasmussen BB, Miller SL, et al. Timing of amino acid-carbohydrate ingestion alters anabolic response of muscle to resistance exercise. Am J Physiol 2001;281:E197–E206.

27. Chandler RM, Byrne HK, Patterson JG, Ivy JL. Dietary supplements affect the anabolic hormones after weight-training exercise. J Appl Physiol 1994;76:839–845.

28. Kraemer WJ, Volek JS, Bush JA, Putukian M, Sebastianelli. Hormonal responses to consecutive days of heavy-resistance exercise with or without nutritional supplementation. J Appl Physiol 1998;85:1544–1555.

29. Esmarck B, Anderson JL, Olsen S, Richter EA, Mizuno M, Kjaer M. Timing of post-exercise protein intake is important for muscle hypertrophy with resistance training in elderly humans. J Physiol 2001; 535:301–311.

30. Gleeson, M. The scientific basis of practical strategies to maintain immuno-competence in elite athletes. Exerc Immunol Rev 2000;6:75–101.

31. Nieman DC. Immune response to heavy exertion. J Appl Physiol 1997; 82:1385–1394.

32. Calder PC, Jackson AA. Undernutrition, infection and immune function. Nutr Res Rev 2000;13:3–29.

33. Gleeson M. Can nutrition limit exercise-induced immunodepression? Nutr Rev 2006;64:119–131.

34. Chan MA, Koch AJ, Benedict SH, Potteiger JA. Influence of carbohydrate ingestion on cytokine responses following acute resistance exercise. Int J Sport Nutr Exerc Metab 2003;13:454–465.

35. Nieman DC, Nehlsen-Cannarella SL, Fagoaga OR, et al. Effects of mode and carbohydrate on the granulocyte and monocyte response to intensive, prolonged exercise. J Appl Physiol 1998;84:1252–1259.

36. Barr SI, Rideout CA. Nutritional considerations for vegetarian athletes. Nutrition 2004;20:696–703.

37. Position of the American Dietetic Association, Dietitians of Canada, and the American College of Sports Medicine: nutrition and athletic performance. J Am Diet Assoc 2000;100:1543–1556.

38. Venderley AM, Campbell WW. Vegetarian diets: nutritional considerations for athletes. Sports Med 2006;36:293–305.
39. Burke DG, Chilibeck PD, Parise G, Candow DG, Mahoney D, Tarnopolosky M. Effect of creatine and weight training on muscle and performance in vegetarians. Med Sci Sports Exerc 2003;35:1946–1955.

5 Essential and Nonessential Micronutrients and Sport

Kristen M. Beavers and Monica C. Serra

Abstract

The purpose of this chapter is to review the role of micronutrients in sport. Attention is given to the function of micronutrients in the body, examples of quality dietary sources of each micronutrient, and an assessment of the literature examining how the recommended daily intake of a micronutrient may or may not change with exercise. The discussion includes plausible biological mechanisms of proposed performance enhancement and current research to support or negate these claims. Water-soluble vitamins, fat-soluble vitamins, macrominerals, and select microminerals are discussed in detail, and a comprehensive table reviewing all micronutrients recommendations for the athlete is provided. Practical applications for professionals working with athletes conclude the chapter.

Key words

Micronutrients · Vitamins · Minerals · Ergogenic aid · Physical · Performance

1. INTRODUCTION

To maintain normal health, a wide range of vitamins, minerals, and trace elements must be present in adequate amounts in the body. Micronutrients play many important roles in the body including hemoglobin synthesis, maintenance of bone health, adequate immune function, and the protection of body tissues from oxidative damage *(1)*. In addition, and to interest of the athlete, micronutrients are integral in the process of energy metabolism. Whereas it is the macronutrients (carbohydrate, protein, fat) that constitute the

From: *Nutritional Supplements in Sports and Exercise*
Edited by: M. Greenwood, D. Kalman, J. Antonio,
DOI: 10.1007/978-1-59745-231-1_5, © Humana Press Inc., Totowa, NJ

sources of fuel for our bodies, it is the micronutrients that allow the breakdown and use of these fuels.

To help quantify which micronutrients we need and in what amounts, the Food and Nutrition Board, along with the Institute of Medicine, have provided us with reference values, known as dietary reference intakes (DRIs) (Tables 1 and 2), of suggested intake of micronutrients to prevent deficiency *(2)*. The DRIs consist of four categories: recommended dietary allowance (RDA), adequate intake (AI), estimated average requirement (EAR), and tolerable upper intake level (UL). The goal of the RDA is to provide a dietary intake level that is sufficient to meet the requirement for 98% of healthy individuals. The AI is used when no RDA has been determined, and the EAR is used to satisfy the needs of 50% of individuals within a particular group. The UL is the maximum recommended intake that individuals could consume without the risk of adverse effects. The DRIs vary among sex and age groups. These values represent what is needed for the "normal" individual. How, and if, these needs change with increased physical activity is still a matter of debate. Sound biological mechanisms postulate that regular intense exercise training may increase micronutrient requirements by increasing degradation rates or increasing losses from the body *(3)*. Moreover, high intakes of micronutrients may be required to cover increased needs for the repair and maintenance of the lean tissue mass in athletes *(1)*.

The purpose of this chapter is to review the function of micronutrients in the body, provide examples of quality dietary sources of each micronutrient, and assess the literature examining how the recommended daily intake of a micronutrient may or may not change with exercise.

2. VITAMINS

Vitamins are organic compounds naturally found in small amounts in food products. They are designated essential nutrients because they cannot be synthesized by the body in amounts that are necessary to support normal physiological function. Generally, vitamins are classified as either water-soluble or fat-soluble based on the medium needed for their absorption. Water-soluble vitamins

Table 1
2004 DRI Table for Vitamins
Dietary Reference Intakes (DRIs): Recommended Intakes for Individuals, Vitamins
Food and Nutrition Board, Institute of Medicine, National Academies

Life Stage Group	Vit A (μg/d)[a]	Vit C (mg/d)	Vit D (μg/d)[b,c]	Vit E (mg/d)[d]	Vit K (μg/d)	Thiamin (mg/d)	Riboflavin (mg/d)	Niacin (mg/d)[e]	Vit B_6 (mg/d)	Folate (μg/d)[f]	Vit B_{12} (μg/d)	Pantothenic Acid (mg/d)	Biotin (μg/d)	Choline[g] (mg/d)
Infants														
0–6 mo	400*	40*	5*	4*	2.0*	0.2*	0.3*	2*	0.1*	65*	0.4*	1.7*	5*	125*
7–12 mo	500*	50*	5*	5*	2.5*	0.3*	0.4*	4*	0.3*	80*	0.5*	1.8*	6*	150*
Children														
1–3y	300	15	5*	6	30*	0.5	0.5	6	0.5	150	0.9	2*	8*	200*
4–8y	400	25	5*	7	55*	0.6	0.6	8	0.6	200	1.2	3*	12*	250*
Males														
9–13y	600	45	5*	11	60*	0.9	0.9	12	1.0	300	1.8	4*	20*	375*
14–18y	900	75	5*	15	75*	1.2	1.3	16	1.3	400	2.4	5*	25*	550*
19–30y	900	90	5*	15	120*	1.2	1.3	16	1.3	400	2.4	5*	30*	550*
31–50y	900	90	5*	15	120*	1.2	1.3	16	1.3	400	2.4	5*	30*	550*
51–70y	900	90	10*	15	120*	1.2	1.3	16	1.7	400	2.4[i]	5*	30*	550*
>70y	900	90	15*	15	120*	1.2	1.3	16	1.7	400	2.4[i]	5*	30*	550*
Females														
9–13y	600	45	5*	11	60*	0.9	0.9	12	1.0	300	1.8	4*	20*	375*
14–18y	700	65	5*	15	75*	1.0	1.0	14	1.2	400[i]	2.4	5*	25*	400*
19–30y	700	75	5*	15	90*	1.1	1.1	14	1.3	400[i]	2.4	5*	30*	425*
31–50y	700	75	5*	15	90*	1.1	1.1	14	1.3	400[i]	2.4	5*	30*	425*
51–70y	700	75	10*	15	90*	1.1	1.1	14	1.5	400	2.4[h]	5*	30*	425*
>70y	700	75	15*	15	90*	1.1	1.1	14	1.5	400	2.4[h]	5*	30*	425*
Pregnancy														
14–18y	750	80	5*	15	75*	1.4	1.4	18	1.9	600[j]	2.6	6*	30*	450*
19–30y	770	85	5*	15	90*	1.4	1.4	18	1.9	600[j]	2.6	6*	30*	450*
31–50y	770	85	5*	15	90*	1.4	1.4	18	1.9	600[j]	2.6	6*	30*	450*

(Continued)

Table 1
(Continued)

Life Stage Group	Vit A (μg/d)[a]	Vit C (mg/d)	Vit D (μg/d)[b,c]	Vit E (mg/d)[d]	Vit K (μg/d)	Thiamin (mg/d)	Riboflavin (mg/d)	Niacin (mg/d)[e]	Vit B6 (mg/d)	Folate (μg/d)[f]	Vit B12 (μg/d)	Pantothenic Acid (mg/d)	Biotin (μg/d)	Choline[g] (mg/d)
Lactation														
14–18y	1,200	115	5*	19	75*	1.4	1.6	17	2.0	500	2.8	7*	35*	550*
19–30y	1,300	120	5*	19	90*	1.4	1.6	17	2.0	500	2.8	7*	35*	550*
31–50y	1,300	120	5*	19	90*	1.4	1.6	17	2.0	500	2.8	7*	35*	550*

Note: This table (taken from the DRI reports, see www.nap.edu) presents Recommended Dietary Allowances (RDAs) in bold type and Adequate Intakes (AIs) in ordinary type followed by an asterisk(*). RDAs and AIs may both be used as goals for individual intake. RDAs are set to meet the needs of almost all (97 to 98 percent) individuals in a group. For healthy breastfed infants, the AI is the mean intake. The AI for other life stage and gender groups is believed to cover needs of all individuals in the group, but lack of data or uncertainty in the data prevent being able to specify with confidence the percentage of individuals covered by this intake.

[a] As retinol activity equivalents (RAEs). 1 RAE = 1 μg retinol, 12 μg β-carotene, 24 μg α-carotene, or 24 μg β-cryptoxanthin. The RAE for dietary provitamin A carotenoids is twofold greater than retinol equivalents (RE), whereas the RAE for preformed vitamin A is the same as RE.

[b] As cholecalciferol. 1 μg cholecalciferol = 40 IU vitamin D.

[c] In the absence of adequate exposure to sunlight.

[d] As α-tocopherol. α-Tocopherol includes *RRR*-α-tocopherol, the only form of α-tocopherol that occurs naturally in foods, and the 2*R*-stereoisomeric forms of α-tocopherol (*RRR*-, *RSR*-, *RRS*-, and *RSS*-α-tocopherol) that occur in fortified foods and supplements. It does not include the 2*S*-stereoisomeric forms of α-tocopherol (*SRR*-, *SSR*-, *SRS*-, and *SSS*-α-tocopherol), also found in fortified foods and supplements.

[e] As niacin equivalents (NE). 1 mg of niacin = 60 mg of tryptophan; 0–6 months = preformed niacin (not NE).

[f] As dietary folate equivalents (DFE). 1 DFE = 1 μg food folate = 0.6 μg of folic acid from fortified food or as a supplement consumed with food = 0.5 μg of a supplement taken on an empty stomach.

[g] Although AIs have been set for choline, there are few data to assess whether a dietary supply of choline is needed at all stages of the life cycle, and it may be that the choline requirement can be met by endogenous synthesis at some of these stages.

[h] Because 10 to 30 percent of older people may malabsorb food-bound B₁₂, it is advisable for those older than 50 years to meet their RDA mainly by consuming foods fortified with B₁₂ or a supplement containing B₁₂.

[i] In view of evidence linking folate intake with neural tube defects in the fetus, it is recommended that all women capable of becoming pregnant consume 400 μg from supplements or fortified foods in addition to intake of food folate from a varied diet.

[j] It is assumed that women will continue consuming 400 μg from supplements or fortified food until their pregnancy is confirmed and they enter prenatal care, which ordinarily occurs after the end of the periconceptional period—the critical time for formation of the neural tube.

Reprinted with permission from *Dietary Reference Intakes* © (2004) by the National Academy of Sciences, courtesy of the National Academies Press, Washington, DC.

Table 2
2004 DRI Table for Elements
Dietary Reference Intakes (DRIs): Recommended Intakes for Individuals, Elements
Food and Nutrition Board, Institute of Medicine, National Academies

Life Stage Group	Calcium (mg/d)	Chromium (µg/d)	Copper (µg/d)	Fluoride (mg/d)	Iodine (µg/d)	Iron (mg/d)	Magnesium (mg/d)	Manganese (mg/d)	Molybdenu (µg/d)	Phosphorus (mg/d)	Seleniu (µg/d)	Zinc (mg/d)	Potassium (g/d)	Sodium (g/d)	Chloride (g/d)
Infants															
0–6 mo	210*	0.2*	200*	0.01*	110*	0.27*	30*	0.003*	2*	100*	15*	2*	0.4*	0.12*	0.18*
7–12mo	270*	5.5*	220*	0.5*	130*	11*	75*	0.6*	3*	275*	20*	3*	0.7*	0.37*	0.57*
Children															
1–3y	500*	11*	340	0.7*	90	7	80	1.2*	17	460	20	3	3.0*	1.0*	1.5*
4–8y	800*	15*	440	1*	90	10	130	1.5*	22	500	30	5	3.8*	1.2*	1.9*
Males															
9–13y	1,300*	25*	700	2*	120	8	240	1.9*	34	1,250	40	8	4.5*	1.5*	2.3*
14–18y	1,300*	35*	890	3*	150	11	410	2.2*	43	1,250	55	11	4.7*	1.5*	2.3*
19–30y	1,000*	35*	900	4*	150	8	400	2.3*	45	700	55	11	4.7*	1.5*	2.3*
31–50y	1,000*	35*	900	4*	150	8	420	2.3*	45	700	55	11	4.7*	1.5*	2.3*
51–70y	1,200*	30*	900	4*	150	8	420	2.3*	45	700	55	11	4.7*	1.3*	2.0*
>70y	1,200*	30*	900	4*	150	8	420	2.3*	45	700	55	11	4.7*	1.2*	1.8*
Females															
9–13y	1,300*	21*	700	2*	120	8	240	1.6*	34	1,250	40	8	4.5*	1.5*	2.3*
14–18y	1,300*	24*	890	3*	150	15	360	1.6*	43	1,250	55	9	4.7*	1.5*	2.3
19–30y	1,000*	25*	900	3*	150	18	310	1.8*	45	700	55	8	4.7*	1.5*	2.3*
31–50y	1,000*	25*	900	3*	150	18	320	1.8*	45	700	55	8	4.7*	1.5*	2.3*
51–70y	1,200*	20*	900	3*	150	8	320	1.8*	45	700	55	8	4.7*	1.3*	2.0*
>70y	1,200*	20*	900	3*	150	8	320	1.8*	45	700	55	8	4.7*	1.2*	1.8*

(Continued)

Table 2
(Continued)

Life Stage Group	Calcium (mg/d)	Chromium (μg/d)	Copper (μg/d)	Fluoride (mg/d)	Iodine (μg/d)	Iron (mg/d)	Magnesium (mg/d)	Manganese (mg/d)	Molybdenu (μg/d)	Phosphorus (mg/d)	Selenia (μg/d)	Zinc (mg/d)	Potassium (g/d)	Sodium (g/d)	Chloride (g/d)
Pregnancy															
14–18y	1,300*	29*	1,000	3*	220	27	400	2.0*	50	1,250	60	12	4.7*	1.5*	2.3*
19–30y	1,000*	30*	1,000	3*	220	27	350	2.0*	50	700	60	11	4.7*	1.5*	2.3*
31–50y	1,000*	30*	1,000	3*	220	27	360	20*	50	700	60	11	4.7*	1.5*	2.3*
Lactation															
14–18y	1,300*	44*	1,300	3*	290	10	360	2.6*	50	1,250	70	13	5.1*	1.5*	2.3*
19–30y	1,000*	45*	1,300	3*	290	9	310	2.6*	50	700	70	12	5.1*	1.5*	2.3*
31–50y	1,000*	45*	1,300	3*	290	9	320	2.6*	50	700	70	12	5.1*	1.5*	**2.3 ***

Note: This table presents Recommended Dietary Allowances (RDAs) in bold type and Adequate Intakes (AIs) in ordinary type followed by an asterisk (*). RDAs and AIs may both be used as goals for individual intake. RDAs are set to meet the needs of almost all (97 to 98 percent) individuals in a group. For healthy breastfed infants, the AI is the mean intake. The AI for other life stage and gender groups is believed to cover needs of all individuals in the group, but lack of data or uncertainty in the data prevent being able to specify with confidence the percentage of individuals covered by this intake.

Sources: *Dietary Reference Intakes for Calcium, Phosporous, Magnesium, Vitamin D, and Choline (1998)*; *Dietary Reference Intakes for Thiamin, Riboflavin, Niacin, Vitamin B6, folare, Vitamin B12, Pantothenic Acid, Biotin, and Choline (1998)*; *Dietary Reference Intakes for Vitamin C, Vitamin E, Selenium, and Carotenoids (2000)*; *Dietary Reference Intakes for Vitamin A, Vitamin K, Arsenic, Boron, Chromium, Copper, Iodine, Iron, Manganese, Molybdenum, Nickel, Silicon, Vanadium, and Zinc (2001)* and *Dietary Reference Intakes for Water, Potassium, Sodium, Chloride, and Sulfate (2004)*. *These reports may be accessed via http://www.nap.edu*

Copyright 2004 by the National Academy of Sciences. All rights reserved.

Reprinted with permission from *Dietary Reference Intakes* © (2004) by the National Academy of Sciences, courtesy of the National Academies Press, Washington, DC.

include thiamine, riboflavin, niacin, pantothenic acid, pyroxidine, biotin, folic acid, cyanocobalamin, and ascorbic acid. Fat-soluble vitamins include vitamins A, D, E, and K. Each group and its associated vitamins are examined in detail in the paragraphs that follow.

2.1. Water-Soluble Vitamins

As their name suggests, water-soluble vitamins (Table 3) dissolve readily in water and are lost daily in the urine. Therefore, most water-soluble vitamins tend not to be stored in the body, necessitating their regular dietary consumption. The largest contributors to the water-soluble vitamins are the B complex vitamins, including thiamine, riboflavin, niacin, pantothenic acid, vitamin B_6, biotin, folic acid, and cyanocobalamin. Ascorbic acid, or vitamin C, is also a water-soluble vitamin; and it plays a major role as an antioxidant.

2.1.1. VITAMIN B_1 (THIAMINE)

Thiamine monophosphate (TMP), thiamine pyrophosphate (TPP), and thiamine triphosphate (TTP) are the three most studied forms of thiamine. The TPP form accounts for about 80% of the thiamine in the body, and TMP and TTP each account for about 10%, respectively. TPP functions in the metabolism of carbohydrates by serving as a cofactor in the conversion of pyruvate to acetyl-coenzyme A (CoA) and in the transketolase reaction, which synthesizes NADPH, deoxyribose, and ribose sugars in the pentose phosphate pathway. Thiamine also plays a role in branched-chain amino acid metabolism and may serve a role in nerve conduction and transmission. Although found in a variety of animal products and vegetables, an abundance of thiamine is found in only a few foods (Table 3). There are no known side effects associated with thiamine supplementation; therefore, no UL has been set. Thiamine deficiency may lead to cardiac failure, muscle weakness, neuropathy, and gastrointestinal disturbances.

Because of the correlation between thiamine and carbohydrate metabolism, it is logical that requirements would increase as energy requirements and intake increase with exercise *(4)*. Exercise has been suggested to affect thiamine status by decreasing absorption of minerals, increasing turnover and metabolism of the nutrients,

Table 3
Summary of Water-Soluble Vitamins

Nutrient	Function	Recommended intake	Food sources	Comments for the athlete
Thiamine (B₁)	Carbohydrate and amino acid metabolism	*Refer to DRI table* • UL:N/A • Deficiency: weakness, decreased endurance, weight loss	Yeast, pork, fortified grains, cereals, legumes	Studies indicate that there is no need for additional thiamine supplementation above the DRI recommendations with exercise.
Riboflavin (B₂)	Oxidative metabolism, electron transport system	*Refer to DRI table* • UL: N/A • Deficiency: altered skin and mucous membrane and nervous system function	Milk, almonds, liver, eggs, bread, fortified cereals	Athletes who consume adequate levels through the diet do not require supplementation above the DRI.
Niacin (B₃)	Oxidative metabolism, electron transport system	*Refer to DRI table* • UL: 35 mg niacin/day • Deficiency: irritability, diarrhea	Meats, fish, legumes, peanuts, some cereals	All persons should obtain the DRI for niacin intake to ensure adequate intake and performance.

Nutrient	Function	UL / Deficiency	Food Sources	Exercise Notes
Pantothenic acid	Essential to the metabolism of fatty acids, amino acids, and carbohydrates	*Refer to DRI table* • UL: N/A • Deficiency: muscle cramps, fatigue, apathy, malaise, nausea, vomiting	Liver, egg yolk, sunflower seeds, mushrooms, peanuts, brewer's yeast, yogurt, broccoli	Limited research exists on pantothenic acid supplementation and exercise performance.
Vitamin B6	Gluconeogenesis	*Refer to DRI table* • UL: N/A • Deficiency: dermatitis, convulsions	Meats, whole-grain products, vegetables, nuts	Exercise has been shown to increase the loss of vitamin B6.
Biotin	Cofactor in synthesis of fatty acids, gluconeogenesis, and metabolism of leucine	*Refer to DRI table* • UL: N/A • Deficiency: dermatitis, alopecia, conjunctivitis	Liver, egg yolk, soybeans, yeast, cereals, legumes, nuts	Not enough information to make a recommendation regarding supplementation and exercise.
Folate	Hemoglobin and nucleic acid formation	*Refer to DRI table* • UL: 1000 µg/day • Deficiency: anemia, fatigue	Yeast, liver, fresh green vegetables, strawberries,	Exercise does not appear to increase needs.

(Continued)

129

Table 3
(Continued)

Nutrient	Function	Recommended intake	Food sources	Comments for the athlete
Vitamin B_{12}	Hemoglobin formation	*Refer to DRI table* • UL: N/A • Deficiency: snemia, neuologic symptoms	Organ meats, shellfish, dairy products	Supplemental vitamin B12 does not appear to benefit performance unless a nutritional deficit is present
Vitamin C	Antioxidant	*Refer to DRI table* • UL: 2000 mg/day • Deficiency: fatigue, loss of appetite	Citrus fruits, green vegetables, peppers, tomatoes, berries, potatoes	Results of supplementation on performance are equivocal; possible benefits of supplementation include enhanced immune function, antioxidant effects, and decreasing body temperature

DRI, dietary reference intakes; UL, upper intake level

increasing thiamine-dependent mitochondrial enzymes, increasing needs through tissue repair and maintenance, and varying biochemical adaptations through exercise training *(5)*.

Suzuki and Itokawa *(6)* determined that thiamine supplementation at 100 mg/day significantly decreased subjects' reports of fatigue compared to a placebo after 30 minutes on a bicycle ergometer. However, when the effects of a thiamine derivative versus a placebo were examined during maximal cycle ergometry, it was determined that supplementation had no effect on exercise performance *(7)*. Despite a plausible biological mechanism, the literature indicates that there is no need for additional thiamine supplementation above the DRI recommendations with exercise.

2.1.2. VITAMIN B₂ (RIBOFLAVIN)

Riboflavin functions as a catalyst for redox reactions in energy production and many metabolic pathways, mainly as a component of flavin mononucleotide (FMN) and flavin-adenine dinucleotide (FAD) *(8)*. Also, riboflavin is required for conversion of other nutrients to their active forms, including niacin, folic acid, and vitamin B_6. FAD is part of the electron transport chain, which is central to energy production. Signs and symptoms of deficiency include sore throat, cracked and red lips, inflammation of the tongue and lining of the mouth, and bloodshot eyes. Excess riboflavin is eliminated in urine; therefore, no UL has been established. Most plant and animal food sources contain riboflavin.

Nutritional surveys provided to athletes have shown that most athletes consume adequate amounts of riboflavin *(9)*; however, whether current DRI recommendations are adequate for athletes has been debated. The effects of 2.15 mg riboflavin supplementation/day in young females during 20 to 50 minutes of aerobic exercise were observed by Belko *(10)*. Riboflavin status was assessed using riboflavin-dependent erythrocyte glutathione reductase activity (EGRAC). Overall, the authors observed that riboflavin depletion developed during periods of active exercise but improved with increased vitamin intake. Although some evidence suggests that riboflavin supplementation may be necessary during exercise, the current consensus is that individuals who consume adequate levels through the diet do not require supplementation above the DRI. No studies examining the effects on physical performance with riboflavin supplementation have determined an ergogenic benefit.

2.1.3. Vitamin B₃ (Niacin)

Niacin is a water-soluble vitamin whose derivatives (e.g., NADH, NAD, NAD$^+$, NADP), play vital roles in substrate metabolism. At least 200 enzymes are known to be dependent on NAD and NADP; most are involved in catabolic reactions such as the oxidation of fuel molecules (11). Deficiency of this vitamin can result in a condition known as pellagra, a disease characterized by scaly skin sores, diarrhea, inflamed mucous membranes, mental confusion, and delusions. Although pellagra has almost disappeared from industrialized countries, it is still common to regions that exist primarily on a corn-based diet, as corn is a poor source of niacin. (It is now known that treating corn products in an alkali bath, typically lime water, increases the bioavailability of niacin.)

Although critical to the oxidation of fuel sources, and thus exercise metabolism, studies assessing the role of niacin on metabolic responses during acute exercise are limited. One study examined the effect of niacin supplementation (in the form of nicotinic acid) on selected physiological responses during exercise (12). The study hypothesized that nicotinic acid supplementation would blunt the normal increase in plasma free fatty acid (FFA) concentration during exercise, thereby increasing performance. Results showed that although supplementation effectively prevented FFAs from rising above rest values, performance was not improved. Surveys of athletic populations have shown niacin consumption to be adequate, with only those athletes participating in weight-restrictive behaviors falling below recommended levels (13). At present, it is recommended that all persons obtain the DRI for niacin intake to ensure adequate intake and performance.

2.1.4. Vitamin B₅ (Pantothenic Acid)

Pantothenic acid performs multiple roles in cellular metabolism and regulation as an integral part of two acylation factors: coenzyme A and acyl-carrier protein (ACP) (11). In these forms, pantothenic acid is essential to the metabolism of fatty acids, amino acids, and carbohydrates as well as the synthesis of cholesterol, steroid hormones, vitamin A, and vitamin D (14).

Human studies looking at the effect of pantothenic acid and exercise performance are limited. A study by Webster revealed that supplementation with pantethine (a dimeric form of vitamin B₅) did not alter exercise metabolism or exercise performance (15).

2.1.5. Vitamin B₆ (Pyridoxine and Related Compounds)

Vitamin B_6 collectively refers to all biologically active forms of vitamin B_6, although the metabolically active form of the vitamin is pyridoxal phosphate (PLP). Vitamin B_6 is involved in many cellular processes including gluconeogenesis, niacin formation, lipid metabolism, erythrocyte function and metabolism, and hormone modulation *(14)*. All persons, including athletes, have generally been shown to consume adequate amounts of this vitamin *(16)*. Vitamin B_6 is widely distributed in foods (see Table 3 to view foods with the greatest concentrations of vitamin B_6).

Because exercise has been shown to increase the loss of vitamin B_6 as 4-pyridoxic acid *(17)*, and because PLP acts as a cofactor in both gluconeogenesis and glycogenolysis, it has been postulated that supplementation of the vitamin may increase exercise performance. A study by Linderman et al. sought to explore the effects of pyridoxine on short-term maximal exercise *(18)*. Highly trained cyclists were instructed to cycle to exhaustion to determine maximal power output. During the trial, subjects were administered oral tablets of placebo, pyridoxine-α-ketoglutarate (PAK) and placebo, or sodium bicarbonate and placebo. Results showed that for the dosages used in this study administration of PAK was ineffective for increasing short-term maximal exercise capacity, and thus supplementation above the DRI does not appear to improve performance.

2.1.6. Vitamin B₇ (Biotin)

Biotin, also known as vitamin H, plays an important role in the catalysis of many essential metabolic reactions, including the synthesis of fatty acids, gluconeogenesis, and metabolism of leucine. To date, no studies have been conducted looking at the role of biotin in exercise performance in humans.

2.1.7. Folic Acid

Named for the abundance of the vitamin in green, leafy, vegetables—or foliage—folic acid plays several important roles in metabolism. Folic acid is the synthetic form of folate and is needed for DNA production and erythropoiesis. Deficiencies can cause errors in cellular replication, particularly affecting red blood cells; megaloblastic anemia can occur as a result of folate deficiency. The DRI for folate is 400 µg/day for women and men. Owing to its prominent role in cellular growth and differentiation, this value increases to 600 and 500 µg/day during pregnancy and lactation, respectively *(2)*.

Folate is ubiquitous in nature, being found in most natural foods. However, the vitamin is highly susceptible to oxidative damage, and thus the folate content of foods is easily destroyed by heat.

Because of its role in erythrocyte production, the benefit of folate supplementation on athletic performance has been speculated. Results of one study showed that although folate supplementation in an athletic population did significantly increase circulating levels of serum folate this increase did not translate into increased performance *(19)*. The authors of this study speculated that changes in circulating concentrations of folate may not reflect changes in cellular folate status; thus, oversupplementation cannot be justified. At present, the recommended intake of folate follows the DRI for normal individuals.

2.1.8. VITAMIN B_{12} (CYANOCOBALAMIN)

Cyanocobalamin encompasses a group of cobalt-containing compounds. Vitamin B_{12} is involved in fat and carbohydrate metabolism as well as protein synthesis. Additionally, vitamin B_{12} is responsible for the conversion of homocysteine to methionine; and deficiencies in the vitamin have been linked to hyperhomocysteinemia, an independent risk factor for atherosclerotic disease *(20)*. In nature, this vitamin is synthesized by microorganisms, and it is not found in plant foods except when they are contaminated by microorganisms. Small amounts of vitamin B_{12} are found in legumes, which contain microorganisms, and may provide the only dietary source of vitamin B_{12} for strict vegetarians. Deficiency is almost always the result of inadequate ingestion or absorption.

Although older studies have shown that supplemental B_{12} does not benefit performance unless a nutritional deficit is present *(21, 22)*, controlled, well designed studies utilizing modern technology are needed to determine whether vitamin B_{12} is needed in larger amounts by individuals who exercise.

2.1.9. VITAMIN C (ASCORBIC ACID)

The functions of vitamin C are based primarily on its biological reductant capabilities. As such, vitamin C is involved in collagen formation, carnitine biosynthesis, neurotransmitter synthesis, and iron absorption. Vitamin C also has been show to promote resistance to infection through the immunological activity of leukocytes, the production of interferon, the process of inflammatory reaction, and/or the integrity of the mucous membranes *(14)*. When dietary intake of ascorbic acid is insufficient, a set of conditions occur that are collectively known as scurvy.

Important to the athlete, vitamin C has certain biological functions that can influence physical performance. Due to its requirement in the synthesis of carnitine (the enzyme responsible for the transport of long-chain fatty acids into mitochondria), it is thought to play a major role in energy availability. Additionally, vitamin C may indirectly improve physical performance by enhancing immune function, especially in endurance athletes who are prone to upper respiratory tract infections. Although short-term marginal vitamin C deficiency does not appear to affect physical performance *(23)*, long-term deficiency can negatively influence performance *(9)*. Evidence suggesting increased athletic performance with oversupplementation has been mixed at best, with some studies indicating indirect benefits to physical performance [e.g., decreased body temperature *(24)* and enhanced immune function *(25)*] whereas other studies show no benefits to physical performance *(26,27)*.

2.2. Fat-Soluble Vitamins

The fat-soluble vitamins (Table 4) include vitamins A, D, E, and K. Vitamins A and E function as antioxidants, and vitamins D and K play a role in bone metabolism. Because fat-soluble vitamins are stored for extended periods, when consumed in excess they create a greater risk for toxicity than water-soluble vitamins. Disease states affecting the absorption or storage of fat could cause deficiency of these vitamins.

2.2.1. VITAMIN A

The roles of vitamin A in the body are vast, including immune function, vision, growth, and gene expression. Vitamin A comprises a group of compounds, including retinol, retinal, retinoic acid, or retinyl ester. Provitamin A carotenoids (i.e., α-carotene and ß-carotene) are precursors to retinol, one of the most active forms of vitamin A. Some provitamin A caroteinoids have been found to have antioxidant activity, with ß-carotene suggested to be the primary anticancer agent in fruits and vegetables *(28)*.

Table 4
Summary of Fat-Soluble Vitamins

Nutrient	Function	Recommended intake	Food sources	Comments for the Athlete
Vitamin A	Vision, immune response; epithelial cell growth and repair	*Refer to DRI table* • UL:3000 µg preformed vitamin A/day • Deficiency: dry skin, dry hair, broken fingernails, susceptibility to infections	Broccoli, squash, sweet potatoes, pumpkin, cantaloupe liver, milk, eggs	Supplementation of β-carotene is not recommended
Vitamin D	Bone remodeling and maintaining serum calcium and phosphorus concentrations	*Refer to DRI table* • UL: 2000 IU vitamin D/day • Deficiency: osteomalacia, Osteoporosis, heart disease, hypertension	Natural sources: fatty fish, egg yolks Fortified sources: milk, cereals	May influence bone mineralization and help prevent fractures

Vitamin E	Antioxidant	*Refer to DRI table* • UL: 1000 mg α-tocopherol/day • Deficiency: retinopathy, neuropathy, and myopathy	Vegetable oils, unprocessed cereal grains, green leafy vegetables, nuts	Antioxidant properties may be beneficial in decreasing oxidative stress during exercise bouts
Vitamin K	Essential for normal blood clotting	*Refer to DRI table* • UL: N/A • Deficiency: increase in blood clotting time and decrease in bone mineral density	Green leafy vegetables, cereal, organ meats, dairy products, eggs	Supplementation may be needed for formation of bone

Dietary carotenoids are consumed primarily through oils and brightly colored fruits and vegetables, whereas preformed vitamin A is found only in animal products. The RDA requirements for vitamin A are expressed in retinol activity equivalents (RAE). A UL of preformed vitamin A has been recommended because retinol is stored and metabolized in the liver; however, the liver has a protective mechanism for reducing vitamin A metabolites by excreting the metabolites in bile *(29)*. Adverse effects associated with an overdose of vitamin A include acute effects such as vertigo, blurred vision, and nausea as well as chronic effects such as bone loss *(30)* and liver abnormalities *(29)*.

To date, there has been little research on the effects of ß-carotene supplementation on muscular strength or endurance. Although there is potential for vitamin A supplementation to decrease oxidative stresses from exercise, the research, again, is limited because vitamins C and E are known to have greater antioxidant capabilities. Owing to fat-solubility properties, supplementation of ß-carotene is not recommended.

2.2.2. Vitamin D (Calciferol)

Although many forms of calciferol (vitamin D) exist, the two main forms are ergocalciferol (vitamin D_2) and cholecalciferol (vitamin D_3). Vitamin D_2 is produced from ergosterol in the diet, and vitamin D_3 is synthesized by ultraviolet radiation from a precursor of cholesterol in the skin. Both forms are biologically inert and must be converted to the biologically active form, 1,25-dihydroxyvitamin D, or 1,25 $(OH)_2D$. The major source of vitamin D for humans is sunlight. This source is especially important because few dietary sources contain vitamin D naturally (Table 4). In the United States, many dairy foods (i.e., milk, cheese) are fortified with vitamin D to avoid deficiency in latitudes where exposure to sunlight is limited during the winter months. Hypervitaminosis D may cause adverse effects in various organs throughout the body. These effects are thought to be caused, for the most part, by the hypercalcemia that may occur with hypervitaminosis D *(31)*. This is because the function of vitamin D is to maintain serum calcium and phosphorus levels in the body by enhancing their absorption from the gastrointestinal tract and promoting their release from the bones. Active vitamin D plays such a role by working in combination with parathyroid hormone (PTH) to mobilize calcium and, indirectly, phosphorus from the bone to maintain serum concentrations as needed.

A study conducted by Kenny et al. tested the effects of 1000 IU cholecalciferol supplementation on strength and physical performance in individuals without vitamin D deficiency *(32)*. Although no significant gains in strength or physical performance were seen in the groups, the authors reported a significant decrease in PTH with treatment, which may have an influence on increasing bone mineral density and decreasing the risk of fractures.

2.2.3. VITAMIN E

Unlike the other fat-soluble vitamins, vitamin E has no specific metabolic function. Instead, its major function is as an antioxidant of polyunsaturated fatty acids, preventing free radical damage in biological membranes caused by lipid peroxidation. Because vitamin E is absorbed in a manner similar to that of dietary fat, changes in pancreatic function, chylomicron transport of lipids, and bile production impair vitamin E absorption *(9)*. Most vitamin E is stored in adipose tissue, with smaller amounts in the heart, liver, lungs, brain, muscles, and adrenal glands. Eight naturally occurring isomers of vitamin E exist; however, only the α-tocopherol form is maintained in human plasma *(33)*. Adverse effects of vitamin E deficiency include retinopathy, neuropathy, and myopathy; however, deficiency occurs rarely in humans. There are no adverse effects noted with ingestion of naturally occurring vitamin E in foods; as a nutritional supplement, however, side effects include fatigue, gastrointestinal disturbances, and altered lipid concentrations. Nutritional supplements contain either a natural or synthetic form of α-tocopherol *(34)*.

Numerous studies have examined the effects of vitamin E supplementation on oxidative stress during exercise. Vitamin E supplementation has been studied in trained versus untrained participants performing upper and lower body resistance exercises *(35)*. Results showed no difference in oxidative damage between supplemented or placebo groups or between trained and untrained participants with regard to lipid peroxidation. In triathletes consuming 800 IU α-tocopherol/day versus a placebo, race times were not different between groups; however, a marker of antioxidant potential, the ferric reducing ability of plasma

(FRAP), was significantly higher 1.5 hours after exercise in the supplemented group versus the placebo group *(36)*. Despite the increase in antioxidant potential, the large dose of vitamin E also caused significant increases in oxidative stress markers. The authors concluded that α-tocopherol should be avoided during long-duration exercise. In contrast, Itoh et al. supplemented male participants with 1200 IU α-tocopherol/day versus a placebo *(37)*. Participants were required to take the supplement 4 weeks prior to baseline testing and on 6 successive days of endurance training. Significantly lower levels of creatine kinase, lacatate dehydrogenase (LDH), and LDH isozymes (markers of oxidative muscle damage and tissue breakdown) in the serum were observed in the supplemented group following the endurance training. Regarding performance, no significant difference was seen between groups with respect to enhanced aerobic work capacity. Further research should be aimed at determining specific recommendations regarding vitamin E supplementation on exercise performance, oxidative stress, and muscle damage.

2.2.4. Vitamin K

Two forms of vitamin K exist naturally: phylloquinones (vitamin K_1), produced by plants, and menaquinones (vitamin K_2), produced by bacteria in the large intestine. Phylloquinones comprise the predominant form in the diet and are obtained from green leafy vegetables *(29)*. Vitamin K functions as a critical cofactor of γ-carboxylase, an essential posttranslational modification required for the functional activity of coagulation proteins such as prothrombin. Deficiency of vitamin K leads to changes in blood clotting (increased prothrombin time) and a decrease in bone mineral density (increase in the plasma of under-γ-carboxylated osteocalcin) *(29)*.

Although no studies on the effects of vitamin K supplementation and performance exist, supplementation benefits on bone mass have been studied. Vitamin K_1 (10 mg/day) supplementation in endurance athletes showed no effect with vitamin K supplementation on the rate of bone loss *(38)*. It was determined that females who began endurance training at a young age were at risk for larger amounts of bone loss than those who began their training at a later age; however, both groups had a relatively high rate of bone loss when compared to standards for females of the same age.

2.3. Summary

Because both fat- and water-soluble vitamins are essential to human physiological function, examining their effect on exercise—both athletic performance and deficiency avoidance—is popular in sport nutrition research. In summary, the grouping of B vitamins has two major functions directly related to exercise. Thiamine, riboflavin, vitamin B_6, niacin, pantothenic acid, and biotin are involved in energy production during exercise, whereas folate and vitamin B_{12} are required for the production of red blood cells, protein synthesis, and tissue repair and maintenance (1). Vitamin C may play a role in improving immune function and may indirectly benefit athletic performance. Although it is thought that exercise may slightly increase the need for these vitamins, demands can usually be met by the increased energy intakes required of physically active persons to maintain energy balance. The two major functions of the fat-soluble vitamins include antioxidant activity of vitamins A and E and bone formation of vitamins D and K.

In general, benefits of vitamin supplementation in regard to increased exercise needs or improved athletic performance are assumed to be inconclusive unless stated otherwise. More research is needed before micronutrient supplementation above the DRI should be recommended to athletes, either as a requirement for increased needs during exercise or as an ergogenic aid.

2.4. To Supplement, or Not to Supplement?

Current recommendations from the American Dietetic Association suggest consuming a wide variety of foods to avoid chronic disease and micronutrient deficiency. However, trends toward consuming convenience foods may lead to a diet where essential vitamins and minerals are lacking. Many adolescents and adults are using vitamin and mineral supplements to ensure adequate dietary intake. Studies have found that those who use supplements are more likely to be educated, older, physically active women and likely to be at or below normal body weight (76). The Child and Adolescent Trial for Cardiovascular Health (CATCH) study conducted by Reaves et al. examined 2761 twelfth graders across four states to determine if multiple vitamin intake is associated with lifestyle behaviors (76). In all, 25% were found to consume supplements, with 58% of them being female. Supplement users consumed significantly more energy, with a greater proportion coming from carbohydrates and fat than protein. The supplement-consuming group also consumed more healthful intakes of fiber, saturated fat, and cholesterol and participated in a greater amount of physical activity. Although it is ideal to consume nutrients through a balanced diet, supplements may contribute to total nutrient intake in the adolescent population, especially if dietary intakes are inadequate. In addition, the effects of multivitamin and mineral supplementation on physical performance in well trained male volunteers with normal vitamin and mineral status have been examined *(77)*. Overall, they found no apparent effect on measured physiological variables, including maximal oxygen consumption, peak treadmill running speed, and the blood lactate turnpoint. Although some studies have found an ergogenic effect of supplementation, the authors suggest that these studies were poorly controlled; subjects were not always evaluated for nutrient deficiency; and the placebo effect may have affected subjects' athletic performance. Because supplements are concentrated sources of nutrients, it is important for health care professionals to monitor for excess intake.

3. MINERALS

Dietary minerals (Table 5) are chemical agents required by living organisms to maintain physical health. Like vitamins, minerals also regulate macronutrient use. They are classified as either macrominerals or microminerals/trace elements, depending on the daily amount needed. Additionally, minerals play various roles involved in enzyme regulation, maintenance of acid-base balance, nerve function, and cellular growth *(39)*. Because many of these processes are heightened during exercise, the field of exercise nutrition has sought to explore the relation between various mineral needs and physical activity. Such findings, along with the general function, effects of deficiency or oversupplementation, and recommended intake levels are the focus of this section.

3.1. Macrominerals

Macrominerals are required in amounts higher than 100 mg/day and include calcium, phosphorus, magnesium, sulfur, potassium, sodium, and chloride. A comprehensive review of the above listed macrominerals and their role in physical performance can be found in Table 5.

Table 5
Summary of Minerals

Nutrient	Function	Recommended intake	Food sources	Comments for the athlete
Calcium	Structure of the teeth and bone, vascular and muscle contractions, blood coagulation, nerve transmission	*Refer to DRI table* • UL: 2500 mg/day • Deficiency: improper bone mineralization, tetany, muscle pain and spasms	Dairy products, pinto and black beans, spinach, fortified cereal, orange juice	Possible effects of calcium supplementation on body weight and sweat losses during exercise; however, current recommendation is DRI.
Phosphorus	Essential for strong bones and teeth and energy metabolism	*Refer to DRI table* • UL: 4 g/day • Deficiency: anorexia, muscle weakness, bone pain, rickets, confusion	Milk, carbonated cola drinks, eggs, whole wheat bread, almonds, lentils, some fish	Phosphate loading may increase exercise performance; however, supplementation is potentially harmful.

(Continued)

143

Table 5
(Continued)

Nutrient	Function	Recommended intake	Food sources	Comments for the athlete
Magnesium	Energy metabolism, neuromuscular coordination, bone mineralization	*Refer to DRI table* • UL: 350 mg/day • Deficiency: hypocalcemia, tetany, tremors, muscular weakness, confusion	Wheat flour, artichokes, pumpkin seeds, almonds, tuna	Although exercise may increase needs, current recommendation is DRI.
Sulfate	Protein synthesis and formation of disulfide bridges	*Refer to DRI table* • UL: N/A • Deficiency: stunted growth	Meat, poultry, fish, eggs, dried beans, broccoli, cauliflower	Although exercise may increase needs, current recommendation is DRI.
Potassium	Water balance, acid-base balance, electrical potential gradients across membranes	*Refer to DRI table* • UL: N/A • Deficiency: muscle weakness, myalgia, increased risk of hyponatremia	Tomatoes, orange juice, beans, raisins, potatoes, grapefruit	Exercise does not appear to increase needs.
Sodium		*Refer to DRI table* • UL: 2.3 g/day	Processed and cured meats and	Ultra-endurance athletes and those

	Maintain extracellular volume and plasma osmolality	• Deficiency: hyponatremia, muscle cramps, overhydration, hypotension	cheeses, frozen meals	with occupational physical activity and heat exposure may benefit from supplementation.
Chloride	Same as sodium	*Refer to DRI table* • UL: 3.5 g/day • Deficiency: overhydration, hypotension, muscle cramps	Similar to sodium-containing foods	Similar to sodium.
Iron	Transportation of oxygen in the body	*Refer to DRI table* • UL: 45 mg/day • Deficiency: fatigue, lack of stamina, breathlessness, headaches, insomnia	Lean red meats, seafood, beans, leafy green vegetables, molasses	May have beneficial effects on physical performance in those who are iron deficient.
Zinc	Aids in wound healing and is a vital component of many enzymatic reactions	*Refer to DRI table* • UL: 40 mg/day • Deficiency: altered taste, hair loss, diarrhea, fatigue, delayed wound healing	Oysters, wheat germ, ground beef, liver, ricotta cheese	Evidence supporting zinc supplementation in athletes has been equivocal.

(Continued)

Table 5
(Continued)

Nutrient	Function	Recommended intake	Food sources	Comments for the athlete
Chromium	Involved in carbohydrate, protein, and lipid metabolism andfacilitates the action of insulin	*Refer to DRI table* • UL: N/A • Deficiency: weight loss, peripheral neuropathy, impaired glucose utilization, and increased insulin requirements	Eggs, liver, oysters, wheat germ, spinach, broccoli, apples, bananas	Studies suggest that chromium supplementation benefits may occur only in individuals with impaired chromium concentrations.
Boron	Function unknown; proposed functions include metabolism of vitamin D, macromineral metabolism, and immune function	*Refer to DRI table* • UL: 20 mg/day • Deficiency: proposed effects include decreased bone density, mineral metabolism, and cognitive function	Grapes, leafy vegetables, nuts, grains, apples, raisins	Boron supplementation does not appear to effect physical performance.
Copper	Enzyme catalyst, enhancing iron	*Refer to DRI table* • UL: 10,000 μg/day		No known benefits of supplementation on

	absorption, antioxidant	• Deficiency: leukopenia, fatigue, hair loss, anorexia, diarrhea	Shellfish, finfish, beef, table salt, coffee	performance.
Fluoride	Mineralized bones and teeth	*Refer to DRI table* • UL: 10 mg/day • Deficiency: dental caries weakened bone	Fluorinated water, tea, fish, legumes, potatoes	No known benefits of supplementation on performance; however, suboptimal intake may affect bone mineral density.
Iodine	Essential in thyroid hormone function	*Refer to DRI table* • UL: 1100 µg/day • Deficiency: goiter, reduced mental function, hypothyroidism	Iodized table salt, seafood, kelp, dairy	No known benefits of supplementation on performance; however, suboptimal thyroid hormone concentrations may affect performance.
Manganese	Antioxidant; bone formation; metabolism of amino acids, lipids, and carbohydrates	*Refer to DRI table* • UL: 11 mg/day • Deficiency: decreased growth, impaired glucose tolerance, dermatitis	Nuts, leafy vegetables, whole grains, pineapple, teas	No known benefits of supplementation on performance.

(Continued)

Table 5
(Continued)

Nutrient	Function	Recommended intake	Food sources	Comments for the athlete
Molybdenum	Enzymatic cofactor	*Refer to DRI table* • UL: 2000 µg/day • Deficiency: headache, nightblindness, tachycardia, tachypnea	Carrots, cabbage, carrots, legumes, nuts	No known benefits of supplementation on performance.
Selenium	Defend against oxidative stress	*Refer to DRI table* • UL: 400 µg/day • Deficiency: cardiomyopathy muscular weakness, pain	Brazil nuts, seafood,Fish and shellfish, meats, garlic, eggs	No known benefits of supplementation on performance; however, may be beneficial owing to antioxidant effects. Toxic if consumed in excess.
Vanadium	Stimulates cell proliferation and differentiation, regulates phosphate-dependent enzymes, insulin-mimetic activity	*Refer to DRI table* • UL: N/A • Deficiency: heart and kidney disease, reproductive disorders	Black pepper, beer, wine, mushrooms, sweeteners, grains	No known benefits of supplementation on performance.

3.1.1. Calcium

Calcium is the most abundant mineral in the body, totaling approximately 1% to 2% of body weight. In total, 99% of calcium in the body functions in the structure of teeth and bones. The remainder, found in blood, muscle, extracellular fluid, and other tissue, functions in various roles throughout the body, such as in vascular and muscle contractions, blood coagulation, and nerve transmission *(31)*. At a low calcium concentration, absorption depends on the activation of vitamin D; however, passive diffusion becomes more common at higher concentrations *(31)*. In addition to vitamin D, calcitonin and parathyroid hormone (PTH) are hormones that regulate serum calcium concentrations. Calcitonin and PTH increase when calcium concentrations drop, causing calcium to be released from bone, reabsorbed in the kidneys, and absorbed in the intestines *(6)*. A high protein diet, as well as foods containing sodium, phytates, fiber, oxalic acid, and caffeine may decrease the bioavailability and absorption of calcium in the diet.

The two main calcium compounds found in supplements are calcium citrate and calcium carbonate. Calcium carbonate supplements contains 40% calcium, and calcium citrate supplements contain 21% calcium; therefore, more calcium citrate must be taken to equal a similar amount of carbonate *(40)*. Amounts < 500 mg of calcium generally are recommended because absorption decreases as the amount of calcium in the supplement increases. Calcium citrate supplements are typically better absorbed in individuals with decreased stomach acid *(40)*, usually a result of taking the supplement with food. When adolescent females were supplemented with 670 mg of a calcium citrate malate supplement (mean daily intake approximately 1500 mg/day) for 7 years, supplementation positively influenced the gain of bone mass throughout the bone-modeling phase of the pubertal growth spurt, which is when requirements of calcium are determined to be the highest *(41)*. By the beginning of young adulthood, the only significant findings were seen in tall subjects, suggesting that calcium requirements may vary with skeletal size. Positive effects of calcium supplementation were seen at all skeletal regions examined during the young adulthood assessment period.

Although effects of calcium supplementation on physical performance are lacking, possible effects on body weight and sweat losses during exercise warrant review of calcium supplementation. Bergeron et al. *(42)* collected sweat calcium concentration losses on two successive days of exercise in the heat. The authors concluded that the calcium losses could contribute to a negative calcium balance and that the requirement may be higher during exercise training. Men and women consuming < 600 mg, 600 to 1000 mg, and > 1000 mg were divided into groups and assessed for body composition and lipid panels (43). Those consuming < 600 mg had significantly higher values for body weight, body mass index, percent body fat, and waist circumference than those who consumed intakes of > 600 mg. Lipid panels revealed no significant differences between subgroups for women for high density lipoprotein (HDL) cholesterol, low density lipoprotein (LDL) cholesterol, triacylglycerol, or total cholesterol; there were no significant differences in plasma lipoprotein–lipid concentrations between subgroups of men. Calcium supplementation may benefit those involved in sports with weight constrictions. Overall, current research does not support the need for calcium supplementation above the DRI; however, more research is required.

3.1.2. PHOSPHORUS

Phosphorus is essential for all living cells as a component in phospholipid membranes, as well as in nucleic acids and nucleotides. Approximately 85% of total body phosphorus is found in bone, with the remaining 15% in soft tissues. The form of phosphorus most commonly found in nature is phosphate; however, when stored in bone, the main form is hydroxyapatite crystals. Although rare—because phosphorus is well dispersed throughout plant and animal foods—deficiency of serum phosphorus may lead to anorexia, muscle weakness, bone pain, rickets, confusion, and death. High intake of phosphorus may reduce serum calcium concentrations and reduce the formation of active vitamin D, leading to an increase of PTH. Elevated PTH is associated with increased bone loss to maintain serum calcium concentrations. Along with its adverse effects on bone, overconsumption of phosphorus also may cause calcification of soft tissues, especially the kidney.

Athletes often consume excess phosphorus due to "phosphate loading" through consumption of potassium, sodium, and/or calcium phosphate supplements. The aim of the supplementation is to improve tissue oxidation by increasing erythrocyte 2,3-bisphosphoglycerate concentrations. Phosphate loading may result in improved athletic performance in endurance athletes by improving oxygen release in tissues. After study participants were supplemented with a phosphate derivative for 7 days, a 30% increase in plasma phosphate and a 25% increase in erythrocyte 2,3-bisphosphoglycerate concentrations were seen *(44)*. This study did not determine effects on performance. However, another study determined that supplementing 22.2 g of calcium phosphate had no benefit as an ergogenic aid or on the aerobic fitness level when compared to subjects supplemented with calcium carbonate alone *(45)*. Before phosphorus supplementation is recommended among athletes, effects on performance and bone mineralization must be evaluated.

3.1.3. MAGNESIUM

As a required cofactor for more than 300 enzymatic reactions, magnesium plays an important role in aerobic and anaerobic energy generation. Other functions of magnesium include neuromuscular coordination and bone mineralization. Magnesium is important in vitamin D metabolism. It also has a structural role in the body. Altogether, 50% to 60% of body magnesium is stored in bones; and PTH is dependent on magnesium for regulation of calcium in bone. Also, magnesium is required for regulating the outward movement of potassium from myocardial cells and the intracellular concentration of calcium during muscle contractions *(31)*. Serum magnesium depletion may lead to hypocalcemia, tetany, tremors, muscular weakness, and confusion.

In a recent study looking at the effect of magnesium supplementation on physical performance, postmenopausal women were asked to follow a magnesium depletion diet for 93 days *(46)*. After this period, the effect of 200 mg magnesium supplementation on exercise was examined via a submaximal ergocycle exercise test. An increase in heart rate and decrease in peak and cumulative net oxygen uptake were seen with restricted dietary magnesium intake when compared with adequate dietary magnesium intake during supplementation. Overall, the authors suggested that magnesium intake may affect metabolic and cardiovascular response to exercise. Brilla and Haley supplemented young men with 8 mg/kg/day *(47)*. The mean estimated intakes of magnesium for the placebo and supplemented groups were 250 and 507 mg, respectively. The magnesium-supplemented group had significantly greater quadriceps torque with leg press and leg extension than the placebo group after 7 weeks of supplementation. The authors of a 2006 review stated that although exercise has been found to increase magnesium requirements by as much as 10% to 20% studies do not support the need for supplementation of physically active individuals with adequate magnesium status to improve performance *(48)*.

3.1.4. SULFUR

The mineral sulfur is a major constituent of three amino acids: cystine, cysteine, and methionine. Additionally, sulfur is involved in protein synthesis, as it is responsible for formation of disulfide bridges, a necessary component of the tertiary structure of proteins. Dietary sources of sulfur include meat, poultry, fish, eggs, dried beans, broccoli, and cauliflower *(49)*.

At present, research studying the effect of sulfur on athletic performance is limited to amino acids containing the mineral. Although a DRI is currently not available for sulfur, there is no literature to suggest athletes need to consume higher amounts than the average person.

3.1.5. POTASSIUM

As an electrolyte, potassium plays a major role in electrical and cellular body functions. Along with sodium and chloride, potassium is involved in maintaining water balance and distribution, osmotic equilibrium, acid-base balance, and electrical potential gradients across membranes *(39)*. Because nerve and muscle cells have the highest gradients of bodily cells, potassium plays a major role in nerve and muscle function.

Because of its role in muscle function, several studies have looked at the relation between potassium and exercise performance. Prolonged exhaustive exercise has been shown to impair potassium transport processes in exercising muscle *(50)*. This impairment can lead to a rise in extracellular potassium concentration in skeletal muscle, which is thought to play an important role in the development of fatigue during intense exercise *(51)*.

Although few studies have looked at the effects of potassium supplementation on exercise performance, one interesting study suggested that potassium phosphate supplementation may mediate perceived exertion *(52)*. In a double-blind, placebo-controlled study, eight highly trained endurance runners were asked to provide a rating of perceived exertion (RPE) during maximal graded exercise tests. Results showed that, overall, RPE was lower with supplementation, thus encouraging prolonged activity. Despite these findings, additional studies are warranted before exercise-specific recommendations can be made.

3.1.6. SODIUM AND CHLORIDE

The cation sodium and the anion chloride are normally found together in most foods as sodium chloride (i.e., salt), with highest concentrations found in prepared, cured, or pickled food products. In the body, sodium and chloride are required to maintain extracellular volume and plasma osmolality. Healthy adults should consume 1.5 g of sodium and 2.3 g of chloride each day, or 3.8 g of salt, to replace the amount lost in sweat. Sweat is produced by our bodies as a by-product of thermoregulation. Should the capacity for sweat production be hindered, a rise in core temperature and resultant heat illness could result. For the athlete, conditions such as extreme heat or exercise intensity can significantly elevate sweat losses above what is considered normal, and resultant dietary adjustments must be made.

Given the critical need to maintain fluid homeostasis, the American College of Sports Medicine has put forth guidelines for proper hydration (Table 6). Inclusion of sodium chloride in rehydration beverages has been shown to reduce urinary water loss, leading to a more rapid recovery of fluid balance, with some experts now recommending sodium concentrations of 20 to 50 mmol/L in beverages consumed during physical activity *(53)*.

Recently, there have been reports of hyponatremia among individuals who tend to over-ingest water during exercise lasting more than four hours. Again, inclusion of sodium chloride in the fluid replacement beverage is often suggested as a potential means of reducing risk of hyponatremia. Although hyponatremia is not likely to be a major risk factor for the general population, ultra-endurance athletes and people with occupational physical activity and heat exposure may benefit from these recommendations *(53)*.

3.2. Microminerals/Trace Elements

Microminerals, or trace elements, include iron, zinc, copper, selenium, iodine, fluoride, chromium, manganese, molybdenum, boron, and vanadium. In general, these elements are required in amounts of < 100 mg/day. Although 14 trace minerals have been identified as essential for life, there is sufficient information on only four, as related to physical performance. This section provides detailed

Table 6
Fluid Replacement Guidelines for Exercise

Exercise duration	Before exercise	During exercise	After exercise
< 1 Hour	300–500 mL water 30–50 g CHO	500–1000 mL water	Forced water
1–3 Hours	300–500 mL water 30–50 g CHO	500–1000 mL water/hr 6–8% CHO solution 10–20 mEq Na^+ 10–20 mEq Cl^-	Forced water 50 g CHO 30–40 mEq Na^+ 30–40 mEq Cl^- 3–5 mEq K^+
> 3 Hours	300–500 mL water 30–50 g CHO	500–1000 mL water/hr 6–8% CHO solution 10–20 mEq Na^+ 10–20 mEq Cl^-	Forced water 50 g CHO 30–40 mEq Na^+ 30–40 mEq Cl^- 3–5 mEq K^+

From Gisolfi and Duchman (82), with permission.
CHO, carbohydrate

information on four of these trace elements, with Table 5 summarizing recommended dietary intake, food sources, and functional role in the body for all other microminerals.

3.2.1. IRON

Dietary iron is a constituent of hemoglobin, myoglobin, cytochromes, and iron-containing enzymes. As such, iron plays a fundamental role in the transport of oxygen in the body, and adequate stores are necessary for optimal athletic performance.

Dietary iron can be obtained through quality food sources, as well as obtained from foods prepared in cast-iron cookware. The bioavailability of iron in certain foods (particularly vegetables) can be increased by adding an acid during preparation.

Iron deficiency is the most common single-nutrient-deficiency disease, with iron status negatively altered in many populations of chronically exercising individuals *(54)*. Depending on the sport surveyed, anywhere from 20% to 80% of elite athletes have been found to be iron-deficient *(55)*. If left untreated, iron deficiency can cause anemia, a condition in which hemoglobin cannot be formed. Numerous studies have shown the negative impact of iron deficiency anemia on work output and physical performance *(54,56,57)*. Moreover, although reduced oxygen-carrying capacity of the blood is not affected until the anemic stage, reduced tissue oxidative capacity is present with any level of deficiency *(58)*. Iron deficiency without anemia has been shown to impair adaptation in endurance capacity after aerobic training *(59)*, compromise 15 km time trial times *(60)*, and decrease VO_{2max} and oxygen consumption in a maximum accumulated oxygen deficit test *(61)*. In general, reductions in tissue oxidative capacity hinder endurance and energetic efficiency, which translates into impaired athletic performance.

In the athlete, despite claims of blood loss as a result of foot striking, gastritis, and menstruation, as well as the pseudoanemia caused by an increase in plasma volume during exercise, the true cause of anemia in athletes can be attributed to a diet inadequate in iron *(62)*; thus, efforts should be placed on improvement of dietary quality. Moreover, because some individuals carry a gene for increased iron absorption, oversupplementation is not advised. Iron is a powerful oxidant and is toxic at high concentrations; it is for this reason that iron supplementation should be reserved for individuals who are deficient. In general, female athletes, vegetarians, and endurance athletes are considered at greater risk for iron deficiency than the typical athlete. However, proper diagnosis of the condition by a medical provider is necessary before supplementation should be considered.

3.3. The Vegetarian Athlete

In 2003, approximately 2.8% of U.S. adults reported following a vegetarian diet *(78)*. Although this percentage may seem small, it translates to more than 5 million people. With the U.S. market for vegetarian foods doubling since 2001, this trend shows no sign of stopping.

According to the Vegetarian Resource Group, vegetarian diets can be classified into four major groups.

- *Vegans:* do not eat meat, fish, or poultry; additionally, do not use other animal products or by-products, such as eggs, dairy products, honey, leather, fur, silk, wool, cosmetics, and soaps derived from animal products

- *Lacto-vegetarians:* do not eat meat, fish, poultry, or eggs; do consume dairy products
- *Ovo-vegetarians:* do not eat meat, fish, poultry, or dairy products; do consume egg products
- *Lacto/ovo-vegetarians:* do not eat meat, fish, or poultry; do consume dairy products and eggs

Additionally, some persons self-describe themselves as vegetarians if they are occasional meat eaters who predominantly practice a vegetarian diet.

In addition to the numerous health benefits associated with a vegetarian diet *(79–81)*, the high carbohydrate nature of a vegetarian diet can be beneficial for the athlete during heavy training, when maximizing body glycogen stores is a must.

Although the benefits of following a vegetarian diet are numerous, appropriate nutrition education and planning are necessary to ensure that dietary needs are being met. Certain nutrients are either not present or are not as easily absorbed in plant products as they are in animal products. Specifically, vegetarians need to be mindful of their intake of iron, calcium, vitamin B_{12}, and vitamin D, as good sources of these nutrients are mostly of animal origin.

Table 7 provides vegetarian-friendly food sources of the nutrients that are most likely to be lacking in a vegetarian diet.

3.3.1. Zinc

The mineral zinc primarily serves a structural role in thousands of proteins. Zinc is also involved as a cofactor in many enzyme reactions and plays a vital role in tissue repair. It has been suggested that athletes generally consume less than the recommended amounts *(63)*. In athletes, zinc deficiency can lead to anorexia, significant loss in body weight, latent fatigue with decreased endurance, and a risk of osteoporosis *(64)*. Although zinc can be lost in sweat and urine, exercise has not been shown to cause significant losses in the athlete when dietary zinc intake is sufficient *(34)*.

To date, evidence supporting zinc supplementation in athletes has been equivocal. In a recent study by Kilic et al., a 4-week zinc supplementation study was shown to affect hematological parameters positively in athletes *(65)*. Additionally, zinc supplementation

Table 7
Good Food Sources of Specific Micronutrients for the Vegetarian

Nutrient	Good food choices
Iron	Legumes, nuts and seeds, whole/enriched grains, enriched/fortified cereals and pasta, leafy green and root vegetables, dried fruits
Calcium	Calcium-set tofu; calcium-fortified beverages (orange juice, soy milk); kale, collard, and mustard greens; tahini; blackstrap molasses
Vitamin D	Fortified foods (some breakfast cereals, soy/rice milk, margarine) Sun exposure ($\sim$10–15 minutes two or three times per week)
Vitamin B_{12}	Redstar brand nutritional yeast, fortified foods (cereal, soy, dairy products), meat analogues, some brands of margarine

Adapted from Aarsby et al. *(83)*.

has been shown to improve erythrocyte deformability, decrease the exercise-induced acute increase in blood viscosity, and improve exercise tolerance *(66)*. Despite these findings, there are data that suggest zinc intake does not play a role in exercise performance. Lukaski et al. found that neither zinc supplementation nor a restricted zinc intake was found to have any effect of maximal oxygen uptake over a 4-month period *(67)*.

Although the ergogenic potential for zinc supplementation is debatable, the effects of oversupplementation are not. In the body, an intake of zinc of $> 50\,\mathrm{mg/day}$ has been shown to inhibit copper absorption *(68)*. Additionally, zinc intake 10 times greater than the RDA has been shown to decrease immune function, reduce HDL cholesterol, and increase LDL cholesterol *(69)*. For these reasons, zinc supplements exceeding 15 mg/day are not recommended.

3.3.2. CHROMIUM

The two most common forms of chromium are chromium III and chromium VI, with chromium III being the form most often found in foods because of its greater stability. Chromium VI is recognized as carcinogenic if inhaled or ingested, whereas chromium III is important in carbohydrate, lipid, and protein metabolism. Chromium helps facilitate the action of insulin, ultimately increasing insulin sensitivity and decreasing the need for insulin. Chromium is well dispersed throughout many food sources. Side effects associated with chromium deficiency include weight loss, peripheral neuropathy, impaired glucose utilization, and increased insulin requirements. Although nephritis, hepatic dysfunction and rhabdomyolysis (skeletal muscle injury) are possible effects of high chromium intake, at present no upper limit for chromium has been set.

Because of chromium's role in energy metabolism, numerous studies examining the effects of chromium supplementation and exercise have been performed. Volek et al. determined that with 11 µmol chromium III supplementation, no effects on glycogen synthesis occur during recovery from high intensity cycle erogometry in overweight adult men on a high carbohydrate diet *(70)*. Likewise, no significant effects on body composition or strength gains in young males during resistance training were observed with either 3.3 µmol chromium in the form of chromium picolinate or 3.5 µmol in the form of chromium chloride, both versus a placebo *(71)*. Transferrin saturation was decreased with chromium picolinate supplementation versus chromium chloride or the placebo, suggesting that chromium supplementation may dispose an individual to iron deficiency, depending on the dose and duration of supplementation. Multiple studies have hypothesized that benefits of supplementation may occur only in individuals with impaired chromium concentrations *(71,72)*.

3.3.3. BORON

The physiological role of boron in the body is not clearly understood. Proposed functions include metabolism of vitamin D, macromineral metabolism, and immune function *(29)*. Because of a lack of evidence regarding boron, no DRI has been established.

Meacham et al. have conducted two studies to determine if supplementing 3 mg of boron versus a placebo in athletic versus sedentary participants has an effect on minerals, namely phosphorus, magnesium, and calcium, affecting bone mineral density (BMD) *(73,74)*. The first study found that athletes supplemented with boron had lower serum magnesium concentrations than the sedentary subjects, but no differences were seen among activity groups receiving the placebo *(74)*. Plasma calcium did not differ between any groups, and serum phosphorus concentrations were significantly lower than baseline values among all groups. The second study not only looked at blood mineral concentrations, but bone mineral density (BMD) was analyzed with an absorptiometer *(73)*. Boron supplementation did not appear to influence BMD; however, serum calcium and magnesium increased and phosphorus decreased over time in all subjects. Serum phosphorus concentrations were significantly lower in boron-supplemented subjects with sedentary levels lower than active individuals. Athletic subjects supplemented with boron had lower serum magnesium levels than sedentary individuals. Because of the varied findings on serum mineral concentrations with boron supplementation, more research should be conducted to determine effects on BMD.

3.3.4. OTHER MINERALS

Little research exists on exercise and the following minerals: copper, fluoride, iodine, manganese, molybdenum, selenium, and vanadium. Functions, DRIs, known effects of exercise, and food sources of each nutrient may be found in Table 5. In general, exercise does not appear to increase needs above the DRIs, nor is there conclusive evidence recommending the use of supplementation for increased athletic performance.

3.4. Summary

Athletes should consume a balanced diet in an attempt to obtain adequate amounts of minerals necessary for optimal performance. Mineral supplementation may be recommended for those who do not consume a balanced diet. Research has consistently found iron and calcium to be consumed in low amounts by athletes *(75)*. During strenuous activity or exercise in a hot environment, elevated sweat losses may result in increased dietary requirements of sodium and chloride. Mineral deficiencies, especially iron and chromium, may lead to performance impairment; and deficiencies in calcium, magnesium, and phosphorus may decrease bone health. Overall, when well nourished athletes are supplemented with minerals, including calcium, magnesium, iron, zinc, copper, and selenium, no improvement in athletic performance has been found *(75)*. Phosphorus is the lone mineral for which multiple studies have shown that supplementation may improve performance in athletes without deficiency. However, due to adverse effects with oversupplementation and the need for further controlled research, the current recommended intake remains the DRI.

4. CONCLUSION

The micronutrient needs of the athlete do not appear to differ from that of a healthy individual; that is, the athlete may refer to appropriate DRI tables to gauge nutrient needs. Generally, when dietary intake is adequate, supplementation is unnecessary. If dietary intake is inadequate (e.g., in the case of strict vegans and intake of vitamin B_{12}) or losses through sweat are increased, supplementation may be warranted. However, care should be taken not to exceed the upper limit of the specific micronutrients in question.

4.1. Practical Application

For the professional working with the athlete, caloric intake and expenditure should be properly assessed prior to making dietary prescription recommendations. This involves an evaluation of current dietary habits (i.e., analysis of a 4-day food record or 24-hour dietary recall) and exercise status (i.e., type, frequency, duration, and intensity of exercise). Additionally, age, sex, and environmental factors (i.e., temperature and terrain) should be considered when making dietary recommendations. Professionals should emphasize consuming a well balanced diet before recommending supplementation. Special attention should be given to female and vegetarian athletes who may present with low calcium and iron levels. Tables 3 through 5 serve as a quick reference for recommended intakes and good dietary sources of specific micronutrients, as well as the role each plays in physical performance.

REFERENCES

1. Position of the American Dietetics Association, Dietitians of Canada, and the American College of Sports Medicine: nutrition and athletic performance. J Am Diet Assoc 2000;100:1543–1556.
2. Food and Nutrition Board, Institute of Medicine. Dietary Reference Intakes: Recommended Intakes for Individuals. National Academy Press, Washington, DC, 2004.
3. Maughan RJ. Role of micronutrients in sport and physical activity. Br Med Bull 1999;55:683–690.
4. Sports Nutrition. A Practice Manual for Professionals. 4th ed. American Dietetics Assocation, Chicago, 2006.
5. Manore MM. Effect of physical activity on thiaminee, riboflavin, and vitamin B-6 requirements. Am J Clin Nutr 2000;72:598S–606S.

6. Suzuki M, Itokawa Y. Effects of thiaminee supplementation on exercise-induced fatigue. Metab Brain Dis 1996;11:95–106.

7. Webster MJ, Scheett TP, Doyle MR. Branz M. The effect of a thiamine derivative on exercise performance. Eur J Appl Physiol Occup Physiol 1997;75:520–524.

8. Food and Nutrition Board, Institute of Medicine. Dietary Reference Intakes for Thiamine, Riboflavin, Niacin, Vitamin B_6, Folate, Vitamin B_{12}, Pantotenic Acid, Biotin, and Choline. National Academy Press, Washington, DC, 1998.

9. Lukaski HC. Vitamin and mineral status: effects on physical performance. Nutrition 2004;20:632–644.

10. Belko AZ. Vitamins and exercise—an update. Med Sci Sports Exerc 1987;19:S191–S196.

11. Combs GF Jr, Vitamins. In: Mahan LK. Escott-Stump S (eds) Krause's Food, Nutrition, & Diet Therapy (pp 67–109). Saunders, Philadelphia, 2000.

12. Murray R, Bartoli WP, Eddy DE, Horn MK. Physiological and performance responses to nicotinic-acid ingestion during exercise. Med Sci Sports Exerc 1995;27:1057–1062.

13. Short SH. Surveys of dietary intake and nutrition knowledge of athletes and their coaches. In: Wolinksy I, Hickson JF (ed) Nutrition and Exercise in Sport (p 367). CRC Press, Boca Raton, FL, 1994.

14. Dwyer JT, Shils ME, Olson JA, Shike M. Modern Nutrition in Health and Disease. 8th ed. Williams & Wilkins, Baltimore, 1994.

15. Webster MJ. Physiological and performance responses to supplementation with thiamine and pantothenic acid derivatives. Eur J Appl Physiol Occup Physiol 1998;77:486–491.

16. Manore MM. Vitamin B_6 and exercise. Int J Sport Nutr 1994;4:89.

17. Manore MM. Effect of physical activity on thiaminee, riboflavin, and vitamin B-6 requirements. Am J Clin Nutr 2000;72:598S–606S.

18. Linderman J, Kirk L, Musselman J, Dolinar B, Fahey TD. The effects of sodium bicarbonate and pyridoxine-alpha-ketoglutarate on short-term maximal exercise capacity. J Sports Sci 1992;10:243–253.

19. Matter M, Stittfall T, Graves J et al. The effect of iron and folate therapy on maximal exercise performance in female marathon runners with iron and folate deficiency. Clin Sci (Lond) 1987;72:415–422.

20. Nygard O, Nordrehaug JE, Refsum H, Ueland PM, Farstad M, Vollset SE. Plasma homocysteine levels and mortality in patients with coronary artery disease. N Engl J Med 1997;337:230–236.

21. Montoye HJ, Spata PJ, Pinckney V, Barron L. Effect of vitamin B_{12} supplementation on physical fitness and growth of boys. J Appl Physiol 1955;7:589.

22. Tin-May T, Ma-Win M, Khin-Sann A, Mya-Tu M. Effect of vitamin B_{12} on physical performance capacity. Br J Nutr 1978;40:269.

23. Van der Beek EJ, van Dokkum W, Schrijver J, Wesstra A, Kistemaker C, Hermus RJ. Controlled vitamin C restriction and physical performance in volunteers. J Am Coll Nutr 1990;9:332–339.

24. Kotze HF, van der Walt WH, Rogers GG, Strydom NB. Effects of plasma ascorbic acid levels on heat acclimatization in man. J Appl Physiol 1977;42:711–716.

25. Peters EM, Goetzsche JM, Grobbelaar B, Noakes TD. Vitamin C supplementation reduces the incidence of postrace symptoms of upper-respiratory-tract infection in ultramarathon runners. Am J Clin Nutr 1993;57:170–174.
26. Bryant RJ. Ryder J, Martino P, Kim J, Craig BW. Effects of vitamin E and C supplementation either alone or in combination on exercise-induced lipid peroxidation in trained cyclists. J Strength Cond Res 2003;17:792–800.
27. Connolly DA, Lauzon C, Agnew J, Dunn M, Reed B. The effects of vitamin C supplementation on symptoms of delayed onset muscle soreness. J Sports Med Phys Fitness 2006;46:462–467.
28. Paiva SA, Russell RM. Beta-carotene and other carotenoids as antioxidants. J Am Coll Nutr 1999;18:426–433.
29. Food and Nutrition Board, Institute of Medicine. Dietary Reference Intakes for Vitamin A, Vitamin K, Arsenic, Boron, Chromium, Copper, Iodine, Iron, Manganese, Molybdenum, Nickel, Silicon, Vanadium, and Zinc. National Academy Press, Washington, DC, 2001.
30. Penniston KL, Tanumihardjo SA. The acute and chronic toxic effects of vitamin A. Am J Clin Nutr 2006;83:191–201.
31. Food and Nutrition Board, Institute of Medicine. Dietary Reference Intakes for Calcium, Phosphorus, Magnesium, Vitamin D, and Fluoride. National Academy Press, Washington, DC, 1997.
32. Kenny AM, Biskup B, Robbins B, Marcella G, Burleson JA. Effects of vitamin D supplementation on strength, physical function, and health perception in older, community-dwelling men. J Am Geriatr Soc 2003;51:1762–1767.
33. Food and Nutrition Board, Institute of Medicine. Dietary Reference Intakes for Vitamin C, Vitamin E, Selenium, and Carotenoids. National Academy Press, Washington, DC, 2000.
34. Lukaski HC. Effects of exercise training on human copper and zinc nutriture. Adv Exp Med Biol 1989;258:163–170.
35. Viitala PE, Newhouse IJ, LaVoie N, Gottardo C. The effects of antioxidant vitamin supplementation on resistance exercise induced lipid peroxidation in trained and untrained participants. Lipids Health Dis 2004;3:14.
36. McAnulty SR, McAnulty LS, Nieman DC, et al. Effect of alpha-tocopherol supplementation on plasma homocysteine and oxidative stress in highly trained athletes before and after exhaustive exercise. J Nutr Biochem 2005;16:530–537.
37. Itoh H, Ohkuwa T, Yamazaki Y, et al. Vitamin E supplementation attenuates leakage of enzymes following 6 successive days of running training. Int J Sports Med 2000;21:369–374.
38. Braam LA, Knapen MH, Geusens P, Brouns F, Vermeer C. Factors affecting bone loss in female endurance athletes: a two-year follow-up study. Am J Sports Med 2003;31:889–895.
39. Anderson JJB. Minerals. In: Mahan KL, Escott-Stump S (eds) Krause's Food, Nutrition, & Diet Therapy (pp 110–152). Saunders, Philadelphia, 2000.
40. National Institutes of Health Clinical Center. Dietary Supplement Fact Sheet: Calcium. Version Current. September 23, 2005 http://dietary-supplements.info.nih.gov/factsheets/calcium.asp#en117/, accessed January 30, 2007.

41. Matkovic V, Goel PK, Badenhop-Stevens NE, et al. Calcium supplementation and bone mineral density in females from childhood to young adulthood: a randomized controlled trial. Am J Clin Nutr 2005;81:175–188.

42. Bergeron MF, Volpe SL, Gelinas Y. Cutaneous calcium losses during exercise in the heat: a regional sweat patch estimation technique. Clin Chem 1998;44:A167.

43. Jacqmain M, Doucet E, Despres JP, Bouchard C, Tremblay A. Calcium intake, body composition, and lipoprotein-lipid concentrations in adults. Am J Clin Nutr 2003;77:1448–1452.

44. Bremner K, Bubb WA, Kemp GJ, Trenell MI, Thompson CH. The effect of phosphate loading on erythrocyte 2,3-bisphosphoglycerate levels. Clin Chim Acta 2002;323:111–114.

45. Galloway SD, Tremblay MS, Sexsmith JR, Roberts CJ. The effects of acute phosphate supplementation in subjects of different aerobic fitness levels. Eur J Appl Physiol Occup Physiol 1996;72:224–230.

46. Lukaski HC, Nielsen FH. Dietary magnesium depletion affects metabolic responses during submaximal exercise in postmenopausal women. J Nutr 2002;132:930–935.

47. Brilla LR, Haley TF. Effect of magnesium supplementation on strength training in humans. J Am Coll Nutr 1992;11:326–329.

48. Nielsen FH, Lukaski HC. Update on the relationship between magnesium and exercise. Magnes Res 2006;19:180–189.

49. US Department of Agriculture, Agricultural Research Service. USDA National Nutrient Database for Standard Reference, Release 18. Version Current. November 18, 2005. Nutrient Data Laboratory Home Page (http://www.nal.usda.gov/fnic/foodcomp), accessed January 29, 2007.

50. Leppik JA, Aughey RJ, Medved I, Fairweather I, Carey MF, McKenna MJ. Prolonged exercise to fatigue in humans impairs skeletal muscle Na^+-K^+-ATPase activity, sarcoplasmic reticulum Ca^{2+} release, and Ca^{2+} uptake. J Appl Physiol 2004;97:1414–1423.

51. Nielsen JJ, Mohr M, Klarskov C, et al. Effects of high-intensity intermittent training on potassium kinetics and performance in human skeletal muscle. J Physiol 2004;554:857–870.

52. Goss F, Robertson R, Riechman S, et al. Effect of potassium phosphate supplementation on perceptual and physiological responses to maximal graded exercise. Int J Sport Nutr Exerc Metab 2001;11:53–62.

53. Sharp RL, Role of sodium in fluid homeostasis with exercise. J Am Coll Nutr 2006;25:231S–239S.

54. Beard J, Tobin B. Iron status and exercise. Am J Clin Nutr 2000;72:594S–597S.

55. Hinton PS, Giordano C, Brownlie T, Haas JD. Iron supplementation improves endurance after training in iron-depleted, nonanemic women. J Appl Physiol 2000;88:1103–1111.

56. Perkkio MV, Jansson LT, Brooks GA, Refino CJ, Dallman PR, Work performance in iron deficiency of increasing severity. J Appl Physiol 1985;58:1477–1480.

57. Gardner GW, Edgerton VR, Senewiratne B, Barnard RJ, Ohira Y. Physical work capacity and metabolic stress in subjects with iron deficiency anemia. Am J Clin Nutr 1977;30:910–917.

58. Haas JD, Brownlie T 4th. Iron deficiency and reduced work capacity: a critical review of the research to determine a causal relationship. J Nutr 2001;131: 676S–688S; discussion 688S–690S.

59. Brownlie T 4th, Utermohlen V, Hinton PS, Haas JD. Tissue iron deficiency without anemia impairs adaptation in endurance capacity after aerobic training in previously untrained women. Am J Clin Nutr 2004;79:437–443.

60. Hinton PS, Giordano C, Brownlie T, Haas JD. Iron supplementation improves endurance after training in iron-depleted, nonanemic women. J Appl Physiol 2000;88:1103–1111.

61. Friedmann B, Weller E, Mairbaurl H, Bartsch P. Effects of iron repletion on blood volume and performance capacity in young athletes. Med Sci Sports Exerc 2001;33:741–746.

62. Weight LM, Noakes TD, Labadarios D, Graves J, Jacobs P, Berman PA. Vitamin and mineral status of trained athletes including the effects of supplementation. Am J Clin Nutr 1988;47:186–191.

63. Lukaski HC. Zinc. In: Wolinksy I, Hickson JF (eds) Sports Nutrition, Vitamins and Trace Elements (pp 217–234). CRC Press, Boca Raton, FL, 2006.

64. Micheletti A, Rossi R, Rufini S. Zinc status in athletes: relation to diet and exercise. Sports Med 2001;31:577–582.

65. Kilic M, Baltaci AK, Gunay M. Effect of zinc supplementation on hematological parameters in athletes. Biol Trace Elem Res 2004;100:31–38.

66. Khaled S, Brun JF, Cassanas G, Bardet L, Orsetti A. Effects of zinc supplementation on blood rheology during exercise. Clin Hemorheol Microcirc 1999;20:1–10.

67. Lukaski HC, Bolonchuk WW, Klevay LM, Milne DB, Sandstead HH. Changes in plasma zinc content after exercise in men fed a low-zinc diet. Am J Physiol 1984;247:E88–E93.

68. Hackman RM, Keen, CL. Changes in serum zinc and copper levels after zinc supplementation in running and nonrunning men. In: Katch F (ed) Sport, Health, and Nutrition (pp 89–99). Human Kinetics, Champaign, IL, 1986.

69. Chandra RK. Excessive intake of zinc impairs immune responses. JAMA 1984;252:1443–1446.

70. Volek JS, Silvestre R, Kirwan JP, et al. Effects of chromium supplementation on glycogen synthesis after high-intensity exercise. Med Sci Sports Exerc 2006;38:2102–2109.

71. Lukaski HC, Bolonchuk WW, Siders WA, Milne DB. Chromium supplementation and resistance training: effects on body composition, strength, and trace element status of men. Am J Clin Nutr 1996;63:954–965.

72. Volpe SL, Huang HW, Larpadisorn K, Lesser II. Effect of chromium supplementation and exercise on body composition, resting metabolic rate and selected biochemical parameters in moderately obese women following an exercise program. J Am Coll Nutr 2001;20:293–306.

73. Meacham SL, Taper LJ, Volpe SL. Effects of boron supplementation on bone mineral density and dietary, blood, and urinary calcium, phosphorus, magnesium, and boron in female athletes. Environ Health Perspect 1994;102 (Suppl 7):79–82.

74. Meacham SL, Taper LJ, Volpe SL. Effect of boron supplementation on blood and urinary calcium, magnesium, and phosphorus, and urinary boron in athletic and sedentary women. Am J Clin Nutr 1995;61:341–345.

75. Williams MH. Dietary supplements and sports performance: minerals. JISSN 2005;2:43–49.

76. Reaves L, Steffen LM, Dwyer JT, et al. Vitamin supplement intake is related to dietary intake and physical activity: the Child and Adolescent Trial for Cardiovascular Health (CATCH). J Am Diet Assoc 2006;106:2018–2023.

77. Weight LM, Myburgh KH, Noakes TD. Vitamin and mineral supplementation: effect on the running performance of trained athletes. Am J Clin Nutr 1988;47:192–195.

78. American Dietetic Association, Dietitians of Canada. Position of the American Dietetic Association and Dietitians of Canada: vegetarian diets. Can J Diet Pract Res 2003;64:62–81.

79. Alewaeters K, Clarys P, Hebbelinck M, Deriemaeker P, Clarys JP. Cross-sectional analysis of BMI and some lifestyle variables in Flemish vegetarians compared with non-vegetarians. Ergonomics 2005;48:1433–1444.

80. Barnard ND, Cohen J, Jenkins DJ, et al. A low-fat vegan diet improves glycemic control and cardiovascular risk factors in a randomized clinical trial in individuals with type 2 diabetes. Diabetes Care 2006;29:1777–1783.

81. Robinson F, Hackett AF, Billington D, Stratton G. Changing from a mixed to self-selected vegetarian diet: influence on blood lipids. J Hum Nutr Diet 2002;15:323–329.

82. Gisolfi CV, Duchman SM. Guidelines for optimal replacement beverages for different athletic events. Med Sci Sports Exerc 1992;24:679–687.

83. Aarsby HM, Larson-Meyer DE. Vegetarian athletic diet for exercise, athletic training and performing. Agro Food Industry Hi-tech 2006;17(2):xx–xxiii.

6 Fluid Regulation for Life and Human Performance

Allyn Byars

Abstract

Fluid regulation for life and performance is examined in this chapter. First, however, the nature and function of water in human physiology as it pertains to maintaining fluid balance on a daily basis must be examined. The relation of water to human performance is also examined as the body's requirement for this fluid before, during, and after exercise is explored. In addition, other fluid sources, such as carbohydrate drinks, are examined for their efficacy, including specifically formulated beverages marketed for use during exercise. Finally, recommendations for application in the athletic environment are given.

Key words

Hydration · Fluid replacement · Water

1. NATURE OF WATER IN THE HUMAN BODY

Of all the macronutrients, water is considered the most essential. Death can occur in as few as 3 days without water. Water typically accounts for approximately 40% to 70% of total body composition. Factors affecting this range of water content include sex, age, and body composition content. Specifically in regard to body composition, water content is much greater in muscle tissue (~70%) than in fat tissue (~10%). Based on this relation, body water content is greatly affected by body composition, with lean individuals having greater water content than individuals of the same weight with greater fat mass.

From: *Nutritional Supplements in Sports and Exercise*
Edited by: M. Greenwood, D. Kalman, J. Antonio,
DOI: 10.1007/978-1-59745-231-1_6, © Humana Press Inc., Totowa, NJ

2. FUNCTION OF WATER IN THE HUMAN BODY

Water serves us well through its many functions in the human body. One such example is that water acts as the body's transport system as gases and nutrients travel in a liquid environment and waste products leave the body through the water in urine and feces. Water lubricates the joints and acts as a cushion for organs such as the eyes, intestines, heart, and lungs. It gives structure and form to body tissue. In addition, water helps the body maintain temperature equilibrium as it absorbs considerable heat with only small changes in temperature and, combined with its high heat of vaporization, maintains a relatively stable body temperature during environmental heat stress and the increased internal heat load generated by exercise (1).

Two fluid components comprise the water content in the body. One is intracellular (intracellular fluid, ICF), which represents the fluid inside the cells. The other is extracellular (extracellular fluid, ECF), which includes the fluid that flows interstitially (in the microscopic spaces between cells); the ECF accounts for about one-third of the water and provides cell structure in the human body. ECF also acts as a reactive medium between various tissues, including lymph, saliva, fluid in the eyes, fluid secreted by glands and the digestive tract, fluid that bathes the spinal cord nerves, and fluid excreted from the skin and kidneys. In addition, blood plasma accounts for approximately 20% of the ECF fluid (3–4 liters); it provides most of the fluid lost through sweating (for cooling the body) and assists in maintaining a pH balance. Of the total body water, about two-thirds (26 liters of the body's 42 liters of water for an average 80 kg man) is intracellular, with the other third (~38%) being ECF (1).

These volumes, however, represent only averages from a perpetually dynamic exchange of fluid between components, particularly in physically active men and women (2). Exercise training (in particular resistance training) can increase the amount of water in ICF components because of associated increases in muscle tissue with its large water content. Conversely, an acute bout of exercise momentarily distributes fluid from plasma to interstitial and intracellular spaces because of the increased pressure of fluid in the circulatory system.

3. WATER BALANCE

Considering the many wide-ranging factors (e.g., food, liquids, metabolism) that affect water balance in the human body, water levels remain fairly stable for days, weeks, months, and even years. The activity status of an individually can obviously affect water balance; but, fortunately, with appropriate fluid intake, deficits in the body's water levels can be quickly replenished. Sedentary adults in a temperature-steady environment require about 2.5 liters of water daily, whereas a physically active person in a warm, humid environment obviously has larger water needs (~5–10 liters) daily depending on the amount of activity and the environmental conditions *(1)*.

Water intake includes three sources: foods, liquids, and metabolism. Of the water in foods, roughly 60% comes from fluids ingested into the body (Table 1). This includes not only water but other fluids such as juices, milk, sport drinks, and even caffeinated beverages such as colas and coffee. On any given day, the average adult under typical conditions consumes around 1200 mL or about 41 oz of water per day *(3)*. However, as aforementioned, physical activity and environmental conditions can greatly influence the need to

Table 1
Dietary Reference Intake Values for Total Water

Subjects and age (years)	From foods	From beverages	Total water
Males			
4–8	0.5	1.2	1.7
9–13	0.6	1.8	2.4
14–18	0.7	2.6	3.3
>19	0.7	3.0	3.7
Females			
4–8	0.5	1.2	1.7
9–13	0.5	1.6	2.1
14–18	0.5	1.8	2.3
>19	0.5	2.2	2.7

Adapted from Fink et al. *(3)*.
Data are given as the AI (Adequate Intake) in liters per day

replenish the body's fluid level and can be as high as five or six times the ordinary amount. Extreme conditions can put an even greater demand on the body as reported for an individual who lost 13.6 kg of water weight during a 2 day, 17 hour, 55 mile run across Death Valley, CA *(4)*. Because of suitable fluid replenishment and salt supplementation, the actual body weight loss amounted to only 1.4 kg, although fluid loss and replenishment accounted for almost 4 gallons of liquid.

Another 30% of fluid intake comes from foods (e.g., fruits, vegetables). Fruits and vegetables can contain significant water. In contrast, foods such as butter, oil, dried meats, chocolate, cookies, and cakes have comparatively low water content. For example, peanut butter and shelled peanuts contain only traces of water, whereas walnuts contain 4% water and dried coconuts and pecans around 7%. This percentage increases for foods such as molasses (25%), whole wheat bread (35%), hamburger beef (54%), and whole milk (87%); and the following foods exceed 90% water content: lettuce, raw strawberries, cucumbers, watercress, Swiss chard, boiled squash, green peppers, bean sprouts, boiled collards, watermelon, cantaloupe, canned pumpkin, celery, and raw peaches *(3)*.

Metabolism accounts for the remaining sources of water. Sometimes referred to as metabolic water, this source provides approximately 10% of the daily water requirement of a sedentary person. The breakdown of basic foodstuffs (carbohydrates, fats, protein) in energy metabolism results in the formation of heat, carbon dioxide, and water. The breakdown of glucose releases 55 g of metabolic water with larger amounts of water occurring from protein (100 g) and fat (107 g) catabolism. In addition, each gram of glycogen joins with 2.7 g of water as its glucose unit links together in which glycogen eventually liberates this bound water during its catabolism for energy *(3)*.

The human body loses water via four mechanisms through the skin (evaporation), expired air (water vapor), urine, and feces. Water lost through urine constitutes a main avenue of loss during resting conditions.

Water loss through seepage from the skin, called insensible perspiration, can account for around 350 mL of water per day. Additional water loss comes in the form of perspiration via sweat glands under the skin. Under normal environmental conditions it can amount to as much as 700 mL of sweat per day in an attempt to help maintain a homeostatic temperature through the evaporative process. However, this amount is greatly exceeded in trained athletes and others in extreme conditions of heat and humidity. Insensible water is also lost through small water droplets in expired air in amounts of 250 to 350 mL daily from the complete saturation of inspired air as it travels down the pulmonary airways. Exercise also affects this source of water loss. For physically active persons, the respiratory passages release 2 to 5 mL of water each minute during vigorous exercise, depending on climatic conditions of the environment. Ventilatory water loss is slightest in hot humid weather and greatest in cold temperatures (inspired air contains little moisture) and at high altitude. The latter occurs because inspired air volumes (which require humidification) are significantly larger at sea level than at higher altitudes *(1)*.

Each day, the quantity or volume of urine excreted daily by the kidneys ranges from 1000 to 1500 mL, or about 1.5 quarts per day. This process may become accentuated when substantial quantities of protein are used for energy (as occurs with a high protein/low carbohydrate diet) and may hasten dehydration during exercise *(1)*. Water loss due to intestinal elimination in the form of feces produces 100 to 200 mL of water loss as water accounts for around 70% of fecal material. The remainder of fecal matter is composed of nondigestible matter including bacteria from the digestive process and the residues of digestive juices from the intestine, stomach, and pancreas. With intestinal disturbances and/or vomiting, water loss can be as high as 5000 mL; this can result in fluid and electrolyte imbalance, which is potentially deadly if not corrected *(1)*.

Overall, it is body water lost through excessive sweating during exercise that can result in the most dangerous situations. The intensity of the physical activity, environmental temperature, and relative humidity contribute to the amount of water lost through sweating. However, it is the relative humidity, or the water content of the environmental air, that affects the effectiveness of the sweating mechanism in the autoregulation of normal body temperature. Ambient or environmental air is completely saturated with water vapor at 100% relative humidity, preventing any evaporation of fluid from the skin surface into the air (because the air is completely saturated with water and has no room for any more). This, of course, reduces the contribution of this essential mechanism for body temperature regulation. When extreme environmental conditions such as this exist, sweat beads on the outer layers of the skin and eventually rolls off the body without producing any noticeable cooling effect. On a dry or less humid day or in such environments, ambient air can hold substantially more moisture, and fluid evaporates quickly from the skin into the surrounding air. Thus, the sweat mechanism operates at optimal efficiency and body temperature remains controlled within a narrow range. Also of importance is that the plasma volume decreases when sweating causes a fluid loss of $\geq 2\%$ to 3% of body mass. Water loss from the vascular system puts a strain on circulatory capacity, which eventually impairs exercise performance and temperature regulation. Therefore, monitoring changes in total body weight after voiding the bladder can suitably assess water loss during exercise and/or heat stress. Each 0.45 kg (1 lb) of body weight loss is associated with approximately 450 mL (15 oz) of dehydration *(1)*.

4. EXERCISE AND HYDRATION REQUIREMENTS

Fluid intake that is inadequate and/or dehydration can have detrimentally profound effects on human performance (Fig. 1) *(5,6)*. Dehydration can appear before an activity commences or as the result of the activity itself. Either way, dehydration can significantly affect the metabolism of the muscle, temperature regulation of the body, cardiovascular function, and of course capacity for exercise (with as little as $1\%–2\%$ reduction in body weight) *(7)*. Replacement of fluid or water that is lost is important not only for good health but also for exercise performance. Life-threatening situations can occur as the result of severe dehydration.

Body water loss during a state of euhydration (normal hydrated state) that causes hypohydration (depleted hydrated state) is referred to as *dehydration*. Dehydration can result from body water loss during intense exercise as well as moderate exercise that exceeds 1 hour duration; and it can result in approximately 0.5 to 1.0 liter lost through perspiration. Of course, larger amounts of water are lost as the result of exercise of greater intensity especially when coupled with hot and

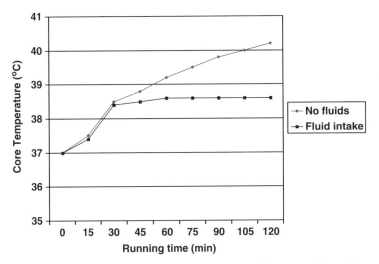

Fig. 1. Effect of fluid intake on core temperature and running time. Adapted from Fink et al. *(3)*.

humid environments. However, even when exercise is conducted in milder environmental conditions, sweat is still produced. Substantial water loss can also occur without the benefit of exercise. Such water loss occurs when athletes concerned with "making weight" (e.g., boxers, wrestlers, weightlifters) seek out rapid weight loss methods such as saunas, steam rooms, or severe diet restriction including fluid, diuretics, and purging. By using these techniques in an attempt to accelerate weight loss, athletes also increase their susceptibility to heat-related illness (exhaustion, stroke).

Hypovolemic conditions or loss of ICF and ECF can have a profound impact on an individual's ability to dissipate heat and can increase the rate of heat storage and cardiovascular load, leading to reductions in the sweating rate and skin blood flow for a given core temperature. This reduced heat tolerance severely compromises cardiovascular function and exercise capacity during intense exercise in hot environments *(8,9)*.

Relative to exercise performance, dehydration and subsequent weight loss have different effects on human performance. For example, dehydration does not alter muscular strength, such as a single bout of power performance up to 60 seconds duration; however, the effect on muscular endurance appears unclear *(10,11)*. Excessive water loss prior to exercise when expressed in relative terms actually improves both parameters because they are expressed relative to body mass in kilograms *(12)*. Performing exercise of high intensity lasting longer than 1 minute has been found to be severely diminished, affecting physiological function as well as training and competing capacity. For example, wrestlers with moderate dehydration of up to 1.5% of the body mass exhibited poorer intermittent overall exercise performance than similar exercise in the euhydrated state *(13)*. Stomach cramps and nausea are also associated with dehydration and body weight, with a concomitant decrease in the gastric emptying rate.

The sweat rate in normal acclimatized individuals usually peaks at around 3 liters per hour during a bout of intense exercise and may reach a total volume of 12 liters in a single day. After several hours of sweating from exercise and climatic conditions, fatigued sweat glands can hinder the regulation of body temperature. Triathletes and other elite endurance athletes can lose approximately ≥ 5 liters of fluid ($\sim$6%–10% of body mass) during competition. For athletes of lower caliber, fluid loss does not usually exceed 500 mL per hour, whereas even in temperate climates of 10°C (50°F) soccer players can lose an average of 2 liters during a 90-minute game *(14)*. However, it must be noted that the ability of the human body to maintain evaporative cooling is dictated, or at least limited, by continued fluid replenishment.

Just as a varied environment can elicit different sweating responses, so can the type sport or activity in which an individual participates. Outside of the aforementioned endurance activities, other sports (e.g., soccer, football, basketball) can produce rather large sweat rates with a concomitant loss in fluid through both practice and competition. In the past (before precompetition weight certification was introduced), it was not uncommon for athletes participating in high school wrestling to lose as much as 13% of their body weight during a match. Most of this type of weight reduction came from excessive sweating and voluntarily choosing to drink minimal portions of water in the days proceeding weigh-in. Past observations have shown than college wrestlers, with the exception of heavyweights, regained an average of 3.7 kg during the 20 hours between the official weigh-in and the actual competition *(15)*.

It was also common for wrestlers (both high school and collegiate wrestlers)—in their quest to make their weight category—to compete in a dehydrated state with a parallel drop in blood and plasma volumes *(16)*. It has also been demonstrate that impaired short-term memory and alterations in mood can result from quick and rapid weight loss in collegiate wrestlers *(17)*.

Dehydration at almost any level negatively affects temperature regulation and physiological functioning. As dehydration worsens, the plasma volume and peripheral blood flow decline; and thus the sweat rate declines in the exercising individual, further complicating the body's ability to maintain thermoregulation. Specifically, dehydration prior to exercise or competition equal to approximately 5% of body weight can result in an increase in the heart rate and core temperature and a reduction in the sweating rate, VO_2max, and exercise capacity compared to a normal euhydrated state *(18)*. In a dehydrated state, this is explained by the reduction in blood volume that results in less ventricular filling pressure, which ultimately causes an increased heart rate and an approximately 25% to 30% reduction in stroke volume. The consequences are reduced cardiac output and arterial blood pressure, leading to reduced aerobic performance *(19)*. Overall, it can be said that an elevation in body core temperature is directly related to reduced systemic blood flow to the skin and therefore a reduction in the amount of fluid lost through sweat. This becomes most pronounced during hot humid climatic conditions in which ambient air is, for the most part, completely saturated and therefore prevents cooling through the evaporative process even though excessive sweating may be present.

Additional studies have demonstrated that during exercise the peripheral blood flow and core temperature are influenced by the hydration level. For example, in one study, body fluid deficits of only 1% body weight increased the core temperature when compared with that seen during equivalent exercise and a normal hydrated state; and for each liter of fluid lost, the heart rate increased approximately 8 beats per minute and there was a 1.0 L/min decline in cardiac output *(20)*. Physical work capacity and physiological function has also been shown to decline with fluid loss of around 4% to 5% body weight *(21)*.

The fluid lost through sweating is predominantly from blood plasma, thereby affecting the circulatory capacity. Coincidentally seen are decreased plasma volume, reduced skin blood flow for a given body temperature, reduced stroke volume, increased heart rate, and a decline in autotemperature regulation and circulatory efficiency *(1)*.

Even modest levels of dehydration can impair exercise performance. One study indicated that dehydration equal to 4.3% of the body mass reduced walking endurance by 48% along with a 22% decrease in VO_2max *(22)*. Also of concern is the use of diuretics to reduce body weight. Dehydration due to these drugs results in a larger percentage of water being drawn from blood plasma than occurs with the normal sweat response. In addition, diuretics may also negatively affect neuromuscular function, including muscle weakness, due to excessive mineral loss from the rapid fluid and weight loss.

5. WATER BALANCE AND EXERCISE

Replacement of water is important for providing consistent blood plasma volume to facilitate the effectiveness of circulation and perspiration. In addition, drinking water or other fluids while participating in exercise allows increased blood flow to the epidermal layers of the skin for evaporative cooling through sweat production. Risks associated with dehydration can be prevented with regimented fluid replenishment.

Although there are some who still believe (erroneously) that fluid replenishment during exercise negatively affects performance, restriction of water expedites heat acclimatization, and that dousing the head with water is all that the athlete needs, research has indicated otherwise. A hydrated individual always functions more effectively than when dehydrated.

5.1. Before Exercise

Prior to exercise, additional fluid intake can provide some initial prevention against dehydration and other heat-related problems affecting performance. This is especially true during exercise in which there is no fluid intake. The preexercise ritual helps facilitate increases in sweating during exercise with a much less pronounced increase in body temperature. A sensible method for increasing preexercise hydration is to ingest around 500 mL of water the night before competition, another 500 mL first thing in the morning, and another 400 to 600 mL of cool water approximately 20 minutes before exercise *(1)*. In a study involving elite soccer players, a regimented intake of fluid of roughly 4.5 liters per day a week before competition helped facilitate better temperature regulation during a warm weather competition creating a 1.1 liter greater total body fluid volume when compared to the athletes' typical daily intake of fluid *(22)*.

5.2. During Exercise

As previously noted, preexercise fluid ingestion does not negate the responsibility of replenishment during the bout of exercise itself. If an athlete does not ingest fluids during an event such as a marathon, the benefits of hyperhydration before competition are soon negated, especially in a hot, humid climate, because it is not possible to match fluid loss (~2000 mL/hr) with fluid intake due to the slower gastric emptying rate (~1000 mL/hr) of the stomach. Thus, preexercise hyperhydration becomes imperative prior to competition, especially long-endurance events *(1)*.

5.3. After Exercise

Because the thirst mechanism is not a good indicator of fluid needed by the body, other mechanisms must be used to ensure adequate hydration of athletes. Water or fluid loss resulting from physical activity and/or exercise can be monitored by weighing athletes prior to and immediately after exercise. Water replacement should equal the amount of water lost through exercise. Monitoring urine output is another method by which the hydration level can be visually monitored as a dark amber/yellow output with subsequent odor is a sign of dehydration whereas a normal euhydrated individual excretes lighter, clearer urine with little or no odor. Approximately 0.5 kg loss in weight is equal to 450 ml (15 fl oz) of dehydration. Coaches and athletic trainers should encourage periodic water breaks during activity to prevent severe dehydration and subsequent loss of performance.

6. ELECTROLYTES REPLACEMENT

Because the water lost through sweat is hypotonic to fluids of the body, replenishment of water appears to be more important than mineral replacement. For example, in a study investigating the impact of potassium- and sodium-laden water in dehydrated individuals found little or no benefit from the additional minerals *(23)*. It is generally accepted that small increases of salt in the daily diet easily replace any sodium lost through the sweating mechanism; and during bouts of extended exercise the body's sodium levels are maintained or conserved via the kidneys. However, as is discussed later, electrolytes and glucose found in fluid replacement beverages bring about more complete rehydration than plain water *(24–26)*.

Restoration of water and electrolyte balance in recovery occurs more rapidly by adding moderate to high amounts of sodium (20–60 mmol/L) to the rehydration drink or combining solid food with appropriate sodium content with plain water *(3)*. Adding a small amount of potassium (2–5 mmol/L) may enhance water retention in the intracellular space and reestablish any extra potassium excretion that accompanies sodium retention by the kidneys *(4)*. The American College of Sports Medicine (ACSM) (see Practical Applications) also recommends that sports drinks contain 0.5 to 0.7 g of sodium per liter of fluid consumed during exercise lasting more than 1 hour. It is also suggested that adherence to rehydration during exercise and recovery may enhanced by beverages that are more palatable for athletes *(27)*.

Restoring fluid balance in the body requires ingesting a fluid volume that exceeds the volume of fluid or water lost through sweat by approximately 25% to 50% because the kidneys are always producing some form of urine independent of body fluid status. When water is ingested by the individual and absorbed and emptied by the stomach, it quickly dilutes plasma sodium, which stimulates urine production concomitantly reducing the effect of the thirst mechanism, all which is counterproductive to replenishment of fluids. Ultimately, without sufficient sodium in the fluid replacement drink, excess fluid intake may increase urine output without fully benefiting the hydration goal *(28)*. Therefore, keeping a consistently high plasma sodium concentration by adding sodium to ingested fluid replacement drinks may sustain the thirst mechanism and promote retention of ingested fluids. Because of the stress of exercising in extreme heat for extended periods of time and the loss of around 13 to 17 g of salt—2.3 to 3.4 g per liter of sweat, which is ~8 g more than is consumed on average—adding salt to the diet (~1/3 tsp of table salt to 1 L of water) would appear to be prudent behavior *(1)*.

Typical activity including exercise generally results in only minimal loss of potassium (partially due to the potassium-conserving mechanism in the kidneys); and at higher intensity exercise levels, losses appear to have no consequences *(29)*. Even under extreme conditions, potassium needs can be met by ingesting foods that contain this mineral (e.g., bananas and potassium-rich citrus fruits).

7. HYPONATREMIA

Excessive fluid intake during some extreme-environment exercise (endurance activities such as marathons and longer triathlons) under certain conditions (heat and high humidity) can result in the serious medical illness hyponatremia, sometimes referred to as "water intoxication." It is the result of low plasma sodium concentration created by overingestion of water. This condition creates an osmotic imbalance across the blood–brain barrier that initiates rapid water influx to the brain resulting in edema of brain tissue. In addition, the following symptoms, from mild to severe, can also be caused by this condition: confusion, headache, malaise, nausea, cramping, seizures, coma, pulmonary edema, cardiac arrest, and death. It has been suggested or recommended that the following steps be taken to prevent development of hyponatremia when planning on competing in prolonged endurance activities *(1)*.

1. Drink 400 to 600 mL (14–22 oz) of fluid 2 to 3 hours before exercise.
2. Drink 150 to 300 mL (5–10 oz) of fluid about 30 minutes before exercise.
3. Drink no more than 1000 mL per hour (33 oz) of plain water spread over 15-minute intervals during or after exercise.
4. Add a small amount of sodium ($\sim$ ¼–½ tsp per 32 oz) to the ingested fluid.
5. Do not restrict salt in the diet.

8. GLYCEROL

It has been proposed that glycerol—a liquid, three-carbon molecule that is the backbone of triglycerides and is part of the storage fat and to some extent body fluids found in all animal and vegetable fats and oils—may serve as a mechanism for retaining water. Ingestion of glycerol in 1 to 2 liters of water has been found to facilitate intestinal water absorption and extracellular fluid retention predominantly in the plasma and interstitial fluid compartments *(30–32)*. Glycerol consumption can increase fluid absorption by the blood and tissues and therefore accentuate hyperhydration for up to 4 hours. Glycerol has also been used clinically to treat cerebral edema and glaucoma as its administration causes a shift in fluid out of the brain and eyes and into the peripheral tissues; however, blurred vision and headaches sometimes ensue *(33)*.

Consumption of glycerol helps hyperhydrate the body. Advocates of glycerol supplementation argue that its effect on hydration reduces overall heat stress during exercise as demonstrated by increased sweating rates. Increased sweating leads to a lower exercise heart rate and body temperature response (for extreme conditions) resulting in enhanced endurance performance *(34,35)*. Research is still unclear about its use; and although it may have more value during exercise under extreme conditions of heat and humidity, the United States Olympic Committee (USOC) currently lists glycerol as a banned substance. It should also be noted that there are currently no tests for the use of supplemental glycerol.

9. CARBOHYDRATE DRINKS AND EXERCISE PERFORMANCE

As previously noted, water by itself may not be enough to replace the deficit created from excessive sweat rates due to exercise of extended duration. If fluid lost during exercise is to be complete restored, it cannot occur without replacing electrolytes, particularly sodium *(36)*.

In addition to sodium, carbohydrate added to fluid replacement solutions can enhance the intestinal absorption of water. It also helps maintain the blood glucose concentration and so delays dependence on muscle glycogen stores, thereby delaying fatigue during bouts of exercise lasting more than 1 hour *(37)*. Aerobic exercise of high intensity for 1 hour or more decreases liver glycogen by about 55%, and a 2-hour strenuous workout almost completely depletes the glycogen content of the liver and active muscle fibers *(1)*. Activities such as football, soccer, hockey, and other similar sports that have repeated supramaximal exercise with intervals of rest can also result in substantial depletion of liver and muscle glycogen. Therefore, it seems prudent to include carbohydrate in drinks used to rehydrate because it is necessary to maintain blood glucose levels during exercise that lasts more than an hour. Repeated ingestion of water and other fluids containing carbohydrate and electrolytes can be an easy method by which to promote optimal performance in athletes.

A carbohydrate concentration of 4% to 8% seems optimal in fluids ingested during exercise. It appears to help replace fluid lost through sweating, maintain blood glucose, and stave off fatigue. This percent solution also contributes to temperature regulation and fluid balance during extreme heat and humidity as effective as plain water and includes an extra 5 calories per minute for maintaining glucose and glycogen levels *(38,39)*.

When it comes to the type of carbohydrate used in the solution, no significant difference exists between glucose, sucrose, and starch. However, the simple sugar fructose should not be used because it has the potential to cause gastrointestinal distress. In addition, because of fructose's low glycemic index, absorption by the gut is slower and promotes less fluid uptake than an equivalent amount of glucose *(1)*.

As noted, carbohydrate ingestion during prolonged activity of an hour or more provides a readily available energy nutrient for active muscles during intense exercise. Studies have found that consuming about 60 g of liquid or solid carbohydrates each hour during exercise benefits high intensity, long duration aerobic exercise and repetitive short bouts of near-maximal effort *(40–43)*. Additionally, supplemental carbohydrate during extended intermittent supramaximal exercise may also provide the same benefits. To improve performance, it has been recommended that athletes should take in 25 to 30 g of carbohydrate (100–200 kcal) every half hour through either food or fluids. However, carbohydrate drinks should be taken every 15 minutes (due to volume and gastric emptying concerns and benefits) which also provide hydration along with the carbohydrates *(44)*. The final result is improved endurance at higher intensities or during intense intermittent exercise as well as increased sprint capacity toward the end of prolonged physical activity such as a marathon in which sustained high energy output and final sprint contribute immensely to the overall and perhaps winning performance *(1)*.

10. PRACTICAL APPLICATIONS

Along with the many aforementioned studies investigating fluid regulation and human performance, the ACSM has made the following recommendations concerning fluid intake before, during, and after exercise *(23)*.

1. It is recommended that individuals consume a nutritionally balanced diet and drink adequate fluids during the 24-hour period before an event—especially during the period that includes the meal prior to exercise—to promote proper hydration before exercise or competition.
2. It is recommended that individuals drink about 500 mL (about 17 oz) of fluid around 2 hours before exercise to promote adequate hydration and allow time for excretion of excess ingested water.
3. During exercise, athletes should start drinking early and at regular intervals in an attempt to consume fluids at a rate sufficient to replace all the water lost through sweating (i.e., body weight loss) or consume the maximum amount that can be tolerate.
4. It is recommended that ingested fluids be cooler than ambient temperature—15° to 22°C (59°–72°F)—and flavored to enhance palatability and promote fluid replacement. Fluids should be readily available and served in containers that allow adequate volumes to be ingested with minimum interruption of exercise.
5. Addition of proper amounts of carbohydrates and/or electrolytes to a fluid replacement solution is recommended for exercise events lasting longer than 1 hour because it does not significantly impair water delivery to the body and may enhance performance. During exercise lasting less than 1 hour, there is little evidence of physiological or physical performance differences after consuming a carbohydrate–electrolyte drink or plain water.
6. During intense exercise lasting longer than 1 hour, it is recommended that carbohydrates be ingested at a rate of 30 to 60 g per hour to maintain oxidation of carbohydrates and delay fatigue. This rate of carbohydrate intake can be achieved without compromising fluid delivery by drinking solutions containing 4% to 8% carbohydrate (grams per 100 mL) at a rate of 600 to 1200 mL per hour. The carbohydrates can be sugars (glucose or sucrose) or starch (e.g., maltodextrin).
7. Including sodium (0.5–0.7 g/L of water) in the rehdydration solution ingested during exercise lasting longer than 1 hour is recommended because it may enhance palatability, promote fluid retention, and possibly prevent hyponatremia (less than normal concentrations of sodium in the blood) in individuals who drink excessive quantities of fluid. There is little physiological basis for the presence of sodium in an oral rehydration solution for enhancing intestinal water absorption so long as sodium is sufficiently available from the previous meal.

In addition to the ACSM position statement, guidelines from other credible sources have been suggested for maintaining fluid levels during practice and competition in hot weather *(45,46)*.

1. Weigh in without your clothes before and after exercise, especially during hot weather. For each pound of body weight lost during exercise, drink 2 cups of fluid.
2. Drink a rehydration beverage containing sodium to replenish lost body fluids quickly. The beverage should also contain 6% to 8% glucose or sucrose.
3. Drink 17 to 20 fluid oz of water or sports drink 2 to 3 hours before practice or competition.
4. Drink 7 to 10 fluid oz of water or sports drink 10 to 20 minutes before the event.
5. Drink 7 to 10 fluid oz of water or sports drink every 10 to 20 minutes during training and competition.
6. Do not restrict fluids before or during the event.
7. Avoid beverages containing caffeine and alcohol because they increase urine production and add to dehydration.

11. CONCLUSION

Fluid regulation for life and performance was examined in this chapter. The nature and function of water in human physiology as it pertains to maintaining fluid balance on a daily basis was examined. The relation of water to human performance as the body's requirement for this fluid was explored before, during, and after exercise. Other fluid sources, such as carbohydrate drinks, were also examined for their efficacy including specifically formulated beverages marketed for use before, during, and recovery after exercise. Finally recommendations were given for practical application.

REFERENCES

1. McArdle WD, Katch FI, Katch VL. Exercise Physiology: Energy, Nutrition, and Human Performance. 6th ed. Lippincott Williams & Wilkins, Philadelphia, 2007.
2. Sawka MN, Coyle EF. Influence of body water and blood volume on thermoregulation and exercise performance in the heat. Exerc Sport Sci Rev 1999;27:167.
3. Fink HH, Burgoon LA, Mikesky AE. Practical Applications in Sports Nutrition. Jones & Barlett, Sudbury, MA, 2006.

4. Robinson S. Cardiovascular and respiratory reactions to heat. In: Yousef MK, et al (eds) Physiological Adaptations. Academic Press, San Diego, 1972.

5. Schoffstall JE, Branch JD, Leutholtz BC, Swain DP. Effect of dehydration and rehydration on the one-repetition maximum bench press of weight trained males. J Strength Cond Res 2001;15:102–108.

6. Webster S, Rutt R, Weltman A. Physiological effects of a weight loss regimen practiced by college wrestlers. Med Sci Sports Exerc 1990;22:229–234.

7. Chandler JT, Brown LE. Conditioning for Strength and Human Performance. Lippincott Williams & Wilkins, Philadelphia, 2007.

8. Montain SJ, Smith SA, Mattot RP, Zientara GP, Jolesz FA, Sawka MN. Hypohydration effects on skeletal muscle performance and metabolism: a ^{31}P-MRS study. J Appl Physiol 1998;84:1889.

9. Sawka MN, Montain SJ, Latzka WA. Hydration effects on temperature regulation. Int J Sports Med 1998;19(Suppl 2):S108.

10. Jacobs I. The effects of thermal dehydration on performance of the Wingate anaerobic test. Int J Sports Med 1980;1:21.

11. Maxwell NS, Gardner F, Nimmo MA. Intermittent running: muscle metabolism in the heat and effect of hypohydration. Med Sci Sports Exerc 1999;31:675.

12. Greiwe JS, Staffey KS, Melrose DR, Narve MD, Knowlton RG. Effects of dehydration on isometric muscular strength and endurance. Med Sci Sports Exerc 1998;30:284.

13. Montain SJ, Laird JE, Latzka WA, Sawka MN. Aldosterone and vasopressin responses in the heat: hydration level and exercise intensity effects. Med Sci Sports Exerc 1997;29:661.

14. Maughan RJ, Leiper JB. Fluid replacement requirements in soccer. J Sports Sci 1994;12(special issue):S29.

15. Scott JR, Horswill CA, Dick RW. Acute weight gain in collegiate wrestlers following a tournament weigh-in. Med Sci Sports Exerc 1994;26:1181.

16. Yankanich J, Kenney W, Fleck S, Kraemer W. Precompetition weight loss and changes in vascular fluid volume in NCAA division I college wrestlers. J Strength Cond Res 1998;12:138.

17. Choma CW, Sforzo GA, Keller BA. Impact of rapid weight loss on cognitive function of collegiate wrestlers. Med Sci Sports Exerc 1998;30:746.

18. Shibasaki M, Kondo N, Crandall CG. Non-thermoregulatory modulation of sweating in humans. Exerc Sport Sci Rev 2003;31:34.

19. Gonzalez-Alonso JR, Mora-Rodriquez ER, Below PR, Coyle EE. Reductions in cardiac output, mean blood pressure and skin vascular conductance with dehydration are reversed when venous return is increased. Med Sci Sports Exerc 1994;26:S163.

20. Coyle EF, Montain SJ. Benefits of fluid replacement with carbohydrate during exercise. Med Sci Sports Exerc 1992;24:S324.

21. Burge CM, Carey MF, Payne WR. Rowing performance, fluid balance, and metabolic function following dehydration and rehydration. Med Sci Sports Exerc 1993;25:1358.

22. Rico-Sanz J, Frontera WR, Rivera MA, Rivera-Brown A, Mole P, Meredith CN. Effects of hyperhydration on total body water, temperature regulation and performance of elite young soccer players in a warm climate. Int J Sports Med 1996;17:85.

23. Costill DL, Coté R, Miller E, Miller T, Wynder S. Water and electrolyte replacement during repeated days of work in the heat. Aviat Space Environ Med 1975;46:795.

24. Rehrer NJ. The maintenance of fluid balance during exercise. Int J Sports Nutr 1996;15:122.

25. Shi X, Sumnmers RW, Schedl HP, Flanagan SW, Chang R, Gisolfi CV. Effects of carbohydrate type and concentration and solution osmolality on water absorption. Med Sci Sports Exerc 1995;27:1607.

26. Wilk B, Bar-Or O. Effect of drink flavor and NaCl on voluntary drinking and hydration in boys exercising in the heat. J Appl Physiol 1996;80:1112.

27. American College of Sports Medicine Position Stand. Exercise and fluid replacement. Med Sci Sports Exerc 1996;28:i–vii.

28. Shirreffs SM, Maughan RJ. Rehydration and recovery of fluid balance after exercise. Exerc Sport Sci Rev 2000;1:27.

29. Cunningham JJ. Is potassium needed in sports drinks for fluid replacement during exercise? Int J Sport Nutr 1997;7:154.

30. Wapnir PA, Sia MC, Fisher SE. Enhancement of intestinal water absorption and sodium transport by glycerol in rats. J Appl Physiol 1996;81:2523.

31. Freund BJ, Montain SJ, Young AJ, et al. Glycerol hyperhydration: hormonal, renal, and vascular fluid responses. J Appl Physiol 1995;79:2069.

32. Riedesel ML, Allen DY, Peake GT, Al-Qattan K Hyperhydration with glycerol solutions. J Appl Physiol 1987;63:2262.

33. Roberts, RA, Griffin, SE. Glycerol: biochemistry, pharmacokinetics, and clinical and practical applications. Sports Med 1998;26:145–167.

34. Montner, P, Stark, DM, Riedesel, ML, et al. Pre-exercise glycerol hydration improves cycling endurance time. Int J Sports Med 1996;17:27–33.

35. Riedesel ML, Allen DY, Peake GT, Al-Qattan K. Hyperhydration with glycerol solutions. J Appl Physiol 1987;63:2262–2268.

36. Takamata A, Mack GW, Gillen CM, Nadel ER. Sodium appetite, thirst, and body fluid regulation in humans during rehydration without sodium replacement. Am J Physiol 1994;266:R1493–R1502.

37. Coggan AR, Coyle EF. Carbohydrate ingestion during prolonged exercise: effects on metabolism and performance. Exerc Sport Sci Rev 1991;19:1–40.

38. Duchman SM, Ryan AJ, Schedl HP, Summers RW, Bleiler TL, Gisolfi CV. Upper limit for intestinal absorption of a dilute glucose solution in men at rest. Med Sci Sports Exerc 1997;29:482.

39. Shi X, Gisolfi CV. Fluid and carbohydrate replacement during intermittent exercise. Sports Med 1998;25:157.

40. Coyle EF, Coggan AR, Hemmert MK, Ivy JL. Muscle glycogen utilization during prolonged strenuous exercise when fed carbohydrate. J Appl Physiol 1986;61:165.

41. Jeukendrup AE, Rouns F, Wagenmakers AJ, Saris WH. Carbohydrate-electrolyte feedings improve 1 h time trial cycling performance. Int J Sports Med 1997;18:125.
42. McConell G, Kloot K, Hargreaves M. Effect of timing of carbohydrate ingestion on endurance exercise performance. Med Sci Sports Exerc 1996;28:1300.
43. Mitchell JB, Costill DL, Houmard JA, Flynn MG, Fink WJ, Belz JD. Effects of carbohydrate ingestion on gastric emptying and exercise performance. Med Sci Sports Exerc 1988;20:110.
44. Coleman E. Carbohydrates: the master fuel. In: Berning JR, Steen SN (eds) Sports Nutrition for the 90's. Aspen, Gaithersburg, MD, 1991.
45. Casa DJ, Armstrong LE, Hillman SK, et al. National Athletic Trainer's Association position statement: fluid replacement for athletes. J Athl Train 2000;25:212–224.
46. Nadel ER. New ideas for rehydration during and after exercise in hot weather. Gatorade Sports Sci Exchange 1988;1(3).

Part III
Specialized Nutritional Supplements and Strategies

7   Muscle Mass and Weight Gain Nutritional Supplements

Bill Campbell

Abstract

There are numerous sports supplements available that claim to increase lean body mass. However, for these sports supplements to exert any favorable changes in lean body mass, they must influence those factors regulating skeletal muscle hypertrophy (i.e., satellite cell activity, gene transcription, protein translation). If a given sports supplement does favorably influence one of these regulatory factors, the result is a positive net protein balance (in which protein synthesis exceeds protein breakdown). Sports supplement categories aimed at eliciting a positive net protein balance include anabolic hormone enhancers, nutrient timing pre- and postexercise workout supplements, anticatabolic supplements, and nitric oxide boosters. Of all the sports supplements available, only a few have been subject to multiple clinical trials with repeated favorable outcomes relative to increasing lean body mass. This chapter focuses on these supplements and others that have a sound theoretical rationale in relation to increasing lean body mass.

Key words

Sports nutrition · Lean body mass · Creatine · Protein supplements · HMB · Nitric oxide · Anabolic · Anticatabolic · Nutrient timing

1. INTRODUCTION

To appreciate fully how certain nutritional supplements increase lean body mass, a thorough understanding of the structural, systemic, and molecular processes that are responsible for such increases in lean body mass is needed. Although there is still a lot to be determined and understood relative to how skeletal muscle is increased, research scientists for the most part have

From: *Nutritional Supplements in Sports and Exercise*
Edited by: M. Greenwood, D. Kalman, J. Antonio,
DOI: 10.1007/978-1-59745-231-1_7, © Humana Press Inc., Totowa, NJ

agreed on several key components that are absolutely necessary for such adaptations to occur. Some of these components are satellite cell activity, muscle-specific gene transcription, protein translation, and nutrient (amino acid) transport into the skeletal muscle. In addition, growth factors/anabolic hormones including testosterone, growth hormone, insulin-like growth factor-1 (IGF-1), and insulin are also necessary for increases in skeletal muscle mass.

Although some sports supplements have been repeatedly observed to increase skeletal muscle mass, they are at best utilized as a complement to an optimally periodized resistance training program. In fact, the greatest stimulus for muscle hypertrophy is mechanical stress in the form of resistance training. The main question to ask relative to sports supplements is whether a given supplement is able to augment the stimulus of resistance training so muscle hypertrophy is maximized. An overview of how skeletal muscle hypertrophy is regulated follows.

2. IMPORTANCE OF NET PROTEIN BALANCE

The functional component of skeletal muscle is comprised of two primary proteins: actin and myosin. Of the two, myosin is the primary protein that increases in size. Hence, when exercise biochemists observe changes in muscle mass from a cellular frame of reference, myosin is often the protein of interest. From a general perspective, muscle hypertrophy can be summarized by the status of *net protein balance*. Net protein balance is equal to muscle protein synthesis minus muscle protein breakdown. For skeletal muscle hypertrophy to occur, the net protein balance must be positive (synthesis must exceed breakdown). At rest, in the absence of an exercise stimulus and nutrient intake, the net protein balance is negative *(1–4)*. As previously stated, resistance training is essential for creating the stimulus necessary for skeletal muscle hypertrophy to occur. However, when resistance training is performed alone, in the absence of nutritional and supplemental interventions, net protein balance still does not increase to the point of becoming anabolic. Specific nutrients and supplements noted later in the chapter are needed in conjunction with the resistance training for the net protein balance to become positive.

3. ROLE OF GENES IN SKELETAL MUSCLE HYPERTROPHY

As already mentioned, net protein balance has two components: synthesis and breakdown. To understand how sports supplements can increase protein synthesis, an understanding of the biochemical process is needed. The center of all bodily functions (including the addition of skeletal muscle) is at the level of genes. Specific to hypertrophy, it is the genes that must be expressed as the proteins in skeletal muscle (i.e., myosin and actin). For instance, once muscle-specific genes are activated, they are copied into messenger RNA (mRNA), which is specific to certain proteins in cells. Once mRNA is transcribed, it is then translated into actual proteins. Using myosin as an example, what must first happen is an increase in activation of the myosin gene. Once the myosin gene is activated, is it copied into myosin mRNA. It is this myosin mRNA that then directs the process of changing amino acids into polypeptides and ultimately a functional myosin protein that is added to the existing matrix of the sarcomere. The point at which the myosin protein is synthesized (from the addition of amino acids under the direction of mRNA), the myosin gene is said to be "expressed." The reason why all of this tedious biochemical information is necessary is that any supplement that claims to increase muscle mass must in some way influence one of these aforementioned variables. For instance, some supplements increase growth hormone, which has been associated with increases in IGF-1. IGF-1, in turn, increases the activity of certain cell-signaling pathways, which may increase muscle-specific (i.e., myosin) gene transcription. Other sports supplements may increase lean body mass by increasing the rate at which amino acids are synthesized into muscle proteins under the direction of mRNA. Yet other sports supplements may increase the delivery of nutrients (i.e., amino acids, glucose) to contracting skeletal muscle, which conceivably results in greater substrate from which lean body mass is acquired. Each of these mechanisms and the supplements that may enhance these contributions to skeletal muscle hypertrophy are discussed. Following is a discussion of one of the best and traditional sports supplements—protein.

4. PROTEIN SUPPLEMENTS

When attempting to increase lean body mass, an essential component equal to a sound resistance training program is protein consumption. Not only is protein intake required for skeletal muscle hypertrophy, protein is also needed to repair damaged cells and tissue and for a variety of metabolic and hormonal activities. Protein is the only macronutrient that contains nitrogen. Given the importance of attaining a positive nitrogen balance, it is vitally important that protein be ingested on a daily (and meal-to-meal) basis. When discussing protein as a nutritional supplement, two main questions arise: 1) How much protein is required for an individual engaging in resistance training? 2) What are the types of protein supplements and which are the best sources of protein?

4.1. Protein Requirements

One of the most controversial subjects in the science of sports nutrition has been protein intake. The main controversy and divided opinions have focused on the safety and effectiveness of protein intake currently recommended by the recommended daily allowance (RDA). Currently, the RDA for protein in healthy adults is 0.8 g/kg body weight per day *(5)*. This recommendation accounts for individual differences in protein metabolism, variations in the biological value of protein, and nitrogen losses in the urine and feces. When determining the amount of protein that needs to be ingested to increase lean body mass, many factors must be considered, such as protein quality, energy intake, carbohydrate intake, the amount and intensity of the resistance training program, and the timing of the protein intake. Although 0.8 g/kg/ day may be sufficient to meet the needs of nearly all non-resistance-trained individuals, it is likely not sufficient to provide substrate for lean tissue accretion or for the repair of exercise-induced muscle damage *(6,7)*. In fact, many clinical investigations indicate that individuals who engage in physical activity/exercise require higher levels of protein intake than 0.8 g/kg/day regardless of the mode of exercise (i.e., endurance, resistance) *(8–12)* or training state (i.e., recreational, moderately or well trained) *(13–15)*. So the question that remains: How much protein *is* required for individuals engaging in resistance training and wanting to increase lean body mass? General recommendations for individuals who engage in strength/power exercise range from 1.6 to 2.0 g/kg/day *(6,13–16)*. A protein intake at these levels help ensure that the net protein balance remains positive, a prerequisite for skeletal muscle hypertrophy to occur.

4.2. Types of Protein Supplement

Although protein can be obtained from whole foods, many resistance trained athletes supplement their diet with protein containing supplements (e.g., protein powders, meal replacements drinks, sports bars). Advances in food processing technology have allowed for the isolation of high quality proteins from both animal and plant sources. Other reasons for supplementing the diet with protein supplements include convenience, simplicity, and the fact that protein supplements also have other benefits, such as a longer shelf life than whole food sources in addition to being more cost-effective in many cases.

Ingesting protein at 1.6 to 2.0 g/kg/day is not the only parameter to consider, however, because it is also important to note that not all protein is the same. Different types of protein are composed of varying amounts of amino acids, which serve as the building blocks of protein. There are approximately 20 amino acids that can be used to make proteins (Table 1). There are eight essential amino acids that must be obtained from the diet because the body cannot synthesize these amino acids. There are also approximately six conditionally essential amino acids that the body has difficulty synthesizing, and therefore individuals are primarily dependent on dietary sources for these amino acids. The body can easily synthesize the remaining amino acids, so they are considered nonessential. Not all protein sources contain the same amounts of amino acids. Protein is classified as *complete* or *incomplete* depending on whether it contains adequate amounts of the essential amino acids. Animal sources of protein contain all essential amino acids and are therefore

Table 1
Classification of Amino Acids

Essential amino acids	Conditionally essential amino acids	Nonessential amino acids
Isoleucine[a]	Arginine	Alanine
Leucine[a]	Cysteine (cystine)	Asparagine
Lysine	Glutamine	Aspartic acid
Methionine	Histidine	Glutamic acid
Phenylalanine	Proline	Glycine
Threonine	Tyrosine	Serine
Tryptophan		
Valine[a]		

[a]Branched-chain amino acids

complete sources of protein, whereas plant proteins are missing some of the essential amino acids (i.e., incomplete). Additionally, there are varying levels of quality of protein depending on the amino acid profile of the protein. Complete protein sources that contain larger amounts of essential amino acids generally have higher protein quality.

4.2.1. WHEY PROTEIN

Four of the most common types of protein found in protein supplements are whey, casein, soy, and egg (ovalbumin) proteins. Each of these proteins is a complete protein, and all are classified as high quality proteins. Whey protein, derived from milk protein, is currently the most popular source of protein used in nutritional supplements. Whey proteins are available as whey protein concentrates, isolates, and hydrolysates. The primary differences among these forms are the method of processing and small differences in fat and lactose content, amino acid profiles, and ability to preserve glutamine residues. In comparison to other types of protein, whey protein is digested at a faster rate, has better mixing characteristics, and is often perceived as a higher quality protein. Research has indicated that the rapid increase in blood amino acid levels following whey protein ingestion stimulates protein synthesis to a greater degree than casein *(17,18)*. Theoretically, individuals who consume whey protein frequently throughout the day may optimize protein synthesis. In fact, a study by Dangin and associates *(19)* reported that frequent ingestion of a small amount of whey protein served to increase protein synthesis to a greater degree than less frequent ingestion of various proteins. Overall, whey protein is an excellent source of protein to supplement due to its amino acid content (including high branched-chain amino acid content) and its ability to be rapidly absorbed *(20)*.

4.2.2. CASEIN PROTEIN

Casein, also a milk protein, is often described as a slower-acting protein *(17,19)*. It is considered a slower protein than whey protein because it takes longer to digest and absorb. This is most likely due to fact that casein has a longer transit time in the stomach *(17)*. Although casein stimulates protein synthesis, it does it to a much lesser extent than whey protein *(17)*. Unlike whey, casein helps decrease protein breakdown *(21)*, which has led to the status of casein as having anticatabolic properties. Given the findings that whey protein stimulates protein synthesis and casein helps decrease muscle breakdown, some supplement manufacturers add both whey and casein to their formulations. The effectiveness of combining whey and casein proteins was illustrated in a recent investigation conducted by Kerksick and colleagues *(22)*. In their study, subjects performed a split body part (training the upper body on one day and the lower body on another) resistance training program 4 days a week for 10 weeks. The subjects were given 48 g of carbohydrate *or* 40 g of whey + 8 g of casein *or* 40 g of whey + 5 g of glutamine + 3 g of branched-chain amino acids (BCAAs). After 10 weeks, the group supplemented with combined whey and casein had the largest increase in lean muscle mass.

4.2.3. SOY PROTEIN

Although soy lacks the essential amino acid methionine, it has a relatively high concentration of remaining essential amino acids and is therefore considered a high quality protein. Soy protein is made from soybeans using water or a water–ethanol mixture to extract the protein *(20)*. Soy protein is similar to whey protein in that there is a soy protein concentrate and isolate. Soy contains compounds called isoflavones, which appear to be strong antioxidants and have been implicated in possibly decreasing the risk of developing cardiovascular disease and cancer. In addition to isoflavones, soy proteins contain protease inhibitors. Given these attributes of soy, there is some evidence to suggest that soy may decrease or prevent the exercise-induced damage to muscle seen following a workout *(23)*. At this point, there are few data relative to soy protein ingestion and accretion of lean body mass in conjunction with resistance training; therefore, more research is needed before definitive recommendations can be given.

4.2.4. EGG PROTEIN

Egg protein is also a high quality protein and has the advantage of being a miscible protein (it mixes easily in solution) *(20)*. However, egg protein supplements generally do not taste good and are more expensive than other protein supplements. For these reasons, along with the availability of other high quality proteins such as whey, casein, and soy, egg protein supplementation is not popular among athletes. Despite this, egg protein is still added in small quantities to some meal replacement/protein powders *(20)*.

4.3. Summary

Adequate protein intake consisting of high quality proteins is a prerequisite for the accretion of lean body mass stimulated by a proper resistance training program. Whey, casein, soy, and egg proteins are all high quality proteins and are commonly found in protein supplements marketed to strength-trained athletes. In addition to ingesting the proper amounts and quality of proteins, the timing of protein intake has been a recent area of scientific investigation. A discussion of the importance of this concept, known as "nutrient timing," follows.

5. CARBOHYDRATE–PROTEIN COMBINATIONS

Ingestion of a high quality protein is essential for increasing lean body mass, but equally important is the timing of the protein intake. This category of sports nutrition has been categorized as *nutrient timing*, and there are multiple research studies highlighting the importance of appropriately timing certain meals throughout the day. In summary, the central idea underlying nutrient timing is to time high glycemic carbohydrate and protein ingestion so it encompasses the time frame in which the resistance training bout exerts a hypertrophic stimulus on the trained skeletal muscles. More specifically, stimulated myofibers are "primed" to synthesize protein, but both insulin and amino acid substrate are required to maximize this adaptation in the moments following an acute bout of resistance exercise. This time period following a resistance training session is commonly referred to as the *anabolic window* to emphasize that this time frame has specific anabolic potential.

5.1. Resistance Training in the Absence of Nutritional Intake

Inherent with the term *anabolic window* is the concept of net protein balance. As stated earlier, net protein balance is equal to muscle protein synthesis minus muscle protein breakdown. For skeletal muscle hypertrophy to occur, net protein balance must be positive (synthesis must exceed breakdown). To improve net protein balance, an appropriate stimulus (e.g., resistance training) must be applied to the skeletal muscles. However, when resistance training is performed alone, in the absence of nutritional and supplemental (i.e., protein, carbohydrate) interventions, net protein balance still does not increase to the point of becoming anabolic. Several studies observing the effects of resistance training and acute changes in net protein balance have concluded that net protein balance is improved as a result of the resistance training bout. Although resistance exercise improves the net balance by stimulating muscle protein synthesis, however, nutrient intake is required for the synthesis to exceed the breakdown *(24)*.

As support for this contention, Biolo and colleagues *(1)* assessed rates of protein synthesis and degradation at rest and 3 hours after a resistance training routine in fasted subjects. At 3 hours after exercise, protein synthesis had increased approximately 108% and protein breakdown had increased 51%. Thus, resistance exercise improved the net protein balance by increasing protein synthesis at a greater rate than protein breakdown. Although the net protein balance was improved, it is important to note that it did not improve to the point of becoming positive (anabolic).

Phillips and coworkers *(3)* conducted a similar study in which they recruited two groups of participants (resistance trained and untrained) and had them perform an eccentric-only resistance exercise workout in a fasted state. Rates of protein synthesis and breakdown were measured within 4 hours of completing the resistance training protocol. Following the resistance training bout, muscle protein synthesis rates increased by 118% in the untrained group and by 48% in the resistance trained group. In terms of muscle protein breakdown, there was an increase of 37% in the untrained group and an increase of 15% in the resistance-trained group. Relative to the net protein balance, the resistance training protocol significantly improved this measure in both groups (+37% in the untrained group and +34% in the trained group), but the overall net protein balance was still negative following the bout of resistance training.

Using a larger time frame, this same researcher *(2)* assessed rates of protein synthesis and protein breakdown at rest and at 3, 24, and 48 hours after a resistance training workout in recreationally active (but not previously resistance trained) subjects. Unfortunately, however, the net protein balance was not assessed in the fasted state; rather, each participant ingested food at his own discretion. There was an important nutritional restriction employed: The participants were instructed to eat a meat-free diet during the study (which limited protein intake). In addition, it appears that the 3-hour net protein balance assessment was conducted in the fasted state. Muscle protein synthesis was significantly increased at each time point following the resistance training bout: at 3 hours 112%; at 24 hours 65%; at 48 hours 34%. Muscle protein breakdown was also increased by 31% at 3 hours after exercise and by 18% at 24 hours. Muscle protein breakdown returned to resting levels by 48 hours. One of the novel findings of this study was the observation that muscle protein synthesis was elevated (by 34%) 48 hours after exercise, during which time muscle protein breakdown returned to baseline levels. Despite this finding, at no time point did the net protein balance become positive (likely due to the restrictions on protein intake).

In summary, each of these aforementioned studies indicates that resistance training alone is not enough to elicit positive changes in net protein balance that lead to increases in lean body mass.

5.2. Insulin, Amino Acids, and Protein Synthesis

As stated in the introduction to the chapter, muscle-specific genes must be activated to initiate the process of skeletal muscle hypertrophy. Once these muscle-specific genes are activated, they are copied into messenger RNA (mRNA) which serves as a template for which muscle proteins are then manufactured (translated). Many researchers believe that resistance training acts as the stimulus for activating muscle-specific genes, but once these genes are copied into muscle-specific mRNA transcripts still other factors are needed to convert the muscle-specific mRNA into functional skeletal muscle proteins. Two biological compounds have been shown to be an integral part of this process: insulin and amino acids. In fact, Bolster and coworkers *(25)* stated in a review paper that, "Without question, investigating the singular role of amino acids or insulin in promoting changes in skeletal muscle protein synthesis with resistance exercise is crucial to elucidating mechanisms regulating muscle hypertrophy."

Insulin has several roles relative to improving the net protein balance following resistance exercise, including increasing protein synthesis *(26–28)*, improving the transport of amino acids into skeletal muscle *(27,29,30)*, and decreasing protein breakdown *(30–33)*. Whereas insulin should never be injected (as multiple adverse events are likely to occur) for the purposes of improving net protein balance, insulin can be significantly increased endogenously via the consumption of carbohydrate. As important as insulin concentrations are to anabolic processes, Biolo and Wolfe *(34)* stated that if high levels of insulin are not supported by an exogenous amino acid supply, insulin loses its anabolic capacity in skeletal muscle. This observation has been shared by other investigators as well *(35,36)*.

Relative to protein synthesis, when essential amino acids were ingested after a bout of resistance exercise, the net protein balance was changed from a negative to a positive state *(37)*. Other clinical studies have also demonstrated that the oral ingestion of amino acids are responsible for increasing protein synthesis rates in multiple populations of participants *(38,39)*. Given the importance of insulin and amino acid availability relative to improving net protein balance, ingesting these nutrients simultaneously is recommended. To further this recommendation, by adding a protein source to carbohydrate ingestion it is possible to increase insulin to levels higher than those induced by carbohydrate ingestion alone.

5.3. Importance of Combined Carbohydrate–Protein
Supplements and Timing of Ingestion

Carbohydrate (to elevate insulin) and amino acids are needed to maximize positive shifts in net protein balance, and the time course for which they must be present should be considered. To highlight the importance of timing, note that when 10 g of protein, 8 g of carbohydrate, and 3 g of fat were ingested either immediately or 3 hours after exercise, protein synthesis was increased more than threefold with the supplement ingested immediately versus ingestion 3 hours after exercise (with which there was only a 12% increase) *(40)*. In a study by Rasmussen and coworkers *(41)*, subjects were given an amino acid–carbohydrate drink or a placebo following a resistance exercise session. Not surprisingly, the amino acid–carbohydrate drink elicited an anabolic response compared to the placebo. In another study of protein breakdown, Bird and colleagues *(42)* gave subjects one of four supplements after a bout of resistance exercise: 1) carbohydrate beverage; 2) essential amino acids; 3) combination of carbohydrate and amino acids; 4) placebo. The result of this nutritional intervention revealed that protein degradation (as measured by urinary 3-methylhistidine) was elevated at 24 and 48 hours after exercise in the placebo group. Relative to the carbohydrate and amino acid group, protein degradation was unchanged at 24 hours and actually decreased 48 hours after exercise. Given these findings and the data on the aforementioned studies, properly timed carbohydrate–protein/amino acid supplements not only increase protein synthesis but also seem to attenuate protein degradation. Most of the scientific investigations have looked at carbohydrate–protein supplements during the postresistance exercise period; however, one study looked at the difference of ingesting an amino acid–carbohydrate supplement before versus after resistance training *(43)*. The investigators reported that protein synthesis was greater as a result of the preresistance training intake of the amino acid–carbohydrate supplement, most likely due to increased delivery of amino acids to the stimulated skeletal muscle fibers *(43)*.

Most studies have examined the combination of amino acid–carbohydrate supplements in the time frame that encompasses a resistance training session, but not many have investigated intact protein (e.g., whey, casein) supplementation after resistance exercise and their effects on the net protein balance. Tipton et al. *(44)* studied the ingestion of casein and whey proteins and their effects on muscle anabolism after resistance exercise. They concluded that the ingestion of both proteins (whey and casein) after resistance exercise resulted in similar increases in muscle protein net balance, resulting in net muscle protein synthesis, despite different patterns of blood amino acid responses (a quicker response of blood amino acids for the whey protein and a more sustained response for the casein protein). In a similar study, Tipton and coworkers *(45)* questioned if ingestion of whole proteins before exercise would stimulate a superior response to that with ingestion after exercise. The authors reported that the net amino acid balance switched from negative to positive following ingestion of the whey proteins at both time points. In another study, when whey protein was added to an amino acid–carbohydrate supplement, the authors indicated that there seemed to be an extension of the anabolic effect compared to that seen with amino acid–carbohydrate supplements without additional whey protein *(46)*.

5.4. Summary

A proper postworkout supplement designed to increase lean body mass should contain both carbohydrates and protein and be in a liquid form. The reason these carbohydrate–protein supplements should be in liquid form is that liquid meals are more palatable and digestible. In addition, liquid meals have a fast absorption profile compared to that of whole foods, which allows faster insulin secretion and peak plasma amino acid levels—both of which are essential to take advantage of the anabolic window created by the resistance training session. This section has highlighted some of the clinical investigations and the mechanisms as to how appropriately timed ingestion of carbohydrate–protein supplements exert their effects. A more detailed explanation can be found by reading Chapter 13, on dietary meal and nutrient timing.

6. CREATINE

The sports supplement creatine has been the gold standard against which other nutritional supplements are compared. The reason for this prominent position is that creatine improves performance, increases lean body mass, and has repeatedly been shown to be safe when recommended dosages are consumed. Consequently, creatine has become one of the most popular nutritional supplements marketed to athletes over the past decade and a half. In fact, one of the most consistent side effects of creatine supplementation has been weight gain in the form of lean body mass. This increase has been observed in several cohorts including males, females, and the elderly (47–54).

In most of the studies published on creatine supplementation, the typical dosage pattern was divided into two phases: a loading phase and a maintenance phase. A typical loading phase consists of ingesting 20 g of creatine (or 0.3 g/kg body weight) in divided doses four times per day for 2 to 7 days, followed by a maintenance dose of 2 to 5 g daily (or 0.03 g/kg) for several weeks to months at a time (55). Another consideration relative to creatine dosage is to base the amount on an individual's lean body mass. Burke and coworkers (56) studied this aspect of creatine supplementation by having subjects ingest creatine at a dosage of 0.1 g/kg of lean body mass (this equates to approximately 8 g of creatine for a 200 pound individual at 15% body fat). Hultman and colleagues (57) demonstrated another interesting approach to creatine ingestion. They demonstrated that when creatine was ingested at 3 g/day over an extended training period of at least 4 weeks the skeletal muscle creatine levels rose more slowly, eventually reaching levels similar to those achieved with the loading method.

In summary, a quick way to "creatine load" skeletal muscle requires ingesting 20 g of creatine monohydrate daily for 6 days and then switching to a reduced dosage of 2 g/day (57). If the immediacy of "loading" is not an important consideration, supplementing with 3 g/day for 28 days achieves the same high levels of intramuscular creatine (57).

6.1. Effects on Lean Body Mass

What type of weight gain (in the form of lean body mass) can be expected with this level of creatine supplementation? Many of the studies performed to date indicate that short-term creatine supplementation increases total body mass by approximately 0.7 to 1.6 kg (~1.5–3.5 lb) (16). Longer-term creatine supplementation (~6–8 weeks) in conjunction with resistance training has been shown to increase lean body mass by approximately 2.8 to 3.2 kg (~7 lb) *(58–60)*. Gain in lean body mass has also been observed in women as a result of creatine supplementation. Vandenberghe et al. *(47)* investigated the changes in fat-free mass in females who ingested creatine (20 g/day for the first 4 days followed by 5 g/day for 65 days) in combination with resistance exercise for 10 weeks. The authors reported an increase of 5.7 lb of fat-free mass after 10 weeks of creatine supplementation and resistance exercise. This increase was 60% greater than in the creatine supplementation group compared to the placebo group.

6.2. Physiological Mechanisms for Increasing Lean Body Mass

The exact physiological mechanisms responsible for increasing lean body mass as a result of creatine supplementation remain poorly understood. Early studies investigating creatine supplementation and weight gain led many to the conclusion that increases in body weight were due to water retention. However, several more recent studies suggest that creatine supplementation may help build lean tissue. Volek et al. *(61)* reported that during a 12 week resistance training program, resistance trained males ingesting creatine significantly increased the fat-free mass compared to those ingesting a placebo. Furthermore, it was reported that the subjects given creatine demonstrated significantly greater increases in types I (35% vs. 11%), IIA (36% vs. 15%), and IIAB (35% vs. 6%) muscle fiber cross-sectional areas *(61)*. The percentage increases in cross-sectional area for all fiber types in those subjects ingesting creatine ranged from 29% to 35%—more than twice the increase observed in placebo subjects (6%–15%) *(16)*.

To help elucidate the physiological mechanisms further, Willoughby and Rosene *(62,63)* conducted a series of studies investigating the effects of oral creatine ingestion and the factors involved in gene expression of contractile filaments and myosin heavy-chain protein expression. In the first of these studies, untrained male subjects ingested creatine at 6 g/day or a placebo in conjunction with heavy resistance training for 12 weeks. At the end of the intervention, those ingesting creatine significantly increased their fat-free mass (~7 lb) in comparison with the placebo group (~1 lb). One of the most interesting parameters in this study was the information that was gathered relative to what was occurring at the cellular level of the skeletal muscle. Myofibrillar protein content (a marker of the amount of intracellular protein) was found to be significantly greater in the creatine group than in the placebo group despite the fact that both groups performed identical resistance training programs. More specifically, the authors reported that there were significant increases in the content of two isoforms of myosin heavy-chain protein (the major constituent of contractile skeletal muscle) *(62)*.

In their other study, Willoughby and Rosene *(63)* investigated the effects of ingesting creatine (in conjunction with a resistance training program) on myogenic regulatory factor gene expression. Myogenic regulatory factors (which include Myo-D, myogenin, MRF-4, and Myf5) are proteins that function as transcription activators that regulate gene expression via their binding to DNA, ultimately activating the transcription of muscle-specific genes such as myosin heavy chain, myosin light chain, α-actin, troponin-I, and creatine kinase *(64)*. After 12 weeks of resistance training, the authors reported that the subjects ingesting creatine had significantly greater mRNA expression for myogenin and MRF-4 than the subjects ingesting a placebo. These findings provide an insight into the mechanisms by which creatine supplementation exerts its effects on increasing lean body mass. Taken together, the aforementioned studies seem to indicate that the increases in lean body mass as a result of creatine supplementation are due to augmenting skeletal muscle fiber hypertrophy and not solely water retention.

6.3. Satellite Cell Activity

In addition to increasing muscle fiber cross-sectional areas, myogenic regulatory factors, and specific isoforms of myosin heavy chain, creatine supplementation has been shown to augment an increase in satellite cell number in human skeletal muscle induced by strength training. In addition to muscle-specific transcription and translation, activation of satellite cells is thought to be a major contributing factor to augmenting skeletal muscle hypertrophy. During the process of load-induced muscle hypertrophy, satellite cells are thought to proliferate, differentiate, and then fuse with existing myofibers *(65)*. The way in which satellite cells are thought to be involved in skeletal muscle hypertrophy is summarized in what is termed the *myonuclear domain theory*. This theory suggests that the myonucleus controls the production of mRNA (i.e., transcription) and proteins (i.e., translation) for a finite volume of cytoplasm, such that increases in fiber size must be associated with a proportional increase in myonuclei, which are contributed from the satellite cell populations (66). If this theory is correct, anything that increases satellite cell activity leading to increases in myonuclei sets the stage for increased skeletal muscle hypertrophy.

In a truly original investigation, Olsen and coworkers (67) investigated the influence of creatine and protein supplementation on satellite cell frequency and the number of myonuclei in human skeletal muscle during 16 weeks of resistance training. After the 16 weeks of training, all groups in the clinical trial (creatine, protein, and placebo groups) demonstrated significant increases in the proportion of satellite cells. However, only the creatine-supplemented group demonstrated consistent significant increases of myonuclei per fiber. This finding led the authors to conclude that "creatine supplementation in combination with strength training amplifies the training-induced increase in satellite cell number and myonuclei concentration in human skeletal muscle fibers, thereby allowing an enhanced muscle fiber growth in response to strength training" (67). Given this important finding relative to creatine supplementation and satellite cell activity, additional clinical trials investigating this aspect of creatine supplementation are needed.

7. ANABOLIC HORMONE ENHANCERS

Insulin, growth hormone, testosterone, and insulin-like growth factor-1 (IGF-1) are all considered primary anabolic hormones. We have already discussed insulin and its role in translating muscle-specific mRNA into skeletal muscle proteins, and the effects that carbohydrate–amino acid supplements have on increasing insulin levels. The other three anabolic hormones are believed to exert their effects on the cell-signaling properties of skeletal muscle fibers, which ultimately result in muscle-specific gene expression. IGF-1, however, not only acts in this regard (cell signaling) but also acts similarly to insulin in its role of translating muscle-specific mRNA transcripts into functional skeletal muscle proteins (actin, myosin). A further discussion of IGF-1, growth hormone, and testosterone follows.

7.1. Insulin-Like Growth Factor-1

There are three isoforms of IGF-1 in human muscle *(68)*: IGF-1Ea (similar to the type of IGF-1 synthesized in the liver); IGF-1 Eb; and IGF-1Ec (known as mechano growth factor) *(68)*. IGF-1 is produced primarily by the liver as an endocrine hormone and is stimulated by growth hormone release. One of the isoforms of IGF-1, known as mechano growth factor, is detectable only after mechanical stimulation (e.g., resistance training). Skeletal muscle hypertrophy is regulated by at least three major molecular processes: 1) satellite cell activity; 2) gene transcription; and 3) protein translation. Interestingly, IGF-I can influence the activity of all of these mechanisms *(69)*. That being the case, any increases in IGF-1 could significantly increase the potential for skeletal muscle hypertrophy.

In addition to growth hormone release and mechanical stimulation, are there any nutritional or supplemental means that increase endogenous levels of IGF-1? For the most part, the answer is no, but two studies have reported that supplementation with bovine colostrum resulted in increases in serum IGF-I concentration in athletes during training *(70,71)*. However, owing to the relatively acute duration of these studies, lean body mass indices were not measured. Another study that did measure muscle protein balance and strength after 2 weeks of bovine colostrum supplementation *(72)* found that the bovine colostrum had no effect on either of these variables. Given these findings, at this point it is safe to say that there are no sports supplements that effectively increase endogenous IGF-1 levels resulting in changes in lean body mass.

7.2. Growth Hormone

A quick survey of the literature on growth hormone reveals that the hormone does indeed improve body composition by simultaneously increasing lean body mass and decreasing body fat in diseased populations *(73–75)*. However, what is often not mentioned in the marketing campaigns of sports supplements designed to increase growth hormone is the fact that most of these clinical investigations introduced growth hormone into their subjects via subcutaneous injection. Many sports supplements designed to increase endogenous levels of growth hormone are based on studies showing that specific amino acids are able (inconsistently) to increase growth hormone. The main amino acid that has demonstrated potential to increase growth hormone is arginine. As discussed below, arginine is often combined with other compounds to elicit growth hormone release.

It is well documented that the infusion of arginine stimulates growth hormone secretion from the anterior pituitary *(76,77)*. This increase in growth hormone secretion from arginine infusion has been attributed to the suppression of endogenous somatostatin secretion *(76)*. The amounts of arginine infused to elicit the growth hormone response ranged from 12 to 30 g. The clinical investigations observing oral consumption of arginine and its impact on growth hormone release are equivocal. Relative to the practical oral ingestion of arginine, several studies have shown that such supplementation resulted in significant increases in growth hormone secretion.

One such study *(78)* found that oral arginine supplementation of 5 and 9 g resulted in significant growth hormone response in males. Interestingly, 13 g of oral arginine did not increase growth hormone levels and caused gastrointestinal distress in most of the subjects. A common supplemental regimen that has shown promise as a growth hormone enhancer includes the addition of lysine to arginine. Utilizing this combination, Isidori and colleagues *(79)* provided 1.2 g of arginine (as arginine-2-pyrrolidone-5-carboxylate) and 1.2 g of lysine (as lysine hydrochloride) to young males. Plasma growth hormone concentrations increased eightfold at 90 minutes after ingestion. Similarly, Suminski and associates *(80)* reported that the ingestion of arginine and lysine resulted in a 2.7-fold increase in plasma growth hormone concentrations in resistance trained males.

Another compound commonly added to arginine for the purpose of eliciting an increase in growth hormone is aspartate. Besset and colleagues *(81)* gave male subjects arginine aspartate at a dose of 250 mg/kg/day (approximately 17.5 g of arginine aspartate for a 70 kg male) for 1 week. The results indicated that the sleep-related growth hormone peak was about 60% higher after a week of arginine aspartate administration than in the controls. Colombani et al. *(82)* gave 20 male endurance trained athletes 15 g of arginine aspartate (7.5 g in the morning and 7.5 g in the evening) for 14 days before a marathon run. After 31 km had been completed by the runners, plasma growth hormone levels were 40% greater in the arginine aspartate group. At the end of the marathon, plasma growth hormone levels were 8% greater in the supplemented group than in the placebo group.

Not all studies investigating arginine supplementation and growth hormone responses have been favorable. Walberg-Rankin *(83)* gave resistance-trained males ingesting a hypocaloric diet arginine hydrochloride 100 mg/kg/day (approximately 8 g arginine hydrochloride) for a 10-day period. This supplementation protocol did not result in an increase in growth hormone concentration. Another study also reported no increase in plasma growth hormone concentrations when elderly men ingested 3 g of arginine and 3 g lysine for 14 days *(84)*. In yet another study investigating arginine and growth hormone responses, Marcell et al. *(85)* investigated whether oral arginine (5 g) increases growth hormone secretion in young and old people (male and female) at rest and during resistive exercise. The authors concluded that oral arginine supplementation does not increase growth hormone secretion at rest or in combination with resistive exercise.

There are several reasons for the conflicting results in terms of arginine eliciting an increase in growth hormone production. Some of these reasons could be the type of arginine complex, dosages, and delivery methods used and variations in the subjects themselves. It has also been suggested that the growth hormone response to amino acid ingestion may be reduced in individuals who are exercise trained *(55)*.

Even if certain amino acids do increase growth hormone levels (a statement not supported by all investigations), it does not necessarily lead to the conclusion that they increase lean body mass. In a scientific review on this subject, Chromiak and Antonio *(55)* stated: "There is no evidence based on properly conducted, rigorous scientific studies that oral supplementation of specific amino acids induces growth hormone that, in conjunction with resistance training, increases muscle mass and strength to a greater extent than resistance training alone." At this point, it appears as if specific amino acids, even if they do elicit an increase in growth hormone, do not increase lean body mass via this mechanism.

7.3. Testosterone

Although each of the anabolic hormones (testosterone, growth hormone, insulin, IGF-1) is required to stimulate maximum levels of skeletal muscle hypertrophy, testosterone may be the most anabolic. It is important to recognize that not all of the testosterone in the blood is bioavailable; rather, most of it is bound to proteins such as sex hormone-binding globulin (SHBG) or other carrier proteins. Testosterone that is not bound is referred to as "free" or "bioavailable" testosterone; and it is able to bind to the androgen receptor and exert its anabolic signaling. This is an important distinction because as one attempts to increase testosterone levels (via testosterone-enhancing supplements) in the body, it is only the bioavailable testosterone that exerts anabolic actions. Another important consideration is the avoidance of increasing SHBG to a greater extent than total testosterone increases, which would result in an environment in which there is less bioavailable testosterone present. Therefore, when investigating sports supplements designed to increase testosterone, each of these factors must be considered. Currently, there are a few sports supplements that claim to increase testosterone levels: ZMA, *Tribulus terrestris*, and aromatase inhibitors.

7.3.1. ZMA

The primary ingredients in ZMA supplements are zinc monomethionine aspartate, magnesium aspartate, and vitamin B_6. Zinc and magnesium deficiencies as well as urine and sweat losses of these minerals have been observed in athletes and individuals who are physically active *(86–90)*. Relative to testosterone, there have been two well designed studies investigating the effects of ZMA supplementation and its effects on testosterone levels, with the studies reporting contradictory results *(91,92)*.

The first of these studies gave collegiate football players ZMA (30 mg zinc monomethionine aspartate + 450 mg magnesium aspartate + 10.5 mg of vitamin B_6) over the course of their spring practice season (approximately 8 weeks) *(91)*. Total testosterone and, more importantly, free testosterone were significantly elevated as a result of the ZMA supplementation compared to that of the placebo group. This study is consistently cited as proof of the effectiveness of ZMA to elevate testosterone levels. In the other study *(92)*, researchers gave resistance trained males a ZMA supplement (main ingredients consisting of 30 mg zinc monomethionine aspartate + 450 mg magnesium aspartate + and 11 mg of vitamin B_6) and found no such increases in either total or free testosterone. This investigation *(92)* also assessed changes in the fat-free mass and several strength and performance variables. No significant differences were observed in relation to these variables in subjects taking ZMA. The discrepancies concerning these two studies may be explained by deficiencies of these minerals. Given the role that zinc deficiency plays relative to androgen metabolism and interaction with steroid receptors *(93)*, when there are deficiencies of this mineral, testosterone production may suffer. In the study showing increases in testosterone levels *(91)*, there were observed depletions of both zinc and magnesium in the placebo group over the course of the study. Therefore, the increased testosterone levels could have been attributed to impaired nutritional status rather than a pharmacological effect. Obviously, more research is needed on supplemental ZMA before any concrete recommendations can be made relative to testosterone responses.

7.3.2. TRIBULUS TERRESTRIS

Tribulus terrestris is often marketed as a testosterone-boosting sports supplement. There are relatively few, if any, scientific studies to substantiate these claims. In fact, one clinical investigation demonstrated that *Tribulus terrestris* exerts no effect on increasing testosterone levels *(94)*. In this study, healthy men were instructed to supplement with *Tribulus terrestris* for a 4-week period after which serum levels of testosterone and luteinizing hormone were measured at 1, 3, 10, 17, and 24 days after supplementation. *Tribulus terrestris* supplementation did not increase the levels of either testosterone or luteinizing hormone. Given the unsubstantiated claims of *Tribulus terrestris* relative to increasing testosterone levels, supplemental *Tribulus* is not recommended.

7.3.3. AROMATASE INHIBITORS

Aromatase inhibitors exert their effects by inhibiting the action of the enzyme aromatase, which converts androgens to estrogens by a process called aromatization. Aromatase inhibitor supplements claim to suppress estrogen levels and increase endogenous testosterone levels. In the only published study investigating aromatase inhibitor supplements, Willoughby and colleagues *(95)* instructed their male subjects to ingest an aromatase inhibitor supplement (containing hydroxyandrost-4-ene-6,17 dioxo-3-THP ether and 3,17-diketo-androst-1,4,6-triene) at 72 mg/day for an 8-week period. At the end of the 8 weeks there was a 3-week washout period. Multiple anabolic hormones were assayed during the duration of the study, including total testosterone and free testosterone. In addition, body composition was assessed during the investigation in which the participants were instructed to maintain their normal resistance training programs. There were significant increases in both total and free testosterone levels compared to those with placebos, with the total testosterone having an average increase of 283% and free testosterone an average of 625%. The aromatase inhibitor supplement had also elicited a 3.5% decrease in fat mass in the aromatase inhibitor group at the end of the 8-week period. After the 3-week washout period, total and free testosterone levels decreased to the presupplementation values. Finally, the aforementioned supplementation appeared to be safe and well tolerated by the study participants as measured by blood and urinary clinical safety markers. Although this study appears to support aromatase inhibitor supplementation for the purpose of increasing endogenous testosterone levels, additional studies are needed to replicate these findings. In summary, it appears that of all the nutritional supplements designed to increase testosterone levels aromatase inhibitor supplementation is the most scientifically valid option.

8. ANTICATABOLIC SUPPLEMENTS

Because the net protein balance is equal to muscle protein synthesis minus muscle protein breakdown, eliciting increases in lean body mass can be achieved not only by increasing protein synthesis but also by decreasing protein breakdown (catabolism). Hence, a number of sports supplements are marketed for that endeavor, including glutamine, cortisol inhibitors, β-hydroxyl-β-methylbutyrate (HMB), and α-ketoisocaproic acid. In addition to these specific sports supplements, insulin has been shown repeatedly to suppress protein breakdown *(30,31,96,97)*. Hence, carbohydrate (or carbohydrate + protein) taken after resistance exercise (a period when protein breakdown is elevated) for the purpose of increasing insulin secretion is a recommended practice to suppress protein breakdown *(42)*. Other purported anticatabolic supplements are discussed below.

8.1. α-Ketoisocaproic Acid

α-Ketoisocaproic acid (KIC) is the keto acid of the BCAA leucine. Despite many claims of KIC and its anticatabolic properties, there is only one peer-reviewed study in humans that has investigated the inclusion of KIC (along with HMB) on a specific marker of muscle damage (creatine kinase). When non-resistance-trained males ingested 3 g of HMB and 0.3 g of KIC daily for 14 days prior to a resistance training session, it was reported that the HMB–KIC supplementation attenuated the creatine kinase response compared to that seen with the placebo *(98)*. Although this may be an important finding, creatine kinase is not a direct measure of protein breakdown. Also, the extent to which HMB or KIC alone affected this attenuation of creatine kinase cannot be determined. Another study that is commonly cited as evidence for KIC supplementation and its ability to prevent proteolysis was conducted on isolated rat diaphragm skeletal muscle *(99)*. In this venue, one should be cautious of overextrapolating from rodent data to the human condition. At this point, there are not enough data to conclude that KIC supplementation alone is an effective anticatabolic supplement.

8.2. β-Hydroxy-β-Methylbutyrate

β-Hydroxy-β-methylbutyrate is a metabolite of the BCAA leucine and is often associated with anticatabolic potential. The original research study to highlight HMB's anticatabolic potential was conducted by Nissen and coworkers *(100)*. In this study, untrained subjects ingested one of three levels of HMB (0, 1.5, or 3.0 g/day) and two protein levels (117 or 175 g/day) and resistance trained 3 days per week for 3 weeks. Other markers of muscle damage were assessed, and protein breakdown was assessed by measuring urinary 3-methylhistidine (3-MH). After the first week of the resistance training protocol, urinary 3-MH was increased by 94% in the control group and by 85% and 50% in individuals ingesting 1.5 and 3.0 g of HMB per day, respectively. During the second week of the study, 3-MH levels were still elevated by 27% in the control group but were 4% and 15% below basal levels for the groups on HMB 1.5 and 3.0 g/day. Interestingly, 3-MH measures at the end of the third week of resistance training were not significantly different among the groups *(100)*. Other studies demonstrating an anticatabolic effect or suppressing muscle damage have supported the finding of this study *(98,101)*. A study conducted by van Someren and coworkers *(98)* instructed their male subjects to ingest 3 g of HMB in addition to 0.3 g KIC daily for 14 days prior to performing a single bout of eccentrically biased resistance exercise. This supplemental intervention that included HMB resulted in a significant reduction in the plasma markers of muscle damage.

Although HMB supplementation may suppress protein breakdown and markers of muscle damage, the main question is if this anticatabolic effect leads to gains in lean body mass. The scientific literature on this topic is divided. In a second arm to the study conducted by Nissen and colleagues *(100)*, male subjects ingested 3 g of HMB or a placebo for 7 weeks in conjunction with resistance training 6 days per week. In this study, the fat-free mass increased in the HMB-supplemented group at various times throughout the investigative period but not at the conclusion of the study. Other studies have also reported evidence for HMB supplementation (~3 g/day) relative to increasing lean body mass *(102,103)*. In addition, a meta-analysis conducted by Nissen and Sharp *(104)* stated that only HMB and one other sports supplement (creatine) were found to increase lean body mass significantly.

Not all studies agree with the findings that HMB increases lean body mass, however *(105–107)*. Each of the studies showing no effect of HMB on lean body mass accretion also supplemented their subjects with approximately the same amount of HMB as the studies that demonstrated increases in lean body mass. Although not conclusive, it appears that HMB supplementation does suppress protein breakdown, ultimately leading to increased lean body mass in some individuals. Following carbohydrate supplementation (for the purpose of secreting insulin), HMB is the next best anticatabolic sports supplement.

8.3. *Glutamine*

Another sports supplement commonly marketed as an anticatabolic agent is the amino acid glutamine. Glutamine is the most abundant free amino acid in plasma and skeletal muscle and accounts for more than half of the total intramuscular free amino acid pool *(108)*. The rationale for glutamine's anticatabolic effects is the fact that it is one of the major fuels used by the gut, resulting in a high cellular turnover of glutamine in the gut (intestinal mucosal cells). This high turnover may result in the supply of amino acids (glutamine) to the cells of the gastrointestinal tract at the expense of skeletal muscle protein. By providing supplemental glutamine to the gut, it theoretically spares the glutamine that is available in skeletal muscle and in this way serves as an anticatabolic agent. One clinical investigation that gave supplemental glutamine to individuals engaging in resistance exercise did not demonstrate an anticatabolic potential or result in increases in lean body mass *(109)*. Other studies that have demonstrated anticatabolic potential for glutamine have used critically ill subjects or subjects who underwent surgery *(110, 111)*. Despite a valid theoretical rationale for glutamine supplementation, at this point there are no scientific data demonstrating that glutamine supplementation suppresses protein breakdown in resistance trained individuals.

9. NITRIC OXIDE BOOSTERS

One of the more recent developments in sports supplements has been the introduction of supplements intended to increase nitric oxide production. Relative to biological processes in humans, nitric oxide is synthesized in cells by nitric oxide synthase. One of nitric oxide's primary physiological functions is to relax smooth muscle, and hence it is one of the body's major regulators of blood flow, especially during exercise. Kingwell *(112)* indicated that nitric oxide potentially affects metabolic control during exercise via multiple mechanisms, including:

- Elevation in skeletal muscle and cardiac blood flow and increased delivery of oxygen, substrates, and regulatory hormones (e.g., insulin)
- Preservation of intracellular skeletal muscle energy stores by promoting glucose uptake, inhibiting glycolysis, mitochondrial respiration, and phosphocreatine breakdown

Together, these actions of nitric oxide on blood flow and substrate utilization appear to be directed toward protection from ischemia *(112)*. Also, if adequate amounts of oxygen and substrate are supplied to the skeletal muscle undergoing mechanical stress, the possibility of extending the total workload on each set of a resistance training bout may lead to greater stimulus for muscle fiber hypertrophy. The aforementioned observation, in conjunction with some earlier studies showing nitric oxide as an integral compound relative to improving skeletal muscle's force production and maximal power output *(113,114)*, has provided a rationale for investigating ways to increase endogenous nitric oxide production.

In every sports supplement claiming to augment endogenous nitric oxide production, the amino acid arginine is included in the list of ingredients. This is due to the fact that arginine serves as a precursor for the biosynthesis of nitric oxide *(115)*. In fact, arginine is the only endogenous nitrogen-containing substrate of nitric oxide synthase and thus governs production of nitric oxide. To date, only one study has investigated the effects of an arginine-containing sports supplement aiming at augmenting endogenous nitric oxide production *(116)*. Although this study was well designed, it should be noted that nitric oxide production was not assessed in the clinical investigation. The study investigated the effects of ingesting 12 g of arginine α-ketoglutarate in conjunction with a periodized resistance training program over a period of 8 weeks. Although there was some improvement in some exercise performance variables, the investigators did not observe any increases in lean body mass. Specifically, those ingesting arginine α-ketoglutarate had a significant increase in upper body strength (as measured by the bench press) compared to the subjects ingesting a placebo. In fact, those ingesting the arginine α-ketoglutarate increased their bench press by 19 lb versus an increase of approximately 6 lb in the placebo group. Additionally, the arginine α-ketoglutarate group significantly increased their peak power output (as measured by a 30-second cycle sprint test) in comparison to the placebo group. Whether these improvements in exercise performance can be associated with increases in nitric oxide production is not known, but the fact remains that there were no increases relative to lean body mass observed in this investigation. At this point, there is a theoretical rationale that augmenting nitric oxide production may lead to more intense training and ultimately greater skeletal muscle hypertrophy. However, because clinical investigations have not demonstrated this, it is premature to state emphatically that sports supplements designed to increase nitric oxide lead to greater gains in lean body mass.

10. CONCLUSION

Increasing lean body mass is a goal of many athletes, recreational weight trainers, and those who wish to improve their body composition. When choosing a dietary supplement to augment increases in lean body mass, it is important to consider the way in which the supplement contributes to the highly regulated process of skeletal muscle hypertrophy. The science of sports supplements is relatively new, although certain sports supplements (protein, creatine) have been scientifically investigated and have repeatedly demonstrated their ability to increase lean body mass. Other sports supplements (e.g., anticatabolic agents, anabolic hormone enhancers, nitric oxide boosters) require more rigorous scientific investigation before they can be deemed effective (or not).

REFERENCES

1. Biolo G, Maggi SP, Williams BD, Tipton KD, Wolfe RR. Increased rates of muscle protein turnover and amino acid transport after resistance exercise in humans. Am J Physiol 1995;268(Pt 1):E514–E20.
2. Phillips SM, Tipton KD, Aarsland A, Wolf SE, Wolfe RR. Mixed muscle protein synthesis and breakdown after resistance exercise in humans. Am J Physiol 1997;273(Pt 1):E99–E107.
3. Phillips SM, Tipton KD, Ferrando AA, Wolfe RR. Resistance training reduces the acute exercise-induced increase in muscle protein turnover. Am J Physiol 1999;276(Pt 1):E118–E124.
4. Wagenmakers AJ. Tracers to investigate protein and amino acid metabolism in human subjects. Proc Nutr Soc 1999;58:987–1000.
5. Trumbo P, Schlicker S, Yates AA, Poos M. Dietary reference intakes for energy, carbohydrate, fiber, fat, fatty acids, cholesterol, protein and amino acids. J Am Diet Assoc 2002;102:1621–1630.
6. American College of Sports Medicine, American Dietetic Association, and Dietitians of Canada. Joint position statement: nutrition and athletic performance. Med Sci Sports Exerc 2000;32:2130–2145.
7. Tarnopolsky M. Protein requirements for endurance athletes. Nutrition 2004;20:662–668.

8. Forslund AH, El-Khoury AE, Olsson RM, Sjodin AM, Hambraeus L, Young VR. Effect of protein intake and physical activity on 24-h pattern and rate of macronutrient utilization. Am J Physiol 1999;276(Pt 1):E964–E976.

9. Meredith CN, Zackin MJ, Frontera WR, Evans WJ. Dietary protein requirements and body protein metabolism in endurance-trained men. J Appl Physiol 1989;66:2850–2856.

10. Phillips SM, Atkinson SA, Tarnopolsky MA, MacDougall JD. Gender differences in leucine kinetics and nitrogen balance in endurance athletes. J Appl Physiol 1993;75:2134–2141.

11. Lamont LS, Patel DG, Kalhan SC. Leucine kinetics in endurance-trained humans. J Appl Physiol 1990;69:1–6.

12. Friedman JE, Lemon PW. Effect of chronic endurance exercise on retention of dietary protein. Int J Sports Med 1989;10:118–123.

13. Tarnopolsky MA, Atkinson SA, MacDougall JD, Chesley A, Phillips S, Schwarcz HP. Evaluation of protein requirements for trained strength athletes. J Appl Physiol 1992;73:1986–1995.

14. Lemon PW, Tarnopolsky MA, MacDougall JD, Atkinson SA. Protein requirements and muscle mass/strength changes during intensive training in novice bodybuilders. J Appl Physiol 1992;73:767–775.

15. Lemon PW. Protein and amino acid needs of the strength athlete. Int J Sport Nutr 1991;1:127–145.

16. Antonio J, Stout J (eds). Sports Supplements. Lippincott Williams & Wilkins, Philadelphia, 2001.

17. Boirie Y, Dangin M, Gachon P, Vasson MP, Maubois JL, Beaufrere B. Slow and fast dietary proteins differently modulate postprandial protein accretion. Proc Natl Acad Sci U S A 1997;94:14930–14935.

18. Fruhbeck G. Protein metabolism: slow and fast dietary proteins. Nature 1998;391:843, 845.

19. Dangin M, Boirie Y, Garcia-Rodenas C, et al. The digestion rate of protein is an independent regulating factor of postprandial protein retention. Am J Physiol Endocrinol Metab 2001;280:E340–E348.

20. Driskell J, Wolinsky I (eds) Energy-Yielding Macronutrients and Energy Metabolism in Sports Nutrition. CRC Press, Boca Raton, FL, 2000.

21. Demling RH, DeSanti L. Effect of a hypocaloric diet, increased protein intake and resistance training on lean mass gains and fat mass loss in overweight police officers. Ann Nutr Metab 2000;44:21–29.

22. Kerksick CM, Rasmussen CJ, Lancaster SL, et al. The effects of protein and amino acid supplementation on performance and training adaptations during ten weeks of resistance training. J Strength Cond Res 2006;20:643–653.

23. Nikawa T, Ikemoto M, Sakai T, et al. Effects of a soy protein diet on exercise-induced muscle protein catabolism in rats. Nutrition 2002;18:490–495.

24. Wolfe RR. Effects of amino acid intake on anabolic processes. Can J Appl Physiol 2001;26(Suppl):S220–S227.

25. Bolster DR, Jefferson LS, Kimball SR. Regulation of protein synthesis associated with skeletal muscle hypertrophy by insulin-, amino acid- and exercise-induced signalling. Proc Nutr Soc 2004;63:351–356.

26. Kimball SR, Jurasinski CV, Lawrence JC Jr, Jefferson LS. Insulin stimulates protein synthesis in skeletal muscle by enhancing the association of eIF-4E and eIF-4 G. Am J Physiol 1997;272(Pt 1):C754–C759.

27. Biolo G, Declan Fleming RY, Wolfe RR. Physiologic hyperinsulinemia stimulates protein synthesis and enhances transport of selected amino acids in human skeletal muscle. J Clin Invest 1995;95:811–819.

28. Hillier TA, Fryburg DA, Jahn LA, Barrett EJ. Extreme hyperinsulinemia unmasks insulin's effect to stimulate protein synthesis in the human forearm. Am J Physiol 1998;274(Pt 1):E1067–E1074.

29. Gore DC, Wolf SE, Sanford AP, Herndon DN, Wolfe RR. Extremity hyperinsulinemia stimulates muscle protein synthesis in severely injured patients. Am J Physiol Endocrinol Metab 2004;286:E529–E534.

30. Biolo G, Williams BD, Fleming RY, Wolfe RR. Insulin action on muscle protein kinetics and amino acid transport during recovery after resistance exercise. Diabetes 1999;48:949–957.

31. Gelfand RA, Barrett EJ. Effect of physiologic hyperinsulinemia on skeletal muscle protein synthesis and breakdown in man. J Clin Invest 1987;80:1–6.

32. Heslin MJ, Newman E, Wolf RF, Pisters PW, Brennan MF. Effect of hyperinsulinemia on whole body and skeletal muscle leucine carbon kinetics in humans. Am J Physiol 1992;262(Pt 1):E911–E918.

33. Denne SC, Liechty EA, Liu YM, Brechtel G, Baron AD. Proteolysis in skeletal muscle and whole body in response to euglycemic hyperinsulinemia in normal adults. Am J Physiol 1991;261(Pt 1):E809–E814.

34. Biolo G, Wolfe RR. Insulin action on protein metabolism. Baillieres Clin Endocrinol Metab 1993;7:989–1005.

35. Bell JA, Fujita S, Volpi E, Cadenas JG, Rasmussen BB. Short-term insulin and nutritional energy provision do not stimulate muscle protein synthesis if blood amino acid availability decreases. Am J Physiol Endocrinol Metab 2005;289:E999–E1006.

36. Fujita S, Rasmussen BB, Cadenas JG, Grady JJ, Volpi E. Effect of insulin on human skeletal muscle protein synthesis is modulated by insulin-induced changes in muscle blood flow and amino acid availability. Am J Physiol Endocrinol Metab 2006;291:E745–E754.

37. Tipton KD, Ferrando AA, Phillips SM, Doyle D Jr, Wolfe RR. Postexercise net protein synthesis in human muscle from orally administered amino acids. Am J Physiol 1999;276(Pt 1):E628–E634.

38. Paddon-Jones D, Sheffield-Moore M, Zhang XJ, et al. Amino acid ingestion improves muscle protein synthesis in the young and elderly. Am J Physiol Endocrinol Metab 2004;286:E321–E328.

39. Volpi E, Kobayashi H, Sheffield-Moore M, Mittendorfer B, Wolfe RR. Essential amino acids are primarily responsible for the amino acid stimulation of muscle protein anabolism in healthy elderly adults. Am J Clin Nutr 2003;78:250–258.

40. Levenhagen DK, Gresham JD, Carlson MG, Maron DJ, Borel MJ, Flakoll PJ. Postexercise nutrient intake timing in humans is critical to recovery of leg glucose and protein homeostasis. Am J Physiol Endocrinol Metab 2001;280:E982–E993.

41. Rasmussen BB, Tipton KD, Miller SL, Wolf SE, Wolfe RR. An oral essential amino acid-carbohydrate supplement enhances muscle protein anabolism after resistance exercise. J Appl Physiol 2000;88:386–392.
42. Bird SP, Tarpenning KM, Marino FE. Liquid carbohydrate/essential amino acid ingestion during a short-term bout of resistance exercise suppresses myofibrillar protein degradation. Metabolism 2006;55:570–577.
43. Tipton KD, Rasmussen BB, Miller SL, et al. Timing of amino acid-carbohydrate ingestion alters anabolic response of muscle to resistance exercise. Am J Physiol Endocrinol Metab 2001;281:E197–E206.
44. Tipton KD, Elliott TA, Cree MG, Wolf SE, Sanford AP, Wolfe RR. Ingestion of casein and whey proteins result in muscle anabolism after resistance exercise. Med Sci Sports Exerc 2004;36:2073–2081.
45. Tipton KD, Elliott TA, Cree MG, Aarsland AA, Sanford AP, Wolfe RR. Stimulation of net muscle protein synthesis by whey protein ingestion before and after exercise. Am J Physiol Endocrinol Metab 2007;292:E71–E76.
46. Borsheim E, Aarsland A, Wolfe RR. Effect of an amino acid, protein, and carbohydrate mixture on net muscle protein balance after resistance exercise. Int J Sport Nutr Exerc Metab 2004;14:255–271.
47. Vandenberghe K, Goris M, Van Hecke P, Van Leemputte M, Vangerven L, Hespel P. Long-term creatine intake is beneficial to muscle performance during resistance training. J Appl Physiol 1997;83:2055–2063.
48. Kelly V, Jenkins D. Effect of oral creatine supplementation on near-maximal strength and repeated sets of high intensity bench press exercise. J Strength Cond Res 1998;12:109–115.
49. Van Loon LJ, Oosterlaar AM, Hartgens F, Hesselink MK, Snow RJ, Wagenmakers AJ. Effects of creatine loading and prolonged creatine supplementation on body composition, fuel selection, sprint and endurance performance in humans. Clin Sci (Lond) 2003;104:153–162.
50. Kreider RB, Ferreira M, Wilson M, et al. Effects of creatine supplementation on body composition, strength, and sprint performance. Med Sci Sports Exerc 1998;30:73–82.
51. Branch JD. Effect of creatine supplementation on body composition and performance: a meta-analysis. Int J Sport Nutr Exerc Metab 2003;13:198–226.
52. Gotshalk LA, Volek JS, Staron RS, Denegar CR, Hagerman FC, Kraemer WJ. Creatine supplementation improves muscular performance in older men. Med Sci Sports Exerc 2002;34:537–543.
53. Brose A, Parise G, Tarnopolsky MA. Creatine supplementation enhances isometric strength and body composition improvements following strength exercise training in older adults. J Gerontol A Biol Sci Med Sci 2003; 58:11–19.
54. Chrusch MJ, Chilibeck PD, Chad KE, Davison KS, Burke DG. Creatine supplementation combined with resistance training in older men. Med Sci Sports Exerc 2001;33:2111–2117.
55. Chromiak JA, Antonio J. Use of amino acids as growth hormone-releasing agents by athletes. Nutrition 2002;18:657–661.

56. Burke DG, Smith-Palmer T, Holt LE, Head B, Chilibeck PD. The effect of 7 days of creatine supplementation on 24-hour urinary creatine excretion. J Strength Cond Res 2001;15:59–62.

57. Hultman E, Soderlund K, Timmons JA, Cederblad G, Greenhaff PL. Muscle creatine loading in men. J Appl Physiol 1996;81:232–237.

58. Stout J, Eckerson J, Noonan D. Effects of 8 weeks of creatine supplementation on exercise performance and fat-free weight in football players during training. Nutr Res 1999;19:217–225.

59. Earnest CP, Snell PG, Rodriguez R, Almada AL, Mitchell TL. The effect of creatine monohydrate ingestion on anaerobic power indices, muscular strength and body composition. Acta Physiol Scand 1995;153:207–209.

60. Kreider RB, Klesges R, Harmon K, et al. Effects of ingesting supplements designed to promote lean tissue accretion on body composition during resistance training. Int J Sport Nutr 1996;6:234–246.

61. Volek JS, Duncan ND, Mazzetti SA, et al. Performance and muscle fiber adaptations to creatine supplementation and heavy resistance training. Med Sci Sports Exerc 1999;31:1147–1156.

62. Willoughby DS, Rosene J. Effects of oral creatine and resistance training on myosin heavy chain expression. Med Sci Sports Exerc 2001;33:1674–1681.

63. Willoughby DS, Rosene JM. Effects of oral creatine and resistance training on myogenic regulatory factor expression. Med Sci Sports Exerc 2003;35:923–929.

64. Lowe DA, Lund T, Alway SE. Hypertrophy-stimulated myogenic regulatory factor mRNA increases are attenuated in fast muscle of aged quails. Am J Physiol 1998;275(Pt 1):C155–C162.

65. Schultz E, McCormick KM. Skeletal muscle satellite cells. Rev Physiol Biochem Pharmacol 1994;123:213–257.

66. Hawke TJ. Muscle stem cells and exercise training. Exerc Sport Sci Rev 2005;33:63–68.

67. Olsen S, Aagaard P, Kadi F, et al. Creatine supplementation augments the increase in satellite cell and myonuclei number in human skeletal muscle induced by strength training. J Physiol 2006;573(Pt 2):525–534.

68. Hameed M, Orrell RW, Cobbold M, Goldspink G, Harridge SD. Expression of IGF-I splice variants in young and old human skeletal muscle after high resistance exercise. J Physiol 2003;547(Pt 1):247–254.

69. Spangenburg EE. IGF-I isoforms and ageing skeletal muscle: an 'unresponsive' hypertrophy agent? J Physiol 2003;547(Pt 1):2.

70. Mero A, Miikkulainen H, Riski J, Pakkanen R, Aalto J, Takala T. Effects of bovine colostrum supplementation on serum IGF-I, IgG, hormone, and saliva IgA during training. J Appl Physiol 1997;83:1144–1151.

71. Mero A, Kahkonen J, Nykanen T, et al. IGF-I, IgA, and IgG responses to bovine colostrum supplementation during training. J Appl Physiol 2002;93:732–739.

72. Mero A, Nykanen T, Keinanen O, et al. Protein metabolism and strength performance after bovine colostrum supplementation. Amino Acids 2005;28:327–335.

73. Fernholm R, Bramnert M, Hagg E, et al. Growth hormone replacement therapy improves body composition and increases bone metabolism in elderly patients with pituitary disease. J Clin Endocrinol Metab 2000;85:4104–4112.

74. Thoren M, Hilding A, Baxter RC, Degerblad M, Wivall-Helleryd IL, Hall K. Serum insulin-like growth factor I (IGF-I), IGF-binding protein-1 and -3, and the acid-labile subunit as serum markers of body composition during growth hormone (GH) therapy in adults with GH deficiency. J Clin Endocrinol Metab 1997;82:223–228.

75. Ahmad AM, Hopkins MT, Thomas J, Ibrahim H, Fraser WD, Vora JP. Body composition and quality of life in adults with growth hormone deficiency; effects of low-dose growth hormone replacement. Clin Endocrinol (Oxf) 2001;54:709–717.

76. Alba-Roth J, Muller OA, Schopohl J, von Werder K. Arginine stimulates growth hormone secretion by suppressing endogenous somatostatin secretion. J Clin Endocrinol Metab 1988;67:1186–1189.

77. Merimee TJ, Rabinowitz D, Riggs L, Burgess JA, Rimoin DL, McKusick VA. Plasma growth hormone after arginine infusion: clinical experiences. N Engl J Med 1967;276:434–439.

78. Collier SR, Casey DP, Kanaley JA. Growth hormone responses to varying doses of oral arginine. Growth Horm IGF Res 2005;15:136–139.

79. Isidori A, Lo Monaco A, Cappa M. A study of growth hormone release in man after oral administration of amino acids. Curr Med Res Opin 1981;7:475–481.

80. Suminski RR, Robertson RJ, Goss FL, et al. Acute effect of amino acid ingestion and resistance exercise on plasma growth hormone concentration in young men. Int J Sport Nutr 1997;7:48–60.

81. Besset A, Bonardet A, Rondouin G, Descomps B, Passouant P. Increase in sleep related GH and Prl secretion after chronic arginine aspartate administration in man. Acta Endocrinol (Copenh) 1982;99:18–23.

82. Colombani PC, Bitzi R, Frey-Rindova P, et al. Chronic arginine aspartate supplementation in runners reduces total plasma amino acid level at rest and during a marathon run. Eur J Nutr 1999;38:263–270.

83. Walberg-Rankin J, Hawkins C, Fild D, Sebolt D. The effect of oral arginine during energy restriction in male weight trainers. J Strength Cond Res 1994;8:170–177.

84. Corpas E, Blackman MR, Roberson R, Scholfield D, Harman SM. Oral arginine-lysine does not increase growth hormone or insulin-like growth factor-I in old men. J Gerontol 1993;48:M128–M133.

85. Marcell TJ, Taaffe DR, Hawkins SA, et al. Oral arginine does not stimulate basal or augment exercise-induced GH secretion in either young or old adults. J Gerontol A Biol Sci Med Sci 1999;54:M395–M399.

86. Lukaski HC. Micronutrients (magnesium, zinc, and copper): are mineral supplements needed for athletes? Int J Sport Nutr 1995;5(Suppl):S74–S83.

87. Lukaski HC. Magnesium, zinc, and chromium nutriture and physical activity. Am J Clin Nutr 2000;72(Suppl):585S–593S.

88. Kikukawa A, Kobayashi A. Changes in urinary zinc and copper with strenuous physical exercise. Aviat Space Environ Med 2002;73:991–995.

89. Nielsen FH, Lukaski HC. Update on the relationship between magnesium and exercise. Magnes Res 2006;19:180–189.

90. Buchman AL, Keen C, Commisso J, et al. The effect of a marathon run on plasma and urine mineral and metal concentrations. J Am Coll Nutr 1998;17:124–127.

91. Brilla L, Conte V. Effects of a novel zinc-magnesium formulation on hormones and strength. J Exerc Physiol Online 2000;3:26–36.

92. Wilborn C, Kerksick CM, Campbell B, et al. Effects of zinc magnesium aspartate (ZMA) supplementation on training adaptations and markers of anabolism and catabolism. J Int Soc Sports Nutr 2004;1:12–20.

93. Om AS, Chung KW. Dietary zinc deficiency alters 5 alpha-reduction and aromatization of testosterone and androgen and estrogen receptors in rat liver. J Nutr 1996;126:842–848.

94. Neychev VK, Mitev VI. The aphrodisiac herb Tribulus terrestris does not influence the androgen production in young men. J Ethnopharmacol 2005;101:319–323.

95. Willoughby DS, Wilborn C, Taylor L, Campbell B. Eight weeks of aromatase inhibition using the nutritional supplement Novedex XT: effects in young, eugonadal men. Int J Sport Nutr Exerc Metab 2007;17:92–108.

96. Fryburg DA, Barrett EJ, Louard RJ, Gelfand RA. Effect of starvation on human muscle protein metabolism and its response to insulin. Am J Physiol 1990;259(Pt 1):E477–E482.

97. Fryburg DA, Louard RJ, Gerow KE, Gelfand RA, Barrett EJ. Growth hormone stimulates skeletal muscle protein synthesis and antagonizes insulin's antiproteolytic action in humans. Diabetes 1992;41:424–429.

98. Van Someren KA, Edwards AJ, Howatson G. Supplementation with beta-hydroxy-beta-methylbutyrate (HMB) and alpha-ketoisocaproic acid (KIC) reduces signs and symptoms of exercise-induced muscle damage in man. Int J Sport Nutr Exerc Metab 2005;15:413–424.

99. Tischler M, Goldberg A. Does leucine, leucyl-tRNA, or some metabolite of leucine regulate protein synthesis and degradation in skeletal and cardiac muscle? J Biol Chem 1982;257:1613–1621.

100. Nissen S, Sharp R, Ray M, et al. Effect of leucine metabolite beta-hydroxy-beta-methylbutyrate on muscle metabolism during resistance-exercise training. J Appl Physiol 1996;81:2095–2104.

101. Knitter AE, Panton L, Rathmacher JA, Petersen A, Sharp R. Effects of beta-hydroxy-beta-methylbutyrate on muscle damage after a prolonged run. J Appl Physiol 2000;89:1340–1344.

102. Jowko E, Ostaszewski P, Jank M, et al. Creatine and beta-hydroxy-beta-methylbutyrate (HMB) additively increase lean body mass and muscle strength during a weight-training program. Nutrition 2001;17:558–566.

103. Gallagher PM, Carrithers JA, Godard MP, Schulze KE, Trappe SW. Beta-hydroxy-beta-methylbutyrate ingestion. I. Effects on strength and fat free mass. Med Sci Sports Exerc 2000;32:2109–2115.

104. Nissen SL, Sharp RL. Effect of dietary supplements on lean mass and strength gains with resistance exercise: a meta-analysis. J Appl Physiol 2003;94:651–659.

105. Kreider RB, Ferreira M, Wilson M, Almada AL. Effects of calcium beta-hydroxy-beta-methylbutyrate (HMB) supplementation during resistance-training on markers of catabolism, body composition and strength. Int J Sports Med 1999;20:503–509.

106. Slater G, Jenkins D, Logan P, et al. Beta-hydroxy-beta-methylbutyrate (HMB) supplementation does not affect changes in strength or body composition during resistance training in trained men. Int J Sport Nutr Exerc Metab 2001;11:384–396.

107. Vukovich MD, Stubbs NB, Bohlken RM. Body composition in 70-year-old adults responds to dietary beta-hydroxy-beta-methylbutyrate similarly to that of young adults. J Nutr 2001;131:2049–2052.

108. Curthoys NP, Watford M. Regulation of glutaminase activity and glutamine metabolism. Annu Rev Nutr 1995;15:133–159.

109. Candow DG, Chilibeck PD, Burke DG, Davison KS, Smith-Palmer T. Effect of glutamine supplementation combined with resistance training in young adults. Eur J Appl Physiol 2001;86:142–149.

110. Hammarqvist F, Wernerman J, Ali R, von der Decken A, Vinnars E. Addition of glutamine to total parenteral nutrition after elective abdominal surgery spares free glutamine in muscle, counteracts the fall in muscle protein synthesis, and improves nitrogen balance. Ann Surg 1989;209:455–461.

111. Vinnars E, Hammarqvist F, von der Decken A, Wernerman J. Role of glutamine and its analogs in posttraumatic muscle protein and amino acid metabolism. JPEN J Parenter Enteral Nutr 1990;14(Suppl):125S–129S.

112. Kingwell BA. Nitric oxide-mediated metabolic regulation during exercise: effects of training in health and cardiovascular disease. FASEB J 2000;14:1685–1696.

113. Morrison RJ, Miller CC 3rd, Reid MB. Nitric oxide effects on shortening velocity and power production in the rat diaphragm. J Appl Physiol 1996;80:1065–1069.

114. Morrison RJ, Miller CC 3rd, Reid MB. Nitric oxide effects on force-velocity characteristics of the rat diaphragm. Comp Biochem Physiol A Mol Integr Physiol 1998;119:203–209.

115. Moncada S, Higgs A. The L-arginine-nitric oxide pathway. N Engl J Med 1993;329:2002–2012.

116. Campbell B, Roberts M, Kerksick C, et al. Pharmacokinetics, safety, and effects on exercise performance of L-arginine alpha-ketoglutarate in trained adult men. Nutrition 2006;22:872–881.

8 Weight Loss Nutritional Supplements

Joan M. Eckerson

Abstract

Obesity has reached what may be considered epidemic proportions in the United States, not only for adults but for children. Because of the medical implications and health care costs associated with obesity, as well as the negative social and psychological impacts, many individuals turn to nonprescription nutritional weight loss supplements hoping for a quick fix, and the weight loss industry has responded by offering a variety of products that generates billions of dollars each year in sales. Most nutritional weight loss supplements are purported to work by increasing energy expenditure, modulating carbohydrate or fat metabolism, increasing satiety, inducing diuresis, or blocking fat absorption. To review the literally hundreds of nutritional weight loss supplements available on the market today is well beyond the scope of this chapter. Therefore, several of the most commonly used supplements were selected for critical review, and practical recommendations are provided based on the findings of well controlled, randomized clinical trials that examined their efficacy. In most cases, the nutritional supplements reviewed either elicited no meaningful effect or resulted in changes in body weight and composition that are similar to what occurs through a restricted diet and exercise program. Although there is some evidence to suggest that herbal forms of ephedrine, such as ma huang, combined with caffeine or caffeine and aspirin (i.e., ECA stack) is effective for inducing moderate weight loss in overweight adults, because of the recent ban on ephedra manufacturers must now use ephedra-free ingredients, such as bitter orange, which do not appear to be as effective. The dietary fiber, glucomannan, also appears to hold some promise as a possible treatment for weight loss, but other related forms of dietary fiber, including guar gum and psyllium, are ineffective.

Key words

Dietary supplements · Ephedra · Obesity · Overweight · Complementary medicine · Alternative medicine · Herbal medicine · Chromium · Calcium · Chitosan · Pyruvate · Garcinia · Psyllium · Citrus aurantium · Bitter orange · Guarana · Herbal caffeine · Conjugated linoleic acid

From: *Nutritional Supplements in Sports and Exercise*
Edited by: M. Greenwood, D. Kalman, J. Antonio,
DOI: 10.1007/978-1-59745-231-1_8, © Humana Press Inc., Totowa, NJ

1. INTRODUCTION

Obesity has reached epidemic proportions in the United States, not only for adults but for adolescents and children, as well. According to data from the 2001–2004 National Health and Nutrition Examination Survey released by the Centers for Disease Control, 32% of adults in the United States were obese [body mass index (BMI) $\geq$ 30 kg·m^2], and 66% were considered overweight (BMI $\geq$ 25 kg·m^2). In comparison, the incidences of overweight and obese adults in 1974 were 47.7% and 14.6%, respectively. Although the number of overweight adults has dramatically increased during the last 30 years, the increase in the number of overweight children and adolescents is even more alarming. In 1970, the incidence of overweight children was 4.2% compared to 17.5% in 2004, which represents an increase of more than 400%. Because of the medical implications and health care costs associated with obesity, as well as the negative social and psychological impacts, many individuals turn to nonprescription nutritional weight loss supplements hoping for a quick fix or an answer to their unsuccessful attempts at dieting and exercise to lose weight. In fact, it has been suggested that many individuals prefer dietary supplements and prescription weight loss drugs to making healthier lifestyle changes (1).

The weight loss industry has responded to the rising rates of obesity and, as a result, generates billions of dollars each year through the sale of a variety of products and commercial weight loss businesses. In 2001, Americans spent more than $36 billion on videos, books, low-calorie foods and drinks, sugar substitutes, medical treatments, commercial weight loss chains, over-the-counter (OTC) drugs, and of course nutritional supplements to assist in their "battle of the bulge" (2). In fact, retail sales of OTC nutritional weight loss supplements alone were estimated to be more than $1.3 billion in 2001 (3).

Weight loss supplements are not just used by overweight individuals. Results of a multistate survey conducted in 1998 by Blanck et al. (4) indicated that 7% of adults at that time used OTC weight loss supplements and that the greatest consumers were obese young women (28.4%). Interestingly, however, 7.9% of normal weight women also reported using dietary supplements for weight loss. Given that obesity rates will likely continue to climb over the next several years and, correspondingly, the use of OTC nutritional weight loss supplements by obese and overweight individuals will also continue to increase, it is important for health professionals to understand the physiological mechanisms by which these products are purported to result in weight loss, as well as their safety and efficacy. Most OTC weight loss supplements are purported to work by increasing energy expenditure, modulating carbohydrate or fat metabolism, increasing satiety (feeling of fullness), inducing diuresis, or blocking fat absorption.

2. NUTRITIONAL SUPPLEMENTS THAT INCREASE ENERGY EXPENDITURE

2.1. Ephedra Alkaloids and Herbal Caffeine

Ephedra, also known as *ma huang*, is probably one of the most widely recognized nutritional supplements used for weight loss. Because of several safety concerns—hypertension, arrhythmias, heart attack, stroke, and even death—the U.S. Food and Drug Administration (FDA) removed ephedra products from the U.S. market in 2004 because the risk of its use was deemed greater than its benefits for weight loss.

Ephedra is derived from a shrub that is native to China (*Ephedra sinica*) and has been used for more than 5000 years as a natural treatment for asthma and other conditions *(5)*. The ephedrine alkaloids present in the plant contain sympathomimetic compounds [i.e., central nervous system (CNS) stimulants] and are a source of ephedrine and pseudophedrine, which are used in many decongestants and cold medicines. The history of ephedrine as a weight loss supplement goes back to 1972 when a Danish physician in Elsinore, Denmark who was treating his asthmatic patients with a compound that contained ephedrine and caffeine noticed that they were experiencing unintentional weight loss. His compound became known as the Elsinore pill, and by 1977 it was being used by more than 70,000 patients *(6)*. Before they were taken off the market in 2004, dietary supplements that contained ephedrine alkaloids were regulated under the Dietary Supplements Health and Education Act (DSHEA), but ephedrine and pseudoephedrine are regulated by the government as drugs.

Ephedrine and norephedrine are analogs of methamphetamine and amphetamine, respectively, and therefore are potent CNS stimulants that act as both α- and β-adrenergic agonists, releasing epinephrine from sympathetic neurons *(7)*. The action of ephedra on adrenergic receptors increases thermogenesis and suppresses hunger, which in turn promotes weight loss *(5)*. In animal studies, it has been shown that ephedrine stimulates thermogenesis in brown adipose tissue via the activation of β-receptors; however, because humans have little brown adipose tissue, it is believed that thermogenesis primarily occurs in the skeletal muscle *(6)*.

A recent meta-analysis by Shekelle et al. *(8)* that examined the efficacy and safety of ephedra and ephedrine-containing products found that they have modest short-term benefits (up to 6 months) and are associated with an increased risk of experiencing an adverse event. In their study *(8)*, ephedra resulted in weight loss that was only 0.9 kg (~2 lb) more per month than was achieved with a placebo, but led to a 2.2- to 3.6-fold increase in the odds of experiencing psychiatric, autonomic, or gastrointestinal problems and arrhythmias of the heart.

In an effort to maximize fat burning, ephedrine has been used in combination with caffeine and aspirin, which is often referred to as an ECA stack. It is believed that when these three compounds are taken together (recommended ratio: 60 mg ephedrine/200 mg caffeine/300 mg aspirin), it results in an even greater thermogenic effect than that of ephedra alone and may be more effective for inducing weight loss *(9,10)*. Herbal equivalents that are often used as a substitute for caffeine in nutritional weight loss supplements include guarana, kola nut, yerba mate, green tea, and yohimbe. Similarly, willow bark is often used in place of aspirin. Hydroxycut™ is just one example of a well known OTC product that contained an ECA stack (ma haung, guarana, and willow bark) before the FDA ban in 2004.

There is some evidence to suggest that the combination of ephedra with caffeine and/or aspirin (or their herbal constituents) may indeed be more effective for inducing weight loss than ephedra alone. In an extensive review, Greenway *(6)* examined the safety and efficacy of ephedrine plus caffeine (EC) and reported that the combination was as effective as some prescription weight loss drugs and was associated with fewer side effects. In the studies reviewed by Greenway *(6)*, the acute side effects for EC were considered mild and transient, and after continuous treatment for 4 to 12 weeks the reported side effects were not significantly different from those of a placebo. As a result, Greenway (6) suggested that the benefits of EC outweigh the associated risks and pointed out that many of the serious adverse events that have been reported in the literature are voluntary case reports that have no placebo or control groups for comparison. Therefore, in his opinion (6), the argument for removing herbal products containing ephedrine was somewhat unfounded.

Hutchins (11) also indicated that many of the ephedra alkaloid-related deaths reported in the literature occurred in individuals with preexisting cardiovascular conditions or risks and, therefore, the dangers associated with ephedra use in healthy individuals may be widely speculative. In addition, Dulloo (12) suggested that mixtures of ephedrine and caffeine may offer a viable, cost-effective approach to the treatment of obesity and has recommended that more large-scale clinical trials be conducted to gain a better understanding of the risks and benefits associated with the combination of ephedrine and caffeine.

In two separate studies, Boozer et al. (5,13) also reported that ma huang (herbal ephedra) combined with either guarana (5) or kola nuts (13) was more effective for weight loss in overweight men and women than placebo and resulted in no adverse events and minimal side effects. In the study that examined the effects of ma huang and guarana (5), the treatment was a commercial herbal mixture called Metabolife-356®, which contained an equivalent of 72 mg of ephedrine and 240 mg of caffeine. Following 8 weeks of supplementation, the treatment group lost significantly more body weight (4.0 ± 3.4 kg) and percent body fat (2.1% ± 3.0% fat) versus the placebo group (0.8 ± 2.4 kg and 0.2% ± 2.3% fat, respectively). In the other study by Boozer et al. (13) that examined the effects of ma huang and kola nut on weight loss, the herbal preparation was equivalent to 90 mg of ephedrine and 192 mg of caffeine; when compared to placebo, it resulted in significant decreases in body weight (5.3 ± 5.0 vs. 2.6 ± 3.2 kg) and fat weight (4.3 ± 3.3 vs. 2.7 ± 2.8 kg). In both studies (5,13), the herbal preparations also resulted in significant improvements in the blood lipid profiles of the subjects. Kalman and Minsch (14) also showed that supplementation of an ECA stack (20 mg ephedrine + 200 mg caffeine + 325 mg aspirin) for 6 weeks in overweight men resulted in weight loss that was significantly greater when compared to placebo (4.17 vs. 0.68 kg, respectively). In agreement, Daly et al. *(15)* also reported that an ECA combination (75–150 mg ephedrine + 150 mg caffeine + 330 mg) resulted in modest sustained weight loss (2.2–5.2 kg) in 24 obese individuals compared to placebo and reported that the doses were well tolerated and had no meaningful effects on heart rate, blood pressure, insulin and glucose concentrations, or cholesterol levels.

In somewhat contrast to the studies described above, Vukovich et al. *(16)* reported that acute administration (3 hours) of herbal ephedrine and caffeine at doses of 20 mg and 150 mg, respectively, significantly increased heart rate (22.7% ± 5.5%), systolic blood pressure (9.1% ± 2.2%), and resting energy expenditure (REE) (8.5% ± 2.0%) compared to baseline values. Although the authors *(16)* did not directly examine the effects of the herbal mixture on body weight, they suggested that the increase in REE would be negligible in terms of weight loss.

Based on most of the findings in the literature, it appears that herbal ephedrine combined with caffeine or with caffeine and aspirin is effective for inducing moderate weight loss in overweight adults who are otherwise healthy and have no preexisting cardiovascular or cerebrovascular conditions. Because of safety concerns, and considering that many individuals might be unaware of underlying conditions that may predispose them to an increased risk from herbal preparations that mimic both caffeine and ephedra, consultation with a physician prior to their use may be warranted as a precaution.

2.2. *Bitter Orange* (Citrus Aurantium)

The ban of ephedra by the FDA in 2004 has led to the proliferation of a number of "ephedra-free" dietary supplements. Many of these products contain bitter orange, which is also known as *Citrus aurantium*, zhi shi, Seville orange, or sour orange and refers to a small citrus tree (*C. aurantium*) and its peel and fruit *(17)*. The active components in bitter orange are synephrine and octopamine, which are structurally similar to epinephrine and norepinephrine, respectively. Therefore, these compounds are also chemically related to ephedrine and are believed to affect α-receptors and β_3 receptors, but not β_1 or β_2 receptors. Because synephrine does not affect β_1 or β_2 receptors, it is reportedly less active in the CNS than ephedrine and is theorized to have fewer adverse effects *(17)*.

There is some confusion in the literature, however, regarding the action of synephrine. For example, Fugh-Berman and Myers *(17)* have reported that synephrine (and octopamine) activate β_3 adrenoreceptors but not β_1 or β_2 receptors, which act to stimulate the heart (β_1 and β_2) and result in systemic vasodilation (β_2). However, Bent et al. *(18)* and Penzak et al. *(19)* reported that the extracts contained in bitter orange primarily stimulate α_1-adrenergic receptors because they resemble phenylephrine (a selective α receptor agonist commonly used as a nasal decongestant that is also known as Neo-Synephrine) and would result in vasoconstriction and increased blood pressure.

It appears that much of the confusion lies in the fact that there are several isomers of synephrine including para (*p*)-synephrine, meta (*m*)-synephrine, and ortho (*o*)-synephrine, and that it is not exactly known which of these isomers or combination of isomers are present in nutritional weight loss supplements products. In addition, there is some confusion as to which synephrine alkaloids are actually present in bitter orange itself *(20)*. This information is critical, as these various isoforms exhibit different pharmacological properties. The differences in the studies *(17–19)* mentioned above regarding the exact mechanism of action of bitter orange are related to the different synephrine isoforms that each author suggests is contained in the extract. For example, Penzak et al. *(19)* and Bent et al. *(18)* stated that it contains *m*-synephrine (i.e., phenylephrine), whereas Fugh-Berman and Myers *(17)* specifically stated that *m*-synephrine (phenylephrine) is *not* present in *C. aurantium* and that its most active component is *p*-synephrine. Blumenthal *(21)* has also stated that the type of synephrine in bitter orange peel is *p*-synephrine and that it has been incorrectly characterized as *m*-synephrine by various authors. However, in their technical report of the constituents of bitter orange, Allison et al. *(20)* stated that they were unable to find any convincing data that bitter orange solely contains *m*-synephrine or *p*-synephrine.

Considering for a moment that bitter orange contains only *p*-synephrine and therefore selectively activates β_3 receptors, it seems reasonable to suggest that this isoform would be able to induce weight loss with fewer side effects than other CNS stimulants, including *m*-synephrine. However, because of their selective activation of β_3 receptors, these compounds may be ineffective in humans. Animal studies using fat cells from rats, hamsters, and dogs have shown that β_3 agonists, such as synephrine and octopamine, have potent lipolytic effects; however, they are weak stimulators in human fat cells *(17)*. Human adipocytes respond to activation of β_1 or β_2 receptors and have little expression for β_3 receptors. Therefore, it requires high concentrations of synephrine (0.1–1.0 mM) to stimulate fat cells in humans *(22)*.

As indicated by Allison et al. *(20)*, it is likely that most weight loss products containing bitter orange contain both *p-* and *m-*synephrine. In their technical report, Allison et al. *(20)* analyzed a weight loss supplement called Ultimate Thermogenic Fuel™, which stated on the label that it contained *m-*synephrine from bitter orange. The authors did indeed find that the product contained both the *p-* and *m-* isoforms. Therefore, if bitter orange contains only *p-*synephrine, the manufacturers of these weight loss products are either adding synthetic *m-*synephrine or are including other CNS stimulants (e.g., guarana, caffeine, ma huang, yohimbe) *(20)*. As evidence of this, one needs only to read the label on several other commonly used OTC weight loss supplements to see that most do, in fact, include several CNS stimulants as a proprietary blend to boost the thermogenic effect. For example, in addition to bitter orange, Xenadrine EFX™ (Cytodyne LLC) also includes tyrosine, green tea, yerba mate, and guarana. Similarly, CortiSlim™ (Window Rock Health Laboratory) contains bitter orange, chromium, and green tea, which are all believed to have an effect on weight loss.

Given that these newly reformulated weight loss supplements contain bitter orange, as well as several other botanicals that also have sympathomimetic activity, there is some concern that they pose the same risks as ephedra. To help clarify this point, Haller et al. *(23)* examined the pharmokinetics and cardiovascular effects of two oral dietary weight loss supplements containing bitter orange in 10 healthy adults who ranged in age from 18 to 49 years. The subjects were given a single dose of Advantra Z™ (46.9 mg synephrine), Xenadrine-EFX (5.5 mg synephrine, 5.7 mg octopamine, 239.2 mg caffeine), or a placebo on three occasions that were separated by a 1-week washout period. The results showed that, compared to the placebo, Xenadrine-EFX increased systolic blood pressure (SBP) (9.6±6.2 mM Hg) and diastolic blood pressure (DBP) (9.1±7.8 mM Hg), with peak increases occurring 2 hours after ingestion. Advantra Z, however, had no meaningful effects on blood pressure. Heart rate (HR) was significantly elevated 6 hours after dosing with both Xenadrine-EFX (16.7 ± 12.4 b·min^{-1}) and Advantra Z (11.4 ± 10.8 b·min^{-1}) compared to placebo. The pharmokinetics were similar for the two supplements and showed that both synephrine and octopamine are poorly absorbed and/or rapidly metabolized when taken orally. The time to peak plasma concentration was 1 to 2 hours, and the half-life was approximately 3 hours for both treatments.

The finding that Advantra Z, which contained eight times the amount of synephrine as Xenadrine-EFX, had no effect on blood pressure suggests that it has minimal pharmacological activity *(23)*. However, when bitter orange is combined with other active herbal ingredients, as was the case with Xenadrine-EFX, the blood pressure increased significantly. The finding that both supplements resulted in a short-term increase in HR also suggests that bitter orange may have some β_1-adrenergic activity *(23)*.

In a study that examined the cardiovascular effects of bitter orange—in the form of sour orange juice (SOJ)—Penzak et al. *(19)* reported findings similar to those of Haller et al. *(23)*. The specific purpose of their study *(19)* was to quantify the content of both synephrine and octopamine in SOJ and determine its effects on SBP, DBP, mean arterial pressure (MAP), and HR in 12 normotensive (SBP/DBP $\leq$ 140/90 mmHg) young adults (age range 20–27 years). Each subject served as his or her own control and consumed either 8 oz of freshly squeezed SOJ (~14 mg synephrine) followed by repeat ingestion 8 hours later, or 8 oz of water using the same protocol, with a 1-week washout between the two trials.

The results showed that SOJ had no significant ($p > 0.05$) effect on any of the parameters measured (SBP, DBP, MAP, HR) when compared to water even though the SOJ contained what the authors *(19)* represented as ~28 mg of *m*-synephrine (and no octopamine), which is an amount comparable to that included in commonly used decongestant medications containing phenylephrine. The authors *(19)* also tested regular orange juice from frozen concentrate and found no traces of synephrine or octopamine, which indicates that these compounds are derived only from the fruit of *C. aurantium.*

Penzak et al. *(19)* suggested that their lack of significant findings could have been due to the poor bioavailability of synephrine, which would agree with the findings of Haller et al. *(23)*, as they also reported that synephrine was poorly absorbed. Interestingly, however, Penzak et al. *(19)* did not quantify the isomers (i.e., *p*- and *m*-) of the synephrine contained in the SOJ even though they implied that it contained the *m*- form. Therefore, in this author's opinion, it may be that SOJ primarily contains *p*-synephrine, which would explain in part why there were no significant cardiovascular effects, as it would not activate β_1 or β_2 receptors.

Based upon the findings of Haller et al. *(23)* and Penzak et al. *(19)*, it appears that bitter orange *alone* may be considered safer than ephedra. However, when it is combined with other sympatho-mimetics such as guarana, caffeine, green tea, or tyrosine, it may result in transient increases in blood pressure and HR.

In one of the first clinical studies to investigate the effect of bitter orange on weight loss, Colker et al. *(24)* examined the effect of an herbal mixture that contained 975 mg of *C. aurantium* extract (6% synephrine alkaloid), 900 mg of St. John's wort (3% hypericum), and 528 mg of caffeine on body fat in 20 overweight (BMI > 25 kg/m^2), but otherwise, healthy adults. The subjects were randomly placed in a control group (*n* = 4), a placebo group that ingested maltodextrin (*n* = 7), or the active treatment group (*n* = 9); if applicable, the sub-jects ingested their respective supplement once daily for 6 weeks. All subjects participated in a circuit training exercise program three times a week for 45 minutes per session and received individual counseling from a registered dietitian to comply with an 1800 kcal·d^{-1} diet recom-mended by the American Heart Association.

The change in body weight from baseline to 6 weeks for the group taking the *C. aurantium* was 1.4 kg compared to 0.9 kg in the placebo group, and 0.4 kg in the control group. The change in body fat and percent fat [via bioelectrical impedance analysis (BIA)] for each of the three groups was –3.1 kg/–2.9%, –0.63 kg/0.8%, and –1.8 kg/–2.2%, respectively. The loss in body fat and percent fat was significantly greater in the *C. aurantium* group than in either the placebo or the control group; however, there were no significant differences between the groups in regard to body weight, basal metabolic rate, blood pressure, HR, or electrocardiographic measurements.

Based on these findings, the authors *(24)* suggested that the combination of *C. aurantium* with St. John's wort and caffeine was safe and effective when combined with a diet and exercise program for inducing weight loss and fat loss in healthy, overweight adults. These results should be interpreted with caution, however, because the change in percent body fat (–2.9%) over the 6-week supplementation period was less than the error that is typically associated with BIA (~3%–5%), and this technique is not typically regarded as a criterion measure of body composition *(25)*.

More recently, Sale et al. *(26)* examined the acute effects of Xenadrine-EFX on metabolism and substrate utilization during rest and during treadmill walking in 10 overweight males (> 20% fat). To gather resting data, the subjects ingested the supplement and laid supine for 7 hours. Baseline measurements were taken during the first hour; and expired gases, blood pressure, and HR were measured and a venous blood sample obtained every 30 minutes for the remaining 6 hours. During the exercise arm of the study *(26)* the subjects ingested the supplement or a placebo, and at 1 hour after ingestion they exercised on a treadmill for 1 hour at 60% of their estimated HR reserve. Venous blood was analyzed for nonesterfied fatty acids (NEFAs), glycerol, glucose, and lactate; and expired gases were used to calculate energy (ATP) production and substrate utilization from carbohydrate and NEFAs for both the resting and exercise conditions.

The results showed that there was no significant effect of the supplement on total ATP utilization during the 6 hours of rest or during the 60 minutes of treadmill walking. However, there was a shift in ATP production and substrate utilization during both phases of the study, which demonstrated an increase in ATP production from carbohydrate and a decrease in NEFAs, as well as an increase in carbohydrate oxidation. In fact, the increase in carbohydrate oxidation at rest was shown to be as high as 30%. Weight loss supplements are typically promoted to increase fat utilization; however, these findings indicate that Xenadrine-EFX stimulates carbohydrate use and actually *decreases* ATP production from fat.

Based on these findings, it appears that this product would not have any favorable effects on body weight. However, more long--term studies designed to examine its effect on weight loss, both with and without exercise, are warranted to verify this statement. One positive finding from the study *(26)* was that Xenadrine-EFX had no effect on resting HR or blood pressure (either SBP or DBP), which is in contrast to the findings of Haller et al. *(23)*. Zenk et al. *(27)* also reported that a commercial weight loss product called The Lean System 7 (iSatory Global Technologies), which contains bitter orange, guarana, dehydroepiandrosterone (DHEA), and yerba mate among its seven ingredients, had no effect on HR or blood pressure (SBP or DBP) following 8 weeks of supplementation in 47 overweight adults who were also following a low-calorie diet and an exercise program. Perhaps more importantly, however, Zenk et al. *(27)* reported that this product had no effect on BMI, body weight, fat weight, or fat-free weight compared to placebo, even though the resting metabolic rate was significantly ($p = 0.03$) increased in the treatment group ($7.2 \pm 1.6 \, \text{kcal} \cdot \text{d}^{-1}$).

Based on the available research, it does not appear that bitter orange is effective for weight loss. Moreover, because safety information is limited, individuals with preexisting cardiac problems and/or hypertension should proceed with caution before using weight loss supplements containing bitter orange because it is most commonly used in herbal mixtures that contain several other CNS stimulants. In addition, future studies are warranted to gain a better understanding of the exact isomers contained in various forms of bitter orange, as the potency varies widely between the different stereo ($\pm$) and positional (*m*- and *p*-) isomers. In short, it appears to be "buyer beware" at the present moment with regard to the different types of synephrine that manufacturers may be including in their weight loss products, particularly as they do not necessarily need to register with the FDA or get FDA approval before producing or selling these products.

2.3. Pyruvate

Pyruvate (PYR), a three-carbon compound synthesized in the body via glycolysis, has been studied for its effects on weight loss since the late 1970s when Stanko et al. *(28)* found that PYR and a related three-carbon compound known as dihydroxyacetone (DHA) reduced the development of fatty livers in rats fed ethanol. Follow-up studies performed by the same research group *(29,30)* using rat and pig models also showed that PYR and DHA supplementation resulted in increased energy expenditure and fat reduction, possibly through the results of increased thermogenesis. Stanko et al. *(31,32)* have also reported significant effects of PYR and DHA on weight loss in humans. In one study *(32)* that examined the effect of PYR ($19 \, \text{g} \cdot \text{d}^{-1}$) and DHA ($12 \, \text{g} \cdot \text{d}^{-1}$) supplementation for 3 weeks in 21 obese women ingesting a low-calorie diet, it was found that the combination of PYR-DHA resulted in significantly greater losses in body weight (6.5 kg) and fat weight (4.3 kg) compared to placebo (5.6 kg and 3.5 kg, respectively), which represented a difference of ~2 lb between groups. In a related study, Stanko et al. *(31)* also found that when obese women who first lost weight on a low-calorie diet were subsequently placed on a high-calorie diet with PYR, they regained weight at a significantly ($p < 0.05$) slower rate compared to a high-calorie diet without PYR. Based on these findings, it was concluded that PYR and PYR-DHA when combined with a low-calorie diet results in greater weight loss versus calorie restriction alone *(32)* and attenuates weight gain during hypercaloric conditions *(31)*.

Although many of the studies conducted by Stanko and colleagues *(31–34)* resulted in positive findings, it is important to note that he owns several patents *(29)* for PYR-DHA; and although many researchers use patents as a means to protect their intellectual property, it is considered by some to represent a conflict of interest. In addition, the dosages recommended for the patients were large (i.e., $22–44 \, \mathrm{g \cdot d^{-1}}$), representing up to 20% of daily energy intake, and were ingested concurrently with a low-calorie diet *(31,33)*.

To gain a better understanding of the efficacy of PYR on weight loss without caloric restriction, Kalman et al. *(35)* determined the effect of low-dose PYR on body composition in 51 overweight (BMI $> 25 \, \mathrm{kg \cdot m^2}$) men and women consuming a $2000 \, \mathrm{kcal \cdot d^{-1}}$ diet. In their study *(35)*, the subjects randomly received a weight loss product that contained PYR ($6 \, \mathrm{g \cdot d^{-1}}$) and DHA ($50 \, \mathrm{mg \cdot d^{-1}}$) ($n = 18$) or a placebo (maltodextrin $6 \, \mathrm{g \cdot d^{-1}}$; $n = 18$) for 6 weeks, and another 15 subjects served as a control group. Subjects in the PYR and PL groups met with a registered dietitian every 2 weeks and received counseling to follow a $2000 \, \mathrm{kcal \cdot d^{-1}}$ diet (50% carbohydrate, 20% protein, 30% fat), and all subjects completed a circuit training protocol three times a week for ~45 minutes at 60% of their predicted maximal HR. Body composition was tested at baseline and every 2 weeks thereafter using BIA. There were no significant differences between the three groups at baseline; and at the end of the 6-week period, the results showed that none of the groups experienced a significant change in body weight. However, PYR resulted in a significant decrease in fat weight (–2.1 kg) and percent fat (–2.6%), as well as a significant increase in fat-free weight (1.5 kg). There were no significant differences in fat weight and fat-free weight from baseline to 6 weeks for the placebo and control groups. It is unknown if any of the differences in body composition were significantly different between groups because it does not appear that the authors *(35)* performed a statistical analysis to determine between-group differences after baseline. The authors *(35)* stated that PYR supplementation of $6\,g\cdot d^{-1}$ for 6 weeks results in modest decreases in fat weight and a concomitant increase in fat-free weight when performed in conjunction with exercise. These results should be interpreted with caution, however, as the body composition was not assessed using a criterion method, and it appears that incomplete statistical analyses were performed.

Kalman et al. *(36)* performed a similar study that was published 1 year later to examine the effects of exercise (45–60 minutes three times a week) and PYR $6\,g\cdot d^{-1}$ for 6 weeks on body composition in 26 overweight men and women compared to placebo. The results of this study *(36)* showed that PYR resulted in a statistically significant ($p < 0.001$) decrease in body weight (1.2 kg), fat weight (–2.5 kg), and percent fat (–3.0%), whereas the placebo group experienced no changes over the 6 weeks of supplementation and training.

In contrast to the findings of Kalman et al. *(35,36)*, a more recent study *(37)* that examined the effect of PYR during training on body composition reported no significant effects compared to placebo. In this study by Koh-Banerjee et al. *(37)*, 23 untrained women were assigned to receive either PYR ($10\,g\cdot d^{-1}$) or a placebo for 30 days while participating in a supervised exercise program. Prior to and following supplementation, body weight, fat weight, and percent fat were assessed using underwater weighing, which is a commonly accepted criterion method for assessing body composition. Although the PYR group gained less weight (PYR 0.3 ± 0.3 kg vs. placebo 1.2 ± 0.3 kg), lost more fat weight (PYR -0.4 ± 0.5 kg vs. placebo 1.1 ± 0.5 kg), and lost a greater percentage of body fat (PYR $-0.65\% \pm 0.6\%$ vs. placebo $0.1\% \pm 0.5\%$), the results were not statistically significant ($p = 0.16$) when compared to placebo. These findings *(37)* are in agreement with those of Stone et al. *(38)*, who also reported that PYR supplementation ($\sim 9\,g\cdot d^{-1}$) for 5 weeks had no significant effect on body composition or training adaptations in college football players.

Although several studies have shown that PYR, both with and without DHA, results in positive effects on body weight and body composition *(31–36)*, most of these studies were performed in the same laboratory, and supplementation occurred in conjunction with extremely low calorie diets *(30–33)*. In contrast, the results of other more recent well controlled studies in which subjects maintained their normal diet *(37,38)* showed no effect of PYR when compared to placebo. In addition, considering that the subjects used in most of the studies that showed positive results were overweight or obese, the weight loss induced by PYR (~2–3 lb) was what may be considered modest at best. Therefore, future studies conducted by several independent laboratories are necessary to gain a better understanding of the efficacy PYR on body composition before it can be recommended with confidence as a treatment for weight loss.

3. SUPPLEMENTS THAT MODIFY CARBOHYDRATE AND FAT METABOLISM

3.1. *Chromium Picolinate*

Chromium is a popular weight loss supplement that has been on the market for a number of years. From 1996 through 2003, sales of chromium in the United States increased from $65 million to $106 million *(39)*. Chromium is an essential trace mineral that enhances insulin activity and, therefore, is involved in carbohydrate, protein, and fat metabolism. Most dietary weight loss supplements (80% of all chromium sales) contain chromium picolinate (CrP), an organic compound of trivalent chromium and picolinate, which is a derivative of tryptophan *(39,40)*. Because CrP facilitates the action of insulin, it is believed to decrease body fat, increase lean mass, and increase basal metabolism *(39,40)*. However, most studies that have examined the effect of CrP on body weight and lean mass using humans have not shown many positive effects.

Pittler et al. *(40)* conducted a meta-analysis of 10 randomized clinical trials that examined the effect of CrP on body weight reduction. The results showed that body weight decreased 1.1 to 1.2 kg (0.08–0.2 kg·wk^{-1}) compared to placebo during an intervention period ranging from 6 to 14 weeks, with daily dosages ranging from 188 to 924 µg. In addition, no adverse events were reported in the studies that also examined the potential adverse side effects of chromium supplementation *(18,41)*.

More recently, Lukaski et al. *(39)* examined the effect of CrP on body weight and body composition (fat weight and fat-free weight) via dual-energy x-ray absorptiometry (DEXA) in 83 women (age range 19–50 years) with a BMI range of 18–30 kg·m². The subjects in the treatment group (n = 27) ingested an equivalent of 200 μg of CrP for 12 weeks, and the other subjects either randomly received 1720 μg picolinic acid (n = 27) or a placebo (n = 29). All subjects were counseled by a registered dietitian and maintained a 2000 kcal diet (51% carbohydrate, 18% protein, 31% fat) to control for chromium intake during the 12-week study.

The results showed that body weight [−1.0 kg (placebo) to −1.3 kg (CrP)) and fat-weight (−1.1 kg (CrP) to −1.4 kg (placebo)] significantly decreased over the 12-week intervention. However, there were no significant differences among the three groups, which indicates that the calorie restriction, not CrP, was responsible for the weight loss. There were no significant effects across time for any of the groups for fat-free weight or bone mineral density determined from DEXA. These findings support those of Pittler et al. (40) and suggest that CrP is a safe, but ineffective dietary weight loss supplement.

3.2. Chitosan

Chitosan, a positively charged polysaccharide that is a polymer of glucosamine, is derived from the shells of crustaceans (i.e., crabs, shrimp, lobster) *(3,42)*. It is purported to block fat absorption; and although there are some data on animals to suggest that it might be an effective weight loss supplement, there is little support for its use in humans. In one study *(43)*, it was found that chitosan prevented an increase in weight in mice that were fed a high-fat diet for 9 weeks, and another study using hamsters *(44)* reported that chitosan decreased food intake and body weight.

However, in their review article, Pittler and Ernst *(42)* performed a meta-analysis on five randomized clinical trials that examined the effect of chitosan on body weight in obese and overweight individuals and reported that, in three of the five studies, chitosan had no significant effect on body weight compared to a placebo. The dosages used in the studies ranged from 0.48 to 3.1 $g \cdot d^{-1}$ and were 4 to 12 weeks in duration. However, in the study that used the lowest dose, chitosan was combined with *Garcinia cambogia* (1.1 $g \cdot d^{-1}$), which is also commonly used in many dietary weight loss supplements. When examining the other four studies that used chitosan alone, three resulted in no significant findings. All five studies used in the meta-analysis reported minor adverse gastrointestinal effects following chitosan supplementation, including constipation, flatulence, bloating, nausea, and heartburn *(42)*.

As previously mentioned, chitosan is a positively charged polymer that is believed to block, or "trap," the absorption of fat by binding with negatively charged fat molecules in the lumen of the intestine *(3)*. If malabsorption of fat does result from products containing chitosan, it would be expected that there would be an increase in fecal fat excretion. To test this hypothesis, Gades and Stern *(45)* quantified the fecal fat content in subjects who supplemented with an OTC product called Absorbitol (Natrol, Chatsworth, CA, USA). Fifteen males (age 26.3 ± 5.9 years; BMI 25.6 ± 2.3 $kg \cdot m^2$) consumed five meals per day containing 15 g fat (total 75 $g \cdot d^{-1}$) for 12 days. The first 4 days served as a control period, followed by a 4-day supplementation period in which the subjects received chitosan 4.5 $g \cdot d^{-1}$. The subjects were then placed back on the control diet for another 4 days. All fecal matter on days 2 to 12 was collected to be analyzed for fat content.

The results showed that chitosan supplementation significantly ($p = 0.02$) increased fecal fat excretion by 1.1 ± 1.8 $g \cdot d^{-1}$ (from 6.1 ± 1.2 to 7.2 ± 1.8 $g \cdot d^{-1}$), which the authors *(45)* considered clinically negligible. They stated that the product would have no meaningful effect on energy balance or weight loss. Based on these findings *(45)* and those of Pittler and Ernst *(42)*, chitosan does not appear to be effective for reducing body weight in humans and is associated with gastrointestinal discomfort, including constipation, flatulence, and bloating.

3.3. Garcinia cambogia *(Hydroxycitric Acid)*

Garcinia cambogia, also known as brindelberry, is a tropical tree native to India that bears yellowish, pumpkin-shaped fruit. Both the natural fruit and rind contain hydroxycitric acid (HCA), which has been shown to decrease fat synthesis (lipogenesis), spare

carbohydrate, suppress food intake, and attenuate body weight gain
(46). The extract of *G. cambogia* is a component of many dietary
supplements (i.e., Citrimax^{TM}, Citrilean^{TM}), including a bottled
drinking water called "Skinny Water®." Most of these products
claim that when ingested 30 minutes prior to a meal, they suppress
appetite and block carbohydrate absorption.

In general, the results for HCA as a dietary weight loss supplement for humans are not promising. In an article that examined the
effect of HCA as a potential antiobesity agent, Heymsfield et al.
(46) reviewed seven other studies that had examined the effect of
HCA alone or in combination with other ingredients on body weight
and fat weight in overweight humans, including two peer-reviewed
articles, four abstracts, and one open-label study from an industrial
publication. Of the seven studies reviewed, five reported significant
($p < 0.05$) effects of HCA alone or in combination with another
weight loss agent on weight loss and fat loss; one study that failed
to include statistics reported that subjects who ingested HCA +
chromium for 8 weeks lost ~7 pounds more than subjects on a
placebo (HCA 11.14 lb vs. placebo 4.20 lb). However, Heymsfield
et al. *(46)* noted that in five of the studies HCA was taken in
combination with other active ingredients that could also potentially result in weight loss (i.e., CrP, L-carnitine, chitosan, herbal
forms of caffeine); therefore, the studies offered little insight into the
specific weight loss effects of HCA. Other limitations included a lack
of a placebo group or double blinding in one study and the use of
near-infrared interactance to assess body composition in another,
which is considered by many to be an invalid method for measuring
body composition *(46)*.

To gain a better understanding of the effect of HCA alone on body
composition, as well as overcome some of the limitations previously
reported in the literature, Heymsfield et al. *(46)* conducted their
own 12-week double-blind placebo-controlled study using overweight
subjects (ages 18–65 years; BMI > 27–38 kg·m^2) who were either
randomized to receive 1500 mg·d^{-1} HCA (3×500 mg·d^{-1} 30 minutes
prior to meals) ($n = 42$) or a placebo ($n = 42$). Body weight and fat
weight (via DEXA) were assessed at baseline and at 12 weeks.
All subjects were provided with 5040 kJ/day diet plan (i.e.,
~1200 kcal) with 50%, 30%, and 20% as carbohydrate, protein, and
fat, respectively.

The results showed that subjects in both groups lost a significant amount of weight over the 12 weeks; however, there was no significant difference between the two groups (placebo 4.1 ± 3.9 kg vs. HCA 3.2 ± 3.3 kg; $p = 0.14$). In addition, there was no significant difference in percent body fat lost between the HCA and placebo groups ($2.16\% \pm 2.06\%$ vs. $1.44\% \pm 2.15\%$, respectively). Based on these results, HCA was deemed no more effective than placebo for reducing body weight and fat weight in overweight individuals.

Kovacs et al. *(47)* also found that supplementation with 500 mg·d^{-1} HCA alone or in combination with medium-chain triglycerides for 2 weeks did not result in increased satiety, fat oxidation, 24-hour REE, or body weight loss (-1.0 kg for placebo vs. -1.5 kg for HCA alone) compared to placebo in 11 overweight male subjects (BMI 27.4 ± 8.2 kg·m^2). However, in a follow-up study that was also 2 weeks in duration but used a larger dose, investigators from this same laboratory *(48)* reported that 900 mg·d^{-1} HCA reduced energy intake by 15% to 30%, but did not result in any significant changes in satiety or body weight. The authors *(48)* suggested that HCA may not primarily serve as a weight loss agent, but could be effective for preventing weight regain in overweight individuals, since it resulted in reduced energy intake.

In another related study, Kriketos et al. *(49)* examined the effect of HCA on fat oxidation and the metabolic rate in 10 sedentary men (ages 22–38 years; BMI 22.4–37·6 kg·m^2) using a double-blind, crossover design. The subjects visited the laboratory on four occasions to examine the effects of HCA 3.0 g·d^{-1} for 3 days and a placebo on metabolic parameters both with and without moderately intense exercise (30 minutes at 40% VO$_2$max followed by 15 minutes at 60% VO$_2$max). The results showed that HCA had no effect on the respiratory exchange ratio or energy expenditure at rest or during exercise, indicating that it had no effect on fat oxidation. Although their study did not directly examine the effect of HCA on body weight loss, the authors *(49)* commented that the lack of metabolic changes suggest that HCA would be ineffective for inducing weight loss in individuals consuming a typical mixed diet. In agreement, van Loon et al. *(50)* also reported that acute HCA supplementation (18 g·d^{-1}) had no significant effects on total carbohydrate or fat oxidation in 10 trained cyclists at rest or during submaximal exercise.

Although the results of several early studies reviewed by Heymsfield et al. *(46)* suggested that HCA in combination with other ingredients, including CrP and herbal caffeine, enhanced weight loss, more rigorous well controlled studies that used HCA alone suggest that it is not an effective dietary weight loss supplement compared to placebo. Interestingly, however, a recent study *(51)* using rats suggests that there definitely is a difference in efficacy between products and that the dosages used in human studies might be too low to result in significant findings.

In their study, Louter-van de Haar et al. *(51)* compared the effects of three HCA-containing products on Wistar rats (Regulator^{TM}, Citrin K^{TM}, CitriMax) at doses corresponding to 150 and 200 mg·kg^{-1} on food intake and body weight after both single dose and repeated dose (4 days) administration. Regulator and Citrin K significantly reduced food intake following both single and repeated dose administration, whereas similar doses of CitriMax showed little effect on food intake. Repeated administration of Regulator and Citrin K also reduced body weight. Based up the results, the authors *(51)* suggested that because Citrimax has low efficacy, human studies that used CitriMax *(47,48,50)* likely found no results because the dose was too low. Even with these findings, however, it is important to keep in mind that the results of studies using animal models that have shown positive effects of HCA on food intake, fat oxidation, and body weight loss have not necessarily been reflected in human studies. Although more research that examines the effect of HCA on weight loss and body composition using higher doses and for longer periods of time is warranted, most of the current research suggests that it is an ineffective dietary weight loss supplement when ingested alone.

3.4. Conjugated Linoleic Acid

Conjugated linoleic acid (CLA) is a collective term used to describe a group of positional and geometrically conjugated (i.e., alternating single and double bonds) isomers of linoleic acid [a fatty acid (FA)] that are naturally found in animal fat, dairy, and partially hydrogenated vegetable oils *(52,53)*. CLA is also sold commercially as a dietary weight loss supplement. Most CLA products sold OTC have a 40%:40% content of *cis*-9,*trans*-11 (c9t11) FA and *trans*-10, *cis*-12 (t10c12) FA; the remaining 20% is typically composed of other conjugated FA (~1%–4%) and other nonconjugated FA (~15%–19%). The c9t11 CLA isomer accounts for 85% to 90% of the total natural CLA content in the diet, whereas dietary intake of the t10c12 isomer is negligible *(54)*. When these two CLA isomers are combined in approximately equal amounts, several animal studies have shown strong evidence that CLA has anticarcinogenic and antiatherogenic effects, as well as positive effects on body composition and blood lipid profiles *(41,52,53,55)*. The mechanisms of action that may explain how CLA promotes weight loss are not well understood, but it has been theorized that it may elicit positive effects on body composition by: 1) inhibiting hormone-sensitive lipoprotein lipase which, in turn, inhibits lipogenesis (fat synthesis); 2) promoting adipocyte apoptosis (programmed cell death); 3) preventing triglcyceride (TG) accumulation in adipocytes; 4) downregulating the expression of leptin; and 5) modulating glucose and fat metabolism to increase energy expenditure *(56)*. It is well documented that CLA elicits favorable effects on body composition and lipid profiles in animals; however, studies on the effect of CLA on body weight and body composition in humans have produced conflicting findings.

Tricon and Yaqoob *(53)* reviewed 18 studies that examined the effect of body weight and body composition using human subjects and found that the results were much less promising than those found for animal studies using mice and pigs. The amount used in the studies reviewed ranged from 0.7 to $6.8\,\mathrm{g\cdot d^{-1}}$, most of which contained a 1:1 mixture of c9t11 and t10c12 CLA isomers. Of the 18 studies, 4 demonstrated modest reductions in fat weight, 2 studies were deemed inconsistent because a dose-response relation was not found, and the remaining 12 studies showed no effect of CLA on body composition *(53)*. Although four studies reviewed by Tricon and Yaqoob *(53)* did show positive results for CLA on body composition, upon closer examination some of the findings are less than dramatic. For example, Riserus et al. *(57)* examined the effect of CLA $4.2\,\mathrm{g\cdot d^{-1}}$ for 4 weeks on abdominal fat and cardiovascular risk factors in 25 obese men with syndrome X type symptoms (i.e., abdominal obesity, hypertension, dyslipidemia, impaired fasting glucose) and reported a significant ($p = 0.04$)

decrease in sagittal abdominal diameter compared to placebo. However, the change was < 1 cm (–0.57 cm), and there were no significant differences between groups for body weight, waist circumference, waist-to-hip ratio, or blood lipid concentrations [cholesterol, high density lipoprotein (HDL), low density lipopotein (LDL), TG] following supplementation.

In another study showing positive results, Gaullier et al. *(56)* reported that supplementation with two isomers of CLA for 1 year reduced body fat in overweight (BMI 25–30 kg·m^2), but otherwise healthy adult men and women ($n = 180$). In their study *(18)*, 60 subjects ingested TG-CLA 4.5 g·d^{-1} ($n = 60$), FA-CLA 4.5 g·d^{-1} ($n = 61$), or a placebo (olive oil 4.5 g·d^{-1}) ($n = 59$). After 1 year of supplementation, both CLA groups demonstrated significant decreases in body weight (ΔFA-CLA, –1.1 ± 3.7 kg; ΔTG-CLA, –1.8 ± 3.4 kg) compared to placebo (Δ 0.2 ± 3.0 kg), which represents an approximate weight loss of only 2.4 to 4.0 lb over the entire year (0.20–0.33 lb per month). The fact that the standard deviation is greater than the mean change value for each group also suggests that there may have been considerable variation in body weight exhibited by the subjects during the 1-year period, with some subjects likely demonstrating notable increases.

In contrast, Tricon et al. *(55)* used a crossover design to examine the effect of three doses of high-grade c9t11 (0.59, 1.19, or 2.38 g·d^{-1}) and t10c12 CLA (0.63, 1.26, or 2.52 g·d^{-1}) for three consecutive 8-week periods separated by a 6-week washout period. The results showed that there was no significant effect of either isomer on body weight, BMI, or body composition (via skinfold measurements and BIA) at any dose. In agreement, Malpuech-Brugere et al. *(52)* also reported no effect of high grade c9t11 or t10c12 CLA supplementation at dosages of 1.5 or 3.0 g·d^{-1} administered in a dairy drink for 18 weeks on body weight or body composition (via DEXA) in 81 overweight men.

Although animal studies have provided strong evidence to indicate that CLA has positive effects on body composition, its potential as an antiobesity treatment for humans is much less promising. Brown and McIntosh *(41)* suggested that the conflicting findings in human research may be due, in part, to the fact that the mechanism of action of CLA is isomer-specific and that the dosages used in human trials are much less than those used in animal studies. Research from their own laboratory *(41)* has shown that the t10c12 CLA isomer decreases human adipocyte TG content and differentiation, whereas the c9t11 isomer increases TG accumulation and adipocyte-specific gene expression in human fat cells. In a related study, Tricon et al. *(55)* showed that these same two CLA isomers also had opposing effects on blood lipid profiles in healthy adults, with t10c12 resulting in increases in the LDL:HDL cholesterol and total HDL cholesterol ratios, whereas c9t11 resulted in a decrease in these ratios. Given that the c9t11 and t10c12 CLA isomers appear to have opposite effects on adiposity, Brown and McIntosh *(41)* speculated that the inconclusive findings in human studies could be due to the use of mixed isomers for supplementation, which may negate each other and, thus, result in no significant change in body fat. Although future studies are warranted to examine the effect of isomer-specific doses of CLA on body composition, there is no conclusive evidence, to date, to suggest that supplementation with either a mixture of CLA isomers or single CLA isomers results in any meaningful effects on body composition in humans.

3.5. Calcium

Calcium is typically known for its role in maintaining bone density and bone mineral homeostasis. However, evidence now suggests that calcium may also play a role in adipocyte lipid kinetics to help decrease body fat. The potential of calcium to induce weight loss was first reported during the late 1980s when Metz et al. *(58)* demonstrated a reduction in fat weight in hypertensive rats that were ingesting high amounts of calcium and sodium; and Bursey et al. *(59)* found that increasing dietary calcium from 0.1% to 2.0% resulted in less weight gain in both lean and obese Zucker rats. Interestingly, it was not until 1999 with the publication of two abstracts *(60,61)* that more research began to focus on this unexpected relationship between calcium and body fat.

The exact mechanism of action by which calcium exerts its effect on body weight is not entirely clear, however, Zemel et al. *(62)* proposed that low dietary calcium intake stimulates dihydroxyvitamin D and parathyroid hormone (PTH) which, in turn, simulate the uptake of calcium into adipocytes. This influx of intracellular calcium results in increased lipogenesis (i.e., fat synthesis) and decreased lipolysis (i.e., fat breakdown), with the net result being an increase in body fat. In contrast, high dietary intake of calcium has the opposite effect and inhibits vitamin D and PTH, which decreases the uptake of calcium into the adipocytes and, in turn, increases lipolysis and decreases lipogenesis, resulting in weight loss (62).

The discovery of an apparent inverse relationship between dietary calcium intake and body weight has led to several original investigations and the reevaluation of previously published data to examine the efficacy of both dietary and supplemental calcium on body composition. For example, Davies et al. (63) reanalyzed five clinical trials that included 780 women between their third and eighth decades of life and found that in each of the five studies the calcium/protein ratio negatively predicted BMI and/or change in body weight. In contrast, Shapses et al. (64) combined data from three 25-week randomized, double-blind placebo-controlled trials to reexamine the effect of calcium supplementation (1000 mg·d^{-1}, calcium citrate) on body weight and fat weight in pre- and postmenopausal women ($N = 100$). They found no significant differences for calcium versus placebo and no significant action of calcium supplementation on menopausal status. The results from several original investigations (65–69) designed specifically to examine the effect of calcium on body composition have also resulted in conflicting findings.

Zemel et al. (69) placed 32 obese subjects (ages 18–60 years; BMI 30.0–39.9 kg·m^2) on a calorie-restricted diet (500 kcal·d^{-1} deficit) for 24 weeks and randomized them to a standard diet (400–500 mg dietary calcium per day supplemented with placebo), a high-calcium diet (standard diet with calcium carbonate 800 mg·d^{-1}), or a high dairy diet (dietary calcium 1200–1300 mg·d^{-1} in the form of dairy products supplemented with placebo). The results showed that all subjects lost body weight as a result of caloric restriction; however, those assigned to the high calcium (supplemental) and high dairy diet lost 8.58 ± 1.1 kg and 11.07 ± 1.63 kg of body weight, respectively, which was greater than the loss demonstrated by the subjects on the standard diet (6.60 ± 2.58 kg). A similar trend also occurred for fat weight, with the high calcium and high dairy groups experiencing greater losses (5.61 ± 0.98 kg and 7.16 ± 1.22 kg, respectively) compared to those on the standard diet (4.81 ± 1.22 kg), with a significant portion of the fat loss in each group occurring in the trunk region. Although the sample size in the study by Zemel et al.

(69) was relatively small, their findings indicate that calcium significantly augments weight loss secondary to calorie restriction and that dairy products have a greater effect than supplemental forms of calcium. Based on these results, as well as their finding that calcium regulates lipogenesis and lipolysis in the adipocyte *(62)*, Zemel and colleagues submitted a patent for treating obesity with calcium *(70)*.

In contrast to the findings of Zemel et al. (69), several studies with a larger number of subjects *(65–68)* have failed to find a significant relationship between calcium intake and body composition. Gunther et al. *(67)* determined the effect of long-term (1 year) supplementation with dairy calcium on 135 healthy, normal weight women (ages 18–30 years). They found no significant changes in body weight or fat weight between groups that consumed a control diet ($< 800\,\mathrm{mg\cdot d^{-1}}$), a moderate dairy diet (1000–$1100\,\mathrm{mg\cdot d^{-1}}$), or a high dairy diet (1300–$1400\,\mathrm{mg\cdot d^{-1}}$). In agreement, Haub et al. *(68)* showed that a calcium-fortified beverage ($1125\,\mathrm{mg\cdot d^{-1}}$) supplemented over 1 year had no effect on body weight, fat weight, or abdominal fat in 37 postmenopausal women (ages 48–75 years) even though supplementation more than doubled the calcium/protein ratio. Barr et al. *(65)* also showed that three servings of milk per day had no effect on weight loss or other metabolic risk factors in older adults; and in a 1-year study using 178 preschool children (ages 3–5 years), DeJongh et al. *(66)* reported that there were no significant correlations between percent body fat and fat mass changes and dietary calcium intake, and no significant differences between calcium-supplemented groups ($1000\,\mathrm{mg\cdot d^{-1}}$) versus placebo.

There is an overwhelming amount of evidence that calcium has beneficial effects on bone health and metabolism. However, its role as a weight loss supplement has not been firmly established, as retrospective studies and original investigations have resulted in conflicting findings. Therefore, more large-scale clinical trials are necessary to help gain a better understanding of the role of calcium on adiposity using a wide range of subject populations. Although many individuals who are attempting to lose weight tend to eliminate dairy products, none of the studies reviewed herein reported a significant increase in body weight following increased calcium intake. Therefore, so long as energy intake is less than energy expenditure, it does not appear that the addition of calcium-rich dairy sources to the diet would lead to any substantial increases in body weight and may help many individuals meet their daily calcium requirements.

4. SUPPLEMENTS THAT INCREASE SATIETY

4.1. Glucommannan, Guar Gum, and Psyllium

A number of dietary weight loss supplements contain water-soluble fiber, which is theorized to absorb water in the gut, thereby decreasing feelings of hunger and, in turn, reducing food intake, which ultimately leads to weight loss. Examples of some of the most common sources of fiber in these products are glucomannan, psyllium, and guar gum.

Glucomannan (*Amorphophallus konjac*) is a highly viscous dietary fiber native to Asia that is derived from the konjac root (also known as elephant yam); it is composed of a polysaccharide chain of glucose and mannose (42). An early study by Walsh et al. *(71)* that examined the effect of glucomannan ($3 \, \text{g·d}^{-1} \times 8$ weeks taken before meals) on obese patients ($\geq 20\%$ of ideal body weight) found that it was more effective than placebo for inducing weight loss, resulting in a mean loss of 2.49 kg ($\sim$5.5 lb). A more recent review by Keithly and Swanson *(72)* also reported that at doses of 2 to $4 \, \text{g·d}^{-1}$ glucomannan resulted in significant weight loss in overweight individuals by increasing satiety. Moreover, it appears to be well tolerated.

The dietary fiber guar gum, derived from the Indian cluster bean *Cyamposis tetragonolobus*, is found in a number of natural weight loss preparations. To determine its effectiveness for reducing body weight, Pittler and Ernst *(73)* conducted a meta-analysis of 11 randomized controlled trials that ranged from 3 weeks to 6 months. They found that there was no significant difference in overweight subjects who received guar gum ($7.5–21.0 \, \text{g·d}^{-1}$) compared to those who received placebo (mean difference between guar gum and placebo was –0.04 kg). Of the 11 studies reviewed, 7 also reported several adverse gastrointestinal effects from guar gum, including flatulence, diarrhea, gastric pain, and nausea *(73)*. Because of the related adverse events and the fact that it resulted in minimal weight loss compared to placebo, Pittler and Ernst *(73)* did not recommend guar gum as a dietary treatment for weight loss.

The results of a study by Birketvedt et al. *(74)* that examined the effect of three dietary fiber supplements containing glucomannan and/or guar gum on weight loss were consistent with the findings of the studies described above. In their study *(74)*, 176 men and women randomly received one of three dietary fiber supplements or a placebo for 5 weeks while consuming a balanced diet of $1200 \, \text{kcal} \cdot \text{d}^{-1}$. The fiber supplements all contained glucomannan and included Chrombalance® (glucomannan), Appe-Trim® (glucomannan and guar gum), and Glucosahl® (glucomannan, guar gum, and alginate). The results showed that all three fiber supplements in conjunction with the diet resulted in significantly more weight loss than did the placebo and diet alone; however, there were no significant differences between the three treatments (3.8 ± 0.9, 4.4 ± 2.0, and $4.1 \pm 0.6 \, \text{kg}$ for Chrombalance, Appe-Trim, and Glucosahl, respectively), which resulted in an average weight loss of approximately $0.8 \, \text{kg} \cdot \text{wk}^{-1}$. These findings suggest that glucomannan is effective for inducing body weight loss and that guar gum and alginate have no added effect *(74)*.

Psyllium, derived from the husks of ripe seeds from the plant *Plantago ovata* or *Plantago psyllium (26,42)*, is the active ingredient in many nonprescription laxatives and fiber supplements including the well known brand Metamucil®. Although psyllium has been shown to be effective for lowering total cholesterol and LDL cholesterol, it does not appear to be as effective as other fiber types for reducing body weight and is associated with gastrointestinal disturbances, including bloating and flatulence *(75)*.

Dietary fiber is well known for its benefits on colon health and is commonly used for the treatment of high cholesterol levels *(75)*. Although, there is some evidence to suggest that fiber supplements may be useful for increasing satiety and inducing weight loss, there appears to be a difference in the efficacy of the various dietary fiber supplements, with glucomannan showing the most promising effects. Therefore, further studies are warranted to examine the safety and efficacy of glucomannan as an adjuvant treatment for weight loss to help decrease the high prevalence of overweight and obesity in the United States.

5. OTHER DIETARY SUPPLEMENTS

Other dietary supplements that have been purported to induce weight loss as either their primary or secondary action include herbal preparations such as dandelion, bladderwrack, sunflower, germander, and St. John's wort; the prohormone DHEA; ß-hydroxy-ß-methylbutyrate, which is a metabolite of the amino acid leucine; and yerba mate, which would have actions similar to those of ma haung, citrus aurantium, and guarana *(3,9)*. Still others include licorice, vitamin B$_5$, medium-chain triglycerides, and L-carnitine *(3)*. Research regarding the efficacy of these supplements for weight loss has primarily resulted in nonsignificant findings and, in some cases, supplementation has resulted in serious adverse effects *(3,9)*. Therefore, the use of these dietary supplements as a treatment for obesity appears to be largely unwarranted.

6. CONCLUSION

Although the use of dietary supplements for weight loss is widespread, the evidence for their efficacy and safety is not overwhelmingly convincing. In most cases, these products either elicit no meaningful effect despite the manufacturer's claims, or result in changes in body weight and composition that are comparable to those that occur through a restricted diet and exercise program, which can result in weight loss of up to 1.5 to 2.5 kg·wk^{-1} *(76)*. There is some evidence to suggest that herbal forms of ephedrine, such as ma huang, combined with caffeine or as an ECA stack is effective for inducing moderate weight loss in overweight adults. However, because of the recent ban on ephedra, manufacturers must now use "ephedra-free" ingredients, such as bitter orange, which do not appear to be as effective and may still possess the potential for adverse effects, particularly in individuals with preexisting cardiovascular or cerebrovascular conditions. The dietary fiber glucomannan also appears to hold some promise as a possible treatment for weight loss; however, more research is warranted to examine its efficacy, particularly when consumed alone, to confirm the findings of preliminary research.

It is important to remember that dietary supplements are not regulated like drugs, which must undergo rigorous clinical testing using both animal and human models before entering the market. Therefore, the manufacturers of dietary weight loss supplements can make claims regarding the effectiveness of their products without necessarily conducting clinical research trials first. In fact, in January 2007, the Federal Trade Commission received a $25 million settlement from the manufacturers of four well-known dietary weight loss supplements for deceptive advertising: Xenadrine

EFX, CortiSlim, Trim Spa™, and One-A-Day WeightSmart™. When making the decision to use dietary supplements, it is therefore critical to rely on evidence-based research and/or dietetic professionals versus infomercials and advertisements from fitness and body building magazines.

The future of dietary supplements for weight loss depends on well designed, large scale clinical studies and more stringent regulation of the dietary supplement industry. Currently, however, there are few dietary supplements designed for weight loss that can be recommended with much confidence.

REFERENCES

1. Radimer KL, Subar AF, Thompson FE. Nonvitamin, nonmineral dietary supplements: issues and findings from NHANES III. J Am Diet Assoc 2000;100:447–454.
2. Bryant J. Fat is a $34 billion business. Atlanta Business Chronicle September 24, 2001. http://atlanta.bizjournals.co/atlanta/stories/2001/09/24/story4.html/. Accessed May 11, 2007.
3. Saper RB, Eisenberg DM, Phillips RS. Common dietary supplements for weight loss. Am Fam Physician 2004;70:1731–1738.
4. Blanck HM, Khan LK, Serdula MK. Use of nonprescription weight loss products: results from a multistate survey. JAMA 2001;286:930–935.
5. Boozer CN, Nasser JA, Heymsfield SB, Wang V, Chen C, Solomon JL. An herbal supplement containing ma huang-guarana for weight loss: a randomized, double-blind trial. Int J Obes 2001;25:316–324.
6. Greenway FL. The safety and efficacy of pharmaceutical and herbal caffeine and ephedrine use as a weight loss agent. Obes Rev 2001;2:199–211.
7. Young R, Glennon RA. Discriminative stimulus properties of (-)ephedrine. Pharmacol Biochem Behav 1998;60:771–775.
8. Shekelle PG, Hardy ML, Morton SC, et al. Efficacy and safety of ephedra and ephedrine for weight loss and athletic performance: a meta-analysis. JAMA 2003;289:1537–1545.
9. Allison DB, Fontaine KR, Heshka S, Mentore JL, Heymsfield SB. Alternative treatments for weight loss: a critical review. Crit Rev Food Sci Nutr 2001;41:1–28.
10. Kovacs EMR, Mela DJ. Metabolically active functional food ingredients for weight control. Obes Rev 2006;7:59–78.
11. Hutchins GM. Dietary supplements containing ephedra alkaloids. N Engl J Med 2001;344:1095–1097.
12. Dulloo AG. Herbal stimulation of ephedrine and caffeine in the treatment of obesity. Int J Obes 2002;26:590–592.

13. Boozer CN, Daly PA, Homel P, et al. Herbal ephedra/caffeine for weight loss: a 6-month randomized safety and efficacy trial. Int J Obes 2002;26:593–604.

14. Kalman CM, Minsch A. Ephedrine, caffeine, and aspirin enhance fat loss under nonexercising conditions [abstract]. J Am Coll Nutr 1997;16(Suppl 2):S149.

15. Daly PA, Krieger DR, Dulloo AG, Young JB, Landsberg L. Ephedrine, caffeine, and aspirin: safety and efficacy for treatment of obesity. Int J Obes 1993;17(Suppl 1):S73–S78.

16. Vukovich, MD, Schoorman R, Heilman C, Jacob P, Benowitz NL. Caffeine-herbal ephedra combination increases resting energy expenditure, heart rate, and blood pressure. Clin Exp Pharmacol Physiol 2005;32:47–53.

17. Fugh-Berman A, Myers A. Citrus aurantium, an ingredient of dietary supplements marketed for weight loss: current status of clinical and basic research. Exp Biol Med 2004;229:698–704.

18. Bent S, Padula A, Neuhaus J. Safety and efficacy of citrus aurantium for weight loss. Am J Cardiol 2004;94:1359–1361.

19. Penzak SR, Jann MW, Cold JA, Hon YY, Desai HD, Gurley BJ. Seville (sour) orange juice: synephrine content and cardiovascular effects in normotensive adults. J Clin Pharmacol 2001;41:1059–1063.

20. Allison DB, Cutter G, Poehman, ET, Moore DR, Barnes S. Exactly which synephrine alkaloids does citrus aurantium (bitter orange) contain? Int J Obes 2005;29:443–446.

21. Blumenthal M. Bitter orange peel and synephrine. Part 1. HerbalGram 2005;66. American Botanical Council. http://www.herbalgram.org/herbalgram/articleview.asp?a = 2833&p = Y/. Accessed May 11, 2007.

22. Carpene C, Galitzky J, Fontana E, Atgie C, Lafontan M, Berlan M. Selective activation of beta3-adrenoreceptors by octopamine: comparative studies in mammalian fat cells. Naunyn Schmiedebergs Arch Pharmacol 1999;359:310–321.

23. Haller CA, Benowitz NL, Jacob P III. Hemodynamic effects of ephedra-free weight-loss supplements in humans. Am J Med 2005;118:998–1003.

24. Colker CM, Kalman DS, Torina GC, Perlis T, Street C. Effects of Citrus aurantium, caffeine, and St. John's wort on body fat loss, lipid levels, and mood states, in overweight healthy adults. Curr Ther Res 1999;60:145–152.

25. Heyward VH, Stolarczyk LM. Applied Body Composition Assessment. Human Kinetics, Champaign, IL, 1996, pp 47–54.

26. Sale C, Harris RC, Delves S, Corbett J. Metabolic and physiological effects of ingesting extracts of bitter orange, green tea and guarana at rest and during treadmill walking in overweight males. Int J Obes 2006;30:764–773.

27. Zenk, JL, Liekam SA, Kassen LJ, Kuskowski, MA. Effect of Lean System 7 on metabolic rate and body composition. Nutrition 2005;21:179–185.

28. Stanko RT, Mendelow H, Shinozuka H, Adibi SA. Prevention of alcohol-induced fatty liver by natural metabolites and riboflavin. J Lab Clin Med 1978;91:228–235.

29. Stanko RT. Methods for preventing body fat deposition in mammals. US Patent No. 4,812,479. March 14, 1989.

30. Stanko RT, Adibi SA. Inhibition of lipid accumulation and enhancement of energy expenditure by the addition of pyruvate and dihydroxyacetone to a rat diet. Metabolism 1986;35:182–186.

31. Stanko RT, Arch JE. Inhibition of regain in body weight and fat with addition of 3-carbon compounds to the diet with hyperenergetic refeeding after weight reduction. Int J Obes Relat Metab Disord 1996;20:925–930.

32. Stanko RT, Tietze DL, Arch JE. Body composition, energy utilization, and nitrogen metabolism with a severely restricted diet supplemented with dihydroxyacetone and pyruvate. Am J Clin Nutr 1992;55:771–776.

33. Stanko RT, Reynolds HR, Hoyson R, Jonosky JE, Wolf R. Pyruvate supplementation on a low cholesterol, low fat diet: effects on plasma lipid concentrations and body composition in hyperlipidemic patients. Am J Clin Nutr 1994;59:423–427.

34. Stanko RT, Ferguson TL, Newman CW, Newman RK. Reduction of carcass fat in swine with dietary addition of dihydroxyacetone and pyruvate. J Anim Sci 1989;67:1272–1278.

35. Kalman D, Colker CM, Stark R, Minsch A, Wilets I, Antonio J. Effect of pyruvate supplementation on body composition and mood. Curr Ther Res 1998;59:793–802.

36. Kalman D, Colker CM, Welets I, Roufs JB, Antonio J. The effects of pyruvate supplementation on body composition in overweight individuals. Nutrition 1999;15:337–340.

37. Koh-Banerjee PK, Ferreira MP, Greenwood M, et al. Effects of calcium pyruvate supplementation during training on body composition, exercise capacity, and metabolic responses to exercise. Nutrition 2005;21:312–319.

38. Stone MH, Sanborn K, Smith LL, et al. Effects of in-season (5 weeks) creatine and pyruvate supplementation on anaerobic performance and body composition in American football players. Int J Sports Nutr 1999;9:146–165.

39. Lukaski HC, Siders WA, Penland JG. Chromium picolinate supplementation in women: effects on body weight, composition, and iron status. Nutrition 2007;23:187–195.

40. Pittler MH, Stevinson C, Ernst E. Chromium picolinate for body weight reduction: meta-analysis of randomized trials. Int J Obes 2003;27:522–529.

41. Brown JM, McIntosh MK. Conjugated linoleic acid in humans: regulation of adiposity and insulin sensitivity. J Nutr 2003;133:3041–3046.

42. Pittler MH, Ernst E. Dietary supplements for body-weight reduction: a systematic review. Am J Clin Nutr 2004;79:529–536.

43. Han LK, Kimura Y, Okuda H. Reduction in fat storage during chitin-chitosan treatment in mice fed a high fat diet. Int J Obes 1999;23:174–179.

44. Trautwein EA, Jurgensen U, Erbersdobler HF. Choleserol-lowering and gallstone-preventing action of chitosans with different degrees of deacetylation in hamsters fed cholesterol rich-diets. Nutr Res 1997;17:1053–1065.

45. Gades MD, Stern JS. Chitosan supplementation and fecal fat excretion in men. Obes Res 2003;11:683–688.

46. Heymsfield SB, Allison DB, Vasselli JR, Pietrobelli A, Greenfield D, Nunez C. Garcinia cambogia (HCA) as a potential anti-obesity agent. JAMA 1998;280:1596–1600.

47. Kovacs EMR, Westerterp-Plantega MS, Saris WHM. The effects of 2-week ingestion of (-)-hydroxycitrate and (-)-hydroxycitrate combined with medium-chain triglycerides on satiety, fat oxidation, energy expenditure and body weight. Int J Obes 2001;25:1087–1094.

48. Westerterp-Plantenga MS, Kovacs EMR. The effect of (-)-hydroxycitrate on energy intake and satiety in overweight humans. Int J Obes 2002;26:870–872.

49. Kriketos AD, Thomson HR, Greene H, Hill JO. (-)-Hydroxycitric acid does not affect energy expenditure and substrate oxidation in adult males in a post-absorptive state. Int J Obes 1999;23:867–873.

50. Van Loon LJC, van Rooijen JJM, Niesen B, Verhagen H, Saris WHM, Wagenmakers AJM. Effects of acute (-)-hydroxycitrate supplementation on substrate metabolism at rest and during exercise in humans. Am J Clin Nutr 2000;72:1445–1450.

51. Louter-van de Haar, J, Wielinga PY, Scheurink AJW, Nieuwenhuizen AG. Comparison of the effects of three different (-)-hydroxycitric acid preparations on food intake in rats. Nutr Metab 2005;2:23.

52. Malpuech-Brugere C, Wihelmine PHG, Verboeket-vande V, Mensink RP, et al. Effects of two conjugated linoleic acid isomers on body fat mass in overweight humans. Obes Res 2004;12:591–598.

53. Tricon S, Yaqoob P. Conjugated linoleic acid and human health: a critical evaluation of the evidence. Clin Nutr Metab Care 2006;9:105–110.

54. Meinert Larsen T, Toubro S, Gudmundsen O, Astrup A. Conjugated linoleic acid supplementation for 1 y does not prevent weight or body fat regain. Am J Clin Nutr 2006;83:606–612.

55. Tricon S, Burdge GC, Kew S, et al. Opposing effects of cis-9,trans-11 and trans-10,cis-12 conjugated linoleic acid on blood lipids in healthy humans. Am J Clin Nutr 2004;80:614–620.

56. Gaullier JM, Halse J, Hoye K, et al. Conjugated linoleic acid supplementation for 1 y reduces body fat mass in healthy overweight humans. Am J Clin Nutr 2004;79:1118–1125.

57. Riserus U, Berglund L, Vessby B. Conjugated linoleic acid (CLA) reduced abdominal adipose tissue in obese middle-aged men with signs of the metabolic syndrome: a randomized controlled trial. Int J Obes 2001;25:1129–1135.

58. Metz JA, Karanja N, Torok J, McCaron DA. Modification of total body fat in spontaneously hypertensive rats and Wistar-Kyoto rats by dietary calcium and sodium. Am J Hypertens 1988;1:58–60.

59. Bursey RG, Sharkey T, Miller GD. High calcium intake lowers weight in lean and fatty Zucker rats [abstract]. FASEB J 1989;3137:A265.

60. Teegarden D, Lin YC, Weaver CM, Lyle RM, McCabe GP. Calcium intake related to change in body weight in young women [abstract]. FASEB J 1999;13:A873.

61. Zemel MB, Shi H, Zemel PC, DiRienzo D. Calcium and calcium-rich dairy products reduce body fat [abstract]. FASEB J 1999;12:LB211.

62. Zemel MB, Shi H, Greer B, DiRienzo D, Zemel P. Regulation of adiposity by dietary calcium. FASEB J 2000;14:1132–1138.

63. Davies KM, Heaney RP, Recker RR, et al. Calcium intake and body weight. J Clin Enocrinol Metab 2000;85:4635–4638.

64. Shapses, SA, Stanley H, Heymsfield SB. Effect of calcium supplementation on weight and fat loss in women. J Clin Endocrinol Metab 2004;89:632–637.

65. Barr SI, McCarron DA, Heaney RP, et al. Effects of increased consumption of fluid milk on energy and nutrient intake, body weight, and cardiovascular risk factors in healthy older adults. J Am Diet Assoc 2000;100:810–817.

66. DeJongh ED, Binkley TL, Specker BL. Fat mass gain is lower in calcium-supplemented than in unsupplemented preschool children with low dietary calcium intakes. Am J Clin Nutr 2006;84:1123–1127.

67. Gunther CW, Legowski AP, Lyle RM, et al. Dairy products do not lead to alterations in body weight or fat mass in young women in a 1-y intervention. Am J Clin Nutr 2005;81:751–756.

68. Haub MD, Simons TR, Cook CM, Remig VM, Al-Tamimi EK, Holcomb CA. Calcium-fortified beverage supplementation on body composition in postmenopausal women. Nutr J 2005;4:21.

69. Zemel MB, Thompson W, Milstead A, Morris K, Campbell P. Calcium and dietary acceleration of weight and fat loss during energy restriction in obese adults. Obes Res 2004;12:582–590.

70. Zemel MB, Shi H, Zemel PC. Materials and methods for the treatment or prevention of obesity. US patent no. 6,384,087; May 7, 2002.

71. Walsh DE, Yaghoubian V, Behforooz A. Effect of glucomannan on obese patients: a clinical study. Int J Obes 1984;8:289–293.

72. Keithly J, Swanson B. Glucomannan and obesity: a critical review. Altern Ther Health Med 2005;11:30–34.

73. Pittler MH, Ernst E. Guar gum for body weight reduction: meta-analysis of randomized trials. Am J Med 2001;110:724–730.

74. Birketvedt GS, Shimshi M, Erling T, Florholmen J. Experiences with three different fiber supplements in weight reduction. Med Sci Monit 2005;11:PI5–P18.

75. Pittler MH, Schmidt K, Ernst E. Adverse events of herbal food supplements for body weight reduction: a systematic review. Obes Rev 2005;6:93–111.

76. Cowburn G, Hillsdon, M, Hankey CR. Obesity management by life-style strategies. Br Med Bull 1997;53:389–408.

9

Effective Nutritional Supplement Combinations

Matt Cooke and Paul J. Cribb

Abstract

Few supplement combinations that are marketed to athletes are supported by scientific evidence of their effectiveness. Quite often, under the rigor of scientific investigation, the patented combination fails to provide any greater benefit than a group given the active (generic) ingredient. The focus of this chapter is supplement combinations and dosing strategies that are effective at promoting an acute physiological response that may improve/enhance exercise performance or influence chronic adaptations desired from training. In recent years, there has been a particular focus on two nutritional ergogenic aids—creatine monohydrate and protein/amino acids—in combination with specific nutrients in an effort to augment or add to their already established independent ergogenic effects. These combinations and others are discussed in this chapter.

Key words

Acute · Chronic · Supplementation · Aerobic · Anaerobic · Exercise performance · Resistance training · Protein · Amino acids · Carbohydrate · Creatine monohydrate · Protein balance · Glycogen resynthesis · Sodium · D-Pinotol · HMβ · Sodium bicarbonate · Caffeine · Ephedrine

1. INTRODUCTION

The first documented use of "natural preparations" to enhance athletic prowess were the ancient Greeks (300 BCE). It is probable that ever since that time, athletes have been combining various nutritional compounds in an effort to increase the ergogenic potential of the supplement and enhance performance. Whether it is to outperform the competition or maximize personal potential, athletes are competitive by nature. This drive to succeed and a

From: *Nutritional Supplements in Sports and Exercise*
Edited by: M. Greenwood, D. Kalman, J. Antonio,
DOI: 10.1007/978-1-59745-231-1_9, © Humana Press Inc., Totowa, NJ

growing awareness that nutritional choices can influence athletic performance has fueled an explosion in the interest of nutritional combinations as ergogenic aids: dietary supplement formulations that enhance athletic performance. In the sports supplement industry, companies often market various combinations to consumers based on the assumption that the supplement blend (or stack) will provide greater benefit than any single compound alone. However, few supplement combinations that are marketed to athletes are supported by scientific evidence of their effectiveness. Quite often, under the rigor of scientific investigation, the patented product blend in question is shown to be no more effective than one active (generic) ingredient.

From both a scientific and practical perspective, the focus of this chapter is on supplement combinations and dosing strategies that are documented to be safe and effective at promoting an acute physiological response that may improve/enhance exercise performance or influence the chronic adaptations desired from training. Few studies have linked acute physiological responses to chronic adaptations in the same trial. However, manipulation (type timing and quantity) of some nutritional variables, such as the macronutrients, is shown to alter events that affect chronic adaptations. Therefore, where applicable, well controlled longer-term studies that document enhanced chronic adaptations by certain dietary supplement combinations are featured.

2. SUPPLEMENT COMBINATIONS THAT MAY ENHANCE THE PHOSPHAGEN SYSTEM

Exercise at high intensity is dependent on the maximum rate of adenosine triphosphate (ATP) regeneration, which occurs via the phosphagen [ATP, phosphocreatine-creatine (PCr-Cr)] and glycolysis/glycogenolysis systems. Whereas the ADP-ATP aspect of the phosphogen system is considered a "cofactor" (albeit an essential one), the PCr system (encompassing its site-specific CK isoenzymes) plays a pivotal, multifaceted role in muscle energy metabolism (Fig. 1). The availability of PCr is now generally accepted as most critical to the continuation of muscle force production and performance during repeated, short bouts of powerful activity (1,2)

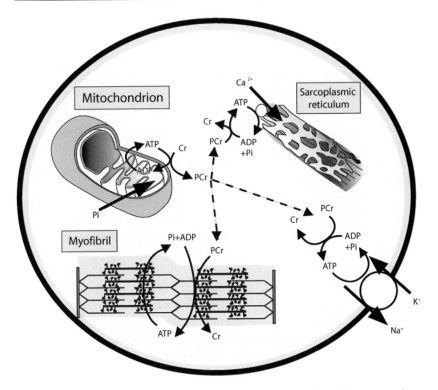

Fig. 1. Main functions of the creatine-phosphocreatine (Cr-PCr) system in a muscle fiber. The first is that of a temporal energy buffer for the regeneration of ATP via anaerobic degradation of PCr to Cr and rephosphorylation of ADP. The second major function of this system is that of a spatial energy buffer or transport system that serves as an intracellular energy carrier connecting sites of energy production (mitochondrion) with sites of energy utilization, such as the Na^+K^+ pump, myofibrils, and the sarcoplasmic reticulum.

as well as aerobic exercise at high intensity *(3,4)*. Since the early 1990s, it has been established that "loading" with creatine mono-hydrate (CrM) (*n*-[aminoiminomethyl]-*N*-methylglycine) (4 × 5 g servings · day^{-1} for 5 days) elevates muscle Cr concentrations (by 15%–40%) *(1,12)* and may enhance athletic performance under a variety of circumstances *(13–19)*. Regular use also appears to enhance chronic adaptations, particularly during strength training *(30,31)* The ergogenic potential of CrM is thought to reside in its ability to augment the phosphagen system (i.e., increased PCr availability) *(2,9,47)*. Chronic use is popular among a variety of athletes and other populations who perform resistance training exercise *(34–40)*. A large scientific body of literature continues to document this supplement's physiological *(23–26)* and performance enhancing *(27–31)* effects as well as dispel concerns of adverse effects *(32,33)*. For these reasons, there is a steadily increasing amount of interest in combining CrM with other compounds to enhance its ergogenic potential.

Other forms of oral Cr have shown limited potential *(20–22)*. Additionally, the presence of phosphatase enzymes in the blood and gut suggests that supplementation with other energy-yielding components of the phosphagen system, such as ATP or PCr, is not a viable option as these enzymes readily cleave the phosphate from the molecule *(41)*. Whereas the Cr portion of the monohydrate form consists of up to 92% Cr, it forms only 50% of the PCr molecule *(20)*. Oral supplementation with CrM enters the circulation intact where active uptake by tissues is facilitated by a Na^+-dependent transporter against a concentration gradient *(42)*. The capacity of CrM to enhance the bioenergetics of the phosphagen system by increasing PCr availability is thought to reside in the extent of Cr accumulation in muscle *(31,47,48)*. To exert a beneficial effect on performance and metabolism, an increase in muscle total Cr (PCr + Cr) content by at least 20 mmol·kg dm-1 appears to be required *(43)*. Although a loading phase is shown consistently to achieve this (as well as increase total Cr concentrations in other tissues with low baseline Cr content), it is also apparent that this response can be highly variable among subjects (see sidebar: Multifaceted Role of the Muscle PCr-Cr System in Exercise Metabolism).

For these reasons, a number of studies have assessed Cr uptake in the presence of other compounds. For example, Cr accumulation in muscle is enhanced by the presence of insulin *(50)* and possibly triiodothyronine *(51)* but may be depressed by the presence of some drugs (e.g., oubain, digoxin) *(52)* or vitamin E deficiency *(53)*. The findings from some investigations suggest that caffeine impairs the advantages of Cr loading *(54)*, whereas other studies have involved administering CrM in caffeine-containing beverages (e.g., tea, coffee) and report significant elevations in muscle Cr and improved athletic performance *(9,13,15)*. Other investigations have reported that muscle Cr uptake is not affected by PCr, creatinine, or cellular concentrations of various amino acids such as glycine, glutamine, alanine, arginine, leucine, and glycine or the sulfur-containing amino acids methionine and cysteine (55,56).

Improved cellular retention of Cr has been attributed to a stimulatory effect of insulin on the Cr transporter protein (50). Carbohydrates (CHO) with a high glycemic index (GI) (e.g., glucose, sucrose) generally evoke a high insulin response (57). Once it had been demonstrated that the presence of insulin (at supraphysiological levels) increased muscle Cr accumulation in humans (43), other investigations that examined the effects of combining CrM with high-GI CHO soon followed. Green et al. were the first to demonstrate reduced urine Cr losses (58) and a 60% increase in muscle Cr accumulation (59) from combining a high dose of glucose (93 g) with each 5 g dose of CrM (4×5 g·day^{-1} for 2 days) compared to CrM alone. Robinson et al. (60) also showed that a high-CHO diet combined with CrM (4×5 g·day^{-1} for 5 days) after exercise provided effective ($p < 0.01$) Cr accumulation in the exercised limb. However, data from subsequent studies suggested that lower doses of CHO (glucose) may also be effective. For instance, Greenwood et al. (61) assessed whole-body Cr retention (via 24-hour urine samples for 4 days) and reported that a 5 g dose of CrM combined with an 18 g dose of glucose ($4 \times$·day^{-1} for 3 days) resulted in significantly greater Cr retention than an equivalent dose of CrM alone or an effervescent Cr supplement (containing sodium and potassium bicarbonate). Along this line, Preen et al. (62) examined the effectiveness of three CrM loading procedures on total Cr accumulation in muscle. A group of 18 physically active males were divided into three equal groups and provided one of three regimens: 1) CrM (4×5 g·day^{-1} for 5 days); 2) the same dose of CrM+glucose (1 g·kg^{-1} twice a day for 5 days); or 3) CrM combined with 60 minutes of daily exercise (repeated sprints) (CrM+E) for 5 days. Results showed that the combination CrM+glucose provided a 7% to 9% greater ($p < 0.05$) elevation in total muscle Cr concentrations than CrM alone or CrM+E (62).

Supplementation with high-GI CHO appears to be effective for promoting Cr uptake, although combining CrM with a protein (PRO) supplement may provide similar benefits. For example, using a group of recreational weightlifters, one study directly compared the effects of two CrM-containing supplements: CrM+CHO (glucose) and CrM+PRO (whey protein isolate)

(1.5 g of supplement·kg^{-1}·day^{-1}) during 11 weeks of resistance training *(63)*. After the 11-week program, the two CrM-containing supplements provided a similar increase in total muscle Cr concentrations (~10%). Additionally, the CrM+CHO and CrM+PRO groups demonstrated greater ($p<0.05$) strength improvements and muscle hypertrophy than an equivalent dose of CHO or PRO *(63)* (see sidebar: Creatine + Protein or Creatine + Carbohydrate for Better Muscle Hypertrophy?) Other studies have reported similar benefits from combining CrM with whey protein *(64)* or CHO (glucose) *(18)* during resistance training, but muscle Cr concentrations were not assessed. Whereas these studies utilized relatively large doses of PRO or CHO (70–100 g or more) in combination with CrM *(63–65)* and reported positive outcomes, the results of a study by Stout et al. *(66)* suggested that a smaller dose of CHO (35 g glucose) with each 5 g dose of CrM is also effective for improving training adaptation. However, no other studies have directly compared the effects of different CrM-containing PRO or CHO supplements on Cr accumulation and training adaptations.

Combining PRO, CHO, and CrM may be the most effective mix for promoting whole-body Cr accumulation, particularly if smaller doses of the macronutrients are desired. Steenge et al. *(43)* reported that ingestion of CrM along with a PRO+CHO supplement (50 g dairy milk protein, 50 g glucose) over 5 days resulted in similar insulin responses and (whole-body) percentage Cr accumulation values (~25%) as the same CrM dose combined with 100 g of glucose *(43)*. Whole-body Cr accumulation is an indirect method assessing CrM uptake by tissues. Percent whole-body Cr retention can be calculated as Cr ingested (g)/urinary Cr excretion (g) × 100 *(43)*. The results obtained by Steenge et al. *(43)* suggested that the combination of PRO and CHO with CrM may be an effective way to improve Cr accumulation, particularly when smaller doses of these macronutrients are desired. This combination may also have important implications for populations where the consumption of large amounts of high-GI CHO is undesirable, such as those with, or at risk of, type 2 diabetes. This combination (PRO-CHO-CrM) has also been used to demonstrate that the timing of supplementation may be important for improving Cr accumulation in muscle and adaptations from training *(67)* (see sidebar: Can Supplement Timing Double Gains in Muscle Mass?). Other studies have shown that CrM supplementation close to exercise promotes muscle Cr uptake *(60)* and increases the girth and thickness of the exercised limb after resistance training *(68)*. Therefore, the use of a CrM-containing PRO-CHO supplement before and after resistance exercise may provide a higher degree of Cr accumulation and muscle anabolism and therefore promote better gains in strength and muscle mass.

To summarize the research in this particular area, co-ingestion of CrM with CHO and/or PRO (i.e., glucose or whey proteins; ~35–100 g) appears to enhance muscle Cr storage, which may result in enhanced performance and better training adaptations. Greater accumulation in muscle appears to be due to a stimulatory effect of insulin on cell Cr transporter. In fact, combining CrM with a PRO and/or CHO supplement seems to reduce the individual variations in muscle Cr accumulation reported previously in studies involving acute loading *(15,43,49)*. Additionally, there is evidence to suggest that the timing of the supplement dose is important. The use of this supplement combination close to exercise (i.e., just before and/or after) appears to promote better Cr accumulation in muscle and influence training adaptations *(60,67)*. Therefore, the use of a CrM-containing PRO-CHO supplement close to the time of exercising represents a simple but highly effective strategy that promotes effective Cr accumulation (to increase PCr availability in muscle) and provides an ergogenic effect during training that results in greater adaptations. Further examination of dose-response data along with the extent of Cr accumulation and adaptations would help define a clearer supplementation prescription.

Aside from the use of macronutrients such as PRO and CHO, some studies have examined the effects of co-ingesting CrM with other compounds that affect insulin secretion and/or tissue sensitivity. For instance, in a single blinded study, Greenwood et al. *(69)* examined whether co-ingestion of D-pinitol (a plant extract with insulin-sensitizing characteristics) *(70)* with CrM affected whole-body Cr retention (determined by 24-hour urine samples for 4 days). Results revealed that whole-body Cr retention (and percentage Cr retention) over the 3-day loading phase was greater ($p < 0.05$) in the two groups given CrM combined with a low dose of D-pinitol (LP) group was given 4×5 g CrM $+ 2 \times 0.5$ g D-pinitol; PreP group was given D-pinitol 2×0.5 g 5 days prior to and during

CrM (4 × 5 g) supplementation compared to an equivalent dose of glucose (placebo) or CrM alone. However, another group given a high dose D-pinitol (4 × 0.5 g) with the same dose of CrM showed no greater Cr retention than in the group given CrM alone *(69)*. Interestingly, the group predosing with D-pinitol (PreP) demonstrated the same results as the LP group, suggesting that no further benefit seems to be gained by taking D-pinitol prior to supplementation *(69)*. The authors concluded that ingesting Cr with D-pinitol may augment whole-body Cr retention in a manner similar to that reported with CHO or CHO + PRO supplementation *(43)*. However, this is the only study that has examined the effects of D-pinitol combined with CrM supplementation on Cr accumulation. Because of the conflicting nature of the results regarding the high versus low doses of D-pinitol, further research is necessary before a clear conclusion can be drawn.

Another compound that has shown potential to enhance Cr uptake and accumulation in muscle is α-lipoic acid (ALA). Supplementation with ALA is shown to increase the expression of glucose transporter proteins (GLUT4) and enhance glucose uptake in muscle *(71,72)*. In light of the fact that Cr uptake is influenced by insulin and that ALA can increase glucose disposal, Burke et al. *(73)* examined the effects of combining ALA with CrM on muscle Cr accumulation. In this study, muscle biopsies were obtained to determine total Cr concentration in 16 male subjects before and after the 5-day supplementation intervention. Results showed a greater increase ($P < 0.05$) in PCr and total Cr in the group given ALA combined with CrM+CHO (CrM 20 g·d^{-1} + sucrose 100 g·d^{-1} + ALA 1000 mg·d^{-1}) compared with a group given the same dose of CrM+CHO or CrM alone. The authors concluded that co-ingestion of ALA with CrM (and a small amount of sucrose) can enhance muscle Cr concentrations compared to an equivalent dose of CrM+CHO or CHO alone *(73)*. However, the authors also acknowledged that a limitation of this study was the high baseline muscle Cr concentrations exhibited by the participants; the groups were ~10% higher than starting values reported in other studies (~135 mmol·kg^{-1} vs. ~125 mmol·kg^{-1}). Initial muscle Cr content is an important determinant of muscle Cr uptake *(48)*. That is, study participants with lower muscle Cr concentrations tend to show the largest increases after supplementation; conversely, those with higher muscle Cr concentrations show little or no increase.

Burke et al. *(73)* suggested that the higher starting values of the participants may have been the reason for the lack of increase in PCr and total Cr experienced by two of the three groups in this study. As is the case with D-pinitol, only one study has examined the effects of ALA on Cr accumulation. The ability of D-pinitol or ALA to affect muscle Cr accumulation during CrM supplementation needs to be confirmed by other investigations. Other compounds such as pyruvate, β-hydroxy-β-methylbutyrate (HMβ), and β-alanine have been examined in combination with CrM. However, these studies did not assess muscle Cr concentrations in response to supplementation, and therefore their results are discussed elsewhere in the chapter.

To summarize this section, supplement combinations that have been shown to increase muscle Cr concentrations successfully are presented in Table 1. The ergogenic potential of CrM and its capacity to enhance the bioenergetics of the phosphagen system are thought to depend on the extent of Cr accumulation in muscle. This has led to increased interest in combining CrM with compounds to improve the uptake and accumulation of Cr in muscle. However, when viewed in comparison to the large body of literature that demonstrates CrM's widespread use, safety, and performance-enhancing effects, a relatively undersized amount of work documents effective strategies and supplement combinations that may improve muscle Cr accumulation in response to supplementation. Probably owing to an insulin-stimulating effect on the cellular Cr transporter, combining each dose of CrM with high-GI CHO and protein ($\sim$50 g of each or a total of $1 \text{ g} \cdot \text{kg}^{-1}$) appears to be a most effective strategy for improving Cr accumulation. The combination of PRO and CHO is particularly effective when smaller doses of these macronutrients are desired. Other compounds that affect insulin secretion and/or tissue sensitivity, such as D-pinotol and ALA, have shown potential to augment muscle Cr accumulation but require further investigation before clear conclusions can be made about their effectiveness.

3. SUPPLEMENT COMBINATIONS TO ENHANCE MUSCLE GLYCOGEN

Along with the phosphagen system, glycolysis and glycogenolysis are considered to be important energy contributors during high intensity exercise. A relation between muscle glycogen concentration

Table 1
Supplement combinations shown to enhance muscle Cr accumulation

Reference	Experimental comparison	Protocol	Supplementation	Change
Green (58)	CrM+CHO vs. CrM only	Muscle [PCr and Cr] before and after supplementation; no exercise	5 g CrM or 5 g CrM + 93 g CHO (glucose) $4 \times \cdot day^{-1}$, 5 days	After 5 days, 60% greater [PCr] from CrM+CHO ($p < 0.01$)
Robinson (60)	High-CHO diet with CrM vs. high-CHO diet without CrM	Muscle [PCr and Cr], one-legged cycle exercise to exhaustion preceded supplementation	20 g CrM $\cdot day^{-1}$ 5 days	After 5 days, 23% greater increase in muscle [total Cr] in exercised limb ($p < 0.01$)
Greenwood (61)	CrM+CHO vs. CrM only and effervescent Cr	Whole-body Cr retention via 24-hr urine samples for 4 days; 3-day supplementation; no exercise	5 g CrM + 18 g CHO (glucose) $4 \times \cdot day^{-1}$, or equivalent dose of CrM, 3 days	After 4 days, greater Cr retention from CrM+CHO compared to other groups ($p < 0.05$) (0%, 60%, 80%, and 60% CrM retained for P, CrM, CrM+CHO, and effervescent Cr, respectively)

Preen (62)	CrM+CHO vs. CrM only and CrM + exercise (E)	Muscle [PCr and Cr] before and after 5-day intervention; one group performed exercise	CrM 20 g·day^{-1} CrM+CHO 20 g·day^{-1} + glucose 1 g·kg^{-1}, 2× · day^{-1} CrM + E 20 g·day^{-1} + 60 min repeated-sprints daily	After 5 days, 9% greater increase in [total Cr] from CrM+CHO ($p < 0.05$).(25% vs. 16% and 18% for CrM+CHO vs. CrM and CrM+E, respectively)
Derave (111)	CrM+PRO compared to CrM only and placebo (P)	Muscle [PCr and Cr] prior to and after 2-week right-leg immobilization followed by 6 weeks of right leg resistance training	CrM: 15 g·day^{-1} during immobilization followed by 2.5 g·day^{-1} during rehabilitation CrM+PRO: CrM dose + 40 g protein and 6 g AA during training	After training, ~30% increase from baseline in [total Cr] (right leg) in both CrM and CrM+PRO vs. P ($p < 0.05$)

(Continued)

Table 1
(Continued)

Reference	Experimental comparison	Protocol	Supplementation	Change
Steenge (43)	CrM+CHO (low and high dose) compared to CrM+CHO + PRO	Insulin and whole-body Cr retention values (24 hr) before and after each supplement trial (all participants completed four trials)	CrM (4 × 5 g) +5 g CHO +50 g CHO, +93 g CHO or +PRO+CHO (50 g each)	After 24 hr PRO+CHO provided similar insulin responses and [total Cr] accumulation values (~25%) as high-dose CHO ($p < 0.05$)
Cribb (63)	Compared CrM+CHO and CrM+PRO to CHO and PRO alone	Muscle [PCr and Cr] before and after 11 weeks of resistance exercise	All groups: 1.5 g of supplement·kg·day⁻¹ for 11 weeks CrM groups: 0.3 g·kg·day⁻¹ 5 days followed by 0.01 g·kg·day⁻¹ for 10 weeks	After 11 weeks, ~10% increase from baseline, [total Cr] after 11 weeks in both CrM+CHO and CrM+PRO groups ($p < 0.05$)

Cribb (67)	Compared supplement-timing; CrM+PRO+CHO before and after resistance exercise to same supplement at times not close to training	Muscle [PCr and Cr] assessed before and after 10 weeks of resistance exercise	Dose: 1 g supplement·kg^{-1} 2×·day^{-1} (CrM 0.01 g·kg·$^{-1}$) Taken immediately before and after workouts *or twice a day* 5 hours outside workouts, 10 weeks	After 10 weeks, 14% greater increase in [PCr] and 18% greater increase in [total Cr] from supplement timing (PCr 16% vs. 2%; total Cr 25% vs. 7%, respectively) ($p < 0.05$).
Greenwood (69)	Compared CrM + D-pinitol high-dose (HP) and low-dose (LP) as well as predosing (PreP) to CrM only and placebo (P)	Whole-body Cr retention via 24-hr urine samples for 4 days, 3 days supplementation, no exercise	CrM 4 × 5 g + 2 × 0.5 g D-pinitol (LP) + 4 × 0.5 g D-pinitol (HP) D-pinitol 2 × 0.5 g D-pinitol 5 days prior to and during CrM (PreP)	After 4 days, whole-body Cr retention was greater in LP and PreP compared to HP, CrM-only, and P ($p < 0.05$). 0%, 61% ± 15%, 83% ± 5%, 61% ± 22%, and 78% ± 9% CrM

(Continued)

Table 1
(Continued)

Reference	Experimental comparison	Protocol	Supplementation	Change
				retained for P, CM, LP, HP, and Pre-P groups, respectively)
Burke (73)	ALA+CrM+CHO vs. CrM+CHO and CrM-only	Muscle [PCr and Cr] assessed before and after 5-day intervention	CrM: 20 g·d^{-1} +100 g·d^{-1} sucrose (CrM+CHO) +1000 mg·d^{-1} ALA (ALA+CrM+CHO)	After 5 days, greater increase in [PCr] and [total Cr] from ALA+CrM+CHO ($p < 0.05$) compared to CrM+CHO and CrM-only (21%, 0%, and 0% increase for [PCr] and 13.8%, 2.0%, and 4.0% [total Cr] for ALA+CrM+CHO, CrM+CHO, and CrM, respectively) ($p < 0.05$)

and exercise performance is well established. That is, the reliance on muscle glycogen during exercise increases with intensity, and a direct relation between fatigue and depletion of muscle glycogen stores has been described *(74–78)*. Furthermore, the increase in endurance after an aerobic training program is associated with increased muscle glycogen storage capacity as well as its more efficient use *(79,80)*. Muscle glycogen is also an essential fuel source for the regeneration of ATP during short-term, high intensity (anaerobic) exercise. For example, during a set of 12 maximum-effort repetitions, just over 82% of ATP demands are estimated to be met by glycogenolysis *(81)*. A single bout of high-intensity resistance exercise characteristically results in a significant (30%–40%) reduction in muscle glycogen *(82–84)*. Muscle glycogen synthesis is affected not only by the extent of depletion but also by the type, duration, and intensity of the preceding exercise *(74–78,85)*. Nevertheless, the rapid restoration of muscle glycogen stores is a critical issue for all athletes who undertake training or competition sessions on the same or successive days. In general, the faster muscle glycogen stores can be replenished after exercise, the faster is the recovery process and the greater the return of performance capacity *(85)*.

Supplementation strategies that may increase the rate of muscle glycogen synthesis have been the focus of extensive investigation. For example, the importance of timing *(86,87)*, frequency *(88)*, and amount *(89,90)* of CHO for postexercise muscle glycogen restoration has been demonstrated. Regarding the effect of various types of CHO that may optimize postexercise glycogen synthesis, some well controlled studies have reported that rapid increases occur during the first 24 hours of recovery with a combination of high-GI CHOs in contrast to low-GI CHOs *(91,92)*. The high-GI sources included glucose and sucrose, and whole foods were also on the list (i.e., white potatoes, rice, pasta). The activation of glycogen synthase (the rate-limiting enzyme for glycogen synthesis) by insulin is well documented *(93,94)*. As high-GI CHOs generally evoke greater blood glucose and insulin levels than low-GI sources, this probably explains the more rapid synthesis of muscle glycogen after exercise from the selection of high-GI CHO sources. [For further reading on the GI of foods and meals, refer to Du et al. *(95)* and Brand-Miller *(57)*]. Aside from the influence of the GI, other efforts to further increase the rate of storage by increasing the amount and frequency of CHO intake or by changing the type and form of CHO supplement used have proved unsuccessful *(86,88,96,97)*.

Rather than focus on a single macronutrient to optimize muscle glycogen stores, a combination would provide a more practical, optimal approach to help meet the complex array of nutritional demands of exercise training. Probably because of the synergistic effect on insulin secretion, the impact of combining PRO with a CHO supplement on muscle glycogen synthesis after exercise has become a topic of interest *(98–105)*. Zawadzki et al. *(101)* were the first to report that the combination of PRO-CHO was more effective than CHO alone in the replenishment of muscle glycogen during the 4 hours immediately after exercise. These authors suggested that the greater rate of muscle glycogen storage from PRO-CHO was the result of a greater plasma insulin response. However, the enhancement of muscle glycogen storage observed by these researchers may have been due to the larger amount of calories provided by the PRO-CHO treatment. Moreover, some evidence suggests that if adequate CHO is provided the addition of PRO has no beneficial effect on muscle glycogen recovery *(106)*. To partially support this notion, some *(85,99,105)* but definitely not all *(98,103,104)* investigations have reported increased glycogen synthesis after the consumption of a PRO-CHO supplement compared to CHO-only of an equivalent dose or caloric content. However, only one study *(85)* has examined the effects of PRO-CHO supplementation compared with CHO supplementation of equal CHO content (LCHO) and equal caloric content (HCHO) in the same trial. Unlike most studies that have assessed glycogen resynthesis with repeated muscle biopsy, Ivy et al. *(85)* utilized natural abundance ^{13}C-nuclear magnetic resonance (NMR) spectroscopy to measure muscle glycogen concentrations. A limitation of repeated muscle biopsies is the number and frequency of measurements that can be obtained as well as the sampling of only a small volume in nonhomogeneous tissue. Because of the noninvasive nature, the NMR technique is purported to provide better time resolution (frequency), repeatability, and precision *(107)*. Using this method to assess muscle glycogen synthesis, Ivy et al. *(85)* reported that the combination of PRO-CHO yielded greater (< 0.05) muscle glycogen storage during the 4 hours immediately after intense exercise compared with both LCHO and HCHO supplements. The percentages of glycogen restored during the 4-hour recovery period were 46.8%, 31.1%, and 28.0% for the PRO-CHO, HCHO, and LCHO treatments, respectively. More recently, Berardi et al. *(105)* utilized the NMR technique and reported a similar result. That is, supplementation with PRO-CHO resulted in greater muscle glycogen resynthesis 6 hours after exercise than an isocaloric dose of CHO ($p < 0.05$).

In general, it appears that the addition of PRO to a CHO supplement increases the rate of muscle glycogen storage during the hours immediately after exercise, particularly if the supplement contains a low to moderate amount of CHO. The mechanism by which protein increases the efficiency of muscle glycogen storage is not known, but there are several possibilities. In brief, the combination of PRO and CHO may accelerate the rate of muscle glycogen storage possibly by activating glycogen synthesis by two mechanisms. First, this combination may raise plasma insulin levels beyond that typical of CHO alone, which may augment muscle glucose uptake and activate glycogen synthase. Insulin stimulates glucose uptake, glycolysis, and glycogen synthesis in muscle via the activation of the PI3–PK-B(Akt)–GSK-3 signaling pathway *(108)*. Second, the increase in plasma amino acids that occur as a result of consuming PRO may activate glycogen synthase through an insulin-independent pathway that has not been clearly identified *(85)*, thereby having an additive effect on the activity of this enzyme. Whereas glucose, sucrose, and glucose polymer supplements as well as high-GI whole-food sources are effective means of replenishing muscle glycogen, the type of protein (or amino acids) that may be best to combine with CHO has received less attention. Studies that have reported a beneficial impact on muscle glycogen stores from the addition of PRO to a CHO supplement have utilized dairy proteins *(85,101,102)*, such as whey isolates *(105)*. A hydrolyzed wheat protein supplement in conjunction with insulin-promoting amino acids (AAs) such as leucine and phenylalanine *(99,100)* has also been shown to have a favorable effect on postexercise glycogen synthesis. In fact, studies that have examined this area directly suggest that the insulin response to PRO-CHO supplementation may be positively correlated with plasma concentrations of AAs such as leucine, phenylalanine, and tyrosine *(100)*. Therefore, the concentration of certain AAs in the supplement may underline its ability to stimulate insulin and therefore muscle glycogen restoration.

Indeed, Yaspelkis and Ivy *(109)* examined the effects of combining CHO with arginine on postexercise muscle glycogen storage following muscle glycogen depletion. Well trained cyclists rode for 2 hours on two occasions to deplete their muscle glycogen stores. At 0, 1, 2, and 3 hours after each exercise bout, the subjects ingested either a CHO supplement (1 g CHO/kg body weight) or a CHO-arginine (CHO/AA) supplement (1 g CHO/kg and 0.08 g arginine HCl/kg). No difference in the rate of glycogen storage was found between the CHO/AA and CHO treatments, although significance was approached. There were also no differences between treatments in regard to plasma glucose, insulin, or blood lactate responses. However, postexercise CHO oxidation during the CHO/AA treatment was significantly reduced compared to that with the CHO treatment. These results suggest that the addition of arginine to a CHO supplement reduces the rate of CHO oxidation after exercise and therefore may increase the availability of glucose for muscle glycogen storage during recovery *(109)*.

As discussed earlier, supplementation with CrM promote an ergogenic effect by enhancing PCr availability in muscle. However, another ergogenic effect of this supplement appears to be its positive impact on muscle glycogen storage. Seven studies have measured muscle glycogen levels in humans after CrM supplementation, and six have reported a stimulatory effect *(60,67,110–113)*. Robinson et al. *(60)* first showed that CrM supplementation in conjunction with a high-CHO diet for 5 days (after a bout of exhaustive exercise) resulted in a 23% greater increase in muscle glycogen than that with a high-CHO diet without CrM. Nelson et al. *(110)* reported that loading with CrM for 5 days enhanced a subsequent 3-day muscle glycogen-loading protocol by 12%. Op 't Eijnde et al. *(111)* demonstrated that supplementation with CrM (20 g daily) had no effect on muscle glycogen stores during 2 weeks of leg immobilization. However, further administration (15 g daily) did enhance muscle glycogen levels (by 46% more than placebo) during 3 weeks of subsequent strength training. In a follow-up study that involved a similar protocol (and a 6-week training phase), these researchers reported that supplementation with PRO (46 g) combined with CrM augmented posttraining muscle glycogen by 35% more than placebo (but not CrM alone) *(112)*. Van Loon et al. *(113)* demonstrated that a 5-day

CrM-loading phase augmented muscle glycogen by 14% compared with no change in the placebo group. Furthermore, this study confirmed a significant correlation between changes in muscle Cr (mean increase of 32%) and muscle glycogen during the loading phase. This substantiates other work *(110–112)* suggesting that significant increases in muscle Cr is a prerequisite for enhanced muscle glycogen storage *(48)*. Supplementation with CrM+CHO or PRO during exercise increases muscle GLUT-4 expression and glycogen storage *(111,112)*. Treatment with CrM has also been shown to increase total body water including intracellular cell volume *(114)*. Changes in cell volume (cellular water content) have been shown to influence glycogen levels *(115)*. Therefore, the ability of CrM to influence GLUT-4 biogenesis and/or regulate cell volume may explain its beneficial impact on muscle glycogen storage. Overall, the findings of these studies suggest that increasing muscle Cr (by ~20%) via supplementation ensures a beneficial impact on glycogen storage.

In summary, when considering all of the research that has been completed on this topic, it appears that the addition of PRO to a CHO supplement increases the rate of muscle glycogen storage during the hours immediately after exercise, particularly if small doses of these macronutrients are desired. However, more work is needed that focuses on different types of protein and/or composition of its amino acids (in combination with CHO) to ascertain what combinations may provide the most beneficial effect on muscle glycogen. Loading with CrM or the addition of CrM to a PRO-CHO supplement not only appears to augment Cr uptake, it is an effective strategy for optimizing muscle glycogen stores. For some athletes, however, careful consideration is needed when contemplating the addition of CrM. For example, loading with CrM characteristically results in a 1- to 3-kg gain in body weight (lean mass) *(13–16,18,19)*. This added mass may offset any potential ergogenic benefit that might be achieved via boosting muscle gylcogen stores. Therefore, in sports where any gain in body weight may disadvantage the athlete, the combination of PRO-CHO (without CrM) maybe a more prudent choice to promote muscle glycogen. However, for all athletes, along with more efficient glycogen restoration, another important advantage of PRO-CHO postexercise supplementation is this combinations' well documented effect on protein synthesis and muscle anabolism.

4. SUPPLEMENT COMBINATIONS TO ENHANCE MUSCLE ANABOLISM

Any adaptive change in muscle mass in response to exercise training must involve alterations in protein turnover. That is, provided the exercise intensity is of sufficient magnitude, muscle protein synthesis and breakdown are acutely stimulated *(116)*. In the absence of nutrient intake, muscle protein degradation exceeds synthesis during the early stages of recovery from exercise, and the net muscle protein balance remains negative (i.e., the muscle is in a catabolic state) *(117)*. Resistance exercise (RE) is incorporated into nearly every athletes' program in an effort to improve either strength, muscle mass, body composition, or the power-to-weight ratio. In addition to athletic populations, others such as older adults and those living with clinical illnesses would benefit from these adaptations. For these reasons, there has been a concentrated focus in the exercise science communities on specific nutritional strategies that affect the acute responses to RE (i.e., enhance muscle protein synthesis, reduce breakdown) and promote a positive protein balance (anabolism). In particular, the stimulation of muscle protein synthesis is thought to be the facilitating process that underlines gains in strength and muscle hypertrophy from training *(118–120)*. Probably for this reason, a number of acute response studies have examined the effects of strategic nutrient supplementation close to RE on muscle protein synthesis in an attempt to stimulate a higher rate and promote a positive net balance after exercise.

The optimal composition of nutrients to maximize muscle protein synthesis (and anabolism) after exercise is not known. However, the acute stimulation of protein synthesis appears to be dependent on the availability of the essential amino acids (EAAs). A positive net protein balance is not achieved unless an exogenous source is provided after exercise *(117)*. It is also clear that the combination of protein and carbohydrate (PRO-CHO) at a time close to exercise (i.e., the hours just before and/or afterward) yields a high anabolic response by altering the acute hormonal and protein turnover response patterns to create an environment that probably helps optimize conditions for recovery. For example, the combination of protein (or EAAs) and RE was initially shown to have a synergistic effect on (thigh) muscle protein synthesis that resulted in a positive net balance *(117,121)*. However, the addition of CHO (glucose 35 g) to EAAs (6 g) at this time amplifies muscle anabolism to a greater extent than when either macronutrient is provided separately after exercise *(122)*. In fact, when this combination was consumed 1 or 3 hours after RE, an increase in synthesis rates of up to 400% above preexercise values was reported, which is the highest ever recorded *(123)*. The same supplement has been shown to promote a similar anabolic effect in muscle when administered just before RE *(124)*. These investigations utilized AA solutions, whereas other studies have confirmed that whole proteins (15–35 g), such as the dairy proteins whey and casein, evoke an acute anabolic response that is similar in magnitude to free-form AA *(87,125–127)*. The finding that doses of whole proteins (e.g., whey, casein) are just as efficient as free-form AA at promoting muscle anabolism is important; in general, whole protein supplements are more economical than free-form AAs and may also provide other health benefits (e.g., additional vitamins, minerals and/or enhanced antioxidant capacity).

It is clear that the strategic intake of nutrients (i.e., consumption of PRO-CHO before and/or after intense exercise) not only augments muscle protein synthesis, most importantly it shifts the net protein balance to a positive state (albeit transiently). This anabolic response can be at least partly attributed to changes in the acute hormonal response pattern. For instance, a novel study by Kraemer et al. *(128)* examined the effects of a high calorie PRO-CHO supplement (total $7.9 \, kcal \cdot kg^{-1}$, 1.3 g glucose polymer $\cdot kg^{-1}$, and 0.7 g dairy proteins $\cdot kg^{-1}(day^{-1})$) consumed 2 hours before and just after resistance exercise for consecutive 3 days of training. Results showed that compared to a low-calorie (non-insulin-stimulating) placebo, the PRO-CHO supplement consistently provided higher blood insulin levels during the hour after exercise *(128)*. This PRO-CHO-induced stimulation of insulin is important; it improves the anabolic response by increasing AA uptake and decreases the rate of muscle protein breakdown *(129)*. Kraemer et al. *(128)* also reported that nutrient timing with the PRO-CHO supplement enhanced acute serum growth hormone (GH) responses for 30 minutes after exercise (on the first day) compared to the noncaloric placebo. Although the reason for this increase is not clear, Chandler et al. *(130)* also reported an increase in serum GH in response to consumption of a similar PRO-CHO supplement immediately and 120 minutes after resistance exercise. In contrast, Williams et al. *(131)* reported no significant effect of PRO-CHO on the GH response to exercise. The regulation of hepatic insulin-like growth fact-1 (IGF-1) is characteristic of GH *(132)*. Kraemer et al. *(128)* also reported that the PRO-CHO supplement elevated serum IGF-1 levels for 30 minutes after exercise on two of three training days. Another investigation reported an increase in (resting) plasma IGF-1 after 6 months of training in response to the daily consumption of a PRO-CHO supplement (42 g PRO, 24 g CHO) in contrast to CHO (70 g) alone *(133)*. Furthermore, Willoughby et al. *(134)* reported that 10 weeks of heavy resistance exercise combined with a similar dose of PRO-CHO before and after each workout was effective for increasing serum IGF-1 and muscle IGF-1 mRNA expression. However, it is important to note that although nutrient timing with PRO-CHO close to RE may increase serum IGF-1 concentrations, the anabolic action of this growth factor on tissue is thought to reside in alterations in its binding proteins *(135)*. Separately, PRO-CHO meals *(136)* and RE *(137)* appear to influence regulation of the IGF-1-binding proteins. However, no studies have examined the impact of combining supplementation and exercise on the IGF-1-binding proteins and muscle anabolism.

Testosterone is an important anabolic hormone thought to augment the synthesis of muscle protein. The intake of a PRO-CHO supplement before and after RE appears to be one of the few strategies shown consistently to affect circulating testosterone levels *(118,127,129,138)*. The nutrient-timing study by Kraemer et al. *(128)* also assessed acute testosterone responses, and these researchers reported an acute increase in circulating testosterone followed by a sharp decrease (to levels that were significantly lower than baseline) with PRO-CHO supplementation. This response was consistently observed on each of the three training days assessed. Chandler et al. *(130)* and Bloomer et al. *(138)* reported a similar response. This rapid decrease in blood testosterone levels in response to supplementation close to exercise may be due to increased metabolic clearance of this hormone, such as increased uptake by muscle. At least one study supported this assumption. Chandler et al. *(130)* showed that a decline in circulating testosterone in response to nutrient timing after RE was not linked to a decrease in luteinizing hormone production. As mentioned previously, nutrient timing with PRO-CHO provides a dramatic increase in muscle protein synthesis in the hours after exercise *(121,124)*. Therefore, the drop in circulating testosterone could be due to increased uptake by muscle to facilitate this process. To further support this contention, Volek *(119)* reported that a postworkout PRO-CHO meal decreased circulating testosterone that corresponded with an increase in muscle androgen receptor content. Along this line, a more recent study (that utilized resistance-trained participants) reported that whereas a PRO-CHO meal after exercise up-regulated androgen receptor content in muscle, the addition of L-carnitine L-tartrate (equivalent to 2 g of L-carnitine(day^{-1} for 3 weeks) resulted in an even greater response *(139)*. Previous work by this research group *(140)* showed that 3 weeks of L-carnitine L-tartrate reduced the amount of exercise-induced muscle tissue damage by 7% to 10% (assessed via magnetic resonance imaging scans of the thigh) as well as increased IGF-1-binding protein (IGFBP-3) concentrations before and up to 180 minutes after acute exercise. Therefore, the addition of L-carnitine L-tartrate to a PRO-CHO postexercise supplementation regimen may improve testosterone uptake and the overall anabolic response from resistance exercise.

At the molecular level, the synergistic effect of a supplement containing PRO-CHO on muscle anabolism is probably due to the activation of insulin-dependent but also insulin-independent pathways. For example, unlike exercise or insulin, amino acids do not appear to stimulate muscle protein synthesis via phosphorylation (activation) of the PI3 and PKB(Akt) signaling proteins *(141,142)*. Human studies *(143)* have confirmed in vitro (144) and in vivo *(145,146)* work that has shown EAAs stimulate muscle protein synthesis directly via the phosphorylation of downstream signaling proteins such as the Raptor–mTOR complex (and its regulatory proteins S6K1 and 4E-BP1) or the eIF–2B complex (the only one of the three regulators of muscle protein synthesis that is not under direct control of mTOR) *(147)*. Additionally, some EAAs, such as the branched-chain amino acids (BCAAs) (leucine, valine, isoleucine) are particularly effective at enhancing muscle protein synthesis via these pathways *(148)*. Consequently, attention has shifted toward examining the effects of combining certain AAs with whole proteins and CHO on postexercise muscle anabolism *(126,149)*. For example, Borsheim et al. *(149)* reported that the combination of whey protein (17.5 g), free-form AAs (4.9 g), and CHO (77.4 g) stimulated net muscle protein synthesis to a greater extent than an isoenergetic CHO supplement after resistance exercise. The authors also concluded that the addition of whole protein to the AA-CHO supplement prolonged the anabolic response observed in previous studies with AA-CHO mixtures.

The BCAA leucine is an established regulator of whole-body and skeletal muscle protein metabolism *(150)*. Supplementation with leucine alone can stimulate muscle protein synthesis, independently of insulin *146,151)*, and may also play a role in minimizing protein breakdown *(152)*. Koopman et al. *(126)* attempted to extend these findings by investigating whether adding leucine to a PRO-CHO supplement could further promote muscle protein anabolism. In this study, eight healthy but untrained male subjects were randomly assigned to three trials in which they consumed drinks containing either CHO (0.3 g·kg⁻¹ h⁻¹), CHO+PRO (0.3 g CHO + 0.2 g whey protein·kg⁻¹ hr⁻¹), or PRO-CHO and free leucine (0.1 g·kg⁻¹) (CHO+PRO+Leu) for 5 hours following 45 minutes of RE *(126)*. Whole-body protein turnover and the fractional synthesis rates in muscle (incorporation of labeled phenylalanine) were assessed. The results obtained suggested that the addition of the leucine significantly increased whole-body net protein balance and provided a higher anabolic response in muscle *(126)*. However, it is worth noting that the total amounts of leucine in the two PRO-CHO supplements were different. The leucine-enriched PRO-CHO supplement provided 9.6 g·hr⁻¹ for an 80 kg person, whereas the PRO-CHO supplement provided only 1.6 g·hr⁻¹ (for an individual of the same weight) *(126)*.

In summary, it is clear that supplementation with PRO-CHO (with or without additional AAs) can alter the acute anabolic response to resistance exercise. However, a more pertinent question is whether repeated metabolic alterations provided by supplementation with PRO-CHO are of sufficient magnitude to alter long-term adaptations to resistance training. As the following section demonstrates, a strong theoretical basis exists for expecting a beneficial effect from supplementation during resistance training, but no studies to date have systematically linked acute physiological responses to chronic adaptations in the same study.

5. COMBINATIONS THAT ENHANCE AEROBIC/ ANAEROBIC PERFORMANCE

Caffeine, a naturally occurring substance, is the most commonly consumed stimulant drug in the world. It produces multiple physiological effects throughout the body including: increased catecholamine release and fat metabolism, resulting in glycogen sparing; increased intracellular Ca^{2+} release; inhibition of cyclic adenosine monophospate (cAMP) phosphodiesterase, and antagonism of adenosine receptors *(153)*. Several studies have demonstrated improved exercise performance in submaximal endurance activities *(153–155)*, but its potential ergogenic effect in acute, high intensity exercise is less clear *(154)*.

Ephedrine was used as a central nervous system (CNS) stimulant in China for centuries before its introduction to Western medicine in 1924 *(156)*. Ephedrine and its related alkaloids (mostly pseudoephedrine) are sympathomimetic agents that stimulate the sympathetic nervous system, increasing circulating catecholamines *(157)*. A number of studies have reported beneficial effects on exercise performance using ephedrine as the supplement *(158,159)*, whereas few studies have reported benefit utilizing the related alkaloids such as pseudoephedrine *(160,161)*. This is most likely due to ephedrine's direct adrenoceptor stimulating actions *(162)*, resulting in it being approximately 2.5-fold more potent than pseudoephedrine *(160)*.

Although both caffeine and ephedrine have demonstrated independent ergogenic effects on exercise performance, research published from Bell and Jacob's laboratory at the Defense and Civil Institute of Environmental Medicine in Canada has indicated that in several instances caffeine–ephedrine mixtures confer a greater ergogenic benefit than either drug alone *(163–167)*. In a series of studies performed by Bell and Jacob *(163–166)*, positive results were observed during various exercise modalities: submaximal steady-state aerobic exercise *(167)*; short- and long-distance running *(164,166)*; and maximal anaerobic cycling *(163)*. The caffeine–ephedrine mixture was normally consumed 1.5 to 2.0 hours prior to exercise at a dosage range of 4 to 5 mg/kg for caffeine and 0.8 to 1.0 mg/kg for ephedrine *(163–166)*. Higher dosages were shown to elicit negative side affects such as vomiting and nausea during the exercise test; thus, Bell and colleagues recommended using the lower dosages of 4 mg/kg for caffeine and 0.8 mg/kg for ephedrine. Importantly, the lower dosage provided an ergogenic effect similar in magnitude to those reported previously using the higher doses *(168)*.

Results from these associated studies showed that caffeine, ephedrine, and the caffeine–ephedrine supplements produced significant effects on a variety of metabolic and cardiovascular responses such as blood glucose, catecholamines, and heart rate during exercise compared to the dietary fiber placebo *(164,166,167,169)*. Despite the independent effects of caffeine and ephedrine on metabolic and cardiovascular responses during exercise, no ergogenic effects on exercise performance was observed. However, when combined, exercise performance was significantly enhanced in a variety of exercise modalities. Researchers suggested that this is most likely due to ephedrine's effect on arousal (i.e., decreasing rating of perceived exertion during exercise) combined with caffeine's ability to enhance muscle metabolism *(164,166,167,169)*.

Although it is clearly evident that the combination of caffeine and ephedrine has a pronounced ergogenic effect on a variety of exercise modalities compared to either supplement alone, it should be noted that these beneficial effects have predominantly been observed in studies involving the Canadian military. Hence, further research is needed to examine the practical application in recreational athletes and untrained individuals. A more important issue is that all dietary supplements containing ephedrine alkaloids are illegal for marketing in the United States *(170)*. Therefore, until ephedrine and ephedrine alakaloids are made legal again, the performance-enhancing effects of the caffeine–ephedrine mixture can be utilized only under research conditions.

The benefits of creatine monohydrate (CrM) to athletes is clear (see earlier). One proposed ergogenic benefit is the capacity of CrM to help maintain normal muscle pH levels during high intensity exercise by consuming excess hydrogen ions during ATP resynthesis and thus possibly delaying fatigue (refer to Multifaceted Role of the Muscle PCr-Cr System in Exercise Metabolism). The intake of sodium bicarbonate ($NaHCO_3$) has also been shown to prevent exercise-induced perturbations in the acid-base balance, which has resulted in enhanced performance *(171–173)*.

Mero and colleagues *(174)* examined the buffering capacity of sodium bicarbonate in combination with CrM on consecutive maximal swims. In a double-blind crossover procedure, competitive male and female swimmers completed, in a randomized order, two treatments (placebo and a combination of CrM + sodium bicarbonate). There was a 30-day washout period between treatments. Both treatments consisted of placebo or CrM supplementation (20 g/day) for 6 days. On the morning of the seventh day, a placebo or sodium bicarbonate supplement (0.3 g/kg body weight) was taken 2 hours prior to the warmup. Two maximal 100-m freestyle swims were performed with a passive recovery of 10 minutes between them. The first swim performances for both treatment groups had similar times. However, the increase in time for the second swim performances was significantly less in the combination group compared to the placebo. Furthermore, the mean blood pH was higher in the combination group compared to the placebo group after supplementation on the test day. The data indicated that simultaneous supplementation of CrM and sodium bicarbonate enhances the buffering capacity of the body and hence the anaerobic performance *(174)*.

In summary, it is evident that when nutritional supplements with complementary independent ergogenic effects are combined, additional benefits can be attained. Research has shown that both caffeine-ephedrine and CrM-sodium bicarbonate supplement mixtures provide an acute physiological response that enhances anaerobic and/or aerobic exercise. However, a number of limitations exist with both supplements. As mentioned, further research is needed to examine the practical application of the caffeine–ephedrine combination in recreational athletes and untrained individuals, as most of the research has been performed in military soldiers. More importantly, however, because all dietary supplements containing ephedrine alkaloids are illegal for marketing in the United States, its use as an ergogenic aid is limited in active individuals. Although sodium bicarbonate is a legal supplement, limited studies have examined its ergogenic effects on exercise performance when combined with CrM, and so further investigation is needed to confirm such observations.

6. CHRONIC ADAPTATIONS: SUPPLEMENT COMBINATIONS THAT PROMOTE MUSCLE HYPERTROPHY AND STRENGTH

In most instances, supplement combination of CrM with protein (PRO) and/or carbohydrate (CHO) has been shown in longer-term trials (6–12 weeks) to enhance the chronic adaptations that are desired from resistance training (i.e., gains in strength and lean body mass and/or improvements in body composition). Kreider and his research group were among the first to examine the effects of CrM-containing PRO and CHO supplements on the development of strength and lean body mass during structured resistance training.

In a study involving 25 National Collegiate Athletic Association (NCAA) division IA football players, Kreider et al. *(18)* demonstrated that 28 days of supplementation with CrM-CHO (containing glucose 99 g·day^{-1} and CrM 15.75 g·day^{-1}) resulted in greater ($p < 0.05$) gains in dual energy x-ray absorptiometry (DEXA)- determined body mass and lean (fat/bone-free) body mass (LBM) compared to an equivalent dose of CHO. Treatment with CrM-CHO also resulted in greater total bench press, squats, and power clean lifting volume as well as sprint performance *(18)*.

Using a group of experienced weightlifters, Burke et al. *(64)* assessed strength and LBM changes after 6 weeks of resistance exercise while ingesting a supplement containing CrM and PRO (whey 1.2 g·kg·day^{-1} and CrM 0.1 g·kg·day^{-1} for 6 weeks) in comparison to a similar dose of PRO (whey) or CHO (maltodextrin) (1.2 g·kg·day^{-1}). LBM increased to a greater extent in the CrM-PRO group than in the PRO- or CHO-alone groups. Bench press strength also increased to a greater extent in the CrM-PRO group than in the PRO- or CHO-only groups, but all other strength/power measures increased to a similar extent *(64)*.

Only one study has directly compared the effects of CrM-CHO and CrM-PRO supplementation (supplement 1.5 g·kg·day^{-1}) on strength, body composition, and muscle hypertrophy during a resistance training program *(63)*. In this study, four groups of matched, recreational bodybuilders were assessed before and after an 11-week program. The groups given the CrM-containing supplements demonstrated greater ($p < 0.05$) strength improvements in all three assessments (1RM bench press, squats, pulldown) and muscle fiber hypertrophy compared to groups given an equivalent dose of CHO or PRO *(63)*. However, there were some subtle but significant differences in body composition changes observed among the groups, and these differences may have implications for different populations (see sidebar: Creatine + Protein or Creatine + Carbohydrate for Better Muscle Hypertrophy?)

Kreider's group were the first to demonstrate that a CrM-containing PRO-CHO supplement (containing glucose $50 \, g \cdot d^{-1}$, dairy protein $50 \, g \cdot d^{-1}$, CrM $15.75 \, g \cdot d^{-1}$) during resistance training can provide greater ($p < 0.05$) gains in strength and LBM than an equivalent dose of PRO-CHO (that does not contain CrM) *(19)*. The effectiveness of adding CrM to a PRO-CHO supplement regarding the development of strength and muscle mass was confirmed some 10 years later in another trial *(175)*. Like Kreider et al. *(19)*, this study utilized experienced lifters (recreational body-builders). However, in this trial the two groups were given the exact same PRO-CHO supplement (50% whey isolate and 50% glucose) (each $1.5 \, g \cdot kg \cdot day^{-1}$) in a double-blind manner with one of the supplements containing a daily serving of CrM ($0.1 \, g \cdot kg \cdot day^{-1}$). A third group was provided with an equivalent dose of PRO only ($1.5 \, g \cdot kg \cdot day^{-1}$). Assessments completed the week before and after the 10-week program included strength (1RM, barbell bench press, squats, pulldown), body composition (determined by DEXA) and vastus lateralis muscle biopsies for determination of muscle fiber type (I, IIa, IIx), cross-sectional area (CSA), and contractile protein content. The most important finding of this investigation was that the CrM-containing PRO-CHO supplement provided greater ($p < 0.05$) gains in 1RM strength (in all three assessments) and muscle hypertrophy compared to supplementation with an equivalent dose of PRO-CHO or PRO *(175)*. Most importantly, a greater ($p < 0.05$) muscle hypertrophic response from the combination of CrM-PRO-CHO was evident at three levels of physiology. That is, this group demonstrated a greater gain in LBM, hypertrophy of the type IIa and IIx fibers, and increased contractile protein *(175)*. This research is particularly relevant as few studies involving exercise and supplementation have confirmed improvements in body composition plus hypertrophic responses at the cellular level (i.e., fiber-specific hypertrophy) and subcellular level (i.e., contractile protein content).

However, not all studies support the hypothesis that a CrM-containing PRO-CHO supplement provides greater adaptations than supplementation with a similar amount of nitrogen and energy. A study by Tarnopolsky et al. *(65)* utilized previously inactive participants and daily supplementation with either CrM (10 g) + CHO (75 g) (1252 KJ or 300 kcal) or protein (10 g) + CHO (75 g) (1420 KJ or 340 kcal) during 10 weeks of resistance training. Results indicated that CrM treatment provided no greater gains in strength, LBM, or muscle fiber hypertrophy *(65)*. One explanation for the discrepancy between these results and those reported by Kreider et al. *(18,19)* and Cribb et al. *(63,176)* may have been the populations used. Whereas Kreider et al. *(18,19)* and Cribb et al. *(63,175)* utilized experienced (trained) participants, Tarnopolsky et al. *(65)* recruited participants who had been inactive prior to the study. Although the influence of training status on the effects of supplementation is unknown, it has been speculated that trained individuals might experience more efficient muscle Cr uptake, as exercise training is associated with improved insulin sensitivity *(30)*. Therefore, resistance-trained individuals may theoretically experience greater adaptations from supplementation *(30)*.

Aside form PRO and CHO, other compounds with purported ergogenic potential have been examined in combination with CrM during resistance training. However, in terms of absolute strength and body composition changes, the benefit of the supplement combination has seldom exceeded the results achieved from CrM treatment alone. For example, when compared with CrM only $(0.22 \, g \cdot kg \cdot day^{-1})$, supplementation with a combination of pyruvate and CrM during 5 weeks of resistance training provided no greater benefit with regard to gains in body mass, LBM, 1RM strength, power output, or force development (vertical jump test) *(176)*. Likewise, studies that have examined the effects of combining CrM with magnesium *(154)* or HMβ *(178,179)* (a leucine metabolite) have shown no greater ergogenic effect than treatment with CrM alone. With regard to HMβ, this is not surprising; research groups outside those involved in the patent of this supplement have been unable to show a consistent beneficial effect from its use. This includes not only strength development but also body composition and a range of symptoms associated with muscle damage *(180–184)*.

One compound that may prove to be an exception is β-alanine. Studies by Hill et al. *(185)* and Harris et al. *(186)* demonstrated that 28 days of β-alanine $(4–6 \, g \cdot kg \cdot day^{-1})$ supplementation increased intramuscular levels of carnosine by approximately 60%. Carnosine appears to serve as a buffer and helps maintain skeletal muscle acid-base homeostasis when a large quantity of H^+ is produced during high-intensity exercise *(187)*. Harris et al. *(188)* also demonstrated improvements in performance during a 4-minute maximal cycle ergometry test in men after supplementation with β-alanine ($3.2\,g\cdot kg\cdot day^{-1}$) for 5 weeks. Others have shown that a similar supplementation protocol can improve submaximal cycle ergometry performance and time-to-exhaustion *(189)*, delay the onset of neuromuscular fatigue during incremental cycle ergometry *(190)*, or increase the amount of work completed during high-intensity exercise (cycling to exhaustion at 110% of estimated power maximum) *(185)*.

The efficacy of combining CrM and β-alanine was examined in regard to strength performance during resistance training. Hoffman et al. completed a 6-week training/supplementation study involving three groups: CrM, CrM+β-alanine, placebo. Both the CrM and CrM+β-alanine groups demonstrated significantly better gains in 1RM strength and LBM than the placebo group, but no differences were detected between the two CrM-treated groups *(191)*. However, there were trends for better gains in LBM in the group given CrM+β-alanine. Additionally, this group tended to show greater (average) training volumes for the bench press and squat exercises. If the study was of longer duration, it is possible that the greater amount of work completed by this group may have had an affect on strength development and lean tissue accruement.

The protein source acutely affects muscle amino acid uptake and net protein balance following resistance exercise. This appears to be related not only to amino acid composition but also to the pattern of amino acid delivery to peripheral tissues. For example, dairy milk proteins are shown to be more effective at supporting protein accretion than soy proteins *(192)*. Whey protein is a collective term that encompasses a range of soluble protein fractions found in dairy milk. In supplement form, whey protein is considered a "fast-absorbing" protein based on studies that showing that consumption (20–30 g) instigates a rapid but transient increase in blood amino acids levels and stimulates a high rate of muscle protein synthesis *(193)*. On the other hand, casein (the other major dairy milk protein) is more slowly absorbed from the gut and manifests a lower but sustained increase in blood amino acids for several hours *(193)*. These attributes suggest that the combination of whey and casein may be most beneficial in supporting muscle protein anabolism and increasing muscle mass during the course of an intense (high-overload) resistance training program.

In young, healthy adults, a blend of whey and casein (30 g) taken after exercise has been shown to result in greater hypertrophy of type I and II muscle fibers and improve muscle performance after 14 weeks of training *(193)*. Kerksick et al. *(194)* examined the effects of supplementation with a combination of whey and casein (40 g and 8 g, respectively) or whey and amino acids (whey 40 g + BCAA 3 g + glutamine 5 g) or a CHO placebo (total 48 g) on performance and training adaptations during 10 weeks of resistance training. Although strength gains were similar among the protein-supplemented groups, the group given the whey-casein combination experienced the greatest ($p < 0.05$) increase in DEXA-determined LBM *(194)*. The whey protein supplements used in these investigations are generally isolates ($\geq 90\%$ protein) and concentrates ($\geq 80\%$ protein). However, the degree of hydrolysis of the material (be it casein or whey) can affect the protein's absorption/digestion kinetics *(195)*.

Although the supplement combination of whey and casein appears to be effective at promoting lean mass during resistance training, no studies have examined what type or ratio is most beneficial. Whether the addition of certain amino acids can optimize the effects of the supplement blend also remains unclear. Nevertheless, a substantial body of evidence now suggests that supplementation with proteins and amino acid mixtures can influence adaptations to training. However, a steadily increasing amount of work suggests that the precise timing of the supplement may enhance the response even further.

The acute response studies discussed earlier clearly demonstrate that oral supplementation with whole proteins (e.g., whey, casein) or essential amino acids immediately before and/or after resistance exercise promotes a better anabolic response (i.e., higher stimulation of protein synthesis and a positive net protein balance) compared to placebo treatments. In young adults, the presence of CHO (e.g., glucose) appears to enhance this response by increasing blood insulin levels. Insulin receptor activation stimulates the PI3K–Akt/PKB–mTOR signaling pathway, which is known to have profound effects on the up-regulation of muscle-specific gene expression and protein synthesis *(196)*. Proteins that contain a high dose of essential amino acids (leucine in particular) are known to up-regulate the activity of mTOR and p70S6 kinase and hyperphosphorylate 4E-BP1 *(198)*. This suggests that amino acids and insulin signaling do not function in isolation but may function cooperatively to optimize the anabolic response in skeletal muscle. For these reasons, it has been suggested that the consumption of a supplement containing PRO and CHO immediately before and after resistance exercise (i.e., supplement timing) may provide the ideal anabolic conditions for muscle growth *(195)*. Indeed, most studies that have assessed chronic adaptations during resistance training have reported greater muscle hypertrophy *(193,199)* or a statistical trend for gains in LBM *(200,201)* from this strategy.

For example, Willoughby et al. *(134)* demonstrated that supplementation with a protein blend (whey 20 g, casein 8 g, and 12 g free amino acids; total 40 g of protein) 1 hour before and immediately after each workout (10 weeks) was more effective than 40 g of CHO (placebo) at increasing muscle strength and mass. Additionally, these researchers reported that the protein blend provided a significant increase in systemic (serum IGF) and local (muscle IGF-1, MHC isoforms mRNA, myofibrillar protein) indicators suggestive of skeletal muscle anabolism and hypertrophy *(133)*. However, there have been some important limitations to these insightful investigations. First, the participants in most studies that have examined the effects of supplement timing were not permitted to consume any nutrients other than the designated supplement for up to 3 hours before and after each workout. Therefore, the results can be attributed to the presence (or absence) of macronutrients but not the supplement per se.

To date, only one study has examined whether supplement timing with PRO and CHO provides greater benefits in terms of muscle hypertrophy or strength development compared to the consumption of the same supplement at other times during the day. This study, by Cribb and Hayes *(67)*, examined the effects of supplement timing with a CrM-containing PRO-CHO supplement during a 10-week resistance exercise training program. The researchers reported that when a CrM-PRO-CHO supplement was consumed immediately before and after each workout this strategy resulted in greater ($p < 0.05$) strength gains (two of three assessments), muscle hypertrophy of type II fibers, and better improvements in body composition *(67)* (see sidebar: Can Supplement Timing Double Gains in Muscle Mass?).

To summarize this section, chronic adaptations that are desired from resistance training (i.e., strength, muscle hypertrophy, and/or lean body mass) are enhanced by the combination of CrM with PRO and/or CHO (up to $1–5\,g\cdot kg\cdot day^{-1}$) appears to be effective. Whether CrM is consumed in combination with PRO or CHO may depend on individual requirements. That is, the additional CHO may be useful to only some athletes. However, as the combination of CrM and PRO appears to provide similar benefits, this combination may be more suited to those in whom a high CHO intake (e.g., glucose) is not desired. The consumption of a supplement containing PRO and CHO before and after resistance exercise (i.e., supplement timing) appears to provide the ideal anabolic conditions for muscle growth. For instance, most studies that have assessed chronic adaptations report significantly greater muscle hypertrophy from this strategy. If smaller doses of these macronutrients are desired, supplement timing with a CrM-containing PRO-CHO supplement ($1\,g\cdot kg\cdot day^{-1}$ containing CrM $0.1\,g^{-1}kg^{-1}day$) has been shown to be a particularly effective strategy for augmenting strength gains and muscle hypertrophy. The incorporation of β-alanine ($3.2\,g\cdot kg\cdot day^{-1}$) may provide a buffer to help maintain skeletal muscle acid-base homeostasis, which may promote greater training volumes during the program. Finally, when considering the protein source, because of their unique digestion/absorption kinetics, the combination of whey and casein proteins appears to be most suitable for promoting muscle anabolism and lean mass during resistance training; however, no studies have examined what type or ratio is most beneficial. Whether the addition of certain amino acids can optimize the effects of this supplement blend also remains unclear.

7. COMBINATIONS SHOWN TO ENHANCE ANAEROBIC/ AEROBIC EXERCISE PERFORMANCE

As mentioned earlier, oral β-alanine supplementation has been shown to improve submaximal cycle ergometry performance and time-to-exhaustion *(189)*, delay the onset of neuromuscular fatigue during incremental cycle ergometry *(190)*, and/or increase the amount of work completed during high-intensity exercise (cycling to exhaustion at 110% of estimated power maximum)

(185). Although carnosine, but more importantly β-alanine supplementation may be an important physiological factor in determining high intensity exercise performance, several studies suggest that it could also potentially enhance the buffering capacity of CrM and thus provide additional ergogenic effects *(185)*. Recently, the potential synergistic effect of β-alanine and CrM supplementation was examined on various indices of cardiorespiratory endurance in healthy males *(202)*. Supplementation groups included CrM only (5.25 g), β-alanine only (1.6 g), CrM+β-alanine (CrM 5.25 g/β-alanine 1.6 g + 34 g dextrose), and dextrose placebo. Following 28 days of supplementation, the CrM and β-alanine groups independently showed improvement in two (power output at ventilatory threshold, time to exhaustion), and one (power output at lactate threshold) of the physiological parameters measured, respectively. However when combined, supplementation resulted in improvements in five of the eight physiological parameters measured (including percent VO_2 peak associated with the lactate threshold and ventilatory threshold) during the incremental cycle ergometry test. Although it is important to reiterate that the improvements were not significant when compared among groups, it was evident by a significant time effect within groups that the combination of CrM and β-alanine was greater at delaying the onset of the fatigue and thus potentially enhancing endurance performance *(202)*. However, with limited research examining the potential synergistic effects of β-alanine and CrM supplementation, further studies are clearly warranted to confirm the beneficial effects of β-alanine and CrM supplementation during exercise performance.

Research has revealed that the combination of specific amino acids (AAs)—particularly BCAAs (leucine, isoleucine, valine), arginine, and glutamine—improves indices of muscle function, damage, and recovery both during and following exercise in college track athletes (middle- and long-distance runners) *(203,204)* and rugby players *(205)*. The AA mixture (% of total protein in grams) used for each study *(203–205)* consisted of L-glutamine (14%), L-arginine (14%), L-leucine, L-isoleucine, L-valine (total BCAA 30%), L-threonine, L-lysine, L-proline, L-methionine, L-histidine, L-phenylalanine, and L-tryptophan, with total protein varying from 2.2 to 7.2 g/day. Ohtani and colleagues *(204)* examined the effects of a daily dose of an AA mixture (mentioned above) on middle- and long distance runners engaging in sustained exercise for 2 to 3 hours/day, 5 days/ week for 6 months. During the 6-month period, subjects received three 1-month dosage treatments (2.2, 4.4, and 6.6 g/day), separated by a washout month between each trial. The 2.2 g/day dose was administered as a single dose at dinner; the 4.4 g/day dose was administered as two 2.2 g doses at breakfast and dinner; and the 6.6 g/day dose was given as three 2.2 g doses, one at each daily meal. Results showed that the AA mixture at the daily dose of 6.6 g had the greatest effect, improving the self-assessment of the physical condition, reducing muscle damage, and enhancing hematopoiesis measures, which suggests improved oxygen-handling capacity *(204)*.

A similar study *(205)* examined the effects of the same AA mixture but at a higher dosage (7.2 g/day), on rugby players for 3 months during a period of intensive physical training. Athletes maintained a regular training schedule with their teammates before, during, and after the 90-day trial period. The subjects were instructed to take a 3.6 g dose of the AA mixture after morning and evening meals each day for 90 days. Results from both studies *(204,205)* suggest that long-term administration of the AA mixture may increase the production of red blood cells, thereby perhaps enhancing the capacity of the blood to carry oxygen. Furthermore, these highly trained athletes reported that long-term intake of the AA mixture produced a favorable effect on their physical fitness. In contrast to trained athletes, another study *(206)* demonstrated significant increases in treadmill time to exhaustion in healthy untrained women following 6 weeks of essential AA supplementation. The essential AA composition per 10 g consisted of L-isoleucine 1.483 g, L-leucine 1.964 g, L-valine 1.657 g, L-lysine 1.429 g, L-methionine 0.699 g, L-phenylalanine 1.289 g, L-threonine 1.111 g, L-tryptophan 0.368 g. Subjects consumed, on average, 128 g of AAs per week, or 18.3 g daily. It is clear from the results of the current study, taken together with the previous studies *(203–205)*, that BCAAs when combined with other essential or nonessential amino acids have a beneficial effect during and after aerobic exercise performance. Although these results are interesting and provide practical application to most athletes when training or competing, a limitation to these studies is that the results were obtained in comparison to an isocaloric sugar (dextrin) placebo and not an equivalent dose of other AAs or protein. Thus, further research is needed to determine whether these specific AA combinations are more advantageous than regular protein supplements at improving indices of muscle function, damage, and recovery during and after exercise.

In summary, research has demonstrated that CrM/β-alanine supplementation and the use of specific AA combinations influence chronic adaptations that enhance exercise performance (predominantly aerobic exercise). However, similar to the combinations mentioned in the section Combinations That Enhance Aerobic/ Anaerobic Performance, there are a number of limitations that exist for both these supplements. First, limited research has proven the beneficial effects of CrM/β-alanine supplementation on exercise performance. Therefore, until further research is conducted, we can only speculate as to whether combining CrM and β-alanine provides benefit additional to that seen when each of the supplements is used alone. Second, although the combination of specific AAs such as BCAAs (leucine, isoleucine, valine), arginine, and glutamine has shown to improve exercise performance, further research is needed to determine whether these specific AA combinations are more advantageous than regular protein supplements, as the results obtained to date were in comparison to an isocalorie sugar (dextrin) placebo, not an equivalent dose of other AAs or protein.

8. CONCLUSION

The focus of this chapter was supplement combinations and dosing strategies that are effective at promoting either an acute physiological response that may improve/enhance exercise performance or influence chronic adaptations desired from training. The main conclusions are as follows.

- Few supplement combinations that are marketed to athletes are supported by scientific evidence of their effectiveness. Quite often, under the rigor of scientific investigation, the patented combination fails to provide any greater benefit than a group given the active (generic) ingredient. One good example is creatine monohydrate (CrM).
- The capacity of CrM to augment the phosphocreatine system and provide an ergogenic benefit under a variety of conditions is well

documented. However, the wide variability with regard to dose responses and muscle uptake among individuals has led to increasing interest in combinations that may improve muscle creatine accumulation in response to supplementation.

- Probably due to an insulin-stimulating effect on the cellular creatine transporter, combining each dose of CrM (5–10 g) with high-GI CHO or dairy proteins (up to $1.5\,g\cdot kg^{-1}\,day^{-1}$) appears to be a highly effective strategy that promotes creatine accumulation. Taking each dose of CrM with PRO and CHO (total 100 g) close to the time of the exercise may be most effective at promoting Cr accumulation.

- Other compounds that show the potential to enhance muscle accumulation and/or the ergogenic effect of CrM are D-pinotol, α-linolic acid, and β-alanine. However, each requires further investigation before clear conclusions can be made regarding their effectiveness.

- The addition of PRO (or amino acids) to a CHO supplement appears to enhance the rate of muscle glycogen storage during the hours following exercise. The combination of CrM, PRO, and CHO not only appears to augment Cr uptake it may optimize muscle glycogen stores as well. It is important to remember that characteristically CrM increases lean mass; therefore, individual requirements should be considered in sports where any gain in body weight may disadvantage the athlete.

- For all athletes, along more efficient glycogen restoration, an important advantage of combining PRO (or essential amino acids) with CHO in a postexercise supplement is this combination's well documented positive effect on protein synthesis and net protein balance, which underlines efficient recovery.

- Chronic adaptations that are desired from resistance training (i.e., increased strength, muscle hypertrophy, lean body mass) appear to be enhanced by the combination of CrM with PRO or CHO (up to $1-5\,g\cdot kg\cdot day^{-1}$). The combination utilized may depend on individual requirements of the athlete. For instance, the additional CHO may be useful to some with high-energy requirements. However, as PRO appears to provide similar benefits, the combination of CrM and PRO may be more suited when high CHO intake (e.g., glucose) is not desired.

- The consumption of a supplement containing PRO and CHO before and after resistance exercise (i.e., supplement timing) appears to provide the ideal anabolic conditions for muscle growth. That is, most resistance training studies that have assessed chronic adaptations report significantly greater muscle hypertrophy from this strategy.

- Additionally, supplement timing with a CrM-containing PRO-CHO supplement [$1\,g\cdot kg^{-1}$ twice a day (CrM $0.1\,g^{-1}kg^{-1}$)] is shown to be a particularly effective strategy for increasing muscle creatine stores and enhancing muscle strength and hypertrophy during resistance training.
- Caffeine-ephedrine and CrM-sodium bicarbonate supplement combinations provide an acute physiological response that enhances anaerobic and/or aerobic exercise, whereas CrM/β-alanine supplementation and the use of specific amino acid combinations influence chronic adaptations that predominantly enhance aerobic exercise performance. However, as mentioned, a number of limitations exist in the research methodology utilized and/or the supplement itself. Thus, the practical application for athletes and recreationally active individuals may require further investigation.

8.1. Multifaceted Role of the Muscle PCr-Cr System in Exercise Metabolism

To appreciate fully the rationale behind the intense research focus on supplements that may enhance the phosphocreatine-creatine (PCr-Cr) system in muscle, one must understand its fundamental, multifaceted roles in relation to exercise metabolism. The PCr-Cr system as a whole integrates all the local pools (or compartments) of adenine nucleotides (i.e., the transfer of energy from mitochondrial compartments to that in myofibrils and cellular membranes as well as the feedback signal transmission from sites of energy utilization to sites of energy production). The availability of PCr is now generally accepted as most critical to the continuation of muscle force production and performance during repeated, short bouts of powerful activity *(1,2)* as well as aerobic exercise at high intensity *(3,4)*.

The main roles of the PCr-Cr system are illustrated in Figure 1. The first is that of a temporal energy buffer for ATP regeneration achieved via anaerobic degradation of PCr to Cr and rephosphorylation of ADP. This energy buffering function is most prominent in the fast-twitch/glycolytic fibers; these fibers contain the largest pool of PCr *(5)*. The ATP required for high intensity exercise is met by the simultaneous breakdown of PCr and anaerobic glycolysis, and the PCr-Cr system provides up to one-third of the total energy required *(6)*. The second major function of the PCr-Cr system is that of a spatial energy buffer (or transport system). In this capacity, the PCr-Cr system serves as an intracellular energy carrier connecting sites of energy production (mitochondria) with sites of energy utilization (Na^+/K^+ pump, myofibrils, sarcoplasmic reticulum) (Fig. 1). To describe the specificity of this system, this system has been coined the creatine-phosphate (Cr-Pi) shuttle *(7)*—Cr literally shuttles energy from the mitochondrion to highly specific sites via compartment-specific creatine kinase (CK) isoenzymes located at each of the energy producing or utilizing sites that transduce the PCr to ATP *(8)* and then returns to regenerate energy exactly the equivalent to its consumption at those sites *(7)*. A third function of the PCr-Cr system is the prevention of a rise in ADP, which would have an inhibitory effect on a variety of ATP-dependent processes, such as cross-bridge cycling. A rise in ADP production would also activate the kinase reactions that ultimately result in the destruction of muscle adenine nucleotides *(2)*. Therefore, the removal of ADP via the CK reaction-induced rephosphorylation serves to reduce the loss of adenine nucleotides while maintaining a high intracellular ATP/ADP ratio at the sites of high energy requirements *(9)*.

The CK reaction during the resynthesis of ATP takes up protons *(8)*. Therefore, another function of this PCr-Cr system is the maintenance of pH in exercising muscle. In a reversible reaction (catalyzed by the site-specific CK), Cr and ATP form PCr and ADP (Fig. 1). The formation of the polar PCr "locks" Cr within the muscle and maintains the retention of Cr because the charge prevents partitioning through biological membranes *(2)*. When pH declines (i.e., during exercise when lactic acid accumulates), the reaction favors the generation of ATP. Conversely, during recovery periods (i.e., periods of rest between exercise sets), when ATP is being generated aerobically, the reaction proceeds toward the right and increases PCr levels. The notion that maintenance of PCr availability is crucial to continued force production and performance during high intensity exercise is further supported by research demonstrating that the rate of PCr utilization is extremely high during the initial seconds of intense contraction—high anaerobic ATP regeneration rates result in a 60% to 80% fall in PCr *(10)*. Not only is the depletion of muscle PCr associated with fatigue *(9)*, the resynthesis of PCr and the restoration of peak performance are shown to proceed in direct proportion to one another despite low muscle pH during recovery *(10)*.

A loading phase with creatine monohydrate (CrM) (4 × 5 g servings·day^{-1} for 5 days) is able to increase Cr concentrations in muscle and other tissues with a low baseline Cr content, such as the brain, liver, and kidney *(4–46)*. Via its accumulation in the cell, CrM enhances the cellular bioenergetics of the PCr-Cr system by increasing PCr availability *(2,9,47)*. The beneficial effect of oral supplementation is thought to be dependent on the extent of Cr accumulation *(31,47,48)*. However, it is also apparent that this response can be highly variable among subjects *(49)*. Large variations in Cr accumulation (0–40 mmol·kg dm^{-1}) in response to supplementation can be partly accounted for by differences in presupplementation muscle concentrations *(48)* and possibly in muscle fiber type distribution *(5)*, but it remains unclear as to why muscle Cr accumulation can vary tremendously (up to sixfold) among individuals with similar presupplementation concentrations *(15,43,49)*. This variability in muscle Cr uptake among some individuals combined with the significance of the PCr-Cr system and CrM's potential to augment this all-important pathway is the underlining rationale of studies that examine the effects of CrM supplementation in combination with other compounds.

8.2. Creatine + Protein or Creatine + Carbohydrate for Better Muscle Hypertrophy?

8.2.1. PAUL J. CRIBB

The combination of creatine monohydrate (CrM) and carbohydrate (CHO) has been shown to provide greater improvements in strength and body composition (i.e., increase lean mass with no increase in fat mass) compared to CHO alone. CrM combined with protein (PRO) (whey protein) has also been shown to augment muscle strength and lean body mass (LBM) when compared to CHO or PRO only. However, prior to this study, no one had compared the effects of different CrM-containing PRO and CHO supplements on muscle Cr accumulation or chronic adaptations during resistance training.

The aim of this study was to examine the effects of combining CrM with CHO and with PRO (whey protein isolate) during an

11-week resistance training program in comparison to PRO and CHO alone. In a double-blind, randomized protocol, resistance-trained males were matched for strength and placed into one of four groups: creatine/carbohydrate (CrCHO), creatine/whey protein isolate (CrWP), WP only, or CHO only (CHO). All participants consumed the supplement ($1.5\,g^{-1}kg^{-1}day^{-1}$) for the duration of the resistance training program while maintaining their habitual daily diet. The CrM-containing supplements (CrCHO, CrWP) protocol included a 1-week loading phase ($0.3\,g^{-1}kg^{-1}day^{-1}$, or 24 g day^{-1}, for an 80 kg individual) that was followed by a maintenance phase ($0.1\,g^{-1}kg^{-1}day^{-1}$ or 8 g day^{-1} for an 80 kg individual) for the duration of the study. All assessments were completed the week before and after the 11-week supervised resistance training program. Assessments included dietary analyses (before and during supplementation), strength (1RM, in the barbell squat, bench press, and cable pulldown), body composition (via DEXA*), and vastus lateralis muscle biopsies for histochemical determination of muscle fiber type (I, IIa, IIx), cross-sectional area (CSA), muscle contractile protein, and Cr content.

Results showed that although there were no differences between the groups at the start of the study and each group consumed a protein-rich diet, the two CrM-treated groups demonstrated greater hypertrophy responses than the WP and CHO-only groups. However, the hypertrophy responses among all groups did vary at the three levels of muscle physiology that were assessed (i.e., LBM, fiber-specific hypertrophy, contractile protein content). For example, the CrCHO and CrWP groups each demonstrated larger gains in LBM (5.5% and 5.0%, respectively) than the CHO (1.1%) and WP (3.7%) groups (Fig. 2). The CrCHO and CrWP groups also demonstrated the largest increases in hypertrophy in type I, IIa, and IIx fibers; but again no difference between the two CrM-treated groups was detected. Additionally, the changes LBM were reflected by the changes in contractile protein content. That is, both CrCHO and CrWP groups demonstrated greater increases in contractile protein content (milligrams per gram of muscle) compared to the CHO and WP groups (Fig. 3). However, there

---

* DEXA (dual x-ray absorptiometry) measures body density and composition via x-rays. Bone, fat and muscle possess different densities and will therefore absorb x-rays at different amounts. This allows researchers then to quantify body composition.

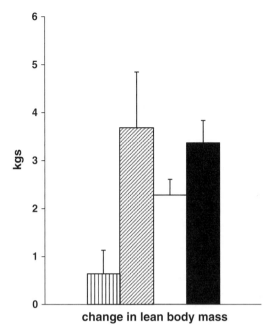

Fig. 2. Change in lean body mass.

was no difference in contractile protein accretion between the two CrM-treated groups. With regard to muscle Cr accumulation, both the CrCHO and CrWP groups demonstrated similar elevations (~10%) in muscle Cr content after the 11-week training/supplementation program.

Based on previous findings of the anabolic effect of whey protein on muscle, an additive effect due to combining CrM and WP on muscle strength and hypertrophy was anticipated in this study. However, no greater effect was observed from combining CrM with whey protein when compared to the CrCHO group. One explanation for this may have been the already high protein intake by all groups (aside from supplementation). For instance, the results of at least one longitudinal study suggested that once dietary protein requirements appear to be met it is the energy content of the diet that has the largest effect on hypertrophy during resistance training *(198)*. In other words, when CrM is consumed in the presence of a high protein diet, the addition of CHO may be

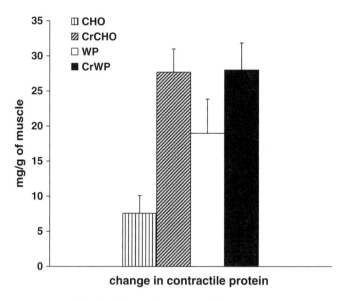

Fig. 3. Change in contractile protein.

more beneficial than extra PRO. However, the results of our study also suggest that the consumption of CrM with PRO provides benefits similar to those of CrM with CHO. This may have important implications for people who cannot consume large amounts of CHO (e.g., glucose) such as those with, or at risk of, type 2 diabetes.

In conclusion, it does appear as though combining CrM with CHO, or PRO can influence the magnitude of chronic adaptations desired from resistance exercise to a greater extent than CHO or PRO alone. The hypertrophic responses from these supplements varied at the three levels assessed (i.e., changes in lean mass, fiber-specific hypertrophy, and contractile protein content). Currently, this is the only study that has compared the effects of different CrM-containing PRO and CHO supplements on muscle Cr accumulation and chronic adaptations during resistance training. Therefore, this topic should continue to receive attention from the scientific community as these results have important implications not only for athletes but also an ageing population and others who have a reduced capacity for exercise.

8.3. Addition of Protein to a Carbohydrate Supplement for Increased Efficiency of Muscle Glycogen Storage

8.3.1. JOHN L. IVY

An essential process in the recovery from exercise is replenishment of muscle glycogen stores. When time is limited between exercise workouts or competitions, it is necessary to maximize the rate of muscle glycogen resynthesis. Research suggests that adding protein to a carbohydrate supplement increases the efficiency by which carbohydrate is converted to muscle glycogen. The mechanism by which protein increases the efficiency of muscle glycogen storage is not known, but there are several possibilities.

Insulin controls two important steps required for muscle glycogen synthesis. First, it activates the transport of glucose across the plasma membrane of the muscle, and second it increases the activity of glycogen synthase, the rate-limiting enzyme in glycogen synthesis. When carbohydrate is consumed, insulin is released from the pancreas to maintain blood glucose homeostasis. Peptides and certain amino acids also stimulate the release of insulin and when combined with carbohydrate the insulin response can be synergistic. This greater insulin response can result in a faster rate of muscle glucose uptake and its conversion to glycogen. The stimulating effect of protein on glycogen synthesis, however, has been observed without a greater insulin response than is typically seen with carbohydrate supplementation alone.

A second possibility is that the amino acids released from protein digestion activate the glycogen synthesis process via a mechanism that is insulin-independent, thus having an additive effect on this process. Glycogen synthase activity is controlled, in part, by glycogen synthase kinase-3, which phosphorylates glycogen synthase, resulting in its inactivation. Inhibition of glycogen synthase kinase-3 results in the dephosphorylation of glycogen synthase and its activation. Glycogen synthase kinase-3 can be inhibited by the protein p70S6K, a downstream target of mTOR (mammalian target of rapamycin), which is activated by essential amino acids. Therefore, an elevation of blood amino acids along with insulin following a carbohydrate–protein supplement may function additively to activate glycogen synthase and increase the rate of glycogen synthesis. Furthermore, certain amino acids, such as leucine, have been found to increase the rate of skeletal muscle glucose transport. This raises the possibility that a rise in blood amino acid levels at the same time blood insulin levels are increasing increases activation of both skeletal muscle glucose transport and glycogen synthase, resulting in an enhanced rate of muscle glycogen synthesis.

8.4. Can Supplement Timing Double Gains in Muscle Mass?

8.4.1. ALAN HAYES

Some studies have reported greater muscle hypertrophy during resistance exercise training from supplement timing (i.e., the strategic consumption of proteins/amino acids and carbohydrates before and/or after each workout). However, prior to this study, no one had examined whether this strategy provided greater muscle hypertrophy or strength development than supplementation at other times during the day. The purpose of this study *(67)* was to examine the effects of supplement timing versus supplementation in the hours not close to the workout on muscle fiber hypertrophy, strength, and body composition during a 10-week resistance exercise program.

Resistance-trained males were matched for strength and randomly placed into one of two groups; group 1 ($n = 8$) consumed a protein-carbohydrate (PRO-CHO) supplement ($1 \, g \cdot kg^{-1}$ twice day) immediately before and after every workout (4 days per week for 10 weeks). Group 2 ($n = 9$) consumed the same dose of the same supplement in the morning and late in the evening. These times were at least 5 hours outside of the workout. The two groups consumed the exact same supplement [0.03 g creatine monophosphate (CrM) + 0.5 g whey isolate + 0.5 g glucose per kilogram body weight] twice each training day, 4 times per week. The only difference was the time of day the supplement doses were consumed. Assessments completed the week before and after the 10-week supervised training program (Max-OT™) included strength (1RM, barbell bench press, squats, dead lifts), body composition (DEXA—see footnote to sidebar Creatine + Protein or Creatine + Carbohydrate for Better Muscle Hypertrophy?), and vastus lateralis muscle biopsies for determination of muscle fiber type (I, IIa, IIx), cross-sectional area (size), contractile protein, and creatine and glycogen content.

Results showed that although both groups demonstrated significant improvements in strength and gains in lean mass, the supplement-timing group showed higher ($p < 0.05$) resting muscle Cr and glycogen concentrations after the training program, greater strength gains (two of three assessments), hypertrophy of type IIa and IIx fibers, and synthesis of contractile protein. Additionally, this group demonstrated a gain in lean body mass that was almost double that of the group that supplemented at times not close to training (2.72 vs 1.45 kg, respectively) (Fig. 4).

There were several aspects of this study that made it unique compared to others that have examined the effects of supplementation close to the time of resistance exercise. First, the changes in body composition were confirmed with hypertrophic responses at the cellular level (i.e., fiber-specific hypertrophy) and the subcellular level (i.e., contractile protein content). Second, this study utilized experienced bodybuilders who characteristically followed regimented eating patterns, and the effects of supplementation were examined in the presence of the participants' normal eating patterns.

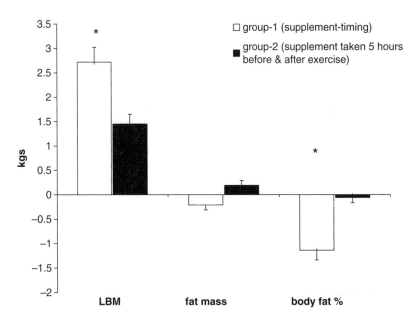

Fig. 4. Body composition changes after 10 weeks of training. *Greater change than that in the group that did not follow supplement timing ($p < 0.05$). From Cribb and Hayes *(67)*, with permission.

Although these results are important, it is the design of this study that makes the findings particularly relevant to a wide sector of the population. Supplement timing with the combination of CrM-PRO-CHO represents a simple but effective strategy that may enhance strength and muscle mass gains during resistance training in healthy adults. However, this protocol may also have important implications for populations that require improvements in strength and body composition but have a reduced capacity for exercise, such as the frail elderly, cardiac rehabilitation patients, and others living with conditions that compromise health such as human immunodeficiency virus infection, cancer, and the various muscular dystrophies.

REFERENCES

1. Balsom PD, Soderlund K, Ekblom B. Creatine in humans with special reference to creatine supplementation. Sports Med 1994;18:268–280.
2. Greenhaff P. The nutritional biochemistry of creatine. J Nutr Biochem 1997;11:610–618.
3. Walsh B, Tonkonogi M, Soderlund K, Hultman E, Saks V, Sahlin K. The role of phosphorylcreatine and creatine in the regulation of mitochondrial respiration in human skeletal muscle. J Physiol 2001;537:971–978.
4. McConell GK, Shinewell J, Stephens TJ, Stathis CG, Canny BJ, Snow RJ. Creatine supplementation reduces muscle inosine monophosphate during endurance exercise in humans. Med Sci Sports Exerc 2005;37:2054–2061.
5. Tesch PA, Thorsson A, Fujitsuka N. Creatine phosphate in fiber types of skeletal muscle before and after exhaustive exercise. J Appl Physiol 1989;66:1756–1759.
6. Greenhaff PL, Bodin K, Söderlund K, Hultman E. Effect of oral creatine supplementation on skeletal muscle phosphocreatine resynthesis. Am J Physiol 1994;266:E725–E730.
7. Bessman SP, Geiger PJ. Transport of energy in muscle: the phosphorylcreatine shuttle. Science 1981;211:448–452.
8. Wallimann T, Wyss M, Brdiczka D, Nicolay K, Eppenberger HM. Intracellular compartmentation, structure and function of creatine kinase isoenzymes in tissues with high and fluctuating energy demands: the "phosphocreatine circuit" for cellular energy homeostasis. Biochem J 1992;281:21–40.
9. Hultman E, Greenhaff PL. Skeletal muscle energy metabolism and fatigue during intense exercise in man. Sci Prog 1991;298:361–370.
10. Bogdanis GC, Nevill NE, Bobbis LH, Lakomy HKA, Nevill MA. Recovery of power output and muscle metabolites following 30 s of maximal sprint cycling in man. J Physiol 1995;482:467–480.

11. Folin O, Denis W. Protein metabolism from the standpoint of blood and tissue analyses: an interpretation of creatine and creatinine in relation to animal metabolism. J Biol Chem 1914;17:493–502.

12. Harris RC, Söderlund K, Hultman E. Elevation of creatine in resting and exercised muscle of normal subjects by creatine supplementation. Clin Sci 1992;83:367–374.

13. Greenhaff PL, Casey A, Short AH, Harris R, Soderlund K, Hultman E. Influence of oral creatine supplementation of muscle torque during repeated bouts of maximal voluntary exercise in man. Clin Sci 1993;84:565–571.

14. Birch R, Noble D, Greenhaff PL. The influence of dietary creatine supplementation on performance during repeated bouts of maximal isokinetic cycling in man. Eur J Appl Physiol 1994;69:268–270.

15. Casey A, Constantin-Teodosiu D, Howell S, Hultman E, Greenhaff PL. Creatine ingestion favourably affects performance and muscle metabolism during maximal exercise in humans. Am J Physiol 1996;271:E31–E37.

16. Earnest CP, Snell PG, Rodriguez R, Almada AL, Mitchell TL. The effect of creatine monohydrate ingestion on anaerobic power indices, muscular strength and body composition. Acta Physiol Scand 1995;153:207–209.

17. Febbraio MA, Flanagan TR, Snow RJ, Zhao S, Carey MF. Effect of creatine supplementation on intramuscular TCr, metabolism and performance during intermittent, supramaximal exercise in humans. Acta Physiol Scand 1995; 155:387–395.

18. Kreider RB, Ferreira M, Wilson M, et al. Effects of creatine supplementation on body composition, strength, and sprint performance. Med Sci Sports Exerc 1998;30:73–82.

19. Kreider RB, Klesges R, Harmon K, et al. Effects of ingesting supplements designed to promote lean tissue accretion on body composition during resistance training. Int J Sport Nutr 1996;6:234–246.

20. Peeters BM, Lantz CD, Mayhew JL. Effect of oral creatine monohydrate and creatine phosphate supplementation on maximal strength indicies, body composition and blood pressure. J Strength Cond Res 1998;13:3–9.

21. Eckerson JM, Stout JR, Moore GA, et al. Effect of creatine phosphate supplementation on anaerobic working capacity and body weight after two and six days of loading in men and women. J Strength Cond Res 2005;19:756–763.

22. Kreider RB, Willoughby D, Greenwood M, Parise G, Payne E, Tarnopolsky MA. Effects of creatine serum on muscle creatine and phosphagen levels. JEPonline 2003;6:24–33.

23. Snow RJ, Murphy RM. Factors influencing creatine loading into human skeletal muscle. Exerc Sport Sci Rev 2003;31:154.

24. Persky AM, Brazeau GA, Hochhaus G. Pharmacokinetics of the dietary supplement creatine. Clin Pharmacokinet 2003;42:557–574.

25. Persky AM, Brazeau GA. Clinical pharmacology of the dietary supplement creatine monohydrate. Pharmacol Rev 2001;53:161–176.

26. Wyss M, Schulze A. Health implications of creatine: can oral creatine supplementation protect against neurological and atherosclerotic disease? Neuroscience 2002;112:243–260.

27. Volek JS, Kraemer WJ. Creatine supplementation: its effect on human muscular performance and body composition. J Strength Cond Res 1996;10:198–203.

28. Casey A, Greenhaff PL. Does dietary creatine supplementation play a role in skeletal muscle metabolism and performance? Am J Clin Nutr 2000;72:607S–6017S.

29. Dempsey RL, Mazzone MF, Meurer LN. Does oral creatine supplementation improve strength? A meta-analysis. J Fam Pract 2002;51:945–951.

30. Rawson ER, Volek JS. The effects of creatine supplementation and resistance training on muscle strength and weightlifting performance. J Strength Cond Res 2003;17:822–831.

31. Kreider RB. Effects of creatine supplementation on performance and training adaptations. Mol Cell Biochem 2003;244:89–94.

32. Poortmans JR, Francaux M. Adverse effects of creatine supplementation: fact or fiction? Sports Med 2000;30:155–170.

33. Farquhar WB, Zambraski EJ. Effects of creatine use on the athlete's kidney. Curr Sports Med Rep 2002;1:103–106.

34. LaBotz M, Smith BW. Creatine use in an NCAA Division I athletic program. Clin J Sport Med 1999;9:167–169.

35. Jacobson BH, Sobonya C, Ransone J. Nutrition practices and knowledge of college varsity athletes: a follow-up. J Strength Cond Res 2001;15:63–68.

36. McGuine TA, Sullivan JC, Bernhardt DA. Creatine supplementation in Wisconsin high school athletes. WMJ 2002;101:25–30.

37. Sundgot-Borgen J, Berglund B, Torstveit MK. Nutritional supplements in Norwegian elite athletes: impact of international ranking and advisors. Scand J Med Sci Sports 2003;13:138–144.

38. Froiland K, Koszewski W, Hingst J, Kopecky L. Nutritional supplement use among college athletes and their sources of information. Int J Sport Nutr Exerc Metab 2004;14:104–120.

39. Morrison LJ, Gizis F, Shorter B. Prevalent use of dietary supplements among people who exercise at a commercial gym. Int J Sport Nutr Exerc Metab 2004;14:481–492.

40. Kristiansen M, Levy-Milne R, Barr S, Flint A. Dietary supplement use by varsity athletes at a Canadian university. Int J Sport Nutr Exerc Metab 2005;15:195–221.

41. Saks VA, Strumia E. Phosphocreatine: molecular and cellular aspects of the mechanism of cardioprotective action. Curr Ther Res 1983;53:565–598.

42. Guimbal C, Kilimann MW. A Na^+-dependent creatine transporter in rabbit brain, muscle, heart, and kidney: cDNA cloning and functional expression. J Biol Chem 1993;268:8418–8421.

43. Steenge GR, Simpson EJ, Greenhaff PL. Protein-and carbohydrate-induced augmentation of whole body creatine retention in humans. J Appl Physiol 2000;89:1165–1171.

44. Dechent P, Pouwels PJ, Wilken B, Hanefeld F, Frahm J. Increase of total creatine in human brain after oral supplementation of creatine-monohydrate. Am J Physiol 1999;277:R698–R704.

45. Leuzzi V, Bianchi MC, Tosetti M, et al. Brain creatine depletion: guanidinoacetate methyltransferase deficiency (improving with creatine supplementation). Neurology 2000;55:1407–1409.

46. Ipsiroglu OS, Stromberger C, Ilas J, Hoger H, Muhl A, Stockler-Ipsiroglu S. Changes of tissue creatine concentrations upon oral supplementation of creatine monohydrate in various animal species. Life Sci 2001;69:1805–1815.

47. Hultman E, Soderlund K, Timmons JA, Cederblad G, Greenhaff PL. Muscle creatine loading in men. J Appl Physiol 1996;81:232–237.

48. Volek JS, Rawson ES. Scientific basis and practical aspects of creatine supplementation for athletes. Nutrition 2004;20:609–614.

49. Lemon PW. Dietary creatine supplementation and exercise performance: why inconsistent results? Can J Appl Physiol 2002;27:663–681.

50. Haughland RB, Chang DT. Insulin effects on creatine transport in skeletal muscle. Proc Soc Exp Biol Med 1975;148:1–4.

51. Odoom JE, Kemp GJ, Radda GK. The regulation of total creatine content in a myoblast cell line. Mol Cell Biochem 1996;158:179–188.

52. Bennett SE, Bevington A, Walls J. Regulation of intracellular creatine in erythrocytes and myoblasts: influence of uraemia and inhibition of Na, K-ATPase. Cell Biochem Funct 1994;12:99–106.

53. Gerber GB, Gerber G, Koszalaka TR, Emmel VM. Creatine metabolism in vitamin E deficiency in the rat. Am J Physiol 1962;202:453–460.

54. Vandenberghe K, Gillis N, Van Leemputte M, Van Hecke P, Vanstapel F, Hespel P. Caffeine counteracts the ergogenic action of muscle creatine loading. J Appl Physiol 1996;80:452–457.

55. Loike JD, Somes M, Silverstein SC. Creatine uptake, metabolism, and efflux in human monocytes and macrophages. Am J Physiol 1986;251:C128–C135.

56. Moller A, Hamprecht B. Creatine transport in cultured cells of rat and mouse brain. J Neurochem 1989;52:544–550.

57. Brand-Miller J. Glycemic index and body weight. Am J Clin Nutr 2005;81:722–723.

58. Green AL, Simpson EJ, Littlewood JJ, Macdonald IA, Greenhaff PL. Carbohydrate ingestion augments creatine retention during creatine feeding in humans. Acta Physiol Scand 1996;158:195–202.

59. Green AL, Hultman E, Macdonald IA, Sewell DA, Greenhaff PL. Carbohydrate ingestion augments skeletal muscle creatine accumulation during creatine supplementation in humans. Am J Physiol 1996;271:E821–E826.

60. Robinson TM, Sewell DA, Hultman E, Greenhaff PL. Role of submaximal exercise in promoting creatine and glycogen accumulation in human skeletal muscle. J Appl Physiol 1999;87:598–604.

61. Greenwood M. Kreider RB, Earnest C, Rasmussen C, Almada A. Differences in creatine retention among three nutritional formulations of oral creatine supplements. JEPonline 2003;6:37–43.

62. Preen D, Dawson B, Goodman C, Beilby J, Ching S. Creatine supplementation: a comparison of loading and maintenance protocols on creatine uptake by human skeletal muscle. Int J Sport Nutr Exerc Metab 2003; 13:97–111.

63. Cribb PJ, Williams AD, Stathis CG, Carey MF, Hayes A. Effects of whey isolate, creatine, and resistance training on muscle hypertrophy. Med Sci Sports Exerc 2007;39:298–307.

64. Burke DG, Chilibeck PD, Davidson KS, Candow DG, Farthing J, Smith-Palmer T. The effect of whey protein supplementation with and without creatine monohydrate combined with resistance training on lean tissue mass and muscle strength. Int J Sport Nutr Exerc Metab 2001;11:349–364.

65. Tarnopolsky MA, Parise G, Yardley NJ, et al. Creatine-dextrose and protein-dextrose induce similar strength gains during training. Med Sci Sports Exerc 2001;33:2044–2052.

66. Stout J, Eckerson J, Noonan D, Moore G, Cullen D. Effects of 8 weeks of creatine supplementation on exercise performance and fat-free weight in football players during training. Nutr Res 1999;19:217–225.

67. Cribb PJ. Hayes A. Effects of supplement timing and resistance exercise on skeletal muscle hypertrophy. Med Sci Sports Exerc 2006;38:1918–1925.

68. Chilibeck PD, Stride D, Farthing JP, Burke DG. Effect of creatine ingestion after exercise on muscle thickness in males and females. Med Sci Sports Exerc 2004;36:1781–1788.

69. Greenwood M, Greenwood L. Kreider RB, Rasmussen C, Almada A, Earnest C. D-Pinitol augments whole body creatine retention in man. JEPonline 2001; 4:41–47.

70. Bates SH, Jones RB, Bailey CJ. Insulin-like effect of pinitol. Br J Pharmacol 2000;130:1944–1948.

71. Estrada E, Ewart H, Tsakiridis T, et al., Stimulation of glucose uptake by the natural coenzyme α-lipoic acid/thioctic acid. Diabetes 1996;45:1798–1804.

72. Kishi Y, Schmelzer J, Yao J, et al. α-Lipoic acid: effect on glucose uptake, sorbitol pathway, and energy metabolism in experimental diabetic neuropathy. Diabetes 1999;48:2045–2051.

73. Burke DG, Chilibeck PD, Parise G, Tarnopolsky MA, Candow DG. Effect of alpha-lipoic acid combined with creatine monohydrate on human skeletal muscle creatine and phosphagen concentration. Int J Sport Nutr Exerc Metab 2003;13:294–302.

74. Bergstrom J, Hultman E. Muscle glycogen synthesis after exercise: an enhancing factor localized to the muscle cells in man. Nature 1966;210:309–310.

75. Bergstrom J, Hermansen L, Hultman E, Saltin B. Diet, muscle glycogen and physical performance. Acta Physiol Scand 1967;71:140–150.

76. Bergstrom J, Hultman E. A study of the glycogen metabolism during exercise in man. Scand J Clin Lab Invest 1967;19:218–228.

77. Ahlborg B, Bergström J, Ekelund LG, Hultman E. Muscle glycogen and muscle electrolytes during prolonged physical exercise. Acta Physiol Scand 1967;70:129–142.

78. Hermansen L, Hultman E, Saltin B. Muscle glycogen during prolonged severe exercise. Acta Physiol Scand 1965;71:334–346.

79. Hickner RC, Fisher JS, Hansen SB, et al. Muscle glycogen accumulation after endurance exercise in trained and untrained individuals. J Appl Physiol 1997;83:897–903.

80. Hurley BF, Nemeth PM, Martin WH III, Hagberg JM, Dalsky GP, Holloszy JO. Muscle triglyceride utilization during exercise: effect of training. J Appl Physiol 1986;60:562–567.

81. MacDougall JD, Ray S, Sale DG, McCartney N, Lee P, Garner S. Muscle substrate utilization and lactate production during weightlifting. Can J Appl Physiol 1999;24:209–215.

82. Robergs RA, Pearson DR, Costill DL, et al. Muscle glycogenolysis during differing intensities of weight-resistance exercise. J Appl Physiol 1991;70:1700–1706.

83. Tesch PA, Ploutz-Snyder LL, Yström L, Castro M, Dudley G. Skeletal muscle glycogen loss evoked by resistance exercise. J Strength Cond Res 1998;12:67–73.

84. Haff GG, Koch AJ, Potteiger JA, et al. Carbohydrate supplementation attenuates muscle glycogen loss during acute bouts of resistance exercise. Int J Sport Nutr Exerc Metab 2000;10:326–339.

85. Ivy JL, Goforth HW Jr, Damon BM, McCauley TR, Parsons EC, Price T. Early postexercise muscle glycogen recovery is enhanced with a carbohydrate-protein supplement. J Appl Physiol 2002;93:1337–1344.

86. Ivy JL, Katz AL, Cutler CL, Sherman WM, Coyle EF. Muscle glycogen synthesis after exercise: effect of time of carbohydrate ingestion. J Appl Physiol 1988;64:1480–1485.

87. Levenhagen DK, Gresham JD, Carlson MG, Maron DJ, Borel MJ, Fakoll PJ. Postexercise nutrient intake timing in humans is critical to recovery of leg glucose and protein homeostasis. Am J Physiol Endocrinol Metab 2001;280:E982–E993.

88. Doyle JA, Sherman WM, Strauss RL. Effects of eccentric and concentric exercise on muscle glycogen replenishment. J Appl Physiol 1993;74:1848–1855.

89. Blom PCS, Høstmark AT, Vaage O, Kardel KR, Mæhlum S. Effect of different post-exercise sugar diets on the rate of muscle glycogen synthesis. Med Sci Sports Exerc 1987;19:491–496.

90. Ivy JL. Glycogen resynthesis after exercise: effect of carbohydrate intake. Int J Sports Med 1998;19:142–146.

91. Cohen PH, Nimmo G, Proud CG. How does insulin stimulate glycogen synthesis? Biochem Soc Symp 1979;43:69–95.

92. Danforth WH. Glycogen synthetase activity in skeletal muscle: interconversion of two forms and control of glycogen synthesis. J Biol Chem 1965;240:588–593.

93. Kiens B, Raben AB, Valeur AK, Richter EA. Benefit of dietary simple carbohydrates on the early post-exercise muscle glycogen repletion in male athletes [abstract]. Med Sci Sports Exerc 1990;22:S88.

94. Burke LM, Collier GR, Hargreaves M. Muscle glycogen storage after prolonged exercise: effect of the glycemic index of carbohydrate feedings. J Appl Physiol 1993;75:1019–1023.

95. Du H, Van der ADL, Feskens EJ. Dietary glycaemic index: a review of the physiological mechanisms and observed health impacts. Acta Cardiol 2006;61:383–397.

96. Keizer HA, Kuipers H, Vankranenburg G, Guerten P. Influence of liquid and solid meals on muscle glycogen resynthesis, plasma fuel hormone response, and maximal physical work capacity. Int J Sports Med 1986;8:99–104.

97. Reed MJ, Brozinick JT, Lee MC, Ivy JL. Muscle glycogen storage postexercise: effect of mode of carbohydrate administration. J Appl Physiol 1989;66:720–726.

98. Van Hall G, Shirreffs SM, Calbet JA. Muscle glycogen resynthesis during recovery from cycle exercise: no effect of additional protein ingestion. J Appl Physiol 2000;88:1631–1636.

99. Van Loon LJ, Saris WH, Kruijshoop M, Wagenmakers AJ. Maximizing post exercise muscle glycogen synthesis: carbohydrate supplementation and the application of amino acid or protein hydrolysate mixtures. Am J Clin Nutr 2000;72:106–111.

100. Van Loon LJ, Saris WH, Verhagen H, Wagenmakers AJ. Plasma insulin responses after ingestion of different amino acid or protein mixtures with carbohydrate. Am J Clin Nutr 2000;72:96–105.

101. Zawadzki KM, Yaspelkis BBD, Ivy JL. Carbohydrate-protein complex increases the rate of muscle glycogen storage after exercise. J Appl Physiol 1992;72:1854–1859

102. Williams MB, Raven PB, Fogt DL, Ivy JL. Effects of recovery beverages on glycogen restoration and endurance exercise performance. J Strength Cond Res 2003;17:12–19.

103. Tarnopolsky MA, Bosman M, Macdonald JR, Vandeputte D, Martin J, Roy BD. Post exercise protein-carbohydrate and carbohydrate supplements increase muscle glycogen in men and women. J Appl Physiol 1997;83:1877–1883.

104. Carrithers JA, Williamson DL, Gallagher PM, Godard MP, Schulze KE, Trappe SW. Effects of post exercise carbohydrate-protein feedings on muscle glycogen restoration. J Appl Physiol 2000;88:1976–1982.

105. Berardi JM, Price TB, Noreen EE, Lemon PW. Postexercise muscle glycogen recovery enhanced with a carbohydrate-protein supplement. Med Sci Sports Exerc 2006;38:1106–1113.

106. Jentjens PG, van Loon LJC, Mann CH, Wagenmakers AJM, Jeukendrup AE. Addition of protein and amino acids to carbohydrates does not enhance postexercise muscle glycogen synthesis. J Appl Physiol 2001;91:839–846.

107. Price TB, Rothman DL, Shulman RG. NMR of glycogen in exercise. Proc Nutr Soc 1999;58:1–9.

108. Kimball SR, Farrell PA, Jefferson LS. Role of insulin in translational control of protein synthesis in skeletal muscle by amino acids or exercise. J Appl Physiol 2002;93:1168–1180.

109. Yaspelkis BB 3rd, Ivy JL. The effect of a carbohydrate-arginine supplement on postexercise carbohydrate metabolism. Int J Sport Nutr 1999;9:241–250.

110. Nelson AG, Arnall DA, Kokkonen J, Day R, Evans J. Muscle glycogen supercompensation is enhanced by prior creatine supplementation. Med Sci Sports Exerc 2001;33:1096.

111. Op 't Eijnde B, Urso B, Richter EA, Greenhaff PL, Hespel P. Effect of oral creatine supplementation on human muscle GLUT4 protein content after immobilization. Diabetes 2001;50:18.

112. Derave W, Eijnde BO, Verbessem P, et al. Combined creatine and protein supplementation in conjunction with resistance training promotes muscle GLUT-4 content and glucose tolerance in humans. J Appl Physiol 2003;94:1910.

113. Van Loon LJ, Murphy R, Oosterlaar AM, et al. Creatine supplementation increases glycogen storage but not GLUT-4 expression in human skeletal muscle. Clin Sci 2004;106:99.

114. Francaux M, Poortmans JR. Effects of training and creatine supplement on muscle strength and body mass. Eur J Appl Physiol Occup Physiol 1999;80:165.

115. Low SY, Rennie MJ, Taylor PM. Modulation of glycogen synthesis in rat skeletal muscle by changes in cell volume. J Physiol 1996;495:299.

116. Phillips SM, Tipton KD, Aarsland A, Wolf SE, Wolfe RR. Mixed muscle protein synthesis breakdown after resistance exercise in humans. Am J Physiol 1997;273:E99–E107.

117. Biolo G, Tipton KD, Klein S, Wolfe RR. An abundant supply of amino acids enhances the metabolic effect of exercise on muscle protein. Am J Physiol 1997;273:E122–E129.

118. Phillips SM, Tipton KD, Ferrando AA, Wolfe RR. Resistance training reduces the acute exercise-induced increase in muscle protein turnover. Am J Physiol 1999;276:E118–E124.

119. Volek JS. Influence of nutrition on responses to resistance training. Med Sci Sports Exerc 2004;36:689–696.

120. Rennie MJ, Wackerhage H, Spangenburg EE, Booth FW. Control of the size of the human muscle mass. Annu Rev Physiol 2004;66:799–828.

121. Tipton KD, Ferrando AA, Phillips SM, Doyle D, Wolfe RR. Postexercise net protein synthesis in human muscle from orally administered amino acids. Am J Physiol 1999;276:E628–E634.

122. Miller SL, Tipton KD, Chinkes DL, Wolf SE, Wolfe RR. Independent and combined effects of amino acids and glucose after resistance exercise. Med Sci Sports Exerc 2003;34:449–455.

123. Rasmussen BB, Tipton KD, Miller SL, Wolf SE, Wolfe RR. An oral amino acid-carbohydrate supplement enhances muscle protein anabolism after resistance exercise. J Appl Physiol 2000;88:386–392.

124. Tipton KD, Rasmussen BB, Miller SL, et al. Timing of amino acid-carbohydrate ingestion alters anabolic response of muscle to resistance exercise. Am J Physiol 2001;281:E197–E206.

125. Tipton KD, Elliott TA, Cree MG, Wolf SE, Sanford AP, Wolfe RR. Ingestion of casein and whey proteins result in muscle anabolism after resistance exercise. Med Sci Sports Exerc 2004;36:2073–2081.

126. Koopman R, Wagenmakers AJ, Manders RJ, et al., Combined ingestion of protein and free leucine with carbohydrate increases postexercise muscle protein synthesis in vivo in male subjects. Am J Physiol Endocrinol Metab 2005;288:E645–E653.

127. Paddon-Jones D, Sheffield-Moore M, Katsanos CS, Zhang XJ, Wolfe RR. Differential stimulation of muscle protein synthesis in elderly humans following isocaloric ingestion of amino acids or whey protein. Exp Gerontol 2006;41:215–219.

128. Kraemer WJ, Volek JS, Bush JA, Putukian M, Sebastianelli WJ. Hormonal responses to consecutive days of heavy-resistance exercise with or without nutritional supplementation. J Appl Physiol 1998;85: 1544–1555.

129. Wolfe RR, Volpi E. Insulin and protein metabolism. In: Jefferson LS, Cherrington AD (eds) Handbook of Physiology (pp 735–757). Vol 7. Oxford University Press, New York, 2001.

130. Chandler RM, Byrne HK, Patterson JG, Ivy JL. Dietary supplements affect the anabolic hormones after weight training exercise. J Appl Physiol 1994;76:839–845.

131. Williams AG, Ismail AN, Sharma A, Jones DA. Effects of resistance exercise volume and nutritional supplementation on anabolic and catabolic hormones. Eur J Appl Physiol 2002;86:315–321.

132. Thissen JP, Ketelslegers JM, Underwood LE. Nutritional regulation of the insulin-like growth factors. Endocr Rev 1994;15:80–101.

133. Ballard TLP, Clapper JA, Specker BL, Binkley TL, Vukovich MD. Effect of protein supplementation during a 6-month strength and conditioning program on insulin-like growth factor I and markers of bone turnover in young adults. Am J Clin Nutr 2005;81:1442–1448.

134. Willoughby DS, Stout J, Wilborn C, Taylor L, Kerksick C. Effects of heavy resistance training and proprietary whey casein leucine protein supplementation on serum and skeletal muscle IGF-1 levels and IGF-1 and MGF mRNA expression. J Int Soc Sports Nutr 2005;2;1–30.

135. Friedlander AL, Butterfield GE, Moynihan S, et al. One year of insulin-like growth factor I treatment does not affect bone density, body composition, or psychological measures in postmenopausal women. J Clin Endocrinol Metab 2001;86:1496–1503.

136. Lee PD, Giudice LC, Conover CA, Powell DR. Insulin-like growth factor binding protein-1: recent findings and new directions. Proc Soc Exp Biol Med 1997;216:319–357.

137. Nindl BC, Kraemer WJ, Marx JO, et al. Overnight responses of the circulating IGF-I system after acute, heavy-resistance exercise. J Appl Physiol 2001;90:1319–1326.

138. Bloomer R J, Sforzo GA, Keller BA. Effects of meal form and composition on plasma testosterone, cortisol, and insulin following resistance exercise. Int J Sport Nutr Exerc Metab 2000;10:415–424.

139. Kraemer WJ, Spiering BA, Volek JS, et al. Androgenic responses to resistance exercise: effects of feeding and L-carnitine. Med Sci Sports Exerc 2006;38;1288–1296.

140. Kraemer WJ, Volek JS, French DN, et al. The effects of L-carnitine L-tartrate supplementation on hormonal responses to resistance exercise and recovery. J Strength Cond Res 2003;17:455–462.

141. Greiwe JS, Kwon G, McDaniel ML, Semenkovich CF. Leucine and insulin activate p70 S6 kinase through different pathways in human skeletal muscle. Am J Physiol 2001;281:E466–E471.

142. Liu Z, Jahn LA, Wei L, Long W, Barrett EJ. Amino acids stimulate translation initiation and protein synthesis through an Akt-independent pathway in human skeletal muscle. J Clin Endocrinol Metab 2002;87:5553–5558.

143. Cuthbertson D, Smith K, Babraj J, et al. Anabolic signalling deficits underlie amino acid resistance of wasting, aging muscle. FASEB J 2005;19:422–424.

144. Christie GR, Hajduch E, Hundal HS, Proud CG, Taylor PM. Intracellular sensing of amino acids in Xenopus laevis oocytes stimulates p70, S6 kinase in a TOR-dependent manner. J Biol Chem 2002;277:9952–9995.

145. Hara K, Yonezawa K, Weng QP, Kozlowski MT, Belham C, Avruch J. Amino acid sufficiency and mTOR regulate p70 S6 kinase and eIF-4E BP1 through a common effector mechanism. J Biol Chem 1998;273:14484–14494.

146. Anthony JC, Yoshizawa F, Anthony TG, Vary TC, Jefferson LS, Kimball SR. Leucine stimulates translation initiation in skeletal muscle of postabsorptive rats via a rapamycin-sensitive pathway. J Nutr 2000;130: 2413–2419.

147. Deldicque L, Theisen D, Francaux M. Regulation of mTOR by amino acids and resistance exercise in skeletal muscle Eur J Appl Physiol 2005;94:1–10.

148. Kimball SR, Jefferson LS. Signalling pathways and molecular mechanisms through which branched-chain amino acids mediate translational control of protein synthesis. J Nutr 2006;136:227S–231S.

149. Borsheim E, Aarsland A, Wolfe RR. Effect of an amino acid, protein, and carbohydrate mixture on net muscle protein balance after resistance exercise. Int J Sport Nutr Exerc Metab 2004;14:255–271.

150. Nair KS, Schwartz RG, Welle S. Leucine as a regulator of whole body and skeletal muscle protein metabolism in humans. Am J Physiol 1992; 263:E928–E934.

151. Anthony JC, Reiter AK, Anthony TG, et al. Orally administered leucine enhances protein synthesis in skeletal muscle of diabetic rats in the absence of increases in 4E-BP1 or S6K1 phosphorylation. Diabetes 2002;51:928–936.

152. Combaret L, Dardevet D, Rieu I, et al. A leucine-supplemented diet restores the defective postprandial inhibition of proteasome-dependent proteolysis in aged rat skeletal muscle. J Physiol 2005;569:489–499.

153. Keisler BD, Armsey TD 2 nd. Caffeine as an ergogenic aid. Curr Sports Med Rep 2006;5:215–219.

154. Greer F, McLean C, Graham TE. Caffeine, performance, and metabolism during repeated Wingate exercise tests. J Appl Physiol 1998;85:1502–1508.

155. Greer F, Friars D, Graham TE. Comparison of caffeine and theophylline ingestion: exercise metabolism and endurance. J Appl Physiol 2000;89:1837–1844.

156. Homsi J, Walsh D, Nelson KA. Psychostimulants in supportive care. Support Care Cancer 2000;8:385–397.

157. Hoffman BB. Adrenoceptor-activating and other sympathomimetic drugs. In: B. G. Katzung BG (eds) Basic and Clinical Pharmacology (pp 120–137). 8th ed. Lange Medical. McGraw-Hill, New York, 2001.

158. Bell DG, Jacobs I, Ellerington K. Effect of caffeine and ephedrine ingestion on anaerobic exercise performance. Med Sci Sports Exerc 2001;33: 1399–1403.

159. Jacobs I, Pasternak H, Bell DG. Effects of ephedrine, caffeine, and their combination on muscular endurance. Med Sci Sports Exerc 2003;35:987–994.

160. Gillies H, Derman WE, Noakes TD, Smith P, Evans A, Gabriels G. Pseudoephedrine is without ergogenic effects during prolonged exercise. J Appl Physiol 1996;81:2611–2617.

161. Hodges AN, Lynn BM, Bula JE, Donaldson MG, Dagenais MO, McKenzie DC. Effects of pseudoephedrine on maximal cycling power and submaximal cycling efficiency. Med Sci Sports Exerc 2003;35:1316–1319.

162. Dulloo AG. Ephedrine, xanthines and prostaglandin-inhibitors: actions and interactions in the stimulation of thermogenesis. Int J Obes Relat Metab Disord 1993;17(Suppl 1):S35–S40.

163. Bell DG, Jacobs I, Zamecnik J. Effects of caffeine, ephedrine and their combination on time to exhaustion during high-intensity exercise. Eur J Appl Physiol Occup Physiol 1998;77:427–433.

164. Bell DG, Jacobs I. Combined caffeine and ephedrine ingestion improves run times of Canadian Forces Warrior Test. Aviat Space Environ Med 1999;70:325–329.

165. Bell DG, Jacobs I, Ellerington K. Effect of caffeine and ephedrine ingestion on anaerobic exercise performance. Med Sci Sports Exerc 2001;33:1399–1403.

166. Bell DG, McLellan TM, Sabiston CM. Effect of ingesting caffeine and ephedrine on 10-km run performance. Med Sci Sports Exerc 2002;34:344–349.

167. Graham TE. Caffeine, coffee and ephedrine: impact on exercise performance and metabolism. Can J Appl Physiol 2001;26(Suppl):S103–S119.

168. Bell DG, Jacobs I, McLellan TM, Zamecnik J. Reducing the dose of combined caffeine and ephedrine preserves the ergogenic effect. Aviat Space Environ Med 2000;71:415–419.

169. Magkos F, Kavouras SA. Caffeine and ephedrine: physiological, metabolic and performance-enhancing effects. Sports Med 2004;34:871–889.

170. Bicopoulos D (ed). AusDI: Drug Information for the Healthcare Professional. 2 nd ed. Pharmaceutical Care Information Services, Castle Hill, NSW, Australia, 2002.

171. Douroudos II, Fatouros IG, Gourgoulis V, et al. Dose-related effects of prolonged $NaHCO_3$ ingestion during high-intensity exercise. Med Sci Sports Exerc 2006;38:1746–1753.

172. Mc Naughton L, Thompson D. Acute versus chronic sodium bicarbonate ingestion and anaerobic work and power output. J Sports Med Phys Fitness 2001;41:456–462.

173. McNaughton L, Backx K, Palmer G, Strange N. Effects of chronic bicarbonate ingestion on the performance of high-intensity work. Eur J Appl Physiol Occup Physiol 1999;80:333–336.

174. Mero AA, Keskinen KL, Malvela MT, Sallinen JM. Combined creatine and sodium bicarbonate supplementation enhances interval swimming. J Strength Cond Res 2004;18:306–310.

175. Cribb PJ, Williams AD, Carey MF, A. Hayes. A creatine-protein-carbohydrate containing supplement enhances responses to resistance training. Med Sci Sports Exerc 2007;39:1960–1968.

176. Stone MH, Sanborn K, Smith LL, et al. Effects of in-season (5 weeks) creatine and pyruvate supplementation on anaerobic performance and body composition in American football players. Int J Sport Nutr 1999;9:146–165.

177. Selsby JT, DiSilvestro RA, Devor ST. Mg^{2+}-creatine chelate and a low-dose creatine supplementation regimen improve exercise performance. J Strength Cond Res 2004;18:311–315.

178. Jowko E, Ostaszewski P, Jank M, et al. Creatine and beta-hydroxy-beta-methylbutyrate (HMB) additively increase lean body mass and muscle strength during a weight-training program. Nutrition 2001;17:558–566.

179. Crowe MJ, O'Connor DM, Lukins JE. The effects of beta-hydroxy-beta-methylbutyrate (HMB) and HMB/creatine supplementation on indices of health in highly trained athletes. Int J Sport Nutr Exerc Metab 2003;13:184–197.

180. Hoffman JR, Cooper J, Wendell M, Im J, Kang J. Effects of beta-hydroxy beta-methylbutyrate on power performance and indices of muscle damage and stress during high-intensity training. J Strength Cond Res 2004;18:747–752.

181. Ransone J, Neighbors K, Lefavi R, Chromiak J. The effect of beta-hydroxy beta-methylbutyrate on muscular strength and body composition in collegiate football players. J Strength Cond Res 2003;17:34–39.

182. Paddon-Jones D, Keech A, Jenkins D. Short-term beta-hydroxy-beta-methylbutyrate supplementation does not reduce symptoms of eccentric muscle damage. Int J Sport Nutr Exerc Metab 2001;11:442–450.

183. Kreider RB, Ferreira M, Wilson M, Almada AL. Effects of calcium beta-hydroxy-beta-methylbutyrate (HMB) supplementation during resistance-training on markers of catabolism, body composition and strength. Int J Sports Med 1999;20:503–509.

184. Slater G, Jenkins D, Logan P, et al. Beta-hydroxy-beta-methylbutyrate (HMB) supplementation does not affect changes in strength or body composition during resistance training in trained men. Int J Sport Nutr Exerc Metab 2001;11:384–396.

185. Hill CA, Harris RC, Kim HJ, Bobbis L, Sale C, Wise JA. The effect of beta-alanine and creatine monohydrate supplementation on muscle composition and exercise performance. Med Sci Sports Exerc 2005;37:S348.

186. Harris RC, Hill CA, Kim HJ, et al. Beta-alanine supplementation for 10 weeks significantly increased muscle carnosine levels. FASEB J 2005;19:A1125.

187. Suzuki Y, Ito O, Mukai N, Takahashi H, Takamatsu K. High level of skeletal muscle carnosine contributes to the latter half of exercise performance during 30-s maximal cycle ergometer sprinting. Jpn J Physiol 2002;52:199–205.

188. Harris RC, Hill C, Wise JA. Effect of combined beta-alanine and creatine monohydrate supplementation on exercise performance. Med Sci Sports Exerc 2003;35:S218.

189. Stout JR, Cramer JT, Zoeller RF, et al. Effects of beta-alanine supplementation on the onset of neuromuscular fatigue and ventilatory threshold in women. Amino Acids 2007;32:381–386.

190. Stout JR, Cramer JT, Mielke M, O'Kroy J, Torok DJ, Zoeller RF. Effects of twenty-eight days of beta-alanine and creatine monohydrate supplementation on the physical working capacity at neuromuscular fatigue threshold. J Strength Cond Res 2006;20:928–931.

191. Hoffman J, Ratamess N, Kang J, Mangine G, Faigenbaum A, Stout J. Effect of creatine and beta-alanine supplementation on performance and endocrine responses in strength/power athletes. Int J Sport Nutr Exerc Metab 2006;16:430–446.

192. Phillips SM, Hartman JW, Wilkinson SB. Dietary protein to support anabolism with resistance exercise in young men. J Am Coll Nutr 2005;24:134S–139S.

193. Andersen LL, Tufekovic G, Zebis MK, et al. The effect of resistance training combined with timed ingestion of protein on muscle fiber size and muscle strength. Metabolism 2005;54:151–156.

194. Kerksick CM, Rasmussen CJ, Lancaster SL, et al. The effects of protein and amino acid supplementation on performance and training adaptations during ten weeks of resistance training. J Strength Cond Res 2006;20:643–653.

195. Manninen AH. Protein hydrolysates in sports and exercise: a brief review. J Sports Sci Med 2004;3:60–63.

196. Bolster DR, Kubica N, Crozier SJ, et al. Immediate response of mammalian target of rapamycin (mTOR)-mediated signalling following acute resistance exercise in rat skeletal muscle. J Physiol 2003;553:213–220.

197. Rozenek R, Ward P, Long S, Garhammer J. Effects of high-calorie supplements on body composition and muscular strength following resistance training. J Sports Med Phys Fitness 2002;42:340–347.

198. Karlsson HK, Nilsson PA, Nilsson J, Chibalin AV, Zierath JR, Blomstrand E. Branched-chain amino acids increase p70S6k phosphorylation in human skeletal muscle after resistance exercise. Am J Physiol 2004;287:E1–E7.

199. Esmarck B, Anderson JL, Olsen S, Richter EA, Mizuno M, Kjaer M. Timing of post-exercise protein intake is important for muscle hypertrophy with resistance training in elderly humans. J Physiol 2001;535:301–311.

200. Chromiak JA, Smedley B, Carpenter W, et al. Effect of a 10-week strength training program and recovery drink on body composition, muscular strength and endurance, and anaerobic power and capacity. Nutrition 2004;20: 420–427.

201. Rankin JW, Goldman LP, Puglisi MJ, Nickols-Richardson SM, Earthman CP, Gwazdauskas FC. Effect of post-exercise supplement consumption on adaptations to resistance training. J Am Coll Nutr 2004;23:322–330.

202. Zoeller RF, Stout JR, O'kroy JA, Torok DJ, Mielke M. Effects of 28 days of beta-alanine and creatine monohydrate supplementation on aerobic power, ventilatory and lactate thresholds, and time to exhaustion. Amino Acids 2007;33:505–510.

203. Ohtani M, Sugita M, Maruyama K. Amino acid mixture improves training efficiency in athletes. J Nutr 2006;136:538S–543S.

204. Ohtani M, Maruyama K, Suzuki S, Sugita M, Kobayashi K. Changes in hematological parameters of athletes after receiving daily dose of a mixture of 12 amino acids for one month during the middle- and long-distance running training. Biosci Biotechnol Biochem 2001;65:348–355.

205. Ohtani M, Maruyama K, Sugita M, Kobayashi K. Amino acid supplementation affects hematological and biochemical parameters in elite rugby players. Biosci Biotechnol Biochem 2001;65:1970–1976.

206. Antonio J, Sanders MS, Ehler LA, Uelmen J, Raether JB, Stout JR. Effects of exercise training and amino-acid supplementation on body composition and physical performance in untrained women. Nutrition 2000;16:1043–1046.

10 Nutritional Supplements for Strength Power Athletes

Colin Wilborn

Abstract

Over the last decade research involving nutritional supplementation and sport performance has increased substantially. Strength and power athletes have specific needs to optimize their performance. Nutritional supplementation cannot be viewed as a replacement for a balanced diet but as an important addition to it. However, diet and supplementation are not mutually exclusive, nor does one depend on the other. Strength and power athletes have four general areas of supplementation needs. First, strength athletes need supplements that have a direct effect on performance. The second group of supplements includes those that promote recovery. The third group comprises the supplements that enhance immune function. The last group of supplements includes those that provide energy or have a direct effect on the workout. This chapter reviews the key supplements needed to optimize the performance and training of the strength athlete.

Key words

Strength · Power · Hypertrophy · Recovery · Performance

1. INTRODUCTION

The supplement industry is a multibillion dollar a year industry. New and more innovative supplements are released on a daily basis. The problem is that in many cases there is little evidence to support claims of their benefit. Thus, we have to determine what a superior sport supplement is and what it is not.

To stay on the cutting edge of nutritional supplementation we must identify what supplements have been shown to be effective

From: *Nutritional Supplements in Sports and Exercise*
Edited by: M. Greenwood, D. Kalman, J. Antonio,
DOI: 10.1007/978-1-59745-231-1_10, © Humana Press Inc., Totowa, NJ

and safe when used properly. Many of the leading experts in the field have identified and classified some of the leading supplements into categories that range from safe and effective to those that have harmful side effects or no supporting literature. If we are to disseminate all of the information available, we have to ask three simple questions in regard to the specific supplement: Is there a sound scientific rationale to support its use? Is there any scientific evidence that the supplement affects health or exercise? Is the supplement legal and safe? Although this book covers many of the supplements out there, it is almost impossible to cover them all. This chapter discusses supplements whose use is supported by strong evidence that they are safe, effective, and legal.

The two main components of becoming a successful strength and power athlete are intense training and superior nutrition. Although we can never devalue the importance of a balanced diet, nutrition supplementation can provide the impetus for huge gains. Nutrition is by far more difficult to take care of than the training. You can have an excellent day from the training standpoint by simply putting in the time. Nutrition involves a 24-hour-a-day 7-day-a-week commitment. Nutritional supplementation is an area where strength athletes can separate themselves from the crowd. Strength and power athletes have many needs from a nutritional standpoint. It would be detrimental to assume that the only supplements that strength athletes need are those that promote protein synthesis (discussed later). The demands of a strength athlete are high, and the body must be put in an advantageous environment to make the appropriate adaptations. Nutrition habits might make the difference between being able to withstand the pounding and physical abuse a strength athlete endures and succumbing to the rapid onset of fatigue. Without providing proper fuel to the body, the athlete cannot operate at maximal ability. This is the main reason why nutrition is such an important part of becoming the "best that you can be."

The supplement needs of strength athletes can be broken down into four general categories. First strength athletes need supplements that have a direct effect on performance. These supplements directly affect protein synthesis and hormonal adaptations, or they have proven effects on strength, power, or hypertrophy. Second, there are supplements that have been categorized to promote recovery. Strength athletes have demanding schedules and intense workloads, and their bodies need optimal recovery. These supplements enhance recovery and support adaptation. Third, certain supplements are touted as having an effective immune function. They go hand in hand with those that promote recovery. Heavy, rigorous strength training can have a negative effect on immune function; so it is necessary to make sure that strength athlete's nutritional standards are high. Finally, a fourth group includes supplements that provide energy or have a direct effect on the workout. These supplements create the optimal environment for adaptation to rigorous training regimens. It is important to note that in many cases these supplements overlap categories. However, it would be redundant to list them more than once. Making the commitment to and perfecting good habits can ensure a drastic improvement in performance, which will no doubt translate into success on the field.

2. REVIEW OF THE LITERATURE

2.1. Supplements that Enhance Strength, Power, or Hypertrophy Directly

2.1.1. CREATINE

Creatine has dominated the supplement market for more than a decade. It has been researched more than any other nutritional supplement on the market today. Creatine is a nitrogenous organic compound obtained predominantly from the ingestion of meat or fish, although it is also synthesized endogenously in the kidney, liver, and pancreas (1,2). Creatine is a naturally occurring amino acid that is derived from the amino acids glycine, arginine, and methionine. When creatine enters the muscle cell, it accepts a high-energy phosphate and forms phosphocreatine (PC). PC is the storage form of high-energy phosphate, which is used by the skeletal muscle cell to regenerate adenosine triphosphate (ATP) rapidly during bouts of maximum muscular contraction (3). The conversion of ATP into adenosine diphosphate (ADP) and a phosphate group generates the energy needed by the muscles during short-term, high-intensity exercise. PC availability in the muscles is vitally important for energy production, as ATP cannot be stored in excessive amounts in muscle and is rapidly depleted during bouts of exhaustive exercise. Oral creatine monohydrate supplementation has been reported to increase muscle creatine and PC content by 15% to 40%, enhance the cellular bioenergetics of the phosphagen system, improve the shuttling of high-energy phosphates between the mitochondria and cytosol via the creatine phosphate shuttle, and enhance the activity of various metabolic pathways *(4)*.

Although creatine research has now expanded to cover its effects on cancer *(5)* and various delivery methods, strength and power athletes are concerned only about creatine's amazing results in athletics. Scientific studies indicate that creatine supplementation is an effective, safe nutritional strategy to promote gains in strength and muscle mass during resistance training *(6–10)*. Creatine has become one of the most popular nutritional supplements for resistance-trained athletes and bodybuilders. In fact, creatine may be the most researched nutritional supplement of the last decade. A recent periodical search yielded more than 1000 hits for creatine in sport and performance.

The energy for all-out maximum-effort exercise lasting up to 10 seconds is primarily derived from limited stores of ATP in the muscle. Creatine, when present in the muscle in sufficient amounts, donates a phosphate group to ADP; the ADP then rapidly transforms to ATP and is immediately available to the muscle to be used as fuel for exercise. During explosive exercise, the phosphate from PC stored in muscle is also cleaved off to provide energy for resynthesis of ATP. This allows the ATP pool to be turned over several dozen times during an all-out maximum-effort exercise bout lasting up to 10 seconds.

Theoretically, the more creatine consumed, the more energy there is for brief high-intensity activity. It does appear, however, that creatine supplementation has an upper limit. Once creatine stores are saturated, excess creatine is not beneficial. Thus, more recent research has been based on the effect of creatine supplementation on body weight or, more specifically, lean body mass. Creatine could be extremely advantageous to strength athletes given its ability to promote strength gains during training. Studies indicate that creatine supplementation during training can increase gains in one repetition maximum (1RM) strength and power.

Peeters et al. *(11)* investigated the effect of creatine monohydrate (CrM) and creatine phosphate (CrP) supplementation on strength, body composition, and blood pressure over a 6-week period. Strength tests performed were the one-repetition maximum (1RM) bench press, 1RM leg press, and maximum repetitions on the seated preacher bar curl with a fixed amount of weight. Subjects were divided into three groups matched for strength: placebo (Pl), CrM, and CrP. All subjects were provided a standardized strength training regimen and ingested a loading dosage of 20 g/day for the first 3 days of the study, followed by a maintenance dose of 10 g/day for the remainder of the 6-week supplementation period. Significant differences were noted between the Pl group and the two Cr groups regarding changes in lean body mass, body weight, and 1RM bench press. These results suggest that oral creatine supplementation results in greater strength and fat-free mass development.

Eckerson et al. *(12)* also studied the effects of 2 and 5 days of creatine loading on anaerobic working capacity (AWC) in women using the critical power (CP) test. Ten physically active women randomly underwent two treatments separated by a 5-week washout period: 1) 18 g dextrose as placebo or 2) 5.0 g Cr + 18 g dextrose taken four times per day for 5 days. The placebo resulted in no significant changes in AWC following supplementation, whereas creatine increased the AWC by 22.1% after 5 days of loading ($p < 0.05$). These results suggest that creatine supplementation is effective for increasing AWC in women following 5 days of loading without an associated increase in BW.

A study by Volek et al. *(10)* investigated the influence of oral supplementation with creatine monohydrate on muscular performance during repeated sets of high-intensity resistance exercise. Fourteen active men were randomly assigned in a double-blind fashion to either a creatine group or a placebo group. Both groups performed a bench press exercise protocol and a jump squat exercise protocol on three occasions separated by 6 days. Creatine supplementation resulted in a significant improvement in peak power output during all five sets of jump squats and a significant improvement in repetitions during all five sets of bench presses. A significant increase in body mass of 1.4 kg was observed after creatine ingestion. One week of creatine supplementation (25 g/day) enhances muscular performance during repeated sets of bench press and jump squat exercise.

Kreider et al. *(13)* conducted a study in which 25 National Collegiate Athletic Association (NCAA) division IA football players were matched-paired and assigned to supplement their diet for 28 days during resistance/agility training with creatine. Subjects performed a maximum repetition test on the isotonic bench press, squat, and power clean; and subjects performed a cycle ergometer sprint test. Gains in bench press lifting volume, the sum of bench press, squat, and power clean lifting volumes, and total work performed during the first five 6-second sprints was significantly greater in the creatine group. The addition of the creatine supplement promoted greater gains in fat/bone-free mass, isotonic lifting volume, and sprint performance during intense resistance/agility training.

Although only a few studies have been reviewed here, dozens of studies have demonstrated an increase in strength and power and improved high-intensity performance.

2.1.2. PROTEIN AND AMINO ACIDS

It has long been believed that excess protein intake was necessary for optimal muscle growth in response to strength training. First, we must understand that skeletal muscle growth is possible only when muscle protein synthesis exceeds muscle protein breakdown. The body is in a continuous state of protein turnover; as old proteins are destroyed or degraded, new ones are being synthesized. Thus, when synthesis of contractile proteins is occurring at a faster rate than degradation the net result is a positive protein balance or, more specifically, myofibrillar hypertrophy. Knowing this, we understand that we need protein or the building blocks of those proteins, amino acids (AAs), available to ensure that we are in a positive balance.

The only problem is that we do not store AAs in our body but draw available AAs from the AA pool. The AA pool is a mixture of AAs available in the cell derived from dietary sources or the degradation of protein. A protein balance is achieved when the dietary intake is balanced by the excretion of urea wastes. AAs enter this pool in three ways: They enter during digestion of protein in the diet, when body protein decomposes, and when carbon sources and NH_3 synthesize nonessential amino acids. In addition, the body attempts to maintain a pattern of constancy in the free AA pool and in the rate of protein turnover. As previously noted, AAs are not stored but are in constant turnover. The AA pool exists to provide individual AAs for protein synthesis and oxidation, and it is replenished only by protein breakdown or AAs entering the body from the diet. Thus, the free AA pool provides the link between dietary protein and body protein in that both dietary protein and body protein feed into the free pool. The effects of dietary protein intake play a vital role in AA pools. It is not only the amount of dietary protein intake that plays a role but also the type of protein/AA and the frequency of ingestion.

It is clear from a physiological perspective that we cannot properly build muscle without sufficient protein intake. Research has clearly defined a vital role of protein and AA supplementation in the development of skeletal muscle. It is interesting to note that previous studies have concluded that acute protein synthesis following resistance training is similar in fed *(14)* and fasted *(15,16)* subjects. Only the rate of turnover changes slightly, although it remains in a negative balance. However, studies have found that with dietary protein or AA supplementation, muscle protein synthesis rate is increased.

Biolo et al. *(17)* evaluated the interactions between resistance training and AA supplementation and the corresponding effects on protein kinetics. Six untrained men were the subjects in this study. Each participant was infused with a mixed (phenylalanine, leucine, lysine, alanine, glutamine) AA solution. Baseline and postresistance training (5×10 sets of leg press and 4×8 sets of nautilus squats, leg curls, and leg extension) samples were taken. The results revealed increased protein synthesis and no change in protein degradation.

Tipton and colleagues *(18)* investigated the effects of orally administered AAs on postexercise net protein synthesis. There were three groups: 1) 40 g carbohydrate (CHO) (Placebo group); 2) 40 g mixed essential and nonessential AAs (EAA+NEAA group); 3) 40 g essential AAs (EAA+arginine). They also sought to determine whether there would be a difference in the anabolic effect of AA supplementation if they used a mixed AA source or essential AAs alone. The findings of this study indicated that postexercise AA supplementation elicits a positive protein balance compared to the negative balance seen with resistance training alone. The study also concluded that supplementation with the essential AAs alone is equivalent to that of a mixed AA supplement.

Although research has concluded that postexercise AA supplementation has a positive effect on protein synthesis, AA supplementation is not always an option. Esmark et al. *(19)* investigated the timing of protein intake after exercise on muscle hypertrophy and strength. This study used a milk and soy protein supplement (containing 10 g protein from skimmed milk and soybean, 7 g carbohydrate, and 3.3 g lipid) instead of an AA mixture. Although protein synthesis was not calculated in this study, hypertrophy was measured. As previously noted, muscular hypertrophy is the result of net protein synthesis. The results of the Esmark study indicated that skeletal muscle hypertrophy was significantly increased after resistance training when a protein supplement was taken. The findings also suggested that when the supplement was taken immediately after the training versus 2 hours later the hypertrophic response was greater.

In the most recent study by Tipton and colleagues *(20)*, they evaluated the effects of casein and whey protein ingestion on protein balance after resistance training. Twenty-three subjects consumed one of three drinks 1 hour after a bout of leg extensions. Subjects consumed either placebo, 20 g of casein protein, or 20 g of whey protein. The results indicated that ingestion of whey or casein protein after a bout of resistance exercise increases net muscle protein synthesis.

In a review, Rennie and colleagues *(21)* concluded that there is no doubt that increasing AA concentrations by intravenous infusion, meal feeding, or ingestion of free AAs increases muscle protein synthesis. They also concluded that during the postexercise period increased availability of AAs enhances muscle protein synthesis. These studies have clearly identified that AA or protein supplementation enhances protein synthesis and suppresses degradation, resulting in net protein synthesis.

Another possibility for AA supplementation is to supplement branched-chain amino acids (BCAAs). The BCAAs include leucine, isoleucine, and valine. BCAAs are the only AAs that are used exclusively for the synthesis of tissue protein and not for other hormones; thus, their value is paramount. The BCAAs cannot be broken down in the liver like the other AAs. Liver does not contain the BCAA aminotransferase enzyme that other tissues (e.g., muscle) do. These AAs are unique in that their amino group can be removed to form α-ketogluturate, which produces branched-chain ketoacids plus the AA glutamate. The amino group on glutamate is transferred to pyruvate, regenerating α-ketoglutarate to form alanine and glutamine. Elevated levels of BCAAs in the free pool contribute to various metabolic pathways and produce many intermediates that could be important to cellular metabolism. Given that BCAAs are involved in the synthesis of glutamine, they may play an important role for strength and power athletes. Glutamine (discussed later) has a beneficial effect on recovery. This is important given the rigor of the training regimens of strength athletes.

Therefore, it is clear from these studies and others that protein or AA supplementation does enhance the physiological adaptations of resistance training. A recent review by Kerksick and Leutholtz *(22)* noted the following conclusions as a synopsis of the current literature on protein and AA supplementation. First, ingestion of AAs after resistance exercise has been shown at many time points in several studies to stimulate increases in muscle protein synthesis, cause minimal changes in protein breakdown, and increase overall protein balance. These authors also concluded that intact proteins or combinations of them that are commonly used in popular protein supplements appear to elicit increases in protein balance after resistance training similar to those reported in other studies using free AAs.

2.1.3. CARNOSINE/β-ALANINE

Carnosine, a dipeptide comprised of the amino acids histidine and β-alanine, occurs naturally in brain, cardiac muscle, kidney, and stomach and in large amounts in skeletal muscles. Carnosine has been widely studied for its contribution to improved wound healing, its antioxidant activity, and its anti-aging properties. Carnosine is found in high concentrations in skeletal muscle, primarily type II muscle fibers, which are the fast-twitch muscle fibers used during explosive movements such as weight training and sprinting. It has also been concluded that carnosine levels are found in higher concentrations in athletes whose performance demands serious anaerobic output.

Carnosine contributes to buffering H^+, thus attenuating a drop in pH associated with anaerobic metabolism. Carnosine is highly effective at buffering the hydrogen ions responsible for producing the ill effects of lactic acid, and it is believed to be one of the primary muscle-buffering substances available in skeletal muscle. In theory, if carnosine could attenuate the drop in pH noted with high-intensity exercise, one could possibly exercise longer. However, via the activity of the enzyme carnosinase, carnosine is rapidly degraded into β-alanine and histidine as soon as it enters the blood. As such, there is no advantage to using direct carnosine supplementation. Thus, β-alanine is believed to be the answer to increasing carnosine in skeletal muscle.

β-Alanine is an amino acid that is not involved in structural proteins, functions to combine with another amino acid, histidine, to form carnosine. However, the synthesis is under some kind of limited control. Using a supplemental form of β-alanine can significantly increase the synthesis. Histidine is already abundant in skeletal muscle, so it is β-alanine that acts as the rate-limiting factor during carnosine conversion. Recent studies conducted by Harris et al. (23) have demonstrated that taking β-alanine orally is effective at increasing carnosine levels. Individuals who take oral β-alanine on a regular basis can expect to increase their muscles' synthesis of carnosine by up to 64%. This study also showed that your body creates only a certain amount of carnosine; it does increase as you work out, but the level eventually plateaus. At this point, the only way to increase your body's production of carnosine is to take β-alanine orally. A recent study (24) supplemented men with β-alanine for 10 weeks. Muscle carnosine was significantly increased by 58.8% and by 80.1% after 4 and 10 weeks of β-alanine supplementation. Carnosine, with an initial 1.71 times higher concentration in type IIa fibers, increased equally in both type I and IIa fibers; there was no increase observed in control subjects.

Researchers including Harris and Stout have begun to do extensive research in the area of β-alanine supplementation for strength athletes. Stout et al. (25) conducted a study that examined the effects of β-alanine supplementation on physical working capacity at fatigue threshold (PWCFT) in untrained young men. The results revealed a significantly greater increase in PWCFT of 9% over placebo. The findings suggest that β-alanine supplementation for 28 days may delay the onset of neuromuscular fatigue. Another study by Stout's group (26) examined the effects of 28 days of β-alanine supplementation on the physical working capacity at fatigue threshold, ventilatory threshold, maximum oxygen consumption, and time-to-exhaustion in women. Results of this study indicated that β-alanine supplementation delays the onset of neuromuscular fatigue and the ventilatory threshold at submaximum workloads and increases the total time-to-exhaustion during maximum cycle ergo-metry performance. These authors concluded that β-alanine supple-mentation appears to improve submaximum cycle ergometry performance and total time-to-exhaustion in young women, per-haps as a result of an increased buffering capacity due to elevated muscle carnosine concentrations.

Recently, several studies have investigated the effects of supple-menting creatine and β-alanine together. This proposed benefit would increase work capacity and decrease time to fatigue. Hoffman et al. *(27)* studied the effects of creatine and creatine + β-alanine on strength, power, body composition, and endocrine changes during a 10-week resistance training program in collegiate football players. The results of this study demonstrated that creatine + β-alanine was effective at enhancing strength performance. Creatine + β-alanine supplementation also appeared to have a greater effect on lean tissue accrual and body fat composition than creatine alone.

Stout et al. *(25)* also investigated the effects of β-alanine and creatine on the onset of fatigue. They found that β-alanine was effective at reducing fatigue, but creatine did not have an additive effect.

The most recent studies that document the effectiveness of β-alanine are just now being published. Many of the studies have been completed but are yet to be published. However, there appears to be enough research currently to evaluate its effectiveness. It appears that not only does β-alanine appear to increase muscle carnosine levels, the changes appear to translate into performance benefit.

2.1.4. ARGININE/α-KETOGLUTURATE

Arginine is one of the amino acids produced in the human body during the digestion or hydrolysis of proteins. Arginine can also be produced synthetically. Because it is produced in the body, it is referred to as nonessential, meaning that no food or supplements are necessary for humans. Arginine compounds can be used to treat people with liver dysfunction because of its role in promoting liver regeneration.

Arginine has several roles in the body: It assists wound healing, helps remove excess ammonia from the body, stimulates immune function, and promotes secretion of several hormones, including glucagon, insulin, and growth hormone. NO_2 is a compound produced from arginine that elicits arteriole vasodilation and assists in nutrient transport/recovery in muscle. This action has been proposed to cause a perpetual muscle pump in users and promote gains in muscle.

Campbell and colleagues *(28)* evaluated the pharmacokinetics, safety, and efficacy of arginine α-ketoglutarate (AAKG) in trained adult men. Subjects participated in two studies that employed a randomized, double-blind, controlled design. In study 1, 10 healthy men fasted for 8 hours and then ingested 4 g of time-released or non–timed-released AAKG. Blood samples were obtained for 8 hours after AAKG ingestion to assess the pharmacokinetic profile of L-arginine. After 1 week, the alternative supplement was ingested. In study 2, which was placebo-controlled, 35 resistance-trained adult men were randomly assigned to ingest 4 g of AAKG (three times a day) or placebo. Participants performed 4 days of periodized resistance training per week for 8 weeks. At 0, 4, and 8 weeks of supplementation, the following tests and examinations were performed: clinical blood markers, one repetition maximum bench press, isokinetic quadriceps muscle endurance, anaerobic power, aerobic capacity, total body water, body composition, and psychometric parameters. In study 1, significant differences were observed in plasma arginine levels in subjects taking non–timed-release and timed-release AAKG. In study 2, significant differences were observed in the AAKG group for 1RM bench press, Wingate peak power, blood glucose, and plasma arginine. No significant differences were observed between groups in terms of body composition, total body water, isokinetic quadriceps muscle endurance, or aerobic capacity. AAKG supplementation appeared to be safe and well tolerated, and it positively influenced 1RM bench press and Wingate peak power performance.

In a study that combined weight training with either arginine + ornithine or placebo, Elam et al. *(29)* found that the amino acid combination produced decreases in body fat, resulting in higher total strength and lean body mass. Moreover, there was less evidence of tissue breakdown after only 5 weeks.

It is well documented that the infusion of arginine stimulates growth hormone secretion from the anterior pituitary *(30–32)*. This increase in growth hormone secretion from arginine infusion has been attributed to the suppression of endogenous somatostatin secretion. However, other studies done in human trials have not had positive results.

Walberg-Rankin and Hawkins *(33)* studied the effects of arginine on growth hormone (GH) and its influence on body composition and muscle function. Male weight trainers were divided into three groups: control, arginine, and placebo. The subjects were given a similar resistance exercise prescription. Subjects in the placebo and arginine groups demonstrated a significant decrease in peak torque for the biceps and quadriceps. Neither supplement acutely affected serum GH or arginine over the 90 minutes after ingestion. There was no significant difference between groups regarding nitrogen balance. Thus, the supplement had no influence on weight loss, fat or lean tissue loss, muscle function, or overall GH status.

At very high intakes (approximately 250 mg per 2.2 pounds of body weight), the amino acid arginine has increased GH levels *(34)*, an effect that has interested bodybuilders owing to the role of GH in stimulating muscle growth. However, at lower amounts recommended by some manufacturers (5 g taken 30 minutes before exercise), arginine failed to increase GH release and may even have impaired the release of GH in younger adults *(35)*. Large amounts of arginine do not appear to raise levels of insulin *(36)*, another anabolic bodybuilding hormone. More modest amounts of a combination of these amino acids have not had measurable effects on any anabolic hormone levels during exercise.

Altough the results in the literature are conflicting, it is important to discuss arginine as a nutritional supplement. Arginine or NO_2 has been one of the hottest supplements over the last few years. There is tons of anecdotal evidence suggesting that arginine is a powerful modulator of strength and performance; however, the current literature is not decisive.

2.1.5. Aromatase Inhibitors

The never-ending pursuit of bigger muscles by athletes and bodybuilders has led to the creation of new and innovative supplements to complement vigorous training. During the last two decades, research in the area of sport nutrition has confirmed the benefit of supplements such as creatine, BCAAs, and whey protein. However, the most potent, anabolic agent is naturally occurring testosterone.

Testosterone is a hormone that is naturally produced by the body. Its anabolic effect includes promotion of protein biosynthesis. It accelerates muscle buildup, increases the formation of red blood cells, speeds regeneration, and speeds the recovery time after injuries or illness. It also stimulates the entire metabolism, which results in the burning of body fat.

Studies by Bhasin et al. *(37–40)* and Sinha-Hikim et al. *(41,42)* have shown that testosterone at supraphysiological levels increases muscle size and strength. Thus, many supplement companies have sought to create ways to increase naturally occurring levels of testosterone. Over the last decade, many supplements (e.g., prohormones, testosterone derivatives, and now aromatase inhibitors) have flooded the market with promises of "testosterone"-like gains in strength and lean mass. The newest of these supplements are the aromatase inhibitors (AIs). Although AIs are not new themselves, they are relatively new to the fitness community. Aromatase is an enzyme involved in the production of estrogen that acts by catalyzing the conversion of testosterone (an androgen) to estradiol (an estrogen) *(43)*.

Aromatase inhibitors have been used in an "athletic" setting for many years. However, their use was thought to be beneficial only to individuals using steroids. Steroids that aromatize heavily are responsible for extreme elevations in estrogen. Higher-than-normal estrogen levels can lead to several physiological problems. Thus, aromatase inhibitors have been used to prevent testosterone conversion and to limit the negative effects of aromatase. Recently the idea has been proposed that AIs may present an advantageous option for individuals not on steroids.

Zmuda et al. *(44)* investigated the effects of an aromatase inhibitor in 14 male subjects. Researchers found that serum testosterone levels increased during all three drug treatments, whereas the estradiol level increased only with testosterone alone, demonstrating that the aromatase inhibitor effectively inhibited testosterone aromatization.

Leder et al. *(45)* investigated the ability of the orally administered aromatase inhibitor anastrozole to increase endogenous testosterone production in 37 elderly men with screening serum testosterone levels of < 350 ng/dL. The findings of this study demonstrated that aromatase inhibition increases serum bioavailable and total testosterone levels to the youthful normal range in older men with mild hypogonadism. Serum estradiol levels decrease modestly but remain within the normal male range. Furthermore, other research has investigated eugonadal boys who were given the aromatase inhibitors exemestane *(46)* and anastrozole *(47)*. They found that there was comparable suppression of estradiol with a parallel increase in the testosterone and free testosterone concentrations.

Now that science has determined that AIs do, in fact, increase testosterone levels, a new wave of AIs have hit the market. Traditionally, AIs are obtained by prescription. However many nutritional supplements are now available that claim to inhibit aromatase. A study on an over-the-counter AI was recently completed *(48)*. This study examined blood hormone responses over a 3-week cycle of 600 mg AI/day in six normal men aged 32 to 40 years. In addition to tracking changes in sex hormone levels it also looked at common indicators of toxicity. Total testosterone levels rose an average of 188%, and free testosterone levels rose an average of 226% over the course of the 3 weeks. A nonsignificant decrease in estradiol was also seen. Furthermore, Numazawa and colleagues *(49,50)* have shown that androst-4-ene-3,6,17-trione (the active ingredient in 6-OXO^{TM}) irreversibly binds to the aromatase enzyme, thereby causing a halt in estradiol production.

A study conducted at Baylor University examined the effects of a newly marketed aromatase inhibitor on body composition and changes in serum hormone levels *(51)*. All subjects were experienced weightlifters with at least 1 year of experience. The 16 participants were equally divided, matched by age and body weight and then assigned an 8-week supplementation protocol consisting of the oral ingestion of either hydroxyandrost-4-ene-6,17-dioxo-3-THP ether + 3,17-diketo-androst-1,4,6-triene—four capsules per day totaling 72 mg/day at bed time—or four capsules per day of placebo (72 mg maltodextrin) also at bed time. Total body mass and body composition as well as various tests on venous blood samples were determined at week 0 and after weeks 4, 8, and 11. Results indicated significant changes in both free and total testosterone. Total endogenous testosterone, free testosterone, and dihydrotestosterone (DHT) had increased an average of 300%, 600%, and 630% from baseline, respectively, at 8 weeks. Whereas there were significant increases in testosterone and DHT, there were no significant changes in estradiol, estrone, estrione, or a variety of serum and urinary clinical safety markers. These increases in testosterone with no change in estrogen led us to believe that the aromatase enzyme was apparently inhibited by the supplement. Furthermore, body fat significantly decreased 1% in the active group, with a 1.5-lb decrease in fat mass. At the end of 8 weeks of supplementation, subjects had a 3-week washout period. After 3 weeks of washout, testosterone and estrogen levels had returned to normal baseline levels. The results of this study also indicated that there were no significant changes in any clinical safety markers, particularly those indicative of hepatic and renal function. Overall, the results suggest that over-the-counter aromatase inhibitors can increase endogenous levels of testosterone without adversely affecting clinical safety markers.

Based on the results of this study, we have concluded that aromatase inhibition can increase the fat-free mass by increasing endogenous testosterone levels. Years of research on aromatase inhibitors have provided results indicating that aromatase inhibitors can decrease estrogen and increase testosterone, thereby helping breast cancer patients and menopause and andropause sufferers. This, in turn, might lead us to believe that these products can provide some benefit to resistance-trained athletes.

2.1.6. ZINC MAGNESIUM ASPARTATE

The biological importance of the minerals magnesium and zinc is revealed in various metabolic and hormonal processes in which they regulate function. Zinc is an essential trace element involved in a range of vital biochemical processes and is required for the activity of more than 300 enzymes. Zinc-containing enzymes participate in many components of macronutrient metabolism, particularly cell replication. The incidence of zinc deficiency has been shown to be higher in athletes and/or individuals who train recreationally (52–54). Singh et al. (53,54) investigated blood (plasma) levels of zinc and other trace minerals in 66 men before and after a 5-day period of sustained physical and psychological stress. They found that zinc levels were decreased by 33%.

These zinc deficiency may also contribute to impaired immune function and decreased performance (52,55–57). A 2004 review by

Gleeson et al. *(58)* concluded that heavy exercise and nutrition exert separate influences on immune function and that these influences appear to be stronger when exercise stress and poor nutrition act synergistically. Dietary deficiencies of energy, protein, and specific micronutrients are associated with depressed immune function and increased susceptibility to infection.

Magnesium is an essential element in human nutrition; it is the cofactor in enzymes of carbohydrate metabolism. Magnesium, a ubiquitous element that plays a fundamental role in many cellular reactions, is involved in more than 300 enzymatic reactions in which food is metabolized and new products are formed. Some important examples include glycolysis, fat and protein metabolism, ATP synthesis, and a second messenger system. Magnesium also serves as a physiological regulator of membrane stability and in neuromuscular, cardiovascular, immune, and hormonal function. It has also been shown to diminish the hormone cortisol *(59)*, which has been shown to be detrimental to strength gains and muscle mass. A 1984 *(60)* study found that 14 days of magnesium supplementation decreased cortisol. Another study *(61)* in 1998 reported similar results, concluding that magnesium supplementation reduced the stress response without affecting competitive potential.

Magnesium supplementation has been reported to improve adaptations to exercise, specifically resistance training *(57,62,63)*. Therefore, supplementation of zinc and magnesium may enhance immune function, increase testosterone, increase strength, and diminish the effects of cortisol, thus having a benefit to resistance training athletes. A 2000 study *(64)* found significant changes in hormones—testosterone, insulin-like growth factor-1 (IGF-1)—and muscle strength when athletes were given a zinc and magnesium supplement.

Zinc and magnesium supplementation has been reported to have positive effects on resistance-training athletes *(62,64,65)*. Subsequently, decreases in zinc and magnesium were associated with loss of strength and muscle mass. However, athletes have been reported to have lower levels of zinc and magnesium possibly due to increased sweating while training or with inadequate intake in their diets *(53,54,66–68)*. A 2000 study was conducted on collegiate football players during 8 weeks of spring practice. The purpose of the study was to assess the effect of a novel zinc, magnesium, and vitamin B_6 formulation (ZMA) on anabolic hormones and muscle function. Twenty-seven subjects successfully followed the nightly supplement regimen over the course of the study and completed the testing sessions. The results of ZMA supplementation on anabolic hormone profile in football players before and after spring football practice indicates that the ZMA group had increased concentrations of total testosterone, free testosterone, and IGF-1 compared to plateaus or decreases in the placebo group. Significant increases in isokinetic torque and power measurements were also seen. The ZMA group increases were significantly different from levels in the placebo group.

Wilborn et al. *(69)* examined whether supplementing the diet with a commercial supplement containing ZMA during training affects zinc and magnesium status, anabolic and catabolic hormone profiles, and/or training adaptations. Forty-two resistance-trained men were matched according to fat-free mass and randomly assigned to ingest in a double-blind manner either a dextrose placebo or ZMA 30 to 60 minutes prior to going to sleep during 8 weeks of standardized resistance training. Subjects completed testing sessions at 0, 4, and 8 weeks that included body composition assessment, 1RM and muscular endurance tests on the bench and leg press, a Wingate anaerobic power test, and blood analysis to assess anabolic/catabolic status and markers of health. Results indicated that ZMA supplementation nonsignificantly increased serum zinc levels by 11% to 17% ($p = 0.12$). However, no significant differences were observed between groups regarding anabolic or catabolic hormone status, body composition, 1RM bench press and leg press, upper or lower body muscular endurance, or cycling anaerobic capacity.

ZMA supplementation has become a popular nutritional practice among resistance-trained athletes. Preliminary research findings have indicated that training decreases zinc and magnesium availability, leading to reductions in testosterone and strength. Zinc and magnesium supplementation has been suggested as a means to maintain zinc and magnesium status and thereby improve training adaptations. However, there is still some discrepancy in the literature.

2.2. Supplements that Promote Recovery

2.2.1. GLUTAMINE

Glutamine is classified as a nonessential amino acid because it can be readily synthesized by various tissues such as the skeletal muscles, liver, and adipose tissue (70). Glutamine contributes a large amount of amino acids to the free pool, as discussed earlier. When the need for glutamine is high, however, the body may not be able to synthesize it a fast enough rate to replenish depletion—hence, the possible need for supplementation. In fact, glutamine is the most abundant amino acid in the body, representing about 60% of the amino acid pool in muscle. In a healthy person, the concentration of glutamine in the blood is three to four times greater than all other amino acids.

Research has shown that glutamine contributes to the prevention of muscle breakdown (71), increase in growth hormone, protein synthesis, improved intestinal health, decrease in the risk of overtraining, and improved immune system function. Glutamine levels are decreased during catabolic conditions in the body and high with an anabolic status (72). Recent reports have shown that plasma glutamine levels decrease acutely after single sessions of high-intensity running (73) and after more extended periods of intensive-running training (74).

Glutamine is also believed to play a large role in enhancement of the immune system. Intense physical training may have a negative effect on the immune system by causing transient suppression of the entire system. The demands on muscle and other organs are so high during intense physical training that the immune system may suffer from a lack of glutamine that temporarily affects its function (75). It has been suggested that because skeletal muscle is the major tissue involved in glutamine production skeletal muscle must thus play a vital role in the process of glutamine utilization in the immune cells. It has been shown that lymphocytes and macrophages utilize glutamine as an energy source at a rate similar to that of glucose. These cells of the immune system have high levels of glutaminase, which is a key enzyme in the catabolism of glutamine. High rates of glycolysis (using glucose) and glutaminolysis (using glutamine) provide powerful supplies of energy for these cells. In addition, glutamine provides nitrogen for the synthesis of nucleotides, which are needed for new

DNA and RNA during cell multiplication of lymphocytes and for mRNA synthesis and DNA repair in macrophages. A decrease in blood glutamine concentration may reduce the maximum rate of lymphocyte cell production and may limit the level of phagocytosis and the rate of cytokine production by macrophages. Because blood glutamine levels decline after heavy exercise, it is possible that exercise-induced immunosuppression is caused in part by glutamine deficiency. A study by Rohde investigated the influence of gluta-mine supplementation on exercise-induced immune system changes for three bouts of bicycle ergometer exercise *(76)*. Oral glutamine supplementation abolished the decrease in plasma glutamine con-centration after exercise without influencing any of the immune system parameters.

Physiological improvement in sports occurs only during the rest period following hard training. This adaptation is in response to maximum loading of the cardiovascular and muscular systems. During recovery periods, these systems build to greater levels to compensate for the stress applied. Proper nutrition and rest are essential for preventing an overtraining effect. Individuals suffering from overtraining also are more susceptible to disease and infections as a result of lowered immunity. One of the results of overtraining is that each day the amount of muscle glutamine gets a little lower. Eventually, the concentration goes below the critical amount of glutamine needed to sustain an anabolic state and the body reverts into a long-term catabolic state. This may be due to the role of glutamine as a primary source of fuel for the cells of the immune system, particularly lymphocytes, macrophages, and killer cells. Thus, glutamine supplementation may enhance performance and adaptation in strength and power athletes.

2.2.2. β-Hydroxy-β-Methylbutyric Acid

β-Hydroxy-β-methylbutyric acid (HMB) is a metabolite of the essential amino acid leucine. HMB is thought to play a role in the regulation of protein breakdown in the body. It helps slow down proteolysis, which is the natural process of breaking down muscle that occurs especially after strenuous activity. It appears that HMB supplementation has a protective effect on muscle and may help the body get a head start on the recovery process by minimizing the amount of protein degradation after exercise. HMB allows the body to stay in an anabolic state longer, which in turn allows it to build more muscle. Thus, theoretically, supplemental HMB might slow the breakdown of protein in the body and thus increase muscle mass and strength.

The body synthesizes a small amount of HMB from α-ketoiso-caproate as a by-product of leucine metabolism. Theoretically, increasing the availability of HMB minimizes protein degradation. In support of this theory, a number of animal studies have reported that adding HMB to the diet enhanced growth rates in pigs, (77) increased muscle mass and decrease body fat in steers (78), improved several markers of immune function in chickens (79,80), and decreased markers of catabolism during training in race horses (81). Based on these findings, it has been hypothesized that HMB supplementation during training in humans inhibits protein degradation, leading to greater gains in strength and muscle mass.

Several studies have evaluated the effects of HMB supplementation on strength and body composition alterations during training in untrained subjects initiating training and in well trained athletes. Nissen et al. (82) studied the effects of dietary supplementation with the leucine metabolite HMB. HMB significantly decreased the exercise-induced rise in muscle proteolysis as measured by urinary 3-methylhistidine during 2 weeks of exercise. The fat-free mass was significantly increased in HMB-supplemented subjects compared with the control group. These authors concluded that supplementation with HMB can partly prevent exercise-induced proteolysis and/ or muscle damage and result in larger gains in muscle function associated with resistance training.

Another study, by Knitter et al. (83), investigated HMB supplementation on muscle damage as a result of intense endurance exercise. Creatine phosphokinase and lactate dehydrogenase (LDH) activities were measured before and after a prolonged run to assess muscle damage. The placebo-supplemented group exhibited a significantly greater increase in creatine phosphokinase activity after a prolonged run than did the HMB-supplemented group. In addition, LDH activity was significantly lower with HMB supplementation compared with the placebo group. These findings support the hypothesis that HMB supplementation helps prevent exercise-induced muscle damage.

Vukovich and Adams *(84)* reported that HMB supplementation (3 g/day for 8 weeks during resistance training) significantly increased muscle mass, reduced fat mass, and promoted greater gains in upper and lower extremity 1RM strength in a group of elderly men and women initiating training. Likewise, Panton and colleagues *(85)* reported that HMB supplementation during 8 weeks of resistance training increased the functional ability to get up, walk, and sit down in a group of elderly subjects. More recently, Gallagher and associates *(86)* evaluated the effects of HMB supplementation (0.38 and 0.76 mg/kg/day) during 8 weeks of resistance training in previously untrained men. The researchers reported that HMB supplementation promoted significantly less muscle creatine kinase excretion and greater gains in muscle mass (in the 0.38 mg/kg/day group only) than subjects taking a placebo. Collectively, these findings support contentions that HMB supplementation may lessen catabolism leading to greater gains in strength and muscle mass.

Kreider et al. *(87)* conducted a study on experienced resistance-trained athletes. Athletes were supplemented with HMB for 28 days during training. Results revealed that although trends were observed HMB supplementation did not significantly affect markers of muscle degradation, muscle mass, or strength. These findings suggest that HMB supplementation does not significantly affect strength and/or muscle mass in well trained subjects.

As with most supplements, there are some conflicting results for HMB. However, there does appear to be sound scientific rationale that HMB supplementation affects catabolism.

2.2.3. Antioxidants

Oxidative stress is the steady-state level of oxidative damage in a cell, tissue, or organ caused by free radicals or the reactive oxygen species (ROS). The ROS, such as free radicals and peroxides, represent a class of molecules that are derived from the metabolism of oxygen and exist inherently in all aerobic organisms. Free radicals are simply electrons that are no longer attached to atoms. There are many sources by which the ROS are generated. Most ROS come from endogenous sources as by-products of normal and essential metabolic reactions, such as energy generation from mitochondria. The level of oxidative stress is determined by the balance between the rate at which oxidative damage is induced and the rate at which it is efficiently repaired and removed. Free radicals interact with other molecules in cells, which can cause oxidative damage to proteins, membranes and genes. Physical exercise induces oxidative stress and tissue damage. Although a basal level ROS is required to drive redox signaling and numerous physiological processes, excess ROS during exercise may have adverse implications on health and performance. High-intensity resistance training has also been shown to increase free radical production. Antioxidant nutrients may be helpful in that regard. Additionally, the overtraining effect can induce increases in ROS.

Antioxidants can be taken to combat the effects of oxidative stress. Antioxidants block the process of oxidation by neutralizing free radicals. That is why there is a constant need to replenish our antioxidant resources. Antioxidants from our diet appear to be of great importance in controlling damage by free radicals. Each nutrient is unique in terms of its structure and antioxidant function.

Vitamin E is a generic term that refers to all entities that exhibit biological activity of the isomer tocopherol. α-Tocopherol, the most widely available isomer, has the highest biopotency, or strongest effect in the body. Because it is fat-soluble, α-tocopherol is in a unique position to safeguard cell membranes, largely composed of fatty acids, from damage by free radicals. Vitamin E is an important intramembrane antioxidant and membrane stabilizer. Vitamin E supplementation has been advocated for athletes in the hope of improving performance, minimizing exercise-induced muscle damage and maximizing recovery.

Meydani et al. *(88)* reported that vitamin E-treated subjects had a decrease in oxidative stress over 12 days following eccentric exercise (downhill running). Cannon et al. *(89)* concluded that supplementation of vitamin E for 48 days reduced the amount of creatine kinase leakage in young and old men during recovery from downhill running bouts. In subsequent studies, Rokitski et al. *(90)* concluded that supplementation of vitamin E for 5 months decreases creatine kinase leakage in aerobic cyclists. These studies indicate that vitamin E supplementation can help reduce muscle damage caused by free radical damage. Hartmann et al. *(91)* noted that DNA damage could occur in white blood cells after exercise. These researchers concluded that a 2400 mg dose of vitamin E resulted in decreased damage to the DNA of white blood cells in exercised individuals.

Vitamin C, also known as ascorbic acid, is a water-soluble vitamin. As such, it scavenges free radicals that are in an aqueous environment inside cells. Vitamin C works synergistically with vitamin E to destroy free radicals. Vitamin C is used in numerous metabolic processes in the body. Theoretically, it may benefit exercise performance by improving metabolism during exercise. Ascorbic acid also may reduce the occurrence of upper respiratory tract infections brought about by intense training *(92)*. Other associations for ascorbic acid involve a role in the synthesis of collagen, which is necessary for strong cartilage, tendons, and bone. Certain hormones such as epinephrine and neurotransmitters need ascorbic acid. Ascorbic acid also aids in the absorption of iron and helps in the formation of red blood cells that carry oxygen to muscle tissues.

Vitamin C does appear to control ROS formed during exercise. If not controlled, these species have the ability to react with cell membranes and damage them. In 1992, Kaminski and Boal *(93)* examined the relation between vitamin C (given to 19 subjects for 3 days before exercise and 7 days afterward) and the muscle damage induced by two bouts of eccentric exercise. The authors concluded that vitamin C reduced muscle damage *(93)*. Vitamin C is also important to a host of numerous other functions in the body.

Vitamins C and E are not the only antioxidants. Vitamin A also has antioxidant properties. Coenzyme Q10 (CoQ10, or ubiquinone), which is essential for energy production, can protect the body from destructive free radicals as well; and uric acid, a product of DNA metabolism, has become increasingly recognized as an important antioxidant. Currently, substances in plants called phytochemicals are being investigated for their antioxidant activity and health-promoting potential.

2.3. Supplements that Enhance Immune Function

2.3.1. VITAMINS AND MINERALS

Immunosuppression in athletes involved in heavy training may increase the athlete's exposure to pathogens and provide optimal conditions for illness and injury. Heavy, prolonged exercise is associated with numerous hormonal and biochemical changes, many of which potentially have detrimental effects on immune function.

Furthermore, improper nutrition can compound the negative influence of heavy exertion on the immune system. For optimal nutrition, one must consume an adequate amount of nutrients. Nutrients are defined as the chemicals taken into the body that are used to produce energy, provide building blocks for new molecules, or function in other chemical reactions. Essential nutrients are the nutrients that must come from the food we ingest because our body cannot make them or is unable to make the required amount of them. Other nutrients can be synthesized from the ingested nutrients. Athletes are not clinically immunodeficient; it is possible that the combined effects of small changes in several immune parameters may compromise resistance to minor illnesses *(94)*. Strategies to prevent immunodeficiencies in athletes include avoiding overtraining, providing adequate rest and recovery during the training, and ensuring adequate nutrition. In a review by Gleeson et al. *(58)* it was concluded that strenuous prolonged exertion and heavy training are associated with depressed immune function. Furthermore, improper nutrition can compound the negative influence of heavy exertion on immune system decrements.

Unlike the macronutrients, the body requires only small amounts of vitamins and minerals. They do not supply any caloric needs, nor do they produce energy. Vitamins are organic compounds that have many important roles in normal functioning, growth, and maintenance of the body. They also help extract energy from the macronutrients. Vitamins are classified into two categories: fat-soluble and water-soluble. Fat-soluble vitamins are vitamins A, D, E, and K. The eight B vitamins and vitamin C make up the water-soluble vitamins. Each vitamin plays a major role in several body functions and may cause health problems if there is a deficiency. Vitamins can be found in just about every food; but the way foods are stored, processed, and cooked often lead to the loss of vitamins *(95)*.

Minerals are elemental atoms that are not destroyed by heat, light, or pH changes. There are seven major minerals: sodium, potassium, chloride, calcium, phosphorus, magnesium, iron. Each mineral has a specific role, from being components of hormones and enzymes to structural function. Significant deficits would lead to severe problems and illnesses. Both plants and animals are good sources of minerals, some examples of which are whole grains, vegetables, milk, meats, and fruit. *(95)*. Although it is impossible to counter the effects of all of the factors that contribute to exercise-induced immunosuppression, it has been shown to be possible to minimize the effects of many factors.

2.3.2. Conjugated Linoleic Acids

Conjugated linoleic acids (CLAs) are essential fatty acids that have been reported to have significant health benefits in animals *(96,97)*. CLA is a naturally occurring fatty acid primarily found in beef and dairy fats. Research has indicated that CLA may have a number of health and performance-enhancing benefits in humans. Some data suggest that CLA supplementation modestly promotes fat loss and/or increases in lean mass *(98–101)*. CLA has been marketed as a supplement that may promote health as well as provide ergogenic value to athletes. Although most research on CLA has been conducted on animals, there have been a number of recent studies that provide greater insight on how CLA may be beneficial to enhance health and performance in humans.

Feeding CLA to animals has been show to lessen markers of catabolism *(102)*, enhance the immune system *(103,104)*, and increase bone content *(105,106)*. A 2005 *(107)* study investigated the effect of dietary CLA supplementation on the immune system and plasma lipids and glucose of healthy human volunteers. Subjects were given 3 g (6 × 500 mg capsules) of CLA per day for 12 weeks. A 12-week washout period followed the intervention period. Levels of plasma immunoglobulin A (IgA) and IgM were increased, and plasma IgE levels were decreased. CLA supplementation also decreased the levels of the proinflammatory cytokines—tumor necrosis factor-α (TNFα) and interleukin-β (IL-1β)—but increased the levels of the antiinflammatory cytokine IL-10. This study showed that CLA could beneficially affect immune function in healthy human volunteers. Another study *(108)* in animals was undertaken to investigate the growth performance and immune responses when supplemented with CLA. There were no significant differences in growth performance among treatments; however, the results indicated that dietary CLA supplementation could enhance the immune response.

Other potential performance benefits of CLA have been shown as well. CLA supplementation has been reported to decrease markers of catabolism. This may help athletes tolerate higher training volumes, leading to greater gains in strength and/or fat free mass over time. Lowery and coworkers *(99)* investigated the effects of CLA supplementation (7.2 g/day) during 6 weeks of resistance training in novice bodybuilders. The researchers reported that CLA supplementation significantly increased arm mass, body mass, and gains in leg press strength. This study provided evidence that CLA supplementation during training may affect training adaptations. Thom and colleagues *(98)* evaluated the effects of CLA supplementation on body composition alterations in 20 healthy male and female subjects. Results revealed no significant changes in body weight. However, subjects ingesting CLA experienced a significant decrease in body fat (–4.3%). Kreider et al. *(109)* evaluated the effects of ingesting 9.2 g/d of CLA for 30-days on body composition, bone density, strength, and markers of catabolism in experienced resistance-trained athletes *(110,112)*. Results revealed that CLA supplementation did not affect body mass, fat mass, or fat-free mass. However, statistical trends were observed indicating that CLA may have lessened markers of catabolism, promoted greater gains in strength, increased bone density, and enhanced immune status. Consequently, results tended to support the theoretical value of CLA supplementation.

Beuker and colleagues *(112)* evaluated the effects of CLA supplementation on body composition, blood lipids, and cycling power output during training. The authors reported that CLA supplementation increased the efficacy of power output and decreased plasma cholesterol by approximately 15%. These findings provide additional support that CLA supplementation affects training adaptations and lipid profiles.

Moreover, CLA appears to have many benefits for the strength athlete. Although there are conflicting results, it appears that there is supporting evidence as to the efficacy of CLA supplementation.

2.4. Supplements that Provide Energy and Enhanced Workouts

2.4.1. CARBOHYDRATE

The amount of carbohydrate that can be stored in the liver and muscle is limited, and it takes time to replenish carbohydrate stores.

When significant amounts of carbohydrate are depleted, it may be difficult to replenish carbohydrate levels fully within 1 day. Consequently, when athletes train once or twice per day over a period of days, carbohydrate levels may gradually decline, leading to fatigue, poor performance, and/or overtraining (113,114). Therefore, it is imperative that active individuals and athletes consume enough carbohydrate in their diet to maintain carbohydrate availability. In addition, certain types of carbohydrate may provide some advantages over others when consumed prior to, during, and/or following exercise.

There is substantial evidence suggesting that the performance of resistance-training exercises can elicit a significant decrement in glycogen stores resulting in decreased performance. Robergs et al. (115) demonstrated that subjects performing six sets of leg extensions at 35% and 70% of 1RM resulted in a decreases in muscle glycogen by 38% and 39%, respectively. In a study by Tesch et al. (116), nine bodybuilders completed five sets each of front squats, back squats, leg presses, and leg extensions to fatigue, comprising 30 minutes of exercise. Biopsies of muscle samples were obtained from the vastus lateralis before and immediately after exercise. The muscle glycogen concentration was 26% lower after exercise. Data from Essen-Gustavsson and Tesch (117) with nine bodybuilders performing the same exercise regimen revealed a 28% decrement in muscle glycogen content as well as a 30% decrease in muscle triglyceride content.

Currently, some scientific evidence suggests that carbohydrate supplementation prior to and during high-volume resistance training results in the maintenance of muscle glycogen concentration, which potentially could result in the maintenance or increase of performance during a training bout. One of the most important nutritional strategies is to ingest adequate amounts of carbohydrate and protein following exercise. Research indicates that athletes who ingest carbohydrate 1.5 g/kg within 2 hours after exercise experience a greater rate of muscle glycogen resynthesis (118).

In a study by Haff et al. (119), six resistance-trained men ingested either a 250-g carbohydrate supplement or placebo during a morning training session, rested 4 hours, and then performed a second session consisting of multiple sets of light-intensity squats (55% 1RM) to exhaustion. During the second training session, the number of sets and repetitions performed were markedly higher with the carbohydrate consumption, and subjects were able to exercise for 30 minutes longer. The authors concluded that athletes engaging in multiple exercise sessions per day would gain a performance advantage with carbohydrate ingestion via maintenance of intramuscular glycogen stores due to greater glycogen resynthesis during recovery.

As discussed earlier, amino acids have a positive effect on protein synthesis. It has been hypothesized that carbohydrate and amino acid supplementation together would result in greater gains. Bird et al. *(120)* investigated chronic alteration of the acute hormonal response associated with liquid carbohydrate (CHO) and/or essential amino acid (EAA) ingestion on hormonal and muscular adaptations following resistance training. Thirty-two untrained young men performed 12 weeks of resistance training twice a week. EAA and CHO ingestion attenuated 3-methylhistidine excretion 48 hours after the exercise bout. CHO + EAA resulted in a 26% decrease, whereas placebo resulted in a 52% increase. In addition muscle cross-sectional area increased across groups for type I, IIa, and IIb fibers, with CHO + EAA producing the greatest gains in cross-sectional area relative to placebo. These data indicate that CHO + EAA ingestion enhances muscle anabolism following resistance training to a greater extent than either CHO or EAA consumed independently.

Based on the current scientific literature, it may be advisable for athletes who are performing high-volume resistance training to ingest carbohydrate supplements before, during, and immediately after resistance training.

2.4.2. CAFFEINE

Caffeine is one of the most widely used stimulants in the world. It occurs naturally in foods and beverages such as coffee, tea, soft drinks, chocolate, and cocoa. The average caffeine consumption in the United States is approximately 200 mg, which is equivalent to two cups of coffee a day. Ten percent of the population ingests more than 1000 mg per day. Caffeine is also added to several over-the-counter medicines such as some weight-loss products, pain medicines, and cold remedies. Caffeine acts as a stimulant on the central nervous system, causing an increase in heart rate and blood pressure.

Most of the research with caffeine and exercise has indicated an improvement in time-to-exhaustion and improved work output in comparison to controls during aerobic exercise (121–125). Ivy and colleagues (125) demonstrated 7.4% improved work output for individuals ingesting caffeine compared to the control group during aerobic cycling bouts. In contrast to several other studies that mainly addressed the ergogenic effects of caffeine in trained individuals, Graham and Spriet (124) showed that the same ergogenic effects of caffeine could be seen on untrained, caffeine-naive subjects. Several studies have indicated that the ergogenic effects of caffeine were best seen when regulated for body weight, with the recommendation for enhancing endurance performance at 80% to 85% VO$_2$max in trained athletes at 9 mg/kg (126–128).

Caffeine has not had the same results for short-term exercise. Doherty and Smith (129) assessed the effects of caffeine ingestion on both aerobic and anaerobic exercise. The results showed that the participants experienced greater improvement during endurance exercise than during graded or short-term exercise. In a related study also conducted by Doherty et al. (130), the participants performed a 1-minute all-out effort on a cycle ergometer. The results suggest that high-intensity cycling performance can be enhanced via an increase in mean power output following moderate caffeine ingestion (5 mg/kg). In a study conducted by Collomp and colleagues (131), the effect of caffeine ingestion (5 mg/kg) on the Wingate anaerobic test was assessed. The results showed that caffeine administration did not significantly change either maximum anaerobic capacity or power. However, there was a significant increase in both catecholamine and blood lactate levels compared to the placebo trials.

Another study (132) was conducted to determine the effect of caffeine on time-to-exhaustion and on associated metabolic and circulatory measures. Eight male subjects ingested either caffeine (5 mg/kg) or a placebo 1 hour prior to exercise at 85% to 90% of maximum workload. Subjects were encouraged to complete three 30-minute intermittent cycling periods at 70 rpm with 5 minutes of rest between each. The exercise was terminated when the subject failed to complete three 30-minute periods or failed to maintain 70 rpm for at least 15 seconds consecutively. Serum free fatty acids, glycerol, blood glucose, lactate, perceived exertion, heart rate, and oxygen cost were measured. The time-to-exhaustion was significantly longer during the caffeine trial than during the placebo trial. Serum free fatty acid levels were significantly different between trials. The decline in blood glucose levels was significantly less during the caffeine trial than during the placebo trial. There were no significant differences between trials for the other measures. It was concluded that caffeine increases time-to-exhaustion when trained subjects cycled intermittently at high intensity.

Although there are conflicting results on the effects of caffeine and performance, there does appear to be benefit. Not only has caffeine demonstrated its capability of enhancing performance directly, it has the potential to have profound effects indirectly. Like other stimulants, caffeine has been advertised and sold as a way to stimulate energy expenditure and possibly result in weight by stimulating both lipolysis and energy expenditure *(133)*. A recent study found that coffee ingestion (200 mg caffeine) resulted in a 7% increase in energy expenditure for 3 hours following ingestion *(134)*. Recent research on the effects of caffeine continues to supports its role in increasing energy expenditure. Another study found that caffeine alone increased energy expenditure by 13% while doubling lipid turnover. It was suggested that these effects were mediated through the sympathetic nervous system *(133)*. It is clear that there is great potential for caffeine as an ergogenic aid.

3. PRACTICAL APPLICATIONS AND CONCLUSION

Efficient sport nutrition is a multifaceted, complex endeavor. With a premium being placed on performance, the strength athlete needs to use every capable tool to perfect his or her performance. Pinnacle nutrition must include a proper balance of carbohydrates, proteins, fats, water, vitamins, and minerals. However, the strength athlete has the benefit of nutrition science to optimize training. Although there is no replacement for a balanced diet, sport supplements can assist in development.

Three supplements have substantial scientific support for their effectiveness. Creatine supplementation has been commonplace in sports for many years. There is no other supplement that has the substantial positive research that creatine has. Creatine has been proven to increase strength, muscle mass, and sprint performance. Protein supplementation is also a necessary component to all strength programs. Protein and amino acids are required for protein synthesis to remain in a positive nitrogen balance. This ensures that strength athletes are getting the most out of their workouts. More recent research suggests that post-workout protein supplementation is a powerful booster to the anabolic process. In conjunction with protein, post-workout carbohydrate supplementation is needed to replace precious glycogen stores that were depleted during the workout. Carbohydrates have received a bad rap in recent years; however, the informed athlete understands the vital important of carbohydrates in optimizing performance. These three supplements are paramount in the resistance-trained athlete.

The next tier of supplements might take the strength athlete to the next level. β-Alanine is the next great supplement. β-Alanine might be capable of increasing time-to-exhaustion and enhancing the workouts necessary to impose substantial overload on the strength athlete. HMB has also been shown to be a powerful anticatabolic supplement. Sparing precious muscle during rigorous training could prove to be the impetus for growth. Coupled with amino acid supplementation, HMB could enhance the hypertrophic effect.

Another group of supplements that might take the strength athlete to the next level comprises the aromatase inhibitors. Aromatase inhibitors have been shown in limited studies to increase endogenous testosterone. The strength athlete understands the importance of testosterone in boosting training adaptations.

Caffeine is a commonplace additive to the diets of millions of athletes and nonathletes alike. However, the stimulatory effects and the boost in metabolism that caffeine induces may enhance workouts and assist in the creation of lean, powerful muscle.

Finally, the supplements that enhance recovery or support immunity could prove to be the difference in peak performance. Endogenous levels of vitamins, minerals, and immune markers have been shown to be low in strength athletes owing to the rigorous nature of their training regimens. This deficiency can be combated by

Table 1
Supplements for Strength and Power Athletes

Nutrient	Theoretical ergogenic value	Summary of research findings/ recommendations
Creatine	Creatine is a naturally occurring amino acid that is derived from the amino acids glycine, arginine, and methionine. When creatine enters the muscle cell, it accepts a high-energy phosphate and forms phosphocreatine (PC). PC is the storage form of high-energy phosphate that is used by the skeletal muscle cell to rapidly regenerate adenosine triphosphate (ATP) during bouts of maximal muscular contraction.	Creatine has been proven to increase strength, muscle mass, and sprint performance. Research supports the use of creatine to increase strength, lean body mass, power, sprint performance, and recovery. No clinical side effects are supported in the literature.
Protein	Protein is the main component of muscles, organs, and hormones. The cells of muscles, tendons, and ligaments are maintained, repaired, and enhanced with protein. Skeletal muscle growth is possible only when muscle protein synthesis	Research has defined a vital role of protein development of muscle mass and hormonal regulation. Post-workout protein supplementation has shown increases in protein synthesis and muscle mass.

(Continued)

Table 1
(Continued)

Nutrient	Theoretical ergogenic value	Summary of research findings/ recommendations
	exceeds muscle protein breakdown; thus, adequate dietary protein is essential.	
Amino acids	Amino acids are organic compounds that combine to form proteins. When proteins are degraded, amino acids are left. The human body requires a number of amino acids to facilitate skeletal muscle growth and repair and for the hormonal deveopment that is necessary for adaptation to stress.	Amino acid supplementation has undergone extensive research, which suggests that pre- and post-workout amino acid supplementation can increase protein synthesis and slow degradation.
β-Alanine	β-Alanine is an amino acid that is not involved in structural proteins. It functions to combine with another amino acid, histidine, to form carnosine. Carnosine is believed to be one of the primary muscle-buffering substances available in skeletal muscle. In theory if carnosine could attenuate the drop in pH noted with high intensity exercise, one could possibly exercise longer.	β-Alanine supplementation appears to improve submaximal cycle ergometry performance and total time to exhaustion. Currently, there appears to be enough research to evaluate its effectiveness. Not only does β-alanine appear to increase muscle carnosine levels, but those changes appear to translate into performance benefits.

Arginine	Arginine is an amino acid that has numerous functions in the body. It is used to make compounds in the body such as nitric oxide, creatine, glutamate, and proline and can be converted to glucose and glycogen if needed. In large doses, arginine also stimulates the release of hormones (e.g., growth hormone, prolactin). Arginine has also been suggested to assist in wound healing, help remove excess ammonia from the body, and stimulate immune function.	There are some conflicting results in the research regarding arginine. However, there is some evidence that arginine supplementation can increase strength, muscle mass, and growth hormone levels.
Aromatase inhibitors	Aromatase is an enzyme involved in the production of estrogen that acts by catalyzing the conversion of testosterone (an androgen) to estradiol (an estrogen). Aromatase inhibitors are believed to inhibit the conversion of testosterone to estrogen, thereby increasing endogenous testosterone levels.	Although aromatase inhibitors as a nutrition supplement are relatively new, there is research suggesting that they can increase endogenous levels of testosterone. However, there is no evidence that these increases in testosterone lead to performance benefits.
Zinc Magnesium, Aspartate	Zinc and magnesium are minerals that are used in a number of metabolic processes and hormonal regulation. Zinc is an essential trace element involved in a range	Preliminary research findings have indicated that training decreases zinc and magnesium availability, leading to reductions in testosterone and strength.

(Continued)

Table 1
(Continued)

Nutrient	Theoretical ergogenic value	Summary of research findings/recommendations
	of vital biochemical processes and is required for the activity of more than 300 enzymes. Magnesium is an essential element in human nutrition; it is the cofactor in enzymes of carbohydrate metabolism and is also involved in several hundred enzymatic reactions in which food is metabolized and new products are formed.	Zinc and magnesium supplementation has been suggested as a means to maintain zinc and magnesium status and thereby improve training adaptations. However, there are still some discrepancies in the literature.
Glutamine	Glutamine is the most abundant amino acid in the body, representing about 60% of the amino acid pool in muscles. Glutamine serves a variety of functions in the body, including cell growth, immune function, and recovery from stress.	Research has shown glutamine to contribute to the prevention of muscle breakdown; it also participates in increased growth hormone levels, protein synthesis, improved intestinal health, decreased risk of overtraining, and improved immune system function.
β-Hydroxy-β-methylbutyric acid	β-Hydroxy-β-methylbutyric acid, or HMB, is a metabolite of the essential amino acid leucine. HMB is thought to play a role in the regulation of protein breakdown	There appear to be sound scientific findings that suggest HMB supplementation may affect catabolism and protein synthesis. However, there is not conclusive evidence

	in the body. It appears that HMB supplementation has a protective effect on muscle and may help the body get a head start on the recovery process by minimizing the amount of protein degradation after exercise.	to suggest that HMB supplementation can increase strength or muscle mass.
Antioxidants	Oxidative stress is the steady-state level of oxidative damage in a cell, tissue, or organ caused by free radicals or the reactive oxygen species (ROS). Reactive oxygen species (e.g., free radicals, peroxides) represent a class of molecules derived from the metabolism of oxygen, which is increased during exercise. Antioxidants block the process of oxidation by neutralizing free radicals.	Research has determined that antioxidants are effective at reducing free radicals. However, their direct effect on the strength and power athlete has yet to be determined.
Vitamins, minerals	Heavy, prolonged exercise is associated with numerous hormonal and biochemical changes, many of which potentially have detrimental effects on immune function. Vitamins are organic compounds that have many important roles for normal functioning, growth, and maintenance of	It is clear that strength trained athletes have been shown to exhibit deficiencies in one or more vitamins and minerals. Thus, the need to raise levels to within a normal range is supported in the literature. However, it is not clear if supplementation in excess of the recommended daily intake

(Continued)

357

Table 1
(Continued)

Nutrient	Theoretical ergogenic value	Summary of research findings/ recommendations
	the body. They also help extract energy from the macronutrients. Each mineral has a specific role—from components of hormones and enzymes to structural function.	(RDI) is advantageous to the strength athlete.
CLA	Conjugated linoleic acids (CLAs) are essential fatty acids that have been reported to have significant health benefits in animals. CLAs are naturally occurring fatty acids primarily found in beef and dairy fats. Research has indicated that CLAs may have a number of health and performance-enhancing benefits.	CLAs appear to have many benefits for the strength athlete. Although there are conflicting results, it appears that there is plenty of supporting evidence regarding the efficacy of CLA supplementation on enhancing immune performance.
Carbohydrate	Carbohydrate serves as the primary fuel for moderate to high intensity exercise. The amount of carbohydrate that can be stored in the liver and muscle, however, is limited; and it takes time to replenish carbohydrate stores. Therefore, it is imperative that	Based on the current scientific literature, it may be advisable for athletes who are performing high volume resistance training to ingest carbohydrate supplements before, during, and immediately after resistance training.

| Caffeine | Caffeine is one of the most widely used stimulants in the world. It occurs naturally in foods and beverages such as coffee, tea, soft drinks, chocolate, and cocoa. Caffeine acts as a central nervous system (CNS) stimulant, which causes the heart rate and blood pressure to increase. Caffeine and exercise is believed to cause an improvement in time-to-exhaustion and improved work output during aerobic exercise. | Caffeine has been shown in the literature to have positive effects on duration of exercise and energy expenditure. Research has also shown an increase in power output. However, further research is needed before conclusive recommendations can be made. |

active individuals and athletes consume enough carbohydrate in their diet to maintain carbohydrate availability.

359

supplementation with multivitamins, antioxidants, glutamine, and CLA. When all of these supplements are combined, coupled with a balanced diet and a rigorous training program, strength and power athletes are sure to maximize their genetic potential.

Each day new supplements hit the market with the promise of enhancing performance, increasing muscle mass, and optimizing strength. However, until valid scientific reports support these claims, there is some risk involved. Table 1 depicts an easy-to-use flow chart regarding macronutrient and micronutrient supplement options for strength and power athletes.

REFERENCES

1. Balsom PD, Soderlund K, Ekblom B. Creatine in humans with special reference to creatine supplementation. Sports Med 1994;18:268–280.
2. Heymsfield SB, Arteaga C, McManus C, Smith J, Moffitt S. Measurement of muscle mass in humans: validity of the 24-hour urinary creatinine method. Am J Clin Nutr 1983;37:478–494.
3. Hirvonen J, Rehunen S, Rusko H, Harkonen M. Breakdown of high-energy phosphate compounds and lactate accumulation during short supramaximal exercise. Eur J Appl Physiol Occup Physiol 1987;56:253–259.
4. Kreider RB. Species-specific responses to creatine supplementation. Am J Physiol Regul Integr Comp Physiol 2003;285:R725–R726.
5. Norman K, Stubler D, Baier P, et al. Effects of creatine supplementation on nutritional status, muscle function and quality of life in patients with colorectal cancer: a double blind randomised controlled trial. Clin Nutr 2006;25:596–605.
6. Greenwood M, Kreider RB, Greenwood L, Byars A. Cramping and injury incidence in collegiate football players are reduced by creatine supplementation. J Athl Train 2003;38:216–219.
7. Kreider RB. Effects of creatine supplementation on performance and training adaptations. Mol Cell Biochem 2003;244:89–94.
8. Kreider RB, Melton C, Rasmussen CJ, et al. Long-term creatine supplementation does not significantly affect clinical markers of health in athletes. Mol Cell Biochem 2003;244:95–104.
9. Stout J, Eckerson J, Ebersole K, et al. Effect of creatine loading on neuromuscular fatigue threshold. J Appl Physiol 2000;88:109–112.
10. Volek JS, Kraemer WJ, Bush JA, et al. Creatine supplementation enhances muscular performance during high-intensity resistance exercise. J Am Diet Assoc 1997;97:765–770.
11. Peeters B, Lantz C, Mayhew J. Effects of oral creatine monohydrate and creatine phosphate supplementation on maximal strength indices, body composition, and blood pressure. J Strength Cond Res 1999;13:3–9.

12. Eckerson JM, Stout JR, Moore GA, Stone NJ, Nishimura K, Tamura K. Effect of two and five days of creatine loading on anaerobic working capacity in women. J Strength Cond Res 2004;18:168–173.

13. Kreider RB, Ferreira M, Wilson M, et al. Effects of creatine supplementation on body composition, strength, and sprint performance. Med Sci Sports Exerc 1998;30:73–82.

14. Phillips SM, Parise G, Roy BD, Tipton KD, Wolfe RR, Tamopolsky MA. Resistance-training-induced adaptations in skeletal muscle protein turnover in the fed state. Can J Physiol Pharmacol 2002;80:1045–1053.

15. Phillips SM, Tipton KD, Ferrando AA, Wolfe RR. Resistance training reduces the acute exercise-induced increase in muscle protein turnover. Am J Physiol 1999;276:E118–E124.

16. Biolo G, Maggi SP, Williams BD, Tipton KD, Wolfe RR. Increased rates of muscle protein turnover and amino acid transport after resistance exercise in humans. Am J Physiol 1995;268:E514–E520.

17. Biolo G, Tipton KD, Klein S, Wolfe RR. An abundant supply of amino acids enhances the metabolic effect of exercise on muscle protein. Am J Physiol 1997;273:E122–E129.

18. Tipton KD, Ferrando AA, Phillips SM, Doyle D Jr, Wolfe RR. Postexercise net protein synthesis in human muscle from orally administered amino acids. Am J Physiol 1999;276:E628–E634.

19. Esmarck B, Andersen JL, Olsen S, Richter EA, Mizuno M, Kjaer M. Timing of postexercise protein intake is important for muscle hypertrophy with resistance training in elderly humans. J Physiol 2001; 535:301–311.

20. Tipton KD, Elliott TA, Cree MG, Wolf SE, Sanford AP, Wolfe RR. Ingestion of casein and whey proteins results in muscle anabolism after resistance exercise. Med Sci Sports Exerc 2004;36:2073–2081.

21. Rennie MJ, Wackerhage H, Spangenburg EE, Booth FW. Control of the size of the human muscle mass. Annu Rev Physiol 2004;66:799–828.

22. Kerksick C, Leutholtz B. Nutrient administration and resistance training. J Int Soc Sport Nutr 2006;2:57–60.

23. Harris R, Dunnett M, Greenhaf P. Carnosine and taurine contents in individual fibres of human vastus lateralis muscle. J Sport Sci 1998;16:639–643.

24. Hill CA, Harris RC, Kim HJ, et al. Influence of beta-alanine supplementation on skeletal muscle carnosine concentrations and high intensity cycling capacity. Amino Acids 2007;32:225–233.

25. Stout JR, Cramer JT, Mielke M, O'Kroy J, Torok DJ, Zoeller RF. Effects of twenty-eight days of beta-alanine and creatine monohydrate supplementation on the physical working capacity at neuromuscular fatigue threshold. J Strength Cond Res 2006;20:928–931.

26. Stout JR, Cramer JT, Zoeller RF, et al. Effects of beta-alanine supplementation on the onset of neuromuscular fatigue and ventilatory threshold in women. Amino Acids 2007;32:381–386.

27. Hoffman J, Ratamess N, Kang J, Mangine G, Faigenbaum A, Stout J. Effect of creatine and beta-alanine supplementation on performance and endocrine

responses in strength/power athletes. Int J Sport Nutr Exerc Metab 2006;16:430–446.

28. Campbell B, Roberts M, Kerksick C, et al. Pharmacokinetics, safety, and effects on exercise performance of L-arginine alpha-ketoglutarate in trained adult men. Nutrition 2006;22:872–881.

29. Elam RP, Hardin DH, Sutton RA, Hagen L. Effects of arginine and ornithine on strength, lean body mass and urinary hydroxyproline in adult males. J Sports Med Phys Fitness 1989;29:52–56.

30. Hembree WC, Ross GT. Arginine infusion and growth-hormone secretion. Lancet 1969;1:52.

31. Merimee TJ, Rabinowitz D, Fineberg S. Arginine-initiated release of human growth hormone: factors modifying the response in normal man. N Engl J Med 1969;280:1434–1438.

32. Merimee TJ, Rabinowitz D, Riggs L, Burgess JA, Rimoin DL, McKusick VA. Plasma growth hormone after arginine infusion: clinical experiences. N Engl J Med 1967;276:434–439.

33. Walberg-Rankin J, Hawkins C. The effect of oral arginine during energy restriction in male weight trainers. J Strength Cond Res 1994;8:170–177.

34. Besset A, Bonardet A, Rondouin G, Descomps B, Passouant P. Increase in sleep related GH and Prl secretion after chronic arginine aspartate administration in man. Acta Endocrinol (Copenh) 1982;99:18–23.

35. Marcell T, Taaffe D, Hawkins S. Oral arginine does not stimulate basal or augment exercise-induced GH secretion in either young or old adults. J Gerontol A Biol Sci Med Sci 1999;54:395–399.

36. Gater DR, Gater DA, Uribe JM, Bunt JC. Effects of arginine/lysine supplementation and resistance training on glucose tolerance. J Appl Physiol 1992;72:1279–1284.

37. Bhasin S, Bremner WJ. Clinical review 85: emerging issues in androgen replacement therapy. J Clin Endocrinol Metab 1997;82:3–8.

38. Bhasin S, Buckwalter JG. Testosterone supplementation in older men: a rational idea whose time has not yet come. J Androl 2001;22:718–731.

39. Bhasin S, Storer TW, Berman N, et al. Testosterone replacement increases fat-free mass and muscle size in hypogonadal men. J Clin Endocrinol Metab 1997;82:407–413.

40. Bhasin S, Woodhouse L, Casaburi R, et al. Testosterone dose-response relationships in healthy young men. Am J Physiol Endocrinol Metab 2001;281:E1172–E1181.

41. Sinha-Hikim I, Artaza J, Woodhouse L, et al. Testosterone-induced increase in muscle size in healthy young men is associated with muscle fiber hypertrophy. Am J Physiol Endocrinol Metab 2002;283:E154–E164.

42. Sinha-Hikim I, Roth SM, Lee MI, Bhasin S. Testosterone-induced muscle hypertrophy is associated with an increase in satellite cell number in healthy, young men. Am J Physiol Endocrinol Metab 2003;285: E197–E205.

43. Brodie A, Inkster S, Yue W. Aromatase expression in the human male. Mol Cell Endocrinol 2001;178:23–28.

44. Zmuda J, Bausserman L, Maceroni D, Thompson P. The effect of supraphysiologic doses of testosterone on fasting total homocysteine levels in normal men. Atherosclerosis 1997;130.
45. Leder BZ, Rohrer JL, Rubin SD, Gallo J, Longcope C. Effects of aromatase inhibition in elderly men with low or borderline-low serum testosterone levels. J Clin Endocrinol Metab 2004;89:1174–1180.
46. Mauras N, Lima J, Patel D, et al. Pharmacokinetics and dose finding of a potent aromatase inhibitor, aromasin (exemestane), in young males. J Clin Endocrinol Metab 2003;88:5951–5956.
47. Mauras N, O'Brien KO, Klein KO, Hayes V. Estrogen suppression in males: metabolic effects. J Clin Endocrinol Metab 2000;85:2370–2377.
48. Inclendon T. The chronic effects of androst-4-ene-3,6,17-trione on endocrine responses in resistance-trained men. Unpublished observation, 2003.
49. Numazawa M, Tsuji M, Mutsumi A. Studies on aromatase inhibition with 4-androstene-3,6,17-trione: its 3 beta-reduction and time-dependent irreversible binding to aromatase with human placental microsomes. J Steroid Biochem 1987;28:337–344.
50. Numazawa M, Mutsumi A, Tachibana M. Mechanism for aromatase inactivation by a suicide substrate, androst-4-ene-3,6,17-trione: the 4 beta, 5 beta-epoxy-19-oxo derivative as a reactive electrophile irreversibly binding to the active site. Biochem Pharmacol 1996;52:1253–1259.
51. Willoughby DS, Wilborn C, Taylor L, Campbell W. Eight weeks of aromatase inhibition using the nutritional supplement Novedex XT: effects in young, eugonadal men. Int J Sport Nutr Exerc Metab 2007;17:92–108.
52. Lukasi H. Micronutrients (magnesium, zinc, and copper): are mineral supplements needed for athletes? Int J Sport Nutr 1995;Jun(Suppl) S74–S83.
53. Singh A, Failla ML, Deuster PA. Exercise-induced changes in immune function: effects of zinc supplementation. J Appl Physiol 1994;76:2298–2303.
54. Singh A, Smoak BL, Patterson KY, LeMay LG, Veillon C, Deuster PA. Biochemical indices of selected trace minerals in men: effect of stress. Am J Clin Nutr 1991;53:126–131.
55. Gleeson M, Lancaster GI, Bishop NC. Nutritional strategies to minimise exercise-induced immunosuppression in athletes. Can J Appl Physiol 2001;26(Suppl):S23–S35.
56. Nieman DC. Nutrition, exercise, and immune system function. Clin Sports Med 1999;18:537–548.
57. Nieman DC, Pedersen BK. Exercise and immune function: recent developments. Sports Med 1999;27:73–80.
58. Gleeson M, Nieman DC, Pedersen BK. Exercise, nutrition and immune function. J Sports Sci 2004;22:115–125.
59. Shephard RJ, Shek PN. Immunological hazards from nutritional imbalance in athletes. Exerc Immunol Rev 1998;4:22–48.
60. Golf SW, Happel O, Graef V, Seim KE. Plasma aldosterone, cortisol and electrolyte concentrations in physical exercise after magnesium supplementation. J Clin Chem Clin Biochem 1984;22:717–721.

61. Golf SW, Bender S, Gruttner J. On the significance of magnesium in extreme physical stress. Cardiovasc Drugs Ther 1998;12(Suppl 2):197–202.
62. Brilla L, Haley T. Effect of magnesium supplementation on strength training in humans. J Am Coll Nutr 1992;11:326–329.
63. Newhouse IJ, Finstad EW. The effects of magnesium supplementation on exercise performance. Clin J Sport Med 2000;10:195–200.
64. Brilla L, Conte V. Effects of a novel zinc-magnesium formulation on hormones and strength. J Exerc Physiol Online 2000;3:26–36.
65. Prasad AS, Mantzoros CS, Beck FW, Hess JW, Brewer GJ. Zinc status and serum testosterone levels of healthy adults. Nutrition 1996;12:344–348.
66. Cordova A, Navas FJ. Effect of training on zinc metabolism: changes in serum and sweat zinc concentrations in sportsmen. Ann Nutr Metab 1998;42:274–282.
67. Khaled S, Brun JF, Micallel JP, et al. Serum zinc and blood rheology in sportsmen (football players). Clin Hemorheol Microcirc 1997;17:47–58.
68. Konig D, Weinstock C, Keul J, Northoff H, Berg A. Zinc, iron, and magnesium status in athletes—influence on the regulation of exercise-induced stress and immune function. Exerc Immunol Rev 1998;4:2–21.
69. Wilborn C, Kerksick C, Campbell B, et al. Effects of zinc magnesium aspartate (ZMA) Supplementation on training adaptations and markers of anabolism and catabolism. J Int Soc Sports Nutr 2004;1:12–20.
70. Newsholme P, Procopio J, Lima MM, Pithon-Curi TC, Curi R. Glutamine and glutamate: their central role in cell metabolism and function. Cell Biochem Funct 2003;21:1–9.
71. Candow DG, Chilibeck PD, Burke DG, Davison KS, Smith-Palmer T. Effect of glutamine supplementation combined with resistance training in young adults. Eur J Appl Physiol 2001;86:142–149.
72. Furst P. Intracellular muscle free amino acids: their measurement and function. Nutr Soc 1983;45:451–462.
73. Parry-Billings M, Budgett R, Koutedakis Y, et al. Plasma amino acid concentrations in the overtraining syndrome: possible effects on the immune system. Med Sci Sports Exerc 1992;24:1353–1358.
74. Keast D, Arstein D, Harper W, Fry RW, Morton AR. Depression of plasma glutamine concentration after exercise stress and its possible influence on the immune system. Med J Aust 1995;162:15–18.
75. Newsholme EA. Biochemical mechanisms to explain immunosuppression in well-trained and overtrained athletes. Int J Sports Med 1994;15(Suppl 3):S142–S147.
76. Rohde T, MacLean DA, Pedersen BK. Effect of glutamine supplementation on changes in the immune system induced by repeated exercise. Med Sci Sports Exerc 1998;30:856–862.
77. Nissen S, Faidley TD, Zimmerman DR, Izard R, Fisher CT. Colostral milk fat percentage and pig performance are enhanced by feeding the leucine metabolite beta-hydroxy-beta-methyl butyrate to sows. J Anim Sci 1994;72:2331–2337.
78. Van Koevering MT, Dolezal HG, Gill DR, et al. Effects of beta-hydroxy-beta-methyl butyrate on performance and carcass quality of feedlot steers. J Anim Sci 1994;72:1927–1935.

79. Peterson AL, Qureshi MA, Ferket PR, Fuller JC Jr. Enhancement of cellular and humoral immunity in young broilers by the dietary supplementation of beta-hydroxy-beta-methylbutyrate. Immunopharmacol Immunotoxicol 1999;21:307–330.

80. Peterson AL, Qureshi MA, Ferket PR, Fuller JC Jr. In vitro exposure with beta-hydroxy-beta-methylbutyrate enhances chicken macrophage growth and function. Vet Immunol Immunopathol 1999;67:67–78.

81. Miller P, Sandberg, L Fuller, JC. The effect of intensive training and β-hydroxy-β-methylbutyrate (HMB) on the physiological response to exercise in horses. FASEB J 1997;11:A1683.

82. Nissen S, Sharp R, Ray M. Effect of leucine metabolite beta-hydroxy-beta-methylbutyrate on muscle metabolism during resistance exercise testing. J Am Physiol 1996;81:2095–2104.

83. Knitter AE, Panton L, Rathmacher JA, Petersen A, Sharp R. Effects of beta-hydroxy-beta-methylbutyrate on muscle damage after a prolonged run. J Appl Physiol 2000;89:1340–1344.

84. Vukovich M, Adams GD. The effect of dietary β-hydroxy-β-methylbutyrate (HMB) on strength gains and body composition. FASEB J 1997; 11:A376.

85. Panton L, Rathmacher J, Fuller J, et al. Effect of β-hydroxy-β-methylbutyrate and resistance training on strength and functional ability in the elderly. Med Sci Sports Exerc 1998;30:S194.

86. Gallagher P, Carrithers JA, Godard MP, Schulze KE, Trappe SW. β-Hydroxy-β-methylbutyrate: supplementation during resistance-training. Med Sci Sports Exerc 1999;31:S402.

87. Kreider RB, Ferreira M, Wilson M, Almada AL. Effects of calcium beta-hydroxy-beta-methylbutyrate (HMB) supplementation during resistance-training on markers of catabolism, body composition and strength. Int J Sports Med 1999;20:503–509.

88. Meydani M, Evans WJ, Handelman G, et al. Protective effect of vitamin E on exercise-induced oxidative damage in young and older adults. Am J Physiol 1993;264:R992–R998.

89. Cannon JG, Meydani SN, Fielding RA, et al. Acute phase response in exercise. II. Associations between vitamin E, cytokines, and muscle proteolysis. Am J Physiol 1991;260:R1235–R1240.

90. Rokitzki L, Logemann E, Sagredos AN, Murphy M, Wetzel-Roth W, Keul J. Lipid peroxidation and antioxidative vitamins under extreme endurance stress. Acta Physiol Scand 1994;151:149–158.

91. Hartmann A, Niess AM, Grunert-Fuchs M, Poch B, Speit G. Vitamin E prevents exercise-induced DNA damage. Mutat Res 1995;346:195–202.

92. Gerster H. The role of vitamin C in athletic performance. J Am Coll Nutr 1989;8:636–643.

93. Kaminski M, Boal R. An effect of ascorbic acid on delayed-onset muscle soreness. Pain 1992;50:317–321.

94. Mackinnon LT. Chronic exercise training effects on immune function. Med Sci Sports Exerc 2000;32:S369–S376.

95. Insel PM, Turner RE, Ross D. Discovering Nutrition. Jones & Bartlett, Sudbury, MA, 2006.

96. Pariza MW, Park Y, Cook ME. Conjugated linoleic acid and the control of cancer and obesity. Toxicol Sci 1999;52:107–110.

97. MacDonald HB. Conjugated linoleic acid and disease prevention: a review of current knowledge. J Am Coll Nutr 2000;19:111S–S118S.

98. Thom E. A pilot study with the aim of studying the efficacy and tolerability of Tonalin CLA on the body composition in humans: technical report. Medstat Research, Lillestram, Norway, July 1997.

99. Lowery LM, Appicelli PA, aleomon PWR. Conjugated linoleic acid enhances muscle size and strength gains in novice bodybuilders. Med Sci Sports Exerc 1998;30:S182.

100. Riserus U, Berglund L, Vessby B. Conjugated linoleic acid (CLA) reduced abdominal adipose tissue in obese middle-aged men with signs of the metabolic syndrome: a randomised controlled trial. Int J Obes Relat Metab Disord 2001;25:1129–1135.

101. Blankson H, Stakkestad JA, Fagertun H, Thom E, Wadstein J, Gudmundsen O. Conjugated linoleic acid reduces body fat mass in overweight and obese humans. J Nutr 2000;130:2943–2948.

102. Miller CC, Park Y, Pariza MW, Cook ME. Feeding conjugated linoleic acid to animals partially overcomes catabolic responses due to endotoxin injection. Biochem Biophys Res Commun 1994;198:1107–1112.

103. Chew BP, Wong TS, Shultz TD, Magnuson NS. Effects of conjugated dienoic derivatives of linoleic acid and beta-carotene in modulating lymphocyte and macrophage function. Anticancer Res 1999;17:1099–1106.

104. Wong MW, Chew BP, Wong TS, Hosick HL, Boylston TD, Shultz TD. Effects of dietary conjugated linoleic acid on lymphocyte function and growth of mammary tumors in mice. Anticancer Res 1997;17:987–993.

105. Li Y, Watkins BA. Conjugated linoleic acids alter bone fatty acid composition and reduce ex vivo prostaglandin E_2 biosynthesis in rats fed n-6 or n-3 fatty acids. Lipids 1998;33:417–425.

106. Watkins BA, Shen CL, McMurtry JP, et al. Dietary lipids modulate bone prostaglandin E_2 production, insulin-like growth factor-I concentration and formation rate in chicks. J Nutr 1997;127:1084–1091.

107. Song HJ, Grant I, Rotondo D, et al. Effect of CLA supplementation on immune function in young healthy volunteers. Eur J Clin Nutr 2005;59:508–517.

108. Zhang H, Guo Y, Yuan J. Conjugated linoleic acid enhanced the immune function in broiler chicks. Br J Nutr 2005;94:746–752.

109. Kreider RB, Ferreira MP, Greenwood M, Wilson M, Almada AL. Effects of conjugated linoleic acid supplementation during resistance training on body composition, bone density, strength, and selected hematological markers. J Strength Cond Res 2002;16:325–334.

110. Ferreira IM, Kreider R, Wilson M, Almada A. Effects of conjugated linoleic acid supplementation during resistance training on body composition and strength. J Strength Cond Res 1997;11:280.

111. Kreider R, Ferreira MP, Greenwood M, Wilson M, Almada AL. Effects of conjugated linoleic acid (CLA) supplementation during resistance-training on bone mineral content, bone mineral density, and markers of immune stress. FASEB J 1998;12.

112. Beuker F, Haak H, Schwietz H. CLA and body styling. Presented at the symposium Vitamine und Zusatzstoffe (pp 229–237), Jena (Thhr.), 1999.

113. Berning JR. Energy intake, diet, and muscle wasting. In: Kreider RB, Fry AC, O'Toole ML (ed) Overtraining in Sport (pp 275–288). Human Kinetics, Champaign, IL, 1998.

114. Sherman WM, Jacobs KA, Leenders N. Carbohydrate metabolism during endurance exercise. In: Kreider RB, Fry AC, O'Toole ML (eds) Overtraining in Sport (289–308). Human Kinetics Champaign, IL, 1998.

115. Robergs RA, Pearson DR, Costill DL, et al. Muscle glycogenolysis during differing intensities of weight-resistance exercise. J Appl Physiol 1991;70:1700–1706.

116. Tesch PA, Colliander EB, Kaiser P. Muscle metabolism during intense, heavy-resistance exercise. Eur J Appl Physiol Occup Physiol 1986;55:362–366.

117. Essen-Gustavsson B, Tesch PA. Glycogen and triglyceride utilization in relation to muscle metabolic characteristics in men performing heavy-resistance exercise. Eur J Appl Physiol Occup Physiol 1990;61:5–10.

118. Kreider RB. Dietary supplements and the promotion of muscle growth with resistance exercise. Sports Med 1999;27:97–110.

119. Haff GG, Lehmkuhl MJ, McCoy LB, Stone MH. Carbohydrate supplementation and resistance training. J Strength Cond Res 2003;17:187–196.

120. Bird SP, Tarpenning KM, Marino FE. Independent and combined effects of liquid carbohydrate/essential amino acid ingestion on hormonal and muscular adaptations following resistance training in untrained men. Eur J Appl Physiol 2006;97:225–238.

121. Clarkson PM. Nutritional ergogenic aids: caffeine. Int J Sport Nutr 1993;3:103–111.

122. Armstrong LE. Caffeine, body fluid-electrolyte balance, and exercise performance. Int J Sport Nutr Exerc Metab 2002;12:189–206.

123. Costill DL, Dalsky GP, Fink WJ. Effects of caffeine ingestion on metabolism and exercise performance. Med Sci Sports 1978;10:155–158.

124. Graham TE, Spriet LL. Performance and metabolic responses to a high caffeine dose during prolonged exercise. J Appl Physiol 1991;71:2292–2298.

125. Ivy JL, Costill DL, Fink WJ, Lower RW. Influence of caffeine and carbohydrate feedings on endurance performance. Med Sci Sports 1979;11:6–11.

126. Powers SK, Dodd S. Caffeine and endurance performance. Sports Med 1985;2:165–174.

127. Falk B, Burstein R, Rosenblum J, Shapiro Y, Zylber-Katz E, Bashan N. Effects of caffeine ingestion on body fluid balance and thermoregulation during exercise. Can J Physiol Pharmacol 1990;68:889–892.

128. Dodd SL, Herb RA, Powers SK. Caffeine and exercise performance: an update. Sports Med 1993;15:14–23.

129. Doherty M, Smith PM. Effects of caffeine ingestion on rating of perceived exertion during and after exercise: a meta-analysis. Scand J Med Sci Sports 2005;15:69–78.

130. Doherty M, Smith P, Hughes M, Davison R. Caffeine lowers perceptual response and increases power output during high-intensity cycling. J Sports Sci 2004;22:637–643.

131. Collomp K, Ahmaidi S, Audran M, Chanal JL, Prefaut C. Effects of caffeine ingestion on performance and anaerobic metabolism during the Wingate test. Int J Sports Med 1991;12:439–443.

132. Trice I, Haymes EM. Effects of caffeine ingestion on exercise-induced changes during high-intensity, intermittent exercise. Int J Sport Nutr 1995;5:37–44.

133. Acheson KJ, Gremaud G, Meirim I, et al. Metabolic effects of caffeine in humans: lipid oxidation or futile cycling? Am J Clin Nutr 2004;79:40–46.

134. Koot P, Deurenberg P. Comparison of changes in energy expenditure and body temperatures after caffeine consumption. Ann Nutr Metab 1995;39:135–142.

11 Nutritional Supplements for Endurance Athletes

Christopher J. Rasmussen

Abstract

Athletes engaged in heavy endurance training often seek additional nutritional strategies to help maximize performance. Specific nutritional supplements exist to combat certain factors that limit performance beginning with a sound everyday diet. Research has further demonstrated that safe, effective, legal supplements are in fact available for today's endurance athletes. Several of these supplements are marketed not only to aid performance but also to combat the immunosuppressive effects of intense endurance training. It is imperative for each athlete to research the legality of certain supplements for their specific sport or event. Once the legality has been established, it is often up to each individual athlete to decipher the ethics involved with ingesting nutritional supplements with the sole intent of improving performance.

Key words

Exercise · Nutrition · Endurance · Supplements · Ergogenics

1. INTRODUCTION

The endurance athlete has special nutritional needs that go above and beyond those of the normal sedentary individual. To optimize performance the endurance athlete needs a solid foundation that includes proper training and nutrition. Proper training should be based on the principles of training and largely depends on the goals of the individual athlete and the point in time within the training cycle (preseason, in-season, postseason). Proper nutrition should focus on a variety of whole foods that adequately meet the demands of the athlete. Supplementing the diet with additional nutrients is

From: *Nutritional Supplements in Sports and Exercise*
Edited by: M. Greenwood, D. Kalman, J. Antonio,
DOI: 10.1007/978-1-59745-231-1_11, © Humana Press Inc., Totowa, NJ

becoming increasingly important for those engaged in heavy endurance training because it is often difficult to obtain the proper amount of macronutrients through whole foods alone. This chapter largely addresses the specific nutritional needs of the endurance athlete and more specifically nutritional supplements for the endurance athlete. Several studies have documented that endurance athletes can improve their training sessions and performance with a combination of proper everyday nutrition and effective supplementation *(1–7)*.

To gain a better understanding of how supplementation can improve performance for the endurance athlete it is important to realize what drives their performance. Therefore, the chapter begins with a description of the factors associated with optimal performance of the endurance athlete and the factors that limit their performance. The second section focuses on specific supplements for endurance athletes and how they can prepare the athlete for action and enhance the training response. The third section discusses how supplementation can aid the immune system to keep it functioning during even the most intense phases of training. The fourth section briefly dives into additional legal and ethical issues. Finally, a brief summary reviews the most important points of the chapter and brings it to a close.

2. FACTORS LIMITING ENDURANCE ATHLETIC PERFORMANCE

Various types of training are utilized by endurance athletes depending on whether they are aiming to improve anaerobic capacity, maximal aerobic power (MAP)—power output when VO_2max is reached—or endurance capability. It is important to point out that the term "endurance athlete" in athletic circles today traditionally refers to those participating in running, cycling, swimming, and/or combinations of these activities, although other activities may certainly be considered. Combinations of training including speed, interval, tempo, resistive, and long slow distance (LSD) are common for the endurance athlete. Manipulating the principles of training depending on the overall goals while coinciding with the respective training season helps optimize training and subsequent performance. The nutritional supplements discussed in this chapter focus primarily on their effectiveness, or lack thereof, in athletes who are involved in running, cycling, and/or swimming events.

Prior to discussing specific supplements for endurance athletes, it is important to understand the factors that limit endurance performance. Dehydration, the depletion of muscle glycogen, and limited blood glucose availability can all play a role in fatigue during long-duration aerobic exercise. The body transfers heat to the environment through conduction, convection, radiation, evaporation, or a combination of these methods. Conduction involves the transfer of heat from one material to another through direct molecular contact, and convection is the transfer of heat from one place to another by the motion of a gas or a liquid across the heated surface. Radiation involves the dissipation of heat in the form of infrared rays, and evaporation is best defined as the evaporation of sweat from the skin's surface.

Evaporation is the primary avenue for heat dissipation during exercise, accounting for roughly 80% of the total heat loss during exercise. With prolonged exercise or exercise in a hot and humid environment, blood volume is reduced by a loss of water through sweat. As exercise continues in a hot, humid environment, a redistribution of blood from the core to the periphery takes place to cool the body. Cardiac filling is reduced as the total blood volume gradually decreases with an increase in the duration of exercise. This leads to a decrease in venous return to the right side of the heart. Subsequently, stroke volume is reduced. The heart rate then tries to compensate for the decreased stroke volume by increasing in an effort to maintain cardiac output. Collectively, these alterations are referred to as the *cardiovascular drift*. The benefit of this phenomenon is that one is able to continue exercising at a low to moderate intensity. The drawback is that the body is unable to compensate fully for the decreased stroke volume at high exercise intensities owing to the fact the heart rate attains its maximum value at a much lower exercise intensity. A loss of body fluid equal to 1% of body weight (approximately 2.0 lb for a 200 lb athlete) can significantly reduce blood volume, placing undue stress on the cardiovascular system and limiting physical performance *(8)*. When dehydration reaches 4%, endurance athletes can experience heat cramps and heat exhaustion *(9)*; and when it reaches upwards of 6%, there may be cessation of sweating, a rise in body temperature, and eventually heat stroke *(10)*.

Collectively the depletion of muscle glycogen and the decline in blood glucose can put a damper on endurance athletic performance. Plasma fatty acids and muscle triglycerides are able to supply the needed energy during low-intensity exercise (i.e., 25% VO_2max) *(11)*. Carbohydrate use is relatively low and comes from blood glucose. Exercise at lower intensities can be maintained for several hours owing to the fact that the liver is able to continually supply glucose to the working muscles. As the intensity of exercise increases, the amount of carbohydrate necessary to keep pace with the increased demand also increases. A combination of blood glucose and muscle glycogen contributes a large percentage of the energy requirements at moderate exercise intensities (i.e., 65% VO_2max) *(11)*. At the beginning of exercise, muscle glycogen is the preferred fuel source; but as these levels decline there is increased dependence on blood glucose by the exercising muscles. Higher-intensity exercise (85% VO_2max) is performed at a level that promotes an even higher rate of muscle glycogen breakdown and carbohydrate oxidation *(11)*. This results in an accelerated rate of lactic acid production and ultimately accumulation in muscle and blood. Contrary to popular belief, lactic acid and lactate—the scapegoats for the pain athletes experience—are not responsible for all the ills for which they are blamed (see sidebar: The Truth About Lactate). Higher-intensity exercise represents the highest level an athlete can maintain for approximately 60 minutes. At these high intensities, carbohydrate oxidation accounts for more than two-thirds of the required energy, with the remainder coming from a combination of plasma fatty acids and intramuscular triglycerides. Given the fact that most competitive endurance athletes often train and compete at a higher intensity, it is easy to reason why the depletion of muscle glycogen and subsequent decline in blood glucose can be such a deterrent to endurance athletic performance. Maintaining adequate hydration and energy supplies is therefore essential to the performance of the endurance athlete.

2.1. The Truth About Lactate

Lactic acid and lactate are not the same compound. Lactate is any salt of lactic acid that enters the blood during high-intensity efforts. Recall that lactic acid is a by-product of glycolysis. Although most people believe that it is responsible for fatigue and exhaustion during all types of exercise, lactic acid accumulates in muscle fibers only during relatively brief high-intensity efforts. Athletes, and endurance athletes in particular, have been told for years that the primary reason they cannot "push" any more during intense exercise is that lactic acid has built up in their muscles. This is not entirely true, and there is evidence to disprove it. It is possible to experience muscle fatigue while the lactic acid concentration in the muscle remains low. Marathon runners, for example, may have near-resting lactic acid levels at the end of the race despite their exhaustion *(12)*. Conversely, there can be an absence of fatigue when the lactic acid concentration in the muscle is high. Exhausting isometric efforts with the quadriceps, for example, can cause fatigue and ultimately terminate the exercise. Minutes after completion the athlete can once again produce the initial force despite the fact that the degree of acidity in the muscles decreases to normal rather slowly. Therefore, it is difficult to accept the idea that an increase in lactic acid in the muscle causes fatigue as a high degree of acidity without fatigue can be observed. Lactate can also serve as an important fuel source by other tissues by converting it to pyruvate and oxidizing it in the mitochondria. For example, during exercise lactate serves as a significant fuel source for the heart *(12)*. Therefore, the fatigue experienced by the endurance athlete is largely caused by a combination of dehydration and inadequate energy supplies, not excess lactic acid.

3. SUPPLEMENTS FOR ENDURANCE ATHLETES

To combat the damaging and depleting effects of intense endurance training and further help the body respond to training, a well planned diet and supplement strategy that meets energy intake needs and incorporates proper timing of essential nutrients is vital. It is important to differentiate between the two popular terms "supplements" and "ergogenic aids" when discussing athletes and nutrition. Although the two are often used interchangeably, there are subtle differences. A nutritional supplement is usually thought of as something that completes or makes an addition to something else. On the other hand, the scientific literature refers to substances that athletes use to help enhance performance as ergogenic aids or sports ergogenics. The term ergogenic is derived from the Greek words *ergon* (work) and *gennan* (to produce). Thus, an ergogenic usually refers to something that produces or enhances work. To avoid confusion, this chapter uses the two terms interchangeably, although specific classifications of the sports ergogenics are mentioned. Countless supplements now exist for athletes of all kinds. This chapter addresses the most popular nutritional supplements available to endurance athletes today.

Nutritional ergogenic aids serve as the foundation for endurance performance and primarily include the macronutrients and micronutrients. Athletes that do not consume enough calories and/or do not consume enough of the right type of calories may hinder training adaptations and subsequent performance. However, athletes who consume a well planned diet during training can help the body adapt to training and more than likely notice improved performance. Furthermore, maintaining a diet that is deficient of the essential macronutrients over time may lead to a loss of body mass and muscle mass, increased susceptibility to illness, and an increase in the symptoms associated with overtraining. Practicing good dietary habits on a daily basis is essential to help optimize training adaptations and subsequent performance.

Preexercise nutrition should consist largely of moderate to low glycemic index (GI) foods/supplements (Table 1) that provide a slow, sustained release of carbohydrates and protein necessary to fuel a workout (see sidebar: What Is the Glycemic Index?). It generally takes about 4 hours for dietary carbohydrate to be digested and begin to be stored as muscle and liver glycogen. Thus, preexercise meals should be consumed about 4 to 6 hours prior to exercise *(13)*. Putting this into an average everyday scenario means that for an athlete who trains in the afternoon breakfast is the most important meal to top off muscle and liver glycogen levels. If the athlete trains first thing in the morning, the meal the evening before is vital. Recent research has also indicated that ingesting a light carbohydrate and protein snack 30 to 60 minutes prior to exercise (e.g., 50 g carbohydrate and 5–10 g protein) further increases carbohydrate availability toward the end of an intense exercise bout owing to the slight increase in glucose and insulin levels *(14,15)*. This can serve to increase the availability of amino acids and decrease

Table 1
Partial list of the glycemic index of foods using glucose as the standard

Low GI		Moderate GI		High GI	
Food	GI	Food	GI	Food	GI
Chana dal	8	Apple juice	40	Life Savers[TM]	70
Peanuts	14	Snickers[TM]	41	White bread	70
Plain yogurt	14	Peach	42	Bagel	72
Soy beans	18	Pudding	43	Watermelon	72
Rice bran	19	Pinto beans	45	Graham crackers	74
Peas	22	Orange juice	46	French fries	75
Cherries	22	Baked beans	48	Total[TM]	76
Barley	25	Strawberry jam	51	Vanilla wafers	77
Grapefruit	25	Sweat potato	54	Gatorade[TM]	78
Kidney beans	27	Pound cake	54	Fava beans	79
Link sausages	28	Popcorn	55	Jelly beans	80
Black beans	30	Brown rice	55	Tapioca pudding	81
Lentils	30	Fruit cocktail	55	Rice cakes	82
Butter beans	31	Pita bread	57	Team Flakes[TM]	82
Soy milk	31	PowerBar[TM]	58	Pretzels	83
Lima beans	32	Honey	58	Corn Chex[TM]	83
Skim milk	32	Blueberry muffin	59	Corn flakes[TM]	84
Split peas	32	Shredded wheat	62	Baked white potato	85
Fettucini	32	Black bean soup	64	Mashed potatoes	86
Chickpeas	33	Macaroni and cheese	64	Dark rye	86
Peanut M&M's[TM]	33	Raisins	64	Instant rice	87
Chocolate milk	34	Canteloupe	65	Crispix[TM]	87
Vermicelli	35	Mars Bar[TM]	65	Boiled Sebago	87
Whole wheat spaghetti	37	Rye bread	65	Rice Chex[TM]	89

(*Continued*)

Table 1
(Continued)

Low GI		Moderate GI		High GI	
Food	*GI*	*Food*	*GI*	*Food*	*GI*
Apple	38	Pineapple	66	Gluten-free bread	90
Pear	38	Grapenuts^{TM}	67	Baked red potato	93
Tomato soup	38	Angel food cake	67	French baguette	95
Ravioli	39	Stoned wheat thins	67	Peeled Desiree	101
Pinto beans	39	Taco shells	68	Dates	103
Plums	39	Whole wheat bread	69	Tofu frozen dessert	115

GI, glycemic index

exercise-induced protein catabolism *(14–16)*. Insulin inhibits protein degradation and apparently offsets the catabolic effects of other hormones (e.g., cortisol) *(17)*. Anabolic actions of insulin appear to be related to its nitrogen-sparing effects and promotion of nitrogen retention *(17)*. The choice of foods and supplements selected is largely up to individual athlete and their personal preferences. It is recommended that the endurance athlete consume something familiar on the day of competition rather than experimenting with a new food or supplement.

3.1. What Is the Glycemic Index?

The glycemic index (GI) is a ranking of foods based on their postprandial blood glucose response compared to a reference food, either glucose or white bread. The GI concept was first developed in 1981 to help determine which foods were best for people with diabetes. The GI of a food is based on several factors, including the physical form of the food, the amylose/amylopectin ratio (two types of starch), sugar content, fiber content, fat content, and the acidity of a food. The index consists of a scale from 0 to 100 with 0 (water) representing the lowest ranking and 100 (pure glucose) the highest ranking. The GI is obtained through use of an oral glucose tolerance test (OGTT) utilizing 50 g of carbohydrate from the test food. Blood samples are then obtained periodically throughout a 2-hour time period, glucose levels are measured, and the area under the curve is calculated *(18)*.

The GI of a carbohydrate has a profound effect on subsequent glucose and insulin responses. High-GI carbohydrates (i.e., dextrose, maltose) produce large increases in glucose and insulin levels. Moderate-GI carbohydrates (i.e., sucrose, lactose) traditionally produce only modest increases in glucose and insulin. Finally, low-GI carbohydrates (i.e., fructose, maltodextrin) have little if any effect on glucose and insulin responses. It has been suggested that manipulating the GI of a sports supplement may optimize carbohydrate availability for exercise, particularly prolonged intense exercise. Caution should be used when applying the GI to whole foods that contain several ingredients. The GI is more accurate for individually packed foods/supplements because of the fewer ingredients and the standardization that exists with the processing of these snacks/supplements. The GI is not as applicable to whole foods/meals and has not been established for many of these whole foods/meals.

Nutrition during an intense endurance training session can aid in the quality of the workout especially if the workout exceeds 60 to 90 minutes. Nutrition during exercise for the endurance athlete usually centers on supplementation more than pre- and postexercise nutrition if for no other reason than the convenience supplements provide. Convenience supplements include glucose-electrolyte solutions (GES), meal replacement powders (MRPs), ready-to-drink supplements (RTDs), energy bars, energy gels, and fitness waters. They are typically fortified with various amounts of vitamins and minerals and differ on the amount of carbohydrate, protein, and fat they contain. The beneficial effects of solid and liquid carbohydrate/protein supplements are similar when thermal stress is not a factor.

Liquid supplements do provide the added benefits of aiding rehydration and tend to digest easier for most athletes while exercising. Cyclists can generally empty from the stomach up to 1000 ml of fluid per hour, and therefore 40 to 60 g of carbohydrate can be easily ingested while consuming a large volume of fluid *(19)*. On the other hand runners generally consume less than 500 ml of fluid per hour *(20)*. This is because of the difficulty of drinking on the run and the potential discomfort of running with a full stomach. Therefore, runners tend to use more concentrated solutions than do cyclists in order to consume adequate amounts of carbohydrate. For cyclists, a carbohydrate solution of 4% to 6% is generally sufficient when fluid replacement is important. For runners, this concentration may have to be 8% to 10% to provide adequate carbohydrate. Glucose concentrations in excess of 10% seem to delay gastric emptying and compromise fluid replacement (Fig. 1). Adequate sodium intake is also important to combat potential electrolyte imbalances during exercise. Most GES solutions contain sufficient quantities of sodium. If sodium is lacking in the supplement, however, electrolyte tablets (i.e., Heat Guard) are also available. The bottom line is that rapid nutrient availability is especially important during a workout to maintain energy levels and training intensity. Thus, high-GI sources should make up most of the supplements ingested during an endurance-type workout. Once again, athletes should experiment with different formulations to find the one that works best for them prior to competition.

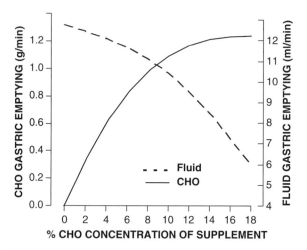

Fig. 1. Effect of carbohydrate (CHO) concentration on the rate of gastric emptying of fluids and carbohydrate from the stomach during exercise. Carbohydrate concentrations of 8% to 10% maximize carbohydrate gastric emptying without substantially reducing fluid delivery. From Gisolfi and Lamb *(100)*, with permission.

It is now well established that with prolonged continuous exercise the time-to-fatigue at moderate submaximum exercise intensities is related to preexercise muscle glycogen concentrations—thus the importance of everyday nutrition along with preexercise nutrition *(21)*. In addition, a GES has been the recommended supplement of choice for decades during exercise to preserve muscle glycogen and maintain blood glucose levels. With short-term, high-intensity exercise, the relation between the availability of muscle glycogen and performance is less clear. One study that utilized 15 high-intensity 6-second bouts on a cycle ergometer concluded that a high carbohydrate regimen over 48 hours helped subjects maintain a higher power output than did the exercise and dietary regimen that included a low carbohydrate content *(22)*. This study demonstrated the importance of a high carbohydrate diet in relation to short-term, high-intensity exercise.

Recent research has shown that the addition of protein can have added benefits to a supplement ingested during exercise by reducing muscle protein degradation and speeding postexercise recovery. Carbohydrate and protein intake significantly alters circulating metabolites and the hormonal milieu (i.e., insulin, testosterone, growth hormone, cortisol) as well as the response of muscle protein and glycogen balance *(23)*. Furthermore, the addition of protein to a carbohydrate supplement enhances the insulin response of a carbohydrate supplement compared to a placebo *(24)*, which can ultimately lead to performance gains *(25)*.

Saunders et al. *(7)* examined the effects of a carbohydrate-protein beverage on cycling endurance and muscle damage. They utilized 15 male cyclists who were randomly administered either a carbohydrate or carbohydrate-protein beverage (4:1 ratio) every 15 minutes during exercise and immediately upon completion of a ride to volitional exhaustion. The carbohydrate-protein beverage produced significant improvement in time-to-fatigue and reduction of muscle damage in the selected endurance athletes. The authors concluded that the benefits observed were the result of a higher total caloric content of the carbohydrate-protein beverage or were due to specific protein-mediated mechanisms.

Controversy exists among numerous studies examining the addition of protein due to the fact that such addition increases the total caloric content of the supplement. Anytime a larger amount of calories is consumed an athlete is likely to perform and recover more rapidly. Therefore, when examining studies that are not based on isocaloric data, one should give them careful consideration.

Whey is the preferred protein to ingest during exercise because of its rapid absorption rates and the fact that is contains all of the essential amino acids (refer to Table 3) as well as a high percentage of leucine and glutamine, which are amino acids the body uses during sustained exercise *(26)*. High-GI carbohydrates (glucose, sucrose, maltodextrin) should be combined with the protein in a 4:1 ratio to provide optimal benefits. Table 2 gives an example of the ideal nutrient composition for a sports drink during exercise *(27)*. Sports drinks such as that shown in Table 2 should be ingested every 20 minutes during an endurance training session to help improve performance and reduce muscle protein breakdown.

Postexercise nutrition for the endurance athlete is vital to restore muscle glycogen stores, enhance skeletal muscle fiber repair and growth, and maintain overall health and wellness. After an intense exercise bout, the body is in a catabolic state (thus key muscle

Table 2
Ideal Nutrient Composition for a Sports Drink During Exercise

Nutrient objectives
  Replace fluids and electrolytes
  Preserve muscle glycogen
  Maintain blood glucose levels
  Maintain hydration
  Minimize cortisol increases
  Set the stage for a faster recovery
  Satisfy thirst

Ideal composition (per 12 oz water)
  High-glycemic carbohydrates (e.g., glucose, sucrose, maltodextrin):
    20–26 g
  Whey protein: 5–6 g
  Vitamin C: 30–120 g
  Vitamin E: 20–60 IU
  Sodium: 100–250 mg
  Potassium: 60–120 mg
  Magnesium: 60–120 mg

Adapted from Ivy and Portman *(27)*, with permission

nutrients are being broken down). However, the opportunity exists to alter the catabolic state into a more anabolic hormonal profile where the athlete begins to rebuild muscle and thus initiates a much faster recovery. Exercise that results in glycogen depletion activates glycogen synthase, the enzyme responsible for controlling the transfer of glucose from UDP-glucose to an amylase chain (28,29). This also happens to be the rate-limiting step of glycogen formation. The degree of glycogen synthase activation is influenced by the extent of glycogen depletion (28). Complete resynthesis of muscle glycogen, however, ultimately depends on adequate carbohydrate intake. Carbohydrates composed of glucose or glucose polymers are the most effective for replenishing muscle glycogen, whereas fructose is most beneficial for replenishing liver glycogen (30,31). Glucose and fructose are metabolized differently. They have different gastric emptying rates and are absorbed into the blood at different rates (29,32). Furthermore, the insulin response to a glucose supplement is generally much greater than that of a fructose supplement (33). The fact that approximately 79% and 14% of total carbohydrate is stored in skeletal muscle and the liver, respectively, is further indication of the importance of consuming glucose or glucose polymers after exercise (13).

Blom et al. (34) found that ingestion of glucose and sucrose was twice as effective as fructose for restoring muscle glycogen. The maximum stimulatory effect of oral glucose intake on post-exercise muscle glycogen synthesis was reached at a dose of 0.70 g/kg taken every second hour following exercise in which the muscle glycogen concentration was reduced by an average of 80%. In addition, the rate of post-exercise muscle glycogen synthesis increases with increasing oral glucose intake, up to a maximum rate of approximately 6 mmol/kg/hr. Blom et al. indicated that the differences between glucose and fructose supplementation were the result of the way the body metabolized these sugars. Fructose metabolism takes place predominantly in the liver, whereas most glucose appears to bypass the liver and is stored or oxidized by muscle (31).

Subsequent research by Burke and associates (35) found that the intake of high-GI carbohydrate foods after prolonged exercise produces significantly more glycogen storage than consumption of low-GI carbohydrate foods 24 hours after exercise. Although the meal immediately after exercise elicited exaggerated blood glucose and plasma insulin responses that were similar for the low-GI and high-GI meals, for the remainder of the 24 hours the low-GI meals elicited lower glucose and insulin responses than the high-GI meals. As previously mentioned, protein has shown additive effects to that of carbohydrate alone in regard to the rate of muscle glycogen resynthesis and overall post-exercise recovery. The addition of protein to a carbohydrate supplement increases insulin levels more than that produced by carbohydrate or protein alone *(36)*. Tarnopolsky et al. showed that post-exercise carbohydrate and carbohydrate-protein-fat nutritional supplements can increase glycogen resynthesis during the first 4 hours after exercise to a greater extent than placebo for both men and women *(37)*. The supplements administered were both isoenergetic and isonitrogenous. Insulin has been demonstrated to have profound anabolic effects on skeletal muscle. In the resting state, insulin has been demonstrated to decrease the rate of muscle protein degradation *(38)*.

To summarize, in addition to the macronutrients selected, the timing of nutrient ingestion can greatly affect the speed of recovery. Research has clearly shown that muscle glycogen resynthesis occurs more quickly if carbohydrate is consumed immediately following exercise in contrast to waiting for several hours *(39)*. Whereas most of the everyday diet for the endurance athlete should be a low- to moderate-GI diet, the post-exercise diet should be centered on moderate- to high-GI sources. This nutritional approach has been found to accelerate glycogen resynthesis and promote a more anabolic hormonal state, which may speed recovery *(37)*.

The increased protein and glycogen synthesis is believed to be due to insulin secretion from the pancreas combined with an increase in muscle insulin sensitivity *(40)*. This was demonstrated in a study (Fig. 2) showing that a carbohydrate-protein combination was 38% more effective in stimulating protein synthesis than a protein supplement and more than twice as effective as a carbohydrate supplement *(41)*. Insulin appears to stimulate biosynthetic pathways that lead to increased glucose utilization, increased carbohydrate and fat storage, and increased protein synthesis. This metabolic pattern is characteristic of the absorptive state. The rise in insulin secretion during this state is responsible for shifting metabolic pathways to net anabolism *(13)*. In contrast, when insulin secretion is low, the opposite effect occurs. The rate of glucose entry into the

Effect of Protein and Carbohydrate Alone and in Combination on Protein Synthesis Following Exercise

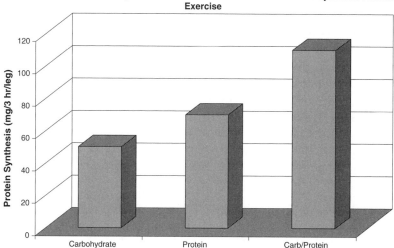

Fig. 2. Effect of protein and carbohydrate alone and in combination on protein synthesis, measured after exercise. From Tarnopolsky et al. *(37)*, with permission.

cells is reduced and net catabolism occurs, rather than net synthesis of glycogen, triglycerides, and protein. This pattern is reminiscent of the postabsorptive state.

In addition to the major macronutrients, phosphorus, an essential mineral distributed widely in foods, may also be classified as a nutritional sports ergogenic. Phosphorus is distributed widely in foods, particularly meat, seafood, eggs, milk, cheese, whole-grain products, nuts, and legumes. The Recommended Dietary Allowance (RDA) is 800 mg for adults and 1200 mg for those 11 to 25 years of age *(42)*. Phosphate salts in both inorganic and organic forms play important roles in human metabolism, particularly as related to sports performance. They may influence all three human energy systems by acting as intracellular buffers. Another theory suggests that phosphate salts increase the formation of 2,3-diphosphoglycerate (2,3-DPG), a compound in the red blood cells (RBCs) that facilitates the release of oxygen to tissues.

An early study on phosphate loading examined highly trained runners who took 1 g sodium phosphate four times daily for 6 days. The phosphate salts increased the concentration of 2,3-DPG in

RBCs by 6.6%, which subsequently resulted in an increase in VO_2max *(43)*. This study also reported decreased production of lactate and reduced sensation of physiological stress. A later study utilizing highly trained crosscountry runners found that 1 g sodium phosphate four times daily for 6 days resulted in a 10% increase in VO_2max and an 11.8% increase in the anaerobic threshold *(5)*. The authors attributed the improvements in exercise performance to increased metabolic efficiency. Well controlled studies support the theory that phosphate salt supplementation may enhance function of the oxygen energy system. Furthermore, studies show that phosphate salt supplementation increases VO_2max and improves performance in endurance exercise tasks, including a greater number of stages completed in a progressive treadmill running test, increased time-to-exhaustion on a bicycle ergometer, and a decreased time to complete a 40-km cycling test *(42)*. However, if phosphate salts are in fact ergogenic, the underlying mechanism has yet to be determined *(42)*. The scientific literature suggests that phosphates may have an ergogenic effect on endurance athletes, but there is little support for its effectiveness during anaerobic exercise *(44)*. More research with tight methodological control is needed to support claims made in regard to phosphorus supplementation in both aerobic and anaerobic events.

Another popular nutritional ergogenic aid is the branched-chain amino acids (BCAAs): leucine, isoleucine and valine. They are prevalent in both protein/meal replacement powders and energy drinks. Table 3 shows the typical amino acid composition of some common protein preparations. The RDA for BCAAs is less than 3 g per day, although supplementation studies frequently utilize 5 to 20 g per day in tablet form and 1 to 7 g per liter in solutions *(42)*. The theory behind BCAA supplements relates to a phenomenon known as central fatigue, which holds that mental fatigue in the brain can adversely affect physical performance in endurance events. The central fatigue hypothesis suggests that low blood levels of BCAAs may accelerate the production of the brain neurotransmitter serotonin, or 5-hydroxytryptamine (5-HTP), and prematurely lead to fatigue *(45)*. Tryptophan, an amino acid that circulates in the blood, is a precursor of serotonin and can be more easily transported into the brain to increase serotonin levels when BCAA levels in the blood are low because high blood levels of

Table 3
Typical amino acid composition of whey, casein, and soy isolates

Amino acid	Whey	Casein	Soy
Alanine	4.6	2.7	3.8
Arginine	2.3	3.7	6.7
Aspartic acid	9.6	6.4	10.2
Cysteine/cystine	2.8	0.3	1.1
Glutamic acid	15.0	20.2	16.8
Glycine	1.5	2.4	3.7
Histidine[a]	1.6	2.8	2.3
Isoleucine[a,b]	4.5	5.5	4.3
Leucine[a,b]	11.6	8.3	7.2
Lysine[a]	9.1	7.4	5.5
Methionine[a]	2.2	2.5	1.1
Phenylalanine[a]	3.1	4.5	4.6
Proline	4.4	10.2	4.5
Serine	3.3	5.7	4.6
Threonine[a]	4.3	4.4	3.3
Tryptophan[a]	2.3	1.1	1.1
Tyrosine	3.3	5.7	3.3
Valine[a,b]	4.5	6.5	4.5

Values are expressed per 100 g of product
[a]Essential amino acid
[b]Branched-chain amino acid

BCAAs can block tryptophan transport into the brain (46). During endurance exercise, as muscle and liver glycogen are depleted for energy the blood levels of BCAAs also decrease, and fatty acid levels increase to serve as an additional energy source (47). The issue with extra fatty acids in the blood is that they need to attach to albumin as a carrier protein for proper transport. In doing so, the fatty acids displace tryptophan from its place on albumin and facilitate the transport of tryptophan into the brain for conversion to serotonin (48). Thus, the combination of reduced BCAAs and elevated fatty acids in the blood causes more tryptophan to enter the brain and more serotonin to be produced, leading to central fatigue (49).

For endurance athletes competing in long races (more than 2 hours), BCAA supplements can help delay central fatigue and maintain mental performance *(50)*. One study looked at BCAA supplementation during a marathon and showed improved performance for slower runners (3.05+ hours) but no effect on faster runners (less than 3.05 hours) *(2)*. Chronic BCAA supplementation (2 weeks) has also been shown to be effective in improving time-trial performance in trained cyclists *(6)*. In addition to their effects on prolonging endurance and delaying central fatigue, BCAA supplements have been associated with a reduced rate of protein and glycogen breakdown during exercise and an inhibition of muscle breakdown following exhaustive endurance exercise *(2,51)*. A number of studies in trained and untrained subjects, however, have shown no effect of BCAA supplements on exercise performance or mental performance *(52)*. In some cases, BCAAs have been compared with carbohydrate supplementation during exercise, with results showing they both delay fatigue to similar degrees *(53)*. The data on BCAA supplements are mixed, but they clearly do not harm endurance performance. Some studies have shown positive adaptations, whereas others have displayed no effect. Biological variations may determine whether BCAAs are effective for the individual athlete and the particular sport or event.

Pharmacological sports ergogenics are drugs designed to function like hormones or neurotransmitter substances that are found naturally in the human body. Like some nutritional sports ergogenics, pharmacological sports ergogenics may enhance physical power by affecting various metabolic processes associated with sport success. The most popular pharmacological sports ergogenics used by endurance athletes today are caffeine and ephedrine.

Caffeine is theorized to enhance endurance performance by first stimulating the central nervous system (CNS) and increasing psychological arousal. Caffeine also stimulates the release of epinephrine from the adrenal glad, which may further enhance physiological processes such as cardiovascular function and fuel utilization. The caffeine-mediated increase in free fatty acid mobilization and sparing of muscle glycogen is the primary theory underlying the ergogenic effects of caffeine on prolonged endurance activities. Lastly, caffeine increases myofilament affinity for calcium and/or increases the release of calcium from the sarcoplasmic reticulum in skeletal muscle, resulting in more efficient muscle contractions.

In one of the first studies conducted on caffeine's ergogenic effect, subjects consumed decaffeinated coffee or decaffeinated coffee combined with 330 mg pure caffeine 60 minutes prior to exercise. Time to exhaustion was more than 19% longer in the caffeine trial than in the decaffeinated trial *(54)*. The authors concluded that the performance increase was more than likely due to the increase in fat oxidation as muscle glycogen was not measured. Another study measured muscle glycogen utilization and found that caffeine prior to exercise reduced muscle glycogen utilization by 30% *(4)*. A later study supported the muscle glycogen-sparing hypothesis by reporting a 55% decrease in muscle glycogenolysis during the first 15 minutes of exercise in the caffeine trial *(55)*. The decrease in glycogenolysis during the initial stages of exercise allowed more glycogen to be available during the final stages, subsequently leading to an increased time to exhaustion. Further support of caffeine supplementation was demonstrated in a study examining the effects of acute caffeine ingestion (6 mg/kg) on prolonged, intermittent sprint performance on a cycle ergometer. The total amount of sprint work performed during the caffeine trial was 8.5% greater than that performed during the placebo trial *(56)*. The authors concluded that acute caffeine ingestion can significantly enhance performance of prolonged, intermittent sprint ability in competitive male, team-sport athletes.

Despite the overwhelming ergogenic evidence of caffeine supplementation, caution should be exercised when considering its use. A recent study showed that in healthy volunteers a caffeine dose corresponding to two cups of coffee (200 mg) significantly decreased blood flow to the heart during exercise by 22%. That percentage increased to 39% for people exercising in a high-altitude chamber, which the researchers used to simulate the way coronary artery disease (CAD) limits the amount of oxygen that gets to the heart *(57)*. Because an increase in blood to the heart is necessary for aerobic activity, the findings theoretically suggest that caffeine could slow the body down. The study's purpose, however, was not to look at whether caffeine could help athletes go faster or farther. Instead, it set out to investigate the effect caffeine has on blood flow to the heart. It can thus be concluded that people with CAD or those at a high risk for heart disease should avoid loading up on caffeine before a run or at least check with their primary care physician first.

A final note in regard to caffeine supplementation is that the International Olympic Committee lists caffeine as a banned substance. Although some amount of caffeine is allowed because of its occurrence in foods, a urinary level that exceeds 12 μg/ml results in a doping violation and possible disqualification or suspension. Therefore, it is recommended to keep the daily caffeine intake to less than 3 mg/kg body weight (i.e., 50–200 mg of caffeine) *(44)*. Table 4 lists typical caffeine content in common beverages, pills, and other products.

Although ephedrine is typically thought of as a weight loss supplement, evidence supports its role in enhancing endurance exercise performance. Ephedrine is a general sympathomimetic agent, which means that it stimulates the CNS by mimicking the effects of many of the body's own sympathetic hormones [e.g., epinephrine (adrenaline), norepinephrine] *(50)*.

Ephedrine is found in various antiasthmatic and cold or cough medications in pill, tablet, or inhaler form. It is also found in herbal teas and dietary supplements containing Ma Huang (Chinese ephedra or herbal ephedrine) as well as in dietary supplements marketed for weight loss and for increasing energy. All cold medications with decongestants are likely to contain prohibited sympathomimetics. Typical doses used in research range from 20 to 25 mg ephedrine *(42)*.

Table 4
Typical Caffeine Content in Common Beverages, Pills, and Other Products

Brewed coffee (cup)^a = 100 mg
Decaffeinated coffee (cup) = 3 mg
Medium-brewed tea (cup) = 50 mg
Cocoa (cup) = 5 mg
Starbucks' Coffee Grande (cup) = 90
Excedrin (one tablet) = 65 mg
No Doz (one tablet) = 100 mg
Vivarin (one tablet) = 200 mg
Guarana (100 mg) = 100 mg
SoBe No Fear (16 oz) = 159 mg
Cola-type soda (can) = 40 mg

Adapted from M.H. Williams *(42)*, with permission
^{a}cup = 5–6 ounces

Ephedrine is commonly combined with caffeine when assessing its ergogenic potential. One study examined ephedrine and caffeine separately and in combination in 12 subjects running a 10-km race. The run times for the ephedrine-containing trials were significantly shorter than those in the caffeine-only and placebo trials. In addition, heart rate and pace for the ephedrine-containing trials were significantly lower than in the caffeine-only and placebo trials *(58)*.

Another study examined the effects of caffeine and ephedrine on running times of nine male subjects performing the Canadian Forces Warrior Tests (WT). This is a standard test that all land forces soldiers must perform within 22 minutes. Time to exhaustion was significantly improved in the treatment group versus the placebo *(1)*. The caffeine and ephedrine enabled the subjects to exercise at a higher percentage of the maximum aerobic power for a longer period of time when compared with the placebo trials. These observations suggest that the improved performance may have been due to the CNS stimulatory effects of both caffeine and ephedrine *(44)*. Ephedrine has a track record of safe use at the recommended amount. Abuse of ephedrine, however, can lead to amphetamine-like side effects, including elevated blood pressure, rapid heart beat, nervousness, irritability, headache, urination disturbances, vomiting, muscle disturbances, insomnia, dry mouth, heart palpitations, and even death caused by heart failure *(59)*. It is important to note that ephedra use is banned by most sports governing bodies (e.g., International Olympic Committee). Additional research on the safety and efficacy of caffeine and ephedrine alone and in combination is warranted.

Physiological sports ergogenics are substances or techniques designed specifically to augment natural physiological processes that generate physical power. Two popular examples are glycerol and creatine. Physiological sports ergogenics are not drugs per se. In a strict sense, however, some may be regarded as drugs because they are prescribed substances.

Glycerol is also known as glycerin and is an alcohol compound that is more commonly found in the diet as a component of fat or triglycerides. It serves as a backbone onto which fatty acid molecules are attached and is marketed as an aid for "hyperhydrating" the body by increasing blood volume and helping to delay dehydration. Thus, glycerol may aid endurance athletes training or competing in hot, humid environments by hydrating tissues, increasing blood volume, and ultimately delaying fatigue and exhaustion associated with dehydration. Glycerin dosages used in research are based on body weight or total body water and have approximated 1 g/kg body weight, with each gram diluted in about 20 to 25 mL of water or similar fluid *(42)*.

Numerous studies support the theory that glycerol added to fluids increase tissue hydration compared with drinking fluid without glycerol added. Following glycerol consumption, the heart rate and body core temperature are lower during exercise in the heat *(60)*, suggesting an ergogenic effect. In endurance type of activities a larger supply of stored water may lead to a delay in dehydration and exhaustion *(61)*. More specifically, one study examined the effect of glycerol (1 g/kg) supplementation on body temperature while exercising on a treadmill (60% VO_2max) at 42°C at 25% relative humidity for 90 minutes 2.5 hours after ingestion of the glycerol. Results showed that the urine volume decreased before exercise, the sweat rate increased, and the rectal temperature was lower during exercise *(62)*. These findings imply that glycerol ingestion was helpful in maintaining normal body temperature during exercise in the heat.

Another study reporting positive results gave 11 fit adults glycerol (1.2 g/kg in a 26 ml/kg body weight solution) or a placebo (26 mg/kg body weight aspartame-flavored solution) 1 hour prior to cycle exercise to exhaustion at 60% of maximum workload (temperature 23.5°–24.5°C, humidity 25%–27%). The heart rate for those taking glycerol was 2.8 bpm lower, and endurance time was 21% longer *(63)*. In a follow-up study, these same researchers wanted to determine whether the same preexercise routine followed by a carbohydrate oral replacement solution during exercise had any further effect. Once again, they found that when glycerol had been taken the endurance time was 25% longer *(63)*.

It is important to note that not all studies show an ergogenic effect and that the benefits—although noted for trained endurance athletes exercising in hot, humid environments—are not necessarily observed in athletes who are less well trained or are exercising in more temperate climates *(61,64)*. These factors should be taken into account when considering glycerol supplements.

Although creatine supplements are typically marketed as bodybuilding and "strength-boosting" supplements, growing evidence suggests that they may prove beneficial for endurance athletes as well. Normally, about 1 to 2 g of creatine daily are produced in the

Table 5
Creatine Content in Select Foods

Food	Creatine content	
	g/lb	g/kg
Cod	1.4	3.0
Beef	2.0	4.5
Herring	3.0–4.5	6.5–10.0
Milk	0.05	0.1
Pork	2.3	5.0
Salmon	2.0	4.5
Shrimp	Trace	Trace
Tuna	1.8	4.0
Plaice	0.9	2.0
Fruits/vegetables	Trace	Trace

Adapted from Williams MH, Kreider RB, Branch D. Creatine: The Power Supplement (p 15). Human Kinetics, Champaign, IL, 1999

body from the amino acids arginine, glycine, and methionine. Dietary sources (Table 5), including meat and fish, add another 1 to 2 g of creatine per day, although overcooking destroys most of the creatine (the 1 g of creatine in an 8-oz steak may fall to zero if that steak is well done).

In the body, creatine plays a vital role in cellular energy production as creatine phosphate (phosphocreatine) in regenerating ATP in skeletal muscle *(50)*. Most studies utilize doses approximating 20 to 30 g/day, consumed in four or five equal doses throughout the day for 5 to 7 days followed by a maintenance dose of 5 g/day *(42)*.

One of the earliest creatine studies utilized well trained distance runners and demonstrated improved cumulative, repeated running times following four 300-m sprints *(65)*. Another study demonstrated significant increases in time to exhaustion during intense cycle ergometry with creatine supplementation *(66)*. Yet another study reported that creatine loading in both male and female athletes resulted in a 12% increase in the anaerobic threshold as well as a decrease in blood lactate during incremental cycle tests *(67)*. One other study showed continued support by delaying the onset of neuromuscular fatigue (a parameter similar to anaerobic threshold) by 13% in highly trained female athletes *(68)*. Most scientists agree that creatine's effectivenss can be attributed to one of two scenarios: 1) Phosphocreatine may aid ATP resynthesis for up to 3 minutes, albeit in a decreasing role with time and intensity of work *(69)*; 2) it may act as an energy shuttle between the mitochondria and muscle fibers, which suggests that creatine may help produce ATP aerobically *(70)*. Regardless, increasing muscle creatine phosphate levels through creatine supplementation may decrease the reliance on anaerobic glycolysis and reduce intramuscular lactate accumulation, thereby delaying the onset of fatigue *(44)*. More research on the effects of creatine supplementation on endurance performance is needed before definitive conclusions can be drawn.

4. IMMUNE SYSTEM AND ENDURANCE PERFORMANCE

Endurance athletes engaged in intense training run the risk of overtraining. Overtraining is usually encountered after several days of intense training and is generally associated with muscle fatigue. During the training period, transient signs and symptoms may occur including changes in the profile of mood state (POMS) where tension, depression, anger, fatigue, and confusion may be present. Other signs include depleted muscle glycogen stores, increased resting heart rate, increased cortisol secretion, decreased appetite, sleep disturbances, head colds, and immunosuppression. Most of the symptoms that result from overtraining, collectively referred to as overtraining syndrome, are subjective and identifiable only after the individual's performance has suffered. Unfortunately, these symptoms can be highly individualized, which can make it difficult for athletes, trainers, and coaches to recognize that performance decrements are brought on by overtraining. The first indication of overtraining syndrome is a decline in physical performance. The athlete can sense a loss in muscle strength, coordination, and maximal working capacity.

The immune system provides a line of defense against invading bacteria, parasites, viruses, and tumor cells. The system depends on the actions of specialized cells and antibodies. Unfortunately, one of the most serious consequences of overtraining is the negative effect it has on the body's immune system. Recent studies confirm that excessive training suppresses normal immune function, increasing the overtrained athlete's susceptibility to infections *(71,72)*. Numerous studies show that short bouts of intense exercise can temporarily impair the immune response, and successive days of heavy training can amplify this suppression *(73)*. Several investigators have reported an increased incidence of illness following a single, exhaustive exercise bout. Also, intense exercise during illness might decrease one's ability to fight off infection and increases the risk of even greater complications. In some cases, supplementation may help attenuate the immunosuppression typically seen with overtraining and in doing so may help prevent overtraining.

To bolster the first line of defense, an endurance athlete must ensure that adequate calories are being consumed. Endurance athletes maintaining heavy volume training often do not consume enough calories to keep up with energy demands because of the suppressive effect exercise can have on the appetite *(74)*. This point can be especially concerning for endurance athletes engaged in prolonged training or competition sessions on the same or successive days. Multiple training sessions on the same day have now become the norm more than the exception for the elite endurance athlete owing to the ever-increasing level of competition and pressure to perform at optimal levels. An example is performing a long, slow, distance (LSD) run in the morning only to follow-up with a strength training session that evening. Although it is unlikely that muscle glycogen stores can be completely resynthesized within a few hours by nutritional supplementation alone, it would behoove all endurance athletes to maximize the rate of muscle glycogen storage after exercise. This ultimately results in faster recovery from training, possibly allowing a greater training volume *(75)*.

To demonstrate the importance of nutrient timing, Flakoll et al. *(76)* provided either a placebo, carbohydrate, or carbohydrate/protein supplement to U.S. Marine recruits immediately after exercise during 54 days of basic training to test the long-term impact of postexercise carbohydrate/protein supplementation on variables such as health, muscle soreness, and function. Compared to the placebo and carbohydrate groups, the combined carbohydrate/protein group had 33% fewer total medical visits, 28% fewer visits due to bacterial/viral infections, 37% fewer visits due to muscle and joint problems, and 83% fewer due to heat exhaustion. Muscle soreness was also reduced immediately after exercise by the carbohydrate/ protein supplement. The authors postulated that post-exercise carbohydrate/protein supplementation not only may enhance muscle protein deposition but also have significant potential to affect health, muscle soreness, and tissue hydration positively during prolonged intense exercise training. This suggests a potential therapeutic approach for the prevention of health problems in severely stressed exercising populations.

Hence, it is essential for the endurance athlete to consume an adequate balance of macro and micronutrients at the proper times throughout the day especially during periods of heavy training. Athletes involved in intense training not only have greater macronutrient needs but, more specifically, greater dietary protein needs than individuals who do not train. Intense training has been shown to decrease the availability of certain essential amino acids. Chronic depletion of essential amino acids may slow the rate of tissue repair and growth. Therefore, athletes involved in intense training need to ingest enough high quality protein in the diet to maintain essential amino acid availability during training. Dietary supplementation of protein has been a common practice among athletes. More recently, advances in food processing have allowed extraction of high quality proteins from food. These proteins have been used in numerous food products and have been marketed to athletes as a convenient means of increasing the quality of protein in their diet.

Several ergogenic aids have been reported to aid the immune system. Colostrum is the clear or cloudy "premilk" that female mammals secrete after giving birth and before producing milk. Colostrum for dietary supplements is usually derived from bovine sources and contains various immunoglobulins (also called antibodies) and antimicrobial factors (i.e., lactoferrin, lactoperoxidase, lysozyme) as well as insulin-like growth factors (e.g., IGF-I, IGF-II). Bovine colostrum is among the highest quality sources of protein.

The most prevalent claims for dietary supplements containing colostrum are in the area of generalized immune function and improved recovery from intense exercise. One study examined the effects of consuming colostrum (60 g/day) or placebo (whey protein) during an 8-week running program, running three times a week for 45 minutes per session (77). Participants conducted two treadmill runs to exhaustion, with 20 minutes of rest between runs, at baseline and 4 and 8 weeks into the study. No differences existed in treadmill running performance at baseline, and at week four both groups had similar improvements in running performance. At week 8, however, the colostrum group ran significantly farther and did more work than the placebo group during the treadmill test. In addition, the colostrum supplemented group exhibited lower serum creatine kinase (CK) levels. CK is an important muscle cell enzyme that some scientists believe can be used as a marker of muscle cell damage. High CK levels often indicate that significant muscle damage has occurred. However, some scientists believe that if CK levels remain fairly normal, the athlete has experienced little muscle trauma *(77)*.

Another study examined rowing performance in a group of elite female rowers. Eight rowers completed a 9-week training program while consuming either colostrum (60 g/day) or whey protein. By week 9, rowers consuming colostrum had greater increases in the distance covered and work done than the whey protein group *(78)*. Additional studies on bovine colostrum consumption suggest that it can deliver some generalized anti-inflammatory benefits *(79,80)* and help prevent and treat the gastric injury associated with non-steroidal anti-inflammatory drugs *(81)*.

Glutamine is another supplement popular in athletic popula-tions today. It is the most abundant amino acid in the body, comprising approximately half of the free amino acids in the blood and muscle. Glutamine is a nonessential glucogenic amino acid and an anaplerotic precursor that has been shown to be a vital fuel for a variety of cells of the immune system *(82)*. In skeletal muscle, glutamine has an inhibitory role on proteolysis and branched-chain amino acid catabolism. Recent evidence has placed emphasis on glutamine as a positive component of immune func-tion *(82,83)*.

Glutamine has two main functions in the body: It is a precursor in the synthesis of other amino acids, and it converts to glucose for energy. Cells of the immune system, small intestine, and kidney are the major consumers of glutamine, making "immune boosting" and "immune maintenance" claims for glutamine supplements quite common. In addition, there are claims for glutamine supplements in maintaining muscle mass, reducing post-exercise catabolism (muscle tissue breakdown), and accelerating recovery from intense exercise. Intense exercise training often exhibited by the endurance athlete results in a drop in plasma glutamine levels. Chronically low glutamine levels have been implicated as a possible contributing factor for athletic overtraining syndrome as well as the transient immunosuppression and increased risk of infections that typically affect competitive athletes during intense training and competition. Under conditions of metabolic stress, the body's need for glutamine may become conditionally essential, meaning that the body cannot produce adequate levels and a dietary source is required to prevent catabolism of skeletal muscle, the primary source of stored glutamine in the body.

A significant body of scientific literature supports the beneficial effects of glutamine supplementation in maintaining muscle mass and immune system function in critically ill patients and in those recovering from extensive burns and major surgery *(84)*. When plasma glutamine levels fall, skeletal muscles may enter a state of catabolism in which muscle protein is degraded to provide free glutamine for the rest of the body. Because skeletal muscle is the major source of glutamine (other than the diet), prolonged deficits in plasma glutamine can lead to a significant loss of skeletal muscle protein and muscle mass. Postsurgical deposition of collagen (a marker for wound healing) can be enhanced by amino acid supplementation containing 14 g of glutamine *(85)*, whereas a lower dose of mixed amino acids containing glutamine (2.9 g) provides no change in athletic performance or on adaptations to cycling training *(86)*.

Studies have also been conducted on glutamine supplementation in athletes, and a strong rationale exists for the efficacy of glutamine supplements in athletic populations. One study found that athletes who consumed glutamine immediately and 2 hours after running a marathon or ultramarathon reported fewer infections than a placebo group *(87)*. The levels of infection were lowest in the middle distance runners and highest in the runners after a marathon or ultramarathon. Glutamine supplements have also been shown to play a role in counteracting the catabolic effects of stress hormones, such as cortisol, which are typically elevated by strenuous exercise. Under conditions of stress-induced protein wasting in adults and children, including burns, surgery, and some forms of cancer, glutamine supplementation has been associated with reduced protein breakdown *(88)*, enhanced lymphocyte function *(89)*, reduced gut permeability *(90)*, and reduced infections *(91)*.

In addition to lowering glutamine levels, prolonged strenuous exercise has been reported to lower zinc levels apparently owing to greater zinc loss during training from sweat and/or inadequate intake of dietary zinc *(92)*. Zinc deficiency has been reported to diminish serum testosterone levels *(93)*, impair immune function *(94)*, and decrease strength *(95)*. Zinc is an essential trace mineral that functions as part of about 300 enzymes. As such, zinc plays a role in numerous biochemical pathways and physiological processes. Because of the varied roles of zinc in the body, claims for zinc-containing dietary supplements are numerous, including those for improved wound healing, general immune system support, and support of various aspects of men's health.

Zinc lozenges have become one of the most popular natural approaches to treating the common cold. The scientific evidence generally supports this use for short periods (1–2 weeks) *(96)*. Zinc lozenges appear to reduce cold symptoms, such as sore throats, hoarseness, and coughing, and may even be able to shorten the duration of colds by a full day.

Exercise performance has been associated with decreased zinc status, especially in athletes who avoid red meat, who concentrate their diets too much on carbohydrates, or who follow an overly restricted dietary regimen. Low zinc intake ($< 3\,\mathrm{mg/day}$) has been linked to reduced activity of a zinc-containing enzyme in red blood cells, carbonic anhydrase, which helps red blood cells transport carbon dioxide from tissues to the lungs to be exhaled. Mild-to-moderate zinc deficiency can lead to significant reductions in the body's ability to take up and use oxygen, remove carbon dioxide, and generate energy during high-intensity exercise *(50)*.

One study examined the effects of daily physical training on serum and sweat zinc concentrations in professional sportsmen between October and December, during the competition season. They discovered a significantly higher zinc concentration in sweat in December than in October. They stated that a daily and maintained exercise routine was probably responsible for the altered zinc metabolism. The authors further concluded that alterations in zinc metabolism with increases in zinc excretion and stress levels lead to a situation of latent fatigue associated with decreased endurance *(97)*.

Another study attempted to investigate how exhaustive exercise affects thyroid hormones and testosterone levels in male wrestlers who were supplemented with oral zinc sulfate (3 mg/kg/day) for 4 weeks. The results demonstrated that exhaustive exercise led to significant inhibition of both thyroid hormones and testosterone concentrations but that 4-week zinc supplementation prevented this inhibition in wrestlers. The authors concluded that physiological doses of zinc administration may benefit performance (98). Another study, however, examined plasma zinc concentrations and markers of immune function in 10 male runners following a moderate increase in training over 4 weeks and found no benefit in zinc concentration, differential leukocyte counts, lymphocyte subpopulations, or lymphocyte proliferation. These authors concluded that athletes are unlikely to benefit from zinc supplementation during periods of moderately increased training volume (99).

Overall, then, studies on colostrum, glutamine, and zinc have provided favorable evidence that they support the immune system and should be considered by the endurance athlete engaged in intense training.

5. ADDITIONAL CONCERNS

Despite the fact that there is solid scientific evidence backing the safety and efficacy of certain nutritional supplements, some remain skeptical as to whether they should be used at all. The two most popular questions raised generally address legal and ethical considerations. Although some drugs may be effective sports ergogenics, their use might also significantly increase health risks. The Medical Commission of the International Olympic Committee (IOC) noted that doping violates the ethics of both sport and medical science and is prohibited. Most athletic governing bodies, such as the IOC, the National Basketball Association (NBA), the National Football League (NFL), The National Collegiate Athletic Association (NCAA), and the United States Olympic Committee (USOC) have developed drug-use policies. Any athlete competing under the jurisdiction of a specific athletic governing body should be aware of its drug rules and regulations before consuming any nutritional supplement designed to aid performance.

What are the ethics of sport? According to Webster's Dictionary, the definition for ethics includes the following: 1) the discipline dealing with what is good and bad and with moral duty and obligation; 2) a theory or system of moral values; 3) the principles of conduct governing an individual or group; and 4) a guiding philosophy. All four definitions can be operational in sports.

The IOC embraces the idea that athletes should succeed through their unaided efforts; in other words, the "Do the best with what you've got" approach. However, the athlete whose primary goal is to win at all costs may be guided by his or her own principles in an attempt to obtain that unfair advantage. With increasing pressures to perform and potential scholarships, jobs, endorsements, and contract extensions on the line, an increasing number of elite athletes participating in both the collegiate and professional ranks are learning to modify their natural ability by techniques that go beyond normal training in order to get an advantage—not necessarily an unfair one—over their competitors. As noted throughout this chapter, safe, effective, legal ergogenics are available. Whether their use is ethical may depend on the individual athlete's code of ethics. Some authorities believe that because a sports ergogenic is legal, its use is ethical. Others may contend that such a sports ergogenic violates the ethics of the IOC antidoping rule. In this situation, the ultimate decision about whether to use a safe, effective, legal ergogenic rests with the individual athlete.

6. CONCLUSION

Proper training, maintaining a positive energy balance, and utilizing effective nutritional supplements form the foundation for optimal performance. Training for the endurance athlete should be based on proper utilization of the principles of training, depending on individual goals and the training season (preseason, in-season, postseason). The nutritional base should focus on an everyday diet that emphasizes a combination of all the nutritional ergogenic aids. The types of macronutrients and micronutrients selected and the timing of their use are of the utmost importance for developing a well refined everyday diet. The use of nutritional supplements that research has shown can help improve energy availability (e.g., sports drinks, carbohydrate) and/or promote recovery (e.g., carbohydrate, protein) can provide additional benefits in certain situations. In addition, a combination of effective pharmacological and physiological ergogenic supplements can further propel performance and support the immune system. Following these training and nutritional recommendations can serve as the foundation for a successful endurance athlete.

7. PRACTICAL APPLICATIONS

- To prepare for competition, endurance athletes should strive to hyperhydrate prior to exercise. Full hydration requires that water along with fluids containing moderate concentrations of carbohydrate and sodium be ingested regularly starting several days before competition.
- Endurance athletes should try to maximize their muscle glycogen stores prior to competition. This can be accomplished during the training taper by consuming 7 to 10 g carbohydrates/kg body weight daily.
- Prior to competition, it is recommended that 500 to 1000 ml of fluid be ingested. It may also be beneficial to consume 200 to 300 g of carbohydrate if supplementation during exercise is limited or not possible.
- During exercise, endurance athletes should try to replace fully any fluid losses that occur, although during hot and humid conditions this may be next to impossible.
- It is best to consume small volumes of fluid frequently (150–250 ml every 15 minutes) rather than consume high volumes of fluid occasionally (e.g., 400 ml every 30 minutes).
- Electrolyte replacement should be considered by the endurance athlete. Most sports drinks are formulated adequately to replace electrolyte losses due to sweating.
- To maximize endurance performance, 45 to 60 g of high glycemic index (GI) carbohydrates should be ingested per hour during exercise. Carbohydrate solutions should not exceed 10% if maintaining hydration is of importance.
- Carbohydrate supplementation can be delayed until about 30 to 40 minutes prior to the onset of fatigue and still be effective. However, if fluid replacement is of concern, it is recommended that supplementation with a dilute carbohydrate solution start as soon as possible and continued periodically throughout exercise.
- The addition of small amounts of protein to a carbohydrate supplement (4:1 ratio) may also increase the effectiveness of the supplement.

- For rapid rehydration following exercise, it is important to start fluid consumption early and replace at least 150% of the fluid lost during exercise with a dilute sodium solution.
- Requirements for the daily recovery of muscle glycogen depend on exercise intensity and duration. If the training duration is moderate and the intensity is low, 5 to 7 g of carbohydrate (CHO)/kg/day should be consumed. If the training duration is moderate and the intensity is high, 7 to 12 g CHO/kg/day should be consumed. If the training is extreme (4–6 hours per day), 10 to 12 g CHO/kg/day should be consumed.
- If there is only limited time to replenish the muscle glycogen stores, 1.0 to 1.2 g CHO/kg/hr should be consumed at frequent intervals starting within the first 30 minutes after exercise.
- The addition of protein to the carbohydrate supplement (4:1 ratio) promotes additional glycogen storage when carbohydrate intake is suboptimal or when frequent supplementation is not possible.
- Post-exercise supplements composed of carbohydrate and protein have the added benefit of limiting muscle tissue damage, stimulating protein accretion, and protecting the immune system from exercise-induced immunosuppression.
- Several pharmacological and physiological ergogenic aids can further propel performance for the endurance athlete and should be considered depending on the individual goal of the athlete.
- There is strong evidence that colostrum, glutamine, and zinc support the immune system. Hence, these supplements should be considered by the endurance athlete engaged in intense training.

REFERENCES

1. Bell DG, Jacobs I. Combined caffeine and ephedrine ingestion improves run times of Canadian Forces Warrior Test. Aviat Space Environ Med 1999;70:325–328.
2. Blomstrand E, Hassem P, Ekblom B, Newsholme EA. Administration of branched-chain amino acids during sustained exercise-effects on performance and plasma concentration of some amino acids. Eur J Appl Physiol 1991;63:83–88.
3. Bridge CA, Jones MA. The effect of caffeine ingestion on 8 km run performance in a field setting. J Sports Sci 2006;24:433–439.
4. Erickson MA. Effects of caffeine, fructose, and glucose ingestion on muscle glycogen utilization during exercise. Med Sci Sports Exerc 1987;19:579–583.
5. Kreider RB, Miller GW, Williams MH, Somma CT, Nasser TA. Effects of phosphate loading on oxygen uptake, ventilatory anaerobic threshold, and run performance. Med Sci Sports Exerc 1990;22:250–256.

6. Mittleman KD, Ricci MR, Bailey SP. Branched-chain amino acids prolong exercise during heat stress in men and women. Med Sci Sports Exerc 1998;30:83–91.

7. Saunders MJ, Kane MD, Todd MK. Effects of a carbohydrate-protein beverage on cycling endurance and muscle damage. Med Sci Sports Exerc 2004;36:1233–1238.

8. Coyle EF, Montain SJ. Benefits of fluid replacement with carbohydrate during exercise. Med Sci Sports Exerc 1992;24:S324–S330.

9. Armstrong LE, Costill DL, Fink WJ. Influence of diuretic-induced dehydration on competitive running performance. Med Sci Sports Exerc 1985;17:456–461.

10. Sutton JR, Bar-Or O. Thermal illness in fun running. Am Heart J 1980;100:778–781.

11. Coyle EF. Substrate utilization during exercise in active people. Am J Clin Nutr 1995;61(Suppl):968S–979S.

12. Thibault G. Ahead of the pack. Training Cond 2006;16:25–31.

13. Kreider R, Fry AC, O'Toole ML (eds). Overtraining in Sport. Human Kinetics, Champaign, IL, 1998.

14. Carli G. Changes in the exercise-induced hormone response to branched chain amino acid administration. Eur J Physiol Occup Physiol 1992;64:272–277.

15. Cade JR. Dietary intervention and training in swimmers. Eur J Physiol Occup Physiol 1997;63:210–215.

16. Kreider RB. Dietary supplements and the promotion of muscle growth with resistance exercise. Sports Med 1999;27:97–110.

17. Komi PV (ed). Strength and Power in Sport. Blackwell Scientific, Cambridge, MA, 1992.

18. Brand-Miller J, Wolever TMS, Colagiuri S, Foster-Powell K (eds). The Glucose Revolution. Marlowe, New York, 1999.

19. Coyle EF, Montain SJ. Carbohydrate and fluid ingestion during exercise: are there trade-offs? Med Sci Sports Ex 1992;24:671–678.

20. Noakes TD, Myburgh KH, Du Plessia J, et al. Metabolic rate, not percent dehydration, predicts rectal temperature in marathon runners. Med Sci Sports Exerc 1991;23:443–449.

21. Bergstrom J, Hermansen L, Hultman E, Saltin B. Diet, muscle glycogen and physical performance. Acta Physiol Scand 1967;71:140–150.

22. Balsom PD, Gaitanos GC, Soderlund K, Ekblom B. High-intensity exercise and muscle glycogen availability in humans. Acta Physiol Scand 1999;165:337–345.

23. Volek JS. Influence of nutrition on responses to resistance training. Med Sci Sports Exerc 2004;36:689–696.

24. Ivy JL, Res PT, Sprague RC, Widzer MO. Effect of a carbohydrate-protein supplement on endurance performance during exercise of varying intensity. Int J Sport Nutr Exerc Metab 2003;13:382–395.

25. Anderson LL, Tufekovic G, Zebis MK, et al. The effect of resistance training combined with timed ingestion of protein on muscle fiber size and muscle strength. Metabolism 2005;54:151–156.

26. Dangin M, Boirie Y, Garcia-Rodenas C, et al. The digestion rate of protein is an independent regulating factor of postprandial protein retention. Am J Physiol Endocrinol Metab 2001;280:E340–E348.
27. Ivy JL, Portman R (eds) The Performance Zone. Basic Health Publications, North Bergen, NJ, 2004.
28. Costill D, Hargreaves M. Carbohydrate nutrition and fatigue. Sports Med 1992;13:86–92.
29. Ivy JL. Muscle glycogen synthesis before and after exercise. Sports Med 1991;11:6–11.
30. Houston M (ed). Biochemistry Primer for Exercise Science. Human Kinetics, Champaign, IL, 1995.
31. Ivy JL. Glycogen resynthesis after exercise: effect of carbohydrate intake. Int J Sports Med 1998;19:S142–S145.
32. Haymond M. Hypoglycemia in infants and children. Endocrinol Metab Clin North Am 1989;18:211–252.
33. Gastelu D, Hatfield F (eds). Nutrition for Maximum Performance. Avery, Garden City Park, NY, 1997.
34. Blom P, Hostmark A, Vaage O, Kardel K, Maehlum S. Effects of different post-exercise sugar diets on the rate of muscle glycogen synthesis. Med Sci Sports Exerc 1987;19:491–496.
35. Burke L, Collier G, Hargreaves M. Muscle glycogen storage after prolonged exercise: effect of the glycemic index of carbohydrate feedings. J Appl Physiol 1993;75:1019–1023.
36. Jentjens RLPG, van Loon LJC, Mann CH, Wagenmakers AJM, Jeukendrup AE. Addition of protein and amino acids to carbohydrates does not enhance post-exercise muscle glycogen synthesis. J Appl Physiol 2001;91:839–846.
37. Tarnopolsky MA, Bosman M, Macdonald JR, Vandeputte D, Martin J, Roy BD. Post-exercise protein-carbohydrate and carbohydrate supplements increase muscle glycogen in men and women. J Appl Physiol 1997;83:1877–1883.
38. Evans WJ. Protein nutrition and resistance exercise. Can J Appl Physiol 2001;26(Suppl):S141–S152.
39. Ivy JL, Katz AL, Dutler CL, Sherman WM, Cyyle EF. Muscle glycogen synthesis after exercise: effect of time of carbohydrate ingestion. J Appl Physiol 1988;64:1480–1485.
40. Garetto OP, Richter EA, Goodman MN, Ruderman NB. Enhanced muscle glucose metabolism after exercise in the rat: the two phases. Am J Physiol 1984;246:E471–E475.
41. Miller SL, Tipton KD, Chinkes DL. Independent and combined effects of amino acids and glucose after resistance exercise. Med Sci Sports Exerc 2003;35:449–455.
42. Williams MH. The Ergogenics Edge. 1st ed. Human Kinetics, Champaign, IL, 1998.
43. Cade R, Conte M. Effects of phosphate loading on 2,3-diphosphoglycerate and maximum oxygen uptake. Med Sci Sports Exerc 1984;16:263–268.

44. Antonio J, Stout J. Supplements for Endurance Athletes. Human Kinetics, Champaign, IL, 2002.

45. Blomstrand E, Celsing F, Newsholme EA. Changes in plasma concentrations of aromatic and branched-chain amino acids during sustained exercise in man and their possible role in fatigue. Acta Physiol Scand 1988;1:115–121.

46. Castell LM, Yamamoto T, Phoenix J, Newsholme EA. The role of tryptophan in fatigue in different conditions of stress. Adv Exp Med Biol 1999;467:697–704.

47. Davis JM. Carbohydrates, branched-chain amino acids, and endurance: the central fatigue hypothesis. Int J Sport Nutr 1995;5(Suppl):S29–S38.

48. Hassmen P, Blomstrand E, Ekblom B, Newsholme EA. Branched-chain amino acid supplementation during 30-km competitive run: mood and cognitive performance. Nutrition 1994;10:405–410.

49. Tanaka H, West KA, Duncan GE, Bassett DRJ. Changes in plasma tryptophan/branched chain amino acid ration in responses to training volume variation. Int J Sport Nutr 1997;18:270–275.

50. Talbott SM, Hughes K. The Health Professional's Guide to Dietary Supplements. 1st ed. Lippincott Williams & Wilkins, Philadelphia, 2007.

51. Gastmann UA, Lehmann MJ. Overtraining and the BCAA hypothesis. Med Sci Sports Exerc 1998;30:1173–1178.

52. Van Hall G, Raaymakers JS, Saris WH, Wagenmakers AJ. Ingestion of branched-chain amino acids and tryptophan during sustained exercise in man: failure to affect performance. J Physiol 1995;486:789–794.

53. Davis JM, Bailey SP, Woods JA, Galiano FJ, Hamilton MT, Bartoli WP. Effects of carbohydrate feedings on plasma free tryptophan and branched-chain amino acids during prolonged cycling. Eur J Appl Physiol Occup Physiol 1992;65:513–519.

54. Costill DL, Dalsky GP, Fink WJ. Effects of caffeine ingestion on metabolism and exercise performance. Med Sci Sports Exerc 1978;10:155–158.

55. Spreit LL, MaClean DA, Dyck DJ, Hultmant E. Caffeine ingestions and muscle metabolism during prolonged exercise in humans. Am J Physiol 1992;262:E891–E898.

56. Schneiker KT, Bishop D, Dawson B, Hackett LP. Effects of caffeine on prolonged intermittent-sprint ability in team-sport athletes. Med Sci Sports Exerc 2006;38:578–585.

57. Namdar M, Koepfli P, Grathwohl R, et al. Caffeine decreases exercise-induced myocardial flow reserve. J Am Coll Cardiol 2006;47:405–410.

58. Bell DG, McLennan TM, Sabiston CM. Effect of ingesting caffeine and ephedrine on 10-km run performance. Med Sci Sports Exerc 2002;34:344–349.

59. Blumenthal M, Busse WR, Goldberg A (eds). The Complete Commission E Monographs: Therapeutic Guide to Herbal Medicines. Integrative Medicine Communications, Boston, 1998.

60. Jimenez C, Melin B, Doulmann N, Allevard AM, Launay JC, Savourey G. Plasma volume changes during and after acute variations of body hydration level in humans. Eur J Appl Physiol Occup Physiol 1999;80:1–8.

61. Wagner DR. Hyperhydrating with glycerol: implications for athletic performance. J Am Diet Assoc 1999;99:207–212.

62. Lyons TP, Riedesel ML, Meuli LE, Chick TW. Effects of glycerol-induced hyperhydration prior to exercise in the heat on sweating and core temperature. Med Sci Sports Exerc 1990;22:477–483.

63. Montner P, Stark DM, Riedesel ML. Pre-exercise glycerol hydration improves cycling endurance time. Int J Sports Med 1996;17:27–33.

64. Arnall DA, Goforth HWJ. Failure to reduce body water loss in cold-water immersion by glycerol ingestion. Undersea Hyperb Med 1993;20:309–320.

65. Harris RC, Viru M, Greenhaff PL, Hultman E. The effect of oral creatine supplementation on running performance during maximal short term exercise in man. J Physiol 1993;467:74P.

66. Smith JC, Stephens DP, Hall EL, Jackson AW, Earnest CP. Effect of oral creatine ingestion on parameters of the work rate-time relationship and time to exhaustion in high-intensity cycling. Eur J Appl Physiol 1998;77:360–365.

67. Nelson A, Day R, Glickman-Weiss E, Hegstad M, Sampson B. Creatine supplementation alters the response to a graded cycle ergometer test. Eur J Appl Physiol 2000;83:89–94.

68. Stout J, Eckerson J, Ebersole K, et al. Effect of creatine loading on neuromuscular fatigue threshold. J Appl Physiol 2000;88:109–112.

69. Bangsbo J, Gollnick PD, Graham TE. Anaerobic energy production and O_2 deficit-debt relationship during exhaustive exercise in humans. J Physiol 1990;422:539–559.

70. Wallimann T, Wyss M, Brdiczka D, Nicolay K, Eppenberger HM. Intracellular compartmentation, structure and function of creatine kinase isoenzymes in tissues with high and fluctuating energy demands: the 'phosphocreatine circuit' for cellular energy homeostasis. Biochem J 1992;281:21–40.

71. Mackinnon LT. Exercise and natural killer cells: what is the relationship? Sports Med 1989;7:141–149.

72. McCarthy DA, Dale MM. The leucocytosis of exercise: a review and model. Sports Med 1988;6:333–363.

73. Brahmi Z, Thomas JE, Park M, Dowdeswell IRG. The effect of acute exercise on natural killer cell activity of trained and sedentary subjects. J Clin Immunol 1985;5:321.

74. Berning JR. Energy intake, diet and muscle wasting. In: Kreider RB, Fry AC, O'Toole ML (eds) Overtraining in Sport (pp 275–288). Human Kinetics, Champaign, IL, 1998.

75. Haff GG, Lehmkuhl MJ, McCoy LB, Stone MH. Carbohydrate supplementation and resistance training. J Strength Cond Res 2003;7:187–196.

76. Flakoll PJ, Judy T, Flinn K, Carr C, Flinn S. Postexercise protein supplementation improves health and muscle soreness during basic military training in marine recruits. J Appl Physiol 2001;96:951–956.

77. Buckley J. Effect of oral bovine colostrum supplement (Intact) on running performance. In: Proceedings of the Australian Conference of Science and Medicine in Sport (p 79).

78. Brinkworth GD, Buckley JD, Bourdon PC, Gulbin JP, David A. Oral bovine colostrum supplementation enhances buffer capacity but not rowing performance in elite female rowers. Int J Sport Nutr Exerc Metab 2002;12:349–365.

79. Bolke E, Jehle PM, Hausmann F, et al. Preoperative oral application of immunoglobulin-enriched colostrum milk and mediator response during abdominal surgery. Shock 2002;17:9–12.

80. Playford RJ, MacDonald CE, Calnan DP, et al. Co-administration of the health food supplement, bovine colostrum, reduces the acute non-steroidal anti-inflammatory drug-induced increase in intestinal permeability. Clin Sci (Lond) 2001;100:627–633.

81. Khan Z, Macdonald C, Wicks AC, et al. Use of the "nutriceutical," bovine colostrum, for the treatment of distal colitis: results from an initial study. Aliment Pharmacol Ther 2002;16:1917–1922.

82. Ardawi MSM, Newsholme EA. Glutamine metabolism in lymphocytes of rats. Biochem J 1983;212:835.

83. Ardawi MSM, Newsholme EA. Metabolism in lymphocytes and its importance in immune response. Essays Biochem 1985;21:1.

84. Carli F, Webster J, Ramachandra V, et al. Aspects of protein metabolism after elective surgery in patients receiving constant nutritional support. Clin Sci (Colch) 1990;78:621–628.

85. Williams JZ, Abumrad N, Barbul A. Effect of a specialized amino acid mixture on human collagen deposition. Ann Surg 2002;236:369–374.

86. Vukovich MD, Sharp RL, Kesl LD, Schaulis DL, King DS. Effects of a low-dose amino acid supplement on adaptations to cycling training in untrained individuals. Int J Sport Nutr 1997;7:298–309.

87. Castell LM, Poortmans JR, Newsholme EA. Does glutamine have a role in reducing infections in athletes? Euro J Appl Physiol 1996;73:488–490.

88. Des Robert C, Le Bacquer O, Piloquet H, Roze JC, Darmaun D. Acute effects of intravenous glutamine supplementation on protein metabolism in very low birth weight infants: a stable isotope study. Pediatr Res 2002;51:87–93.

89. Yoshida S, Kaibara A, Ishibashi N, Shirouzu K. Glutamine supplementation in cancer patients. Nutrition 2001;17:766–768.

90. Klimberg VS, Nwokedi E, Hutchins LF, et al. Glutamine facilitates chemotherapy while reducing toxicity. JPEN J Parenter Enteral Nutr 1992;16:83S–87S.

91. Barbosa E, Moreira EA, Goes JE, Faintuch J. Pilot study with a glutamine-supplemented enteral formula in critically ill infants. Rev Hosp Clin Fac Med Sao Paulo 1999;54:21–24.

92. Khaled S, Brun JF, Micallef JP, et al. Serum zinc and blood rheology in sportsmen (football players). Clin Hemorheol Microcirc 1997;17:47–58.

93. Om A, Chung K. Dietary zinc deficiency alters 5 alpha-reduction and aromatization of testosterone and androgen estrogen receptors in rat liver. J Nutr 1996;126:842–848.

94. Nieman DC. Nutrition, exercise, and immune system function. Clin Sports Med 1999;18:537–548.

95. Van Loan MD, Sutherland B, Lowe NM, Turnland JR, King JC. The effects of zinc depletion on peak force and total work of knee and shoulder extensor and flexor muscles. Int J Sport Nutr 1999;9:125–135.

96. Eby GA. Zinc ion availability: the determinant of efficacy in zinc lozenge treatment of common colds. J Antimicrob Chemother 1997;40:483–493.

97. Cordova A, Navas FJ. Effect of training on zinc metabolism: changes in serum and sweat zinc concentrations in sportsmen. Ann Nutr Metab 1998; 42:274–282.

98. Kilic M, Baltaci AK, Gunay M, Gokbel H, Okudan N, Cicioglu I. The effect of exhaustion exercise on thyroid hormones and testosterone levels of elite athletes receiving oral zinc. Neuro Endocrinol Lett 2006;27:247–252.

99. Peake JM, Gerrard DF, Griffin JF. Plasma zinc and immune markers in runners in response to a moderate increase in training volume. Int J Sports Med 2003;24:212–216.

100. Gisolfi CV, Lamb DR (eds). Fluid Homeostasis During Exercise (Perspectives in Exercise and Sports Medicine). Benchmark Press, Indianapolis, 1990.

12 Nutritional Supplements to Enhance Recovery

Tim N. Ziegenfuss, Jamie Landis, and Mike Greenwood

Abstract

The ability to recover from intense exercise often separates good athletes from great ones. In the past, "recovery" often simply included rest, physical modalities (e.g., massage, hydration therapy) and meeting basic nutritional needs for fluid and energy intake. Today, athletes have a number of additional options to help them recover from high intensity training, one of which includes the judicious use of dietary supplements. This chapter briefly reviews nutritional strategies that have a strong theoretical background for enhancing rehydration/electrolyte balance, replenishing energy reserves, minimizing oxidative damage, and stimulating muscle repair.

Key words

Rehydration · Electrolyte · Antioxidant · Essential amino acids · L-Carnitine

1. EXERCISE RECOVERY

Recovery from exercise is a dynamic process that includes the restoration of body fluids and electrolyte balance, replenishment of energy stores, and repair of damaged tissues. Some aspects of these processes take a few minutes, and others may take days. The ability to continue exercising or competing day after day may be limited by how quickly one's muscles recover after an initial exertion. In general, the more severe the previous exercise stress, the longer and more involved are the body's recovery efforts. By eating a nutrient-dense, well balanced diet that includes adequate fluid intake, the body can generally fully recover from any single bout

From: *Nutritional Supplements in Sports and Exercise*
Edited by: M. Greenwood, D. Kalman, J. Antonio,
DOI: 10.1007/978-1-59745-231-1_12, © Humana Press Inc., Totowa, NJ

of exercise if given enough time. However, by taking proactive steps during the postexercise periods, it is possible to speed the body's recovery from individual bouts while sustaining maximal training and improving exercise performance.

The physical cost of exercise is intimately tied to the intrinsic variables of intensity, duration, and mode. They effectively form the basis of categorizing the stresses on the body that occur during exercise. Consideration of each of the variables requires knowledge of anatomy (extent of muscle mass involved), the kinesiology involved (type of muscle actions), and physiology (e.g., metabolic demands of exercise, hormonal responses to exercise). Other considerations include the ages of the individuals, their current diet, the timing of the exercise bout, and their competitive status, training volume, environmental factors, and overall health status. The obvious consequence of the interplay between these factors is clear: Distinct strategies for nutritionally enhancing recovery are necessary.

When carefully planned based on both the demands of the exercise and the needs of the athlete, we believe that dietary supplementation can help maximize an athlete's recovery from exercise. This chapter provides some basic guidelines for an evidence-based approach to rational, safe, effective nutritional intervention as a means of enhancing exercise recovery.

2. PHASES OF RECOVERY

To address and maximize the benefits of nutritional support properly, two distinct phases of exercise recovery should be recognized: fast (partial) and slow (complete). Fast recovery occurs during the period of approximately 30 to 45 minutes following an exercise bout in a context where the body's metabolic, cardiovascular, respiratory, and hormonal variables rapidly return to within preexercise values. In contrast, the slow phase of recovery can take several days and is only complete when the body maintains a heightened capacity to undertake an additional exercise stimulus.

From a nutritional perspective, the most important characteristics of the rapid phase of recovery involve the cellular transitions from a catabolic (breakdown and mobilization) state to one of anabolism (build-up and storage). In muscle tissue especially, cellular processes shift toward the building of new proteins, the replenishment of energy substrates, reestablishment of optimal fluid and electrolyte levels, and tissue repair. By taking the initiative and maximizing nutritional support during the fast phase of recovery, the complete recovery process is enhanced. Indeed, consider that although a number of recovery processes continue for a day (or even several days) following an intense bout of exercise, the rate of recovery slows appreciably after the first few hours. For example, with no immediate (postexercise) nutritional support, muscle carbohydrate stores may require 24 hours or more for complete restoration, and repair of eccentric contraction-induced muscle damage may require several days. As a general rule, the more severe the exercise duration, intensity, and mode, the longer and more involved are the body's recovery processes. Undertaking a second bout of exercise prior to full recovery complicates the recovery processes. As such, individuals undertaking regular intense training benefit from careful dietary planning for optimal recovery and performance. In short, a well designed nutritional support program can ensure athletes of an advantage for both the rapid and slow phases of recovery.

3. CLASSIFICATION OF RECOVERY SUPPLEMENTS

Table 1 lists five classes of recovery supplements: oral rehydration and electrolyte balance, replenishment of energy reserves, minimizing oxidative damage, muscle repair, immune system support. The remaining sections provide an overview of supplements that may lend support in each of these strata. It should be recognized that full recovery from intense exercise is an integrative process, and all of these strata overlap in many ways.

4. REHYDRATION

Fluid loss is inherent in exercise. Indeed, with prolonged activities such as running or cycling, athletes may lose 1 to 3 liters of body water per hour. These losses may in turn produce a dehydration-associated loss of body weight and impairment of cardiovascular

Table 1
Classification of Recovery Supplements

Oral rehydration and electrolyte balance
- Traditional oral rehydration solutions are made according to World Health Organization guidelines for the prevention and treatment of dehydration during diarrhea or gastroenteritis. For these medical conditions, effective delivery of oral fluid replacement is best achieved by a solution that is low in carbohydrate (2%, or 2 g/100 ml) and provides electrolytes in concentrations of sodium 50–80 mmol/L and potassium 10–30 mmol/L (e.g. Pedialyte®).
- Newer sports drinks have become available with a higher sodium content (~30 mmol/L) to help replace sodium lost in sweat.

Replenishment of energy reserves
Sports drinks
- Usually a carbohydrate-rich solution (6%–8% carbohydrate), containing mainly sodium (10–25 mmol/L) and potassium (3–5 mmol/L). Drinks with additional electrolytes (calcium, chloride, magnesium) and micronutrients are also available.

Sports gels
- A highly concentrated carbohydrate source (65%–70%) in an easily consumed and quickly digested gel form.
- Usually has substantially higher carbohydrate concentration than sports drinks. Use with caution as idiosyncratic gastrointestinal distress may occur.
- Some gels also contain other compounds, such as medium-chain triglycerides (MCTs).

Sports bars
- Compact source of carbohydrate and protein in a bar form. Generally low in both fat and fiber, although there are exceptions. Some are fortified with micronutrients (typically containing about 25%–50% RDA per bar of various vitamins and minerals).

Liquid meal supplements
- Carbohydrate-rich, moderate protein, low-fat powder (or liquid) for mixing with water or milk.
- Generally intended to supply a balance of macronutrients and micronutrients.

Table 1
(Continued)

Minimizing oxidative damage: antioxidants
- Classically, vitamins C and E as well as other micronutrients such as zinc and selenium. Newer additions include L-carnitine, extracts of certain fruits/vegetables, and teas.

Muscle repair

Whole protein and protein hydrolysates
- Whey
- Casein
- Egg
- Soy

Amino acids
- Essential amino acids
- Branched-chain amino acids (leucine, isoleucine, valine)
- Glutamine
- Creatine

Carbohydrates
- See above

Immune system support
- Amino acids (glutamine)
- Vitamins (ascorbic acid)
- Plant extracts *(Andrographis paniculata, Ginseng)*

RDA, recommended daily allowance

and thermoregulatory homeostasis *(1)*. The loss of vascular fluid volume may then result in intracellular water being drawn out of muscle cells *(2)*, an obvious detriment to performance. Clearly, although it is often overlooked as a postexercise dietary supplement or ergogenic, it is of the utmost importance that the appropriate fluid replacement volume be considered along with the attendant need of proper electrolyte balance.

4.1. Water and Sodium

The health benefits of plain water have often been touted and are not disputed here. However, the use of plain water as a sport rehydration beverage may not be ideal. For example, copious water ingestion may cause relative dilution of the serum sodium level *(3)*, which in turn may cause a decrease in thirst sensation and an increase in urine output, stalling the rehydration process. A better option for ensuring a proper plasma volume is to use a sodium/water solution *(4)* In fact, it has been previously demonstrated that when adequate fluid is available, rehydration is achieved more rapidly when the solution sodium content (0.3–0.7 g NaCl per liter of water) is greater than sodium lost via sweat *(5,6)*. This level of rehydration sodium is recommended by both the American College of Sports Medicine *(7)* and the National Athletic Trainers' Association *(8)*. Potassium may also be added to the rehydration solution after exercising, although experimental evidence has not confirmed substantial improvement in intracellular rehydration *(9,10)*.

Because of the combination of exercise-associated water losses and the expected increase in urine production during the postexercise consumption period, it is generally recommended that consumption of 150% or more of the weight lost may be required to achieve euhydration within the short-term postexercise period (i.e., 6 hours) *(5,11,12)*. With such impressive fluid losses and the need for considerable fluid consumption, the palatability of a fluid replacement drink is of no small concern. In fact, research suggests that drinks with a slight citrus taste are especially effective at preventing the quenched sensation following ingestion of water alone *(13)*. It is no coincidence, then, that the popular sports drinks in the American marketplace (e.g., Gatorade, Powerade, Accelerade) all offer a citrus flavor. However, those popular beverages contain sodium concentrations (0.4–0.5 g NaCl per liter of water) at the low end of that recommended for rapid and complete restoration: 0.3 to 0.7 g NaCl per liter of water. When rapid rehydration is the primary goal during the postexercise period, a better choice may be found in a number of specially concentrated solutions (e.g., Endurox R4, Powerbar Recovery). Perhaps also worthy of consideration are the clinical oral rehydration fluids (used to treat diarrhea and dehydration), although a potential drawback of these more concentrated saline solutions is an unpalatable taste, which could affect the total volume consumed. It is also possible to achieve the recommended sodium intake by consuming water with food. Because food consumption slows gastric emptying, the rehydration process is similarly slowed.

In summary, complete and rapid postexercise rehydration requires a special fluid intake plan. Consumption of 150% of fluid losses allows complete fluid restoration. To enhance the process, a supply of palatable drinks should be made available after exercise. Flavored and lightly sweetened drinks are generally preferred and can contribute to achieving carbohydrate intake needs as well. The main drawback of commercial (nonclinical) rehydration beverages lies in their relatively low sodium content. As previously noted, replacement of sodium lost in sweat is important for maximizing retention of ingested fluids. A sodium content equivalent to 0.3 to 0.7 g NaCl per liter water may be necessary for optimal rehydration.

4.2. Carbohydrate-Electrolyte Solutions

With the rise in consumption of carbohydrate-containing sports drinks has come increased scrutiny of their value to the exercising athlete. In particular, there were concerns regarding the potential for slowing the gastric emptying time. However, subsequent studies have demonstrated that dilute solutions of glucose (dextrose), glucose polymers (maltodextrin), and other simple sugars all have suitable gastric emptying characteristics for the delivery of fluid and moderate amounts of carbohydrate substrate *(14)*. As previously noted, most widely available sports drinks provide a low level of sodium (0.3–0.7 g NaCl per liter of water) and thus may not be the ideal rehydration beverage (depending on the athlete's needs) despite their carbohydrate value and palatability.

4.3. Glycerol

Glycerol is a 3-carbon molecule that forms the structural backbone of triglycerides and phospholipids. It is not a carbohydrate, but it can be converted to glucose in the liver without a major change in plasma glucose or insulin levels. Because of this conversion, the direct energy yield of glycerol is similar to that of glucose, although it is not a major energy source when carbohydrates are freely available.

Glycerol has a somewhat unique physiological role in terms of its ability to influence water balance. When glycerol is added to rehydration beverages, it blunts the natural decrease in antidiuretic hormone (ADH) release that occurs with rehydration. Sustained

ADH activity results in greater retention of body fluids and, potentially, the desired enhancement of postexercise rehydration (15). This property has not gone unnoticed clinically, and glycerol is commonly utilized in intravenous rehydration protocols.

The use of glycerol in postexercise rehydration supplements has demonstrated improved short-term hydration but not reliable improvement in athletic performance (15–17). However, the dosage of glycerol is apparently important, as some studies that utilized lowered amounts (0.5 g/kg versus ~1.0 g/kg body weight) did not improve rehydration (18).

In summary, the potential for glycerol-enhanced rehydration is best attempted at a dosage of 1 g/kg body weight with adequate water ingestion as described in the previous section. It is considered a safe, effective supplement; and it has not been shown to elicit negative effects with regard to post-exercise recovery or subsequent exercise performance (19).

4.4. Fluid Balance in American Football Players

4.4.1. DAVID CHORBA

American football players face some unique challenges regarding thermoregulation during preseason training, particularly linemen with relatively high body mass compared to that of other athletes. A small body surface area/body mass ratio and above-average body fat percentage increase the metabolic load of exercise. These athletes are less able to dissipate heat, which is magnified by wearing up to 15 lb of protective gear.

Sweating rates are similar for football athletes and cross-country runners exercising in the same environmental conditions when adjusted for body size. However, longer practice durations, two-a-day preseason sessions, and insulating equipment create a higher overall sweat production in football players (> 2 L/hour, and in some athletes > 3 L/hour) (12). Daily fluid losses from sweating during two-a-day practices can average more than 9 L/day, with some athletes losing more than 14 L/day (12). This amount of fluid loss requires 125% to 150% replacement (during and after practice) to account for urine losses that occur with fluid ingestion (20). Such a large fluid intake can cause mild symptoms associated with hyponatremia (although rare in football) if too much plain water is ingested and dietary sodium intake is inadequate during preseason meals.

With such large fluid losses through sweating, the electrolyte balance must be considered in football athletes. Sodium losses from sweating range from 20 to 80 mmol/L *(21)* and have been documented up to 110 mmol/L in professional football players *(22)*. Normally, sodium and chloride are reabsorbed by sweat glands after excretion. However, this ability does not increase proportionately with the sweating rate, and the electrolyte concentration in sweat increases *(23,24)*. A day's sweat loss of 9.4 L with a sodium concentration of 50 mmol/L equates to > 10 g of sodium depletion *(12)*. Significant sodium depletion can occur with excessive sweating, especially in individuals with poor aerobic conditioning and/or suboptimal heat acclimatization. Some football athletes are prone to large, acute sodium losses during sweating (5.1 g vs. 2.2 g during a 2.5-hour practice) and despite consuming sodium-containing fluids on the field are more likely to experience associated heat cramps *(25)*. These athletes may benefit from ingesting fluids with sodium in addition to making sure they consume sufficient dietary sodium (1.5 g day) *(26)* during meals.

Despite a significantly larger body mass and heat-trapping protective gear, collegiate football athletes during summer preseason two-a-day practices do not experience the continuous core temperature increases seen in distance runners training in the same environment *(27)*. Frequent breaks (averaging 71 seconds, documented for football athletes) cause core temperature fluctuations that may provide the necessary recovery time to dissipate heat. Despite a more intermittent nature of training, however, football athletes were found to be less well hydrated than their endurance-training counterparts. Distance runners may hydrate more frequently throughout the day in preparation for long runs without access to fluids; however, the intermittent nature of football practice should leave adequate opportunity for players to rehydrate.

It is imperative that football athletes properly acclimate to the environment and physical stresses imposed by intense preseason training in order to decrease the risk of adverse health events such as heat exhaustion and heat stroke. Acclimatization requires progressive adaptations to the environment, physical training, and the protective equipment worn by the athlete. In an effort to decrease the risk of heat related injury, the National Collegiate Athletic

Association (NCAA) has implemented a heat-acclimatization protocol during preseason training *(28)*. The protocol requires that protective equipment during practice begins with helmets only and progressively increases to full pads. In this way, exercise intensity is also progressed, as more demanding activities usually require additional protective gear. In addition, exercise duration is progressed to two-a-day practices performed on alternate days, with at least 3 hours of recovery time between sessions. The new NCAA model has been shown to improve exercise-heat tolerance and heat acclimatization in Division I football players *(29)*. An added benefit of proper heat acclimatization is the improved ability to reabsorb sodium and chloride, resulting in a 50% lower sweat sodium concentration for a given sweating rate *(23)*.

Fluid losses greater than 2% of body weight have been shown to impair cognitive/mental function and athletic (aerobic) performance in temperate to hot conditions *(30–32)*, and losses greater than 3% increase the risk of heat-related illness *(32,33)*. Most individuals replace only half of the fluids lost during exercise voluntarily *(34)*. Therefore, it is important that all athletes engaging in prolonged exercise in the heat are assessed for hydration status and follow a water replacement schedule. Individual hydration protocols should be developed that consider the athlete's sweat rate, sport dynamics, environmental factors, acclimatization state, exercise duration and intensity, and individual preferences *(32,33)*. The American College of Sports Medicine and the National Athletic Trainers' Association provide guidelines to assist in designing customized fluid replacement programs *(32,33)*.

A simple, effective method to assess hydration status involves tracking body weight before and after practices. An athlete should be weighed first thing in the morning after voiding for 3 days consecutively (more days for women due to menstrual cycle changes) to establish the baseline, euhydrated (normal) state *(32)*. General recommendations to ensure appropriate hydration from the National Athletic Trainer's Association *(33)* include consuming 17 to 20 oz of water or a sports drink 2 to 3 hours before exercising and 7 to 10 ounces 20 minutes prior to exercise. Fluid replacement during exercise should attempt to match sweat and urine losses, with no more than 2% total body weight loss. This generally equates to 7 to 10 oz every 10 to 20 minutes of exercise but is highly dependent on the individual athlete, as described above. After exercise, athletes should be weighed (after urination) and then consume 450 ml (15 oz) for 125% to 150% of body weight loss in pounds, ideally within 2 hours after exercise. The addition of 0.3 to 0.7 g NaCl/L is acceptable to stimulate thirst and increase voluntary fluid intake for shorter sessions; it should also be considered during initial training days in hot weather, during exercise lasting more than 4 hours, and when inadequate access to meals is anticipated. For intense sessions or those lasting more than 45 to 60 minutes, consuming a sports drink containing 6% to 8% carbohydrate can speed replenishment of glycogen stores and delay fatigue, as well as contribute to the palatability of the beverage.

4.5. Exercise-Associated Hyponatremia

4.5.1. RITA CHORBA

At the South African Ironman Triathalon, an experienced 34-year-old male ultra-marathoner completing the race in just under 12.5 hours was brought to the medical tent near the finish line with mild confusion, sleepiness, and difficulty concentrating and maintaining conversation *(35)*. His hands and face were swollen, and he complained of feeling ill. Testing revealed an absolute weight *gain* during the race of 3.8 kg (8.4 lb), which, after consideration of fuel oxidation and fluid release during the race, approximates at least 5 kg (11 lb) excess fluid. Blood analysis revealed a serum $[Na^+]$ of 127 mmol/L, which was decreased from a before-race value of 143 mmol/L. The athlete was subsequently diagnosed with exercise-associated hyponatremia (EAH).

Despite medical management including sodium replacement, the athlete had initial difficulty producing enough urine to decrease the fluid overload and required overnight hospitalization until his serum $[Na^+]$ normalized and he had excreted 4.6 liters of urine. Only then was the athlete permitted to resume drinking fluids. It was later discovered that the athlete had ingested 750 ml fluid per hour during the cycling component (more than 6 hours in duration), 750 ml early in the run, and then "as much as possible" throughout the remainder of the run, as he thought his ill feelings were signaling dehydration. This amount of fluid ingestion was contradictory to the briefing given to all athletes before the race, where it was recommended that no more than 500 to 800 ml/hr be ingested during the race. In fact, these new recommendations resulted in only this one athlete being treated for hyponatremia in the South African Ironman (0.2% of entrants), compared to previous races exhibiting rates of 18% and 27%.[2,3] Studies of Ironman athletes have revealed an inverse relation between post-race body mass and [Na$^+$], as those who lost less weight during competition (or gained weight) had the lower plasma [Na$^+$] *(35–37)*.

Commonly known as water intoxication, EAH is a dilutional hyponatremia caused by consuming too much plain water during periods of prolonged exercise *(4)*. Although drinking excessive fluids normally stimulates urine production, this mechanism is impaired during sustained endurance exercise, increasing the potential for hyponatremia *(38,39)*. EAH can occur when excessive ADH is released, causing fluid retention and maintaining decreased serum [Na+] with subsequent cell edema. Although symptomatic EAH was previously described in terms of sodium loss, more recent studies confirm significant fluid overload as the primary independent factor, with sodium loss playing a secondary role in the pathogenesis of the condition *(35,38,40,41)*.

Sodium loss through sweating combined with dilution of extracellular sodium due to excessive ingestion of hypotonic fluids (plain water) causes serum sodium concentrations to fall, as sodium is drawn into the unabsorbed water in the intestines. Whereas moderate loss or dilution (serum sodium < 135 mEq/L) can lead to headaches, confusion, nausea and muscle cramps, concentrations below 125 to 130 mEq/L can create more life-threatening situations *(42–44)*.

Severely decreased serum sodium concentration creates a significant osmotic imbalance, causing water to rush into brain and lung cells. The subsequent swelling may cause seizures, coma, pulmonary edema, or death. A number of factors lead to hyponatremia *(38)*.

- Low body mass index
- Female sex
- Prolonged exercise (> 4 hours)
- Extreme hot or cold environment
- Slow running or performance pace
- Race inexperience, poor conditioning

- Excessive drinking behavior and availability of drinking fluids
- Weight gain during exercise
- Altered renal excretory capacity (NSAIDs, diuretics, renal disease)
- Exercise in a sodium-depleted state (more research is necessary to confirm)

Although continuous, high intensity exercise for more than 1 hour can deplete fluid and sodium levels, athletes at risk for hyponatremia most commonly include ultra-marathoners and triathletes, as well as marathoners who take more than 4 hours to complete the race (45).

Current recommendations for marathon runners include drinking ad libitum (according to thirst) no more than 400 to 600 ml/hour (13–20 oz), the lower limit for slower, lighter persons in cool weather, and the upper limit for faster, heavier athletes in warm weather (38,46). As stated previously, similar guidelines were employed in an Ironman Triathalon of significantly longer duration with a dramatically decreased incidence of hyponatremia (35).

Although it seems prudent to recommend increased sodium intake during prolonged exercise, it might not be necessary and might not be effective in decreasing the risk of EAH. Recent guidelines by the Institute of Medicine suggest that the daily adequate intake of Na^+ (1.5 g/day) is appropriate for physically active people, considering that the sodium intake in a Western diet averages 2.3 to 4.7 g/day (47). In another investigation, during the South African Ironman triathalon (48) ingestion of an additional 3.6 g of sodium (15 tablets, equal to 9 liters of Gatorade) during competition along with usual sports drinks and food was not necessary to preserve serum $[Na^+]$ despite more than 12 hours of high intensity activity, nor did supplementation affect any other performance, physiological, or psychological variables collected. [The one exception was the athlete with hyponatremia described previously (35) who was part of the placebo group for the current study (35). However, he was also the only athlete to demonstrate substantial weight gain during competition from overhydrating.] It is postulated that during acute Na^+ loss, contraction of the extracellular volume or Na^+ release from intracellular stores may act to buffer Na^+ loss until it is replenished during a subsequent meal (47). Consumption of electrolyte-containing sports drinks may be able to decrease the severity of EAH [49], although studies have not shown that their consumption prevents development of EAH in the presence of overdrinking *(50,51)*.

4.6. Electrolyte Balance

Electrolytes are minerals that become ions in solution and acquire the capacity to conduct electricity. Proper electrolyte balance in the body is essential for normal functioning of cells. Loss of body water and subsequent alterations in electrolyte balance can impair cardiovascular and thermoregulatory function. In terms of exercise performance, the most important electrolytes are sodium, chloride, and potassium. Sodium is the primary positive ion (cation) in the extracelluar fluid and helps regulate acid-base balance, nerve conduction, blood pressure, and muscle function. Chloride is the primary negative ion (anion) in the extracellular fluid and works in tandem with sodium to regulate nerve impulse conduction and body water balance. During exercise, the body loses fluids and electrolytes (mainly sodium and chloride) via sweat. The resulting decrease in blood volume increases the relative sodium and chloride concentrations in blood and triggers the thirst mechanism. Potassium is the primary cation in intracellular fluid and helps maintain electrical activity in nerves, skeletal muscles and the heart. Potassium also aids carbohydrate metabolism by enhancing glucose transport and glycogen storage. Finally, calcium and magnesium are cationic electrolytes that play important roles in the regulation of muscle contraction and enzymatic reactions. Because their losses in sweat are minimal, they are usually not included in hydration solutions.

Generally, electrolyte replacement is not needed during short bouts of exercise, as sweat is approximately 99% water and less than 1% electrolytes. Adequate water, in combination with a nutrient-dense, well balanced postexercise diet, can restore normal fluid and electrolyte levels in the body. However, replacing electrolytes may be beneficial during continuous activity of longer than 90 minutes, particularly in a hot, humid environment. Most research shows there is little evidence of physiological or physical performance differences between consuming a carbohydrate-electrolyte drink versus plain water during exercise lasting less than 1 hour *(52)*.

5. MUSCLE FUEL ENERGY

An important goal of the athlete's everyday diet is to provide the muscle with substrates to fuel the training program and achieve optimal performance, recovery, and long-term adaptation. Carbohydrate and fat constitute most of the exercising muscle's energy fuel, with protein typically playing a minor role. For example, even the leanest of athletes store enough fat ($>100,000$ kcal) to fuel exercise; but by comparison the same athlete's carbohydrate stores are limited (~ 2500 kcal). It is important to recognize that the initial concentration of stored muscle carbohydrate can dictate performance in a variety of exercises. Therefore, the restoration of muscle energy fuels, particularly carbohydrate, is a principal concern for optimal recovery and subsequent exercise performance.

5.1. Carbohydrate

Skeletal muscle stores the simple carbohydrate glucose in highly branched glycogen granules. Stored glucose is then released during muscle contraction for both anaerobic and aerobic energy production. Approximately 500 g of carbohydrate are stored in the muscle of an average-sized male with an additional 100 g stored in the liver and available for release into the blood. Muscle glycogen is an essential fuel source for sustained moderate to high intensity aerobic and anaerobic exercise metabolism *(53,54)*. Therefore, replenishment of depleted muscle glycogen levels after strenuous exercise is paramount to complete recovery. Glycogen restoration is of utmost importance for preparation for a subsequent training or competition bout of exercise. Furthermore, failure to restore muscle glycogen between exercise sessions results in compromised training and competition capacity and contributes to symptoms of overtraining *(55)*.

Classic work by Costill et al. found that individuals consuming a high carbohydrate diet could replenish their postexercise muscle glycogen to normal preexercise concentrations within 24 hours *(56)*. These investigators reported that the consumption of 600 g of carbohydrate per day resulted in proportionately greater muscle glycogen restoration during the 24-hour period after exercise. However, consumption of more than 600 g provided no additional benefit. When dietary carbohydrate was inadequate (i.e., <150 g in 24 hours) during successive days of intense exercise, there was a gradual reduction in muscle glycogen stores and deterioration in performance *(55)*. Thus, when insufficient carbohydrates are consumed during the 24-hour postexercise period, suboptimal glycogen levels result, particularly over the course of several successive days of intense training. By applying the appropriate postexercise carbohydrate restoration practices described below, athletes can avoid such a decrement in muscle carbohydrate stores.

5.1.1. GLYCOGEN STORAGE IMMEDIATELY AFTER EXERCISE

Recent research has focused on the most effective means of promoting glycogen replenishment during the early hours of exercise recovery. Postexercise synthesis of glycogen occurs in a biphasic fashion, with the initial, rapid phase occurring in the first 30 minutes and the subsequent, slower phase of resynthesis occurring until muscle glycogen levels are fully restored *(57)*. Fatigued muscles are highly sensitive to nutrient activation following exercise, with the sensitivity declining over time. Therefore, to maximize the rate of muscle glycogen repletion, an athlete must initiate carbohydrate restoration supplementation immediately after the completion of exercise.

Resynthesis of muscle glycogen is twice as rapid if carbohydrate is consumed immediately after exercise in contrast to waiting several hours *(58)*. This rapid rate of synthesis can be maintained if carbohydrate is consumed throughout the hours following exercise. For example, supplementing carbohydrate at 2-hour intervals at a rate of 1.5 g/kg body weight appears to maximize resynthesis for a period of 4 hours after exercise *(58)*. Carbohydrate provided immediately after exercise can be in the form of solid or liquid provided the glycemic index (GI) is high and the presence of other macronutrients minimized so as not to affect the rate of gastric emptying, intestinal absorption, or glycogen storage *(59,60,61)*. Practical considerations, such as the availability and appetite appeal of foods or drinks and gastrointestinal comfort may determine individual carbohydrate choices and intake patterns. Review of appropriate carbohydrate types is provided later in the chapter. Intriguingly, postexercise muscle glycogen synthesis may be enhanced with the addition of protein and certain amino acids. Furthermore, the combination of carbohydrate and protein has the added benefit of stimulating amino acid transport, protein synthesis, and muscle tissue repair. Both of these topics are discussed later.

5.2. Sustained Post-Exercise Glycogen Storage

Maehlum et al. *(58)* and Ivy et al. *(58)* have demonstrated that the postexercise rate of glycogen storage plateaus at carbohydrate ingestion rates of 1 to 2 g/kg. Consumption of more than 2 g/kg provides no additional benefit and may result in gastrointestinal distress. Investigators have determined that providing carbohydrate at 1.5 g/kg every 2 hours (in either liquid or solid form) after exercise resulted in a near-maximal rate of glycogen resynthesis over the first 4 hours of recovery *(58,60)*. Other scientists have also reported sustained, rapid resynthesis of muscle glycogen stores from similar rates of carbohydrate intake (i.e., 1.5 g/kg every 2 hours), particularly when high-GI carbohydrate foods are consumed *(62)*. Over a 24-hour period, this equates to ingesting carbohydrate at approximately 7 to 10 g/kg relative to body weight per day. Again, practical considerations, such as the availability and appetite appeal of foods or drinks and gastrointestinal comfort, may determine ideal carbohydrate choices and intake patterns.

5.3. Glycogen Supercompensation

Because of the paramount importance of muscle glycogen during intense, prolonged exercise as well as exercise of an anaerobic nature, methods for maximizing initial, preexercise levels and replenishing the glycogen stores on a day-to-day basis have been studied extensively *(53,63)*. The provision of adequate carbohydrate during the recovery process can result in "supercompensation" of muscle glycogen storage. In other words, more glycogen can be stored than prior to the previous exercise bout. Postexercise carbohydrate feeding promotes glycogen supercompensation by increasing blood concentrations of glucose and insulin. Insulin promotes muscle carbohydrate storage by increasing blood flow through muscle and by regulating the synthesis of glycogen in muscle. Specifically, insulin regulates the synthesis of glycogen in two steps: first, by increasing the transport and uptake of glucose into muscle cells; second, by increasing the intramuscular enzymes involved in glycogen synthesis and decreasing the enzymes involved in degradation *(64)*. Exercise is important in this process as muscle contraction results in an increase in insulin stimulation of glucose transport into the muscle cell *(65,66)* and indirectly enhances insulin-mediated activation of key enzymes of the glycogen synthesis pathway *(67)*.

Bergstrom and Hultman *(68)* first observed that glycogen synthesis occurred most rapidly in muscle depleted of its glycogen stores. They also found that consumption of a high-carbohydrate diet for 3 days would elevate the glycogen concentration of muscle above normal, and that this phenomenon was observed only in muscle that was previously glycogen-depleted by exercise. Later, Sherman and colleagues provided findings for a modified glycogen supercompensation regimen in which a 5-day workout taper was accompanied by a moderate (50% of caloric intake) dietary carbohydrate intake for the first three taper days followed by 3 days of higher carbohydrate intake (i.e., 70% caloric intake) *(69)*. In a follow-up study, these investigators determined that preexercise muscle glycogen levels could be maintained at high levels despite daily training if carbohydrate was consumed at a rate of 10 g/kg/day. More recently, Bussau and associates described a 1-day glycogen supercompensation regimen *(70)*.

5.3.1. DIFFERENCES BETWEEN SIMPLE CARBOHYDRATES

Fructose, sucrose, and glucose are common dietary carbohydrates. However, the physiological responses to the consumption of these "simple" sugars differ considerably. For example, the rise in blood glucose and insulin following ingestion of fructose is significantly lower than that following ingestion of glucose *(62)*. This is due to the necessary conversion of fructose to the more readily utilized glucose by the liver. Researchers have demonstrated that ingestion of glucose and high-GI carbohydrate are twice as effective for restoring muscle glycogen *(62,71)*. Blom and colleagues suggested that the differences between the glucose and fructose supplements were due to the ways the body handles these sugars *(62)*. Fructose metabolism takes place predominantly in the liver, whereas most glucose appears to bypass the liver to be stored or oxidized by muscle *(54)*. Thus, carbohydrate supplement products containing glucose (dextrose) or glucose polymers are more effective for restoring muscle glycogen after exercise than supplements composed predominantly of fructose. Many postexercise supplement drinks contain glucose or glucose polymers; and, despite its name, the common food sweetener high fructose corn syrup (HFCS) contains a significant percentage of glucose ($\sim 50\%$).

5.4. Protein

Recent research demonstrates that ideal postexercise meals/beverages should contain both carbohydrate and protein. This combination promotes an accelerated rate of muscle glycogen storage possibly by activating the glycogen synthesis pathway by two mechanisms *(72,73)*. First, it raises the plasma insulin level beyond that typical of a carbohydrate supplement, which may augment muscle glucose uptake and activate glycogen synthase, the rate-limiting enzyme for glycogen synthesis. Second, the increase in plasma amino acids that occur as a result of protein consumption may activate glycogen synthase through an insulin-independent pathway, thus having an additive effect on the activity of this enzyme. Continued supplementation at 2-hour intervals can maintain an active glycogen recovery for up to 8 hours after exercise. The combination of carbohydrate and protein provided immediately after exercise is also ideal for stimulating protein synthesis and tissue repair. This is due to the additive effect of insulin and amino acids on the enzymes controlling amino acid transport, protein translation, and protein degradation. Moreover, research has demonstrated that appropriate nutrient supplementation after exercise can have a significant impact on subsequent physical performance *(72)*.

In summary, general guidelines for postexercise carbohydrate and protein intake are the following: 1) During the first 3 hours after exercise, ingest two or three mixed meals providing approximately 400 to 600 kcal (each). 2) For strength/power athletes, a carbohydrate/protein ratio of 2:1 is recommended. 3) For team sport athletes, a 3:1 ratio is recommended. 4) For endurance athletes, a 4:1 ratio is recommended. 5) Make sure the first postexercise meal includes high-GI carbohydrates (dextrose, maltodextrin) and rapidly digesting protein (whey protein concentrates, hydrolysates). 6) Be sure to provide more energy for larger athletes and/or for higher volumes of training. For endurance athletes engaged in repeated high-volume training, the recommended daily target for carbohydrate consumption of 10 to 13 g/kg relative to body weight per day.

When the period between exercise sessions is short (< 12 hours), the athlete should begin carbohydrate-protein intake as soon as possible after the first exercise bout to maximize the effective recovery time between sessions. There may be some advantages in meeting carbohydrate-protein intake targets with a series of snacks during the early recovery phase, but during longer recovery periods (1–2 days) the athlete should plan their timing of high-GI, carbohydrate-rich feedings according to what is most practical and comfortable for their individual situation.

5.5. *Lipids*

Athletes rarely ingest dietary fat to improve their performance and recovery, but there are several specialized lipids that may have unique effects on human physiology. Medium-chain triglycerides (MCTs) are fatty acids constituents of coconut and palm kernel oils. They have an important action as energy-providing molecules, in particular for individuals with intestinal malabsorption syndromes *(74)*. Indeed, from the medial perspective, MCTs have been shown to be helpful for nutritional support in a number of conditions, including the specialized nutritional need of some infants, the chronically ill, and surgical patients *(75–77)*. For almost a decade, MCT's have been promoted as useful for weight loss and improving athletic performance. However, to date, most studies have shown that MCTs are ineffective in this regard. For example, researchers have investigated the role of MCTs in the enhancement of long-chain fatty acid oxidation, preservation of muscle glycogen, and the ability to improve overall exercise performance *(78–80)*. In general, results were unimpressive. More importantly, some studies even reported that exercise performance was impaired in subjects taking high doses of MCTs. This reduction in performance was associated with an increase in gastrointestinal problems, such as intestinal cramping. For these reasons, use of MCTs during the postexercise period cannot be justified at this time.

5.5.1. Conjugated Linoleic Acid

Conjugated linoleic acid (CLA) is a group of isomers of the essential fat linoleic acid. CLA is naturally found in a number of animal meats, eggs, and dairy products. CLA has drawn interest in the sports nutrition community because of a reported ability to increase lean body mass and reduce body fat *(81–83)* and more recently improve immune function *(84)*. Specifically, Song and associates reported that 12 weeks of supplementation (at 3 g/day) with a 50:50 blend of the *cis*-9,*trans*-11 and *trans*-10,*cis*-12 isomers significantly decreased levels of the proinflammatory cytokines tumor necrosis factor-α

(TNFα) and interleukin-1β (IL-1β) while simultaneously increasing levels of the antiinflammatory cytokine IL-10 *(84)*. Although this study needs to be confirmed by other research groups, it indicates that CLA may be useful for enhancing recovery by limiting exercise-induced inflammation and soreness. It should be noted that although the *trans*-10, *cis*-12 isomer appears to be responsible for the fat-lowering effects of CLA *(85,86)*, some studies have reported that supplementation with this purified isomer in high doses (i.e., 2.6 g/day) may increase insulin resistance, lower high density lipoprotein (HDL) cholesterol, and elevate biomarkers of oxidative stress in obese men with metabolic syndrome *(87)*. In contrast, the aforementioned 50:50 blend of CLA does not appear to have these adverse effects, even after prolonged use (i.e., up to 2 years).

5.6. Other Muscle Substrates

5.6.1. CREATINE

First discovered in 1832, creatine is an amino acid-like compound produced in the liver, kidneys, and pancreas from arginine, glycine, and methionine. Dietary sources of creatine include meat, fish, and poultry. Adults and teenagers who regularly consume these foods typically eat 1 to 2 g of creatine per day, an amount roughly equal to the body's rate of creatine degradation. In a normal, healthy adult man, 95% of creatine stores are found in skeletal muscle, with the remaining 5% in the heart, brain, eyes, kidneys, and testes. In general, most (65%–70%) of clinical trials on creatine have found beneficial effects, particularly during short, repeated bursts of high intensity activity. Although responses are quite variable from person to person, subjects ingesting creatine average a 2- to 5-lb greater gain in muscle mass, and 5% to 15% greater increases in muscle strength and power compared to control (or placebo) subjects *(88–90)*. Vegetarians who consume no animal products are likely to experience even greater benefits from creatine use.

A useful analogy is that creatine is to the strength/power and repeated sprint athlete what carbohydrate is to the endurance athlete. In other words, the physiological functions of creatine enhance energy production during intense, repeated muscular activity. More specifically, following its transport and uptake into muscle, creatine is thought to serve at least four vital functions: 1) It serves as an energy capacitor, storing energy that can be used to regenerate adenosine triphosphate (ATP). 2) It enhances the capacity for energy transfer between the mitochondria and muscle fibers. 3) It serves as a buffer against intracellular acidosis during exercise. 4) It activates glycogenolysis (glycogen breakdown) during exercise *(88–90)*. Collectively, these effects allow athletes to perform at higher levels during training, ultimately leading to greater increases in muscle mass, strength, power and high intensity muscle endurance. Recent studies have also focused on the potential benefits of creatine use on certain neurological and cardiovascular diseases *(91,92)*.

Initial recommendations for creatine use stemmed from early research using 5 to 7 days of "loading" with 20 to 30 g per day (divided into four to six equal 5 g doses). Based on new research, refinements have been made to this strategy; and now many athletes consume only one 5 g dose approximately 60 minutes prior to or immediately after training (exercise is known to enhance creatine uptake by about 10%). Because it is also known that the uptake and storage of creatine can be augmented by up to 60% when blood levels of insulin are elevated, many athletes co-ingest creatine with high-GI carbohydrates (usually dextrose, maltodextrin, or fruit juice).

Although still somewhat speculative, creatine cycling (i.e., 4 weeks of use followed by a 4 week break) is supported by studies showing that once muscle stores of creatine are full they can remain elevated (with performance enhanced) for an additional 4 to 5 weeks without supplementing at all. In this regard, athletes who continue to use creatine after their muscle stores are "topped off" may be wasting their money, as creatine that is unabsorbed by the body is simply excreted in the urine. (*Note*: Urinary creatinine levels are commonly used as a marker of kidney function. Individuals who ingest creatine frequently have elevated creatinine levels; this is normal and represents an increased rate of muscle creatine degradation to creatinine, rather than an abnormality of kidney function.)

5.6.2. D-RIBOSE

D-Ribose is a ubiquitously occurring five-carbon sugar found in all living cells. It is sometimes referred to as a "genetic sugar" because of its importance in the nucleosides of RNA as well as ATP.

All foods have some D-ribose, but certain substances, such as brewers yeast, are rich in RNA and thus are thought to be good sources of D-ribose. Clinically, research has been presented that suggests consuming large amounts of D-ribose may offer some degree of cardioprotection for the ischemic heart *(93)*.

The use of supplemental ribose as an energy booster and athletic performance enhancer appears to be mostly without substantive foundation. In studies dealing with exercising patients who have metabolic enzyme deficiency disease (noted to have quite limited exercise tolerance), subtle improvements in muscle cramping and stiffness have been noted *(94–96)*. However, overall improvements in exercise tolerance have not been documented, even at relatively high intake levels (60 g of ribose daily). One study did note improvements in sprint cycle performance, although the effects were small, variable, and inconsistent *(97)*. Thus at present, use of D-ribose cannot be adequately justified.

6. MINIMIZING OXIDATIVE DAMAGE

Over the past decade, a great deal of research has focused on the link between free radical formation and muscle damage. Free radicals are continuously formed as a normal consequence of body processes, including exercise. A free radical is an atom with one unpaired electron in its outermost shell, making it unstable and highly reactive. A free radical sequesters an electron from any cellular component it can but typically from the lipid membrane of a cell, initiating a process called *lipid peroxidation*. The most common radical atom is oxygen, and the process whereby this element gains another electron is termed *oxidation*. The resultant cellular damage caused by these processes is known as *oxidative stress*, and the free radicals formed via oxygen are called reactive oxygen species (ROS). The most common ROS include the superoxide anion (O_2), the hydroxyl radical (OH·), singlet oxygen ($^1O_2$), and hydrogen peroxide (H_2O_2).

Free radicals damage cell membranes and increase protein breakdown; and they are partially to blame for the local inflammation and soreness associated with the postexercise period [98]. Aerobic metabolism increases oxidative stress due to the increased amount of oxygen processed throughout the exercising muscle cell

(99–101). The generation of free radicals is related to the rate of oxygen consumption in working muscles and therefore increases as the intensity of exercise increases *(102,103)*. The body maintains several natural antioxidants whose levels increase with endurance training as an adaptation to the chronic exposure to aerobic metabolism-derived oxidative stress *(104)*. In addition to the body's natural antioxidants, several dietary sources of antioxidants consisting of vitamin and vitamin-like substances are available to aid recovery from exercise.

6.1. Oxidation and Reduction Reactions
6.1.1. JASON THOMAS

Oxidation-reduction or "redox" reactions are usually defined as a transfer of electrons from one chemical species to another. During an oxidation-reduction reaction, one reactant gains one or more electrons from another reactant, which in turn is losing one or more electrons. The process of gaining electrons is called *reduction*, and the process of losing electrons is called *oxidation*. The oxidizing agent, or oxidant, is the substance that causes other chemical species to be oxidized while it is reduced. The reducing agent, or reductant, is the substance that causes other chemical species to be reduced while it is oxidized. Redox reactions are ubiquitous. We can find them in our cell phone batteries, on our cars as rust, and in our bodies during combustion reactions. A common mnemonic for redox reactions is OIL RIG (*oxidation is loss, reduction is gain*).

In biological systems, it is important to recognize that free radicals are powerful oxidizing agents. They are highly reactive chemical species that have an odd number of electrons. These chemicals oxidize other chemicals in an attempt to pair their odd electrons. In these free radical reactions, the free radicals must gain electrons and therefore be reduced. In the body, the antioxidant enzymes superoxide dismutase (SOD), catalase (CAT), and glutathione peroxidase (GPx) are thought to be our most important lines of defense for destroying free radicals.

SOD first reduces (adds an electron to) the radical superoxide (O_2^-) to form hydrogen peroxide (H_2O_2) and oxygen (O_2).

$$2O_2^- + 2\,H \rightarrow SOD \rightarrow H_2O_2 + O_2$$

Catalase and GPx then work simultaneously with the protein glutathione to reduce hydrogen peroxide and ultimately produce water (H_2O).

$$2 H_2O_2^- - CAT \rightarrow H_2O + O_2$$

$$H_2O_2 + 2 \text{ glutathione} \rightarrow GPx \rightarrow \text{Oxidized glutathione} + 2 H_2O$$

Oxidized glutathione is then reduced by another antioxidant enzyme called glutathione reductase.

6.2. Vitamins C and E

The water-soluble vitamin C (also known as ascorbic acid) and the fat-soluble vitamin E are well known antioxidants that minimize cellular damage from oxidative stress. These vitamins are consumed during exercise but not synthesized in the body, so repetitive exercise training can reduce tissue levels of these vital compounds. Vitamin C and E may work in a synergistic fashion providing a greater protective benefit when combined by inhibiting the release of the inflammation-promoting molecule IL-6 from working muscle (105). Indeed, postexercise supplements containing both antioxidant vitamins have proven effective (106,107). However, vitamin E seems to be the more relevant of the two with respect to postexercise applications. Vitamin E protects the cell membrane phospholipids from becoming oxidized from free-radical molecules (108). This vitamin is especially beneficial for preventing muscle soreness in individuals who undergo a vigorous exercise bout to which they are unaccustomed. In addition to protecting membrane phospholipids, vitamin E has been shown to increase immune cell migration to damaged muscle cells and decrease the number of free radicals produced during a given bout of exercise (109).

In contrast to the positive postexercise effects of vitamin E, the findings for its counterpart vitamin C are less compelling. Thompson and colleagues investigated the effects of 200 mg vitamin C supplementation following an intense shuttle-running test (110). These investigators did not observe improved indices of muscle damage, muscle function, or muscle soreness compared to a placebo group. Similar results were reported in a study by Close and colleagues (111). In this study, supplementation with ascorbic acid after downhill running exercise decreased indices of oxidative muscle damage compared to placebo but did not attenuate delayed-onset muscle soreness. In fact, at least two studies have reported that postexercise vitamin C supplementation may delay the recovery process, especially following prolonged eccentric exercise *(110,112)*. Thus, although preexercise vitamin C supplementation has been clearly demonstrated to play a protective role in muscle *(113–115)*, its application to postexercise recovery is less clear.

In terms of supplementation, there are no known adverse effects of increased consumption of either of these antioxidant vitamins at the dosages mentioned earlier *(116)*. However, a daily intake of more than 400 IU vitamin E, particularly in the *d*α form, is ill-advised *(117)*. It is important to note that the term "vitamin E" actually describes a family of eight antioxidants, four tocopherols and four tocotrienols, and the effects of the latter two families of vitamin E are largely unstudied at this time. Although it could be argued that high dosages of certain antioxidants are necessary to observe benefits on exercise recovery, it is important to recognize that at low concentrations ROS have important functional roles in the body (i.e., defense against infectious agents and various cell signaling pathways). Thus, when it comes to antioxidant supplementation, "more is not necessarily better."

6.3. L-Carnitine

L-Carnitine is an amino acid derivative found mainly in skeletal muscles and the heart (95%), with the remaining 5% in the liver, kidneys, and extracellular fluid (ref). Initial interest in L-carnitine stemmed from the observation that it is a substrate for carnitine palmitoyltransferase I (CPT-I), the rate-limiting step in fatty acid oxidation. As such, L-carnitine plays an essential role in the transport and oxidation of fat in heart and skeletal muscle. Although L-carnitine supplementation does not appear to improve body composition, recent studies strongly suggest that doses of 1 to 2 g per day decrease the production of free radicals, reduce tissue damage, and enhance recovery from intense resistance exercise *(118–120)*. L-Carnitine may have additional health benefits and appears to be safe during long-term use *(121–123)*.

7. IMMUNE FUNCTION

7.1. Glutamine

Glutamine is the most abundant amino acid in blood plasma and skeletal muscle *(124)*. Although produced in the body from branched-chain amino acids (BCAAs), under certain circumstances (e.g., injury, surgery, overtraining) the body may not be able to synthesize sufficient glutamine to match demand. As such, glutamine is considered as a "conditionally essential" amino acid *(125)*. Glutamine plays a minor role in most cells' energy metabolism and may be important for muscle hypertrophy (see below), but this carbon source seems to be a vital fuel for some cells of the immune system *(126)*. Typically, glutamine (9 g/day) is released from muscle and is available for lymphocyte and macrophage fuel use *(127)*. However, after prolonged (> 2 hours) exhaustive exercise, there is a significant decrease in the circulating plasma glutamine concentration *(128,129)*. Exercise-associated suppression of circulating glutamine may remain for many hours upon cessation of exercise *(130)*.

Although the significance of postexercise glutamine alterations in the intracellular environment of previously exercised muscle has not been determined, the effects on the cells of the immune system are another matter. Because it is now widely accepted that glutamine is utilized at high rates by isolated cells of the immune system such as lymphocytes, macrophages, and neutrophils *(131,132)*, and considerable demand can be placed on these cells as a result of the elevated stress of exercise, there is concern regarding the possibility of an exercise-associated immunosuppression. For example, it is known that strenuous exercise can decrease plasma glutamine concentrations, which in turn may result in an increased risk of developing infection *(128,132)*. Furthermore, it has previously been documented that the depression of glutamine results in a concomitant drop in lymphocyte activated killer cell activity (LAK) *(133)*. Although more data regarding this potential phenomenon are needed, it seems possible that even small changes in immune function following daily exercise may be amplified with chronic high intensity training. As such, an attempt regarding the dietary remediation of glutamine depletion between exercise sessions seems appropriate.

In recent years several studies have investigated the value and safety of glutamine supplementation. For example, Candow and colleagues indicated that glutamine is well tolerated and not associated with toxicity at dosages up to 0.9 g/kg of lean body mass per day *(134)*. However, with respect to postexercise supplementation, investigators have demonstrated decreased rates of postexercise infection following supplementation of 5 to 12 g of glutamine within the first 2 hours after exercise (in conjunction with adequate rehydration) *(128,134)*. The relations noted in these studies suggest that there is a sound theoretical rationale for the consumption of glutamine immediately after exercise. However, access to glutamine may not be absolutely necessary, as other authors have demonstrated that the postexercise glutamine and immune function decrements may also be improved indirectly by supplementation with BCAAs *(135)*.

8. REPAIR OF MUSCLE DAMAGE

Appropriate resistance exercise leads to significant increases in skeletal muscle mass (hypertrophy). Muscle hypertrophy occurs through an increase in muscle protein synthesis, a decrease in muscle protein degradation (or both, resulting a net protein accumulation), an increase in fluid volume, and through increases in mitochondrial mass. From a dietary perspective, the changes in muscle protein metabolism are the aspects of growth most readily influenced. Additionally, the body's resistance training-induced hormonal milieu also has a major impact on regulation of muscle protein synthesis (Table 2). Postexercise nutrient availability plays a critical factor in regulating the degree of hypertrophy by promoting and supporting net protein accumulation in previously exercised muscle as well as driving the anabolic hormonal environment to support muscle recovery and growth *(136)*.

With respect to postexercise supplement considerations, it is important to identify the key ingredients in muscle hypertrophy. Essentially, blood and muscle tissue amino acids are used by the previously exercised muscle for synthesis of proteins that contribute greatly to the strength and size increases observed in resistance-trained muscle. This elaborate synthetic mechanism is supported by energy from carbohydrates, which have the additional role of glycogen resynthesis following exercise. Increased blood levels of carbohydrate and protein can stimulate insulin release from the pancreas. Insulin may be key in the process of muscular growth, as this highly anabolic hormone increases skeletal muscle blood flow, promotes glycogen and protein synthesis in previously exercised muscle, and may influence the effectiveness of other anabolic hormones such as testosterone and growth hormone during the recovery process (Table 2).

Table 2
Major anabolic hormones and putative and established secretagogues

Insulin
    Carbohydrates[a]
    Proteins[a]
    4-Hydroxyisoleucine[a]
    Chromium
    α-Lipoic acid
Growth hormone/IGF-1
    α-Glycerylphosphorylcholine[a]
    L-DOPA[a]
    L-Arginine[a]
    L-Glutamine
    L-Ornithine
    Glycine
    Ornithine α-ketoglutarate
    Macuna pruriens
Testosterone
    Aromatase inhibitors[a]
    Eurycoma longifolia
    Boron
    Fenugreek
    Avena sativa
    Tribulus terrestris

IGF-1, insulin-like growth factor-1
[a]Established secretagogues

8.1. Protein and Amino Acid Intake

Dietary protein provides the amino acids necessary to rebuild muscle tissue that is damaged during intense, prolonged exercise. The amino acids in dietary protein are subjected to digestion (hydrolysis) in the stomach and intestine, are subsequently absorbed into the blood, and finally can circulate to the muscle for synthesis of muscle protein. Purified amino acids and larger molecules of protein hydrolysates (i.e., predigested proteins) are both effective in increasing the circulating levels of amino acids available to the muscle. Extensively hydrolyzed proteins containing mostly dipeptides and tripeptides (chains of two and three amino acids, respectively) are absorbed more rapidly than isolated amino acids and much more rapidly than intact (nonhydrolyzed) proteins *(137)*. Currently, protein hydrolysate is a preferred form of supplement because it results in a faster increase in blood amino acid concentration (when compared to intact protein) *(138)*, and it strongly stimulates insulin secretion *(139)*.

Several studies have demonstrated the beneficial effects of postexercise protein supplementation *(140)*. The timing of protein supplementation may also be important. For example, when compared to protein consumption several hours after exercise, protein consumption (10 g) immediately following exercise has been shown to enhance the accumulation of muscle protein *(141,142)*. However, because of the enhanced insulin response to supplements containing protein and carbohydrate (also noted above), supplementing with 6 g of essential amino acids and 35 g of carbohydrate after exercise has demonstrated a markedly higher anabolic response when compared to those supplemented on protein alone *(143)*. This finding is supported by other studies as well *(144,145)*.

8.2. Leucine and Muscle Protein Synthesis

The BCAA leucine appears to be the most important amino acid signal for the stimulation of muscle protein anabolism. Leucine affects muscle protein metabolism by decreasing the rate of protein degradation, most likely via increases in circulating insulin. However, leucine also activates key molecules involved in the regulation of protein synthesis even in the absence of an increase in circulating insulin concentration *(146)*. A study by Koopman and colleagues emphasized the potential importance of leucine to muscle growth. The investigation concerned postexercise muscle protein anabolism and whole-body protein balance following the combined ingestion of high-GI carbohdrates, with or without whey protein hydrolysate and/ or leucine *(144)*. The results revealed that the whole-body protein synthesis rates were highest in the carbohydrates + protein hydrolysate + leucine trial. Similarly, muscle anabolism was significantly greater in the carbohydrates + protein hydrolysate + leucine trial compared with the carbohydrates-only trial and the values observed in the carbohydrates + protein hydrolysate trial. The authors concluded that additional ingestion of free leucine in combination with protein and carbohydrate likely represents an effective strategy for increasing muscle anabolism following resistance exercise.

8.3. Carbohydrate and Protein Intake

As was already briefly noted, consumption of protein/amino acids and carbohydrate immediately before and after individual training sessions may augment protein synthesis and muscle glycogen resynthesis and reduce protein degradation. The optimal rate for carbohydrate (choose one with a high GI) ingestion immediately after a training session is 1.2 g/kg/hr at 30-minute intervals for 4 hours. However, the effects of repeated supplementation on long-term adaptations to training are currently unclear.

To shed additional light on this issue, Bird and coworkers examined the effects of postexercise, high-GI carbohydrate and/or essential amino acid supplementation on hormonal and muscular adaptations in untrained young men *(147)*. Subjects followed the same supervised, resistance-training protocol twice a week for 12 weeks. Following resistance exercise, the subjects consumed a high-GI carbohydrate (CHO), an essential amino acids (6 g), a combined high-GI CHO + essential amino acid (EAA) supplement, or a placebo containing only aspartame and citrus flavoring. The results revealed that CHO + EAA supplementation enhances muscular and hormonal adaptations to a greater extent than either CHOs or EAAs consumed independently. Specifically, CHO + EAA ingestion demonstrated the greatest relative increase in type I muscle fiber cross-sectional area. Changes in type II muscle fibers exhibited a similar trend.

Although beyond the scope of this chapter, it is likely that chronic reductions in the exercise-induced cortisol response associated with postexercise carbohydrate-amino acid ingestion positively affect skeletal muscle hypertrophic adaptation to resistance training via reductions in hormone-mediated protein degradation.

8.4. Glutamine

As previously noted, glutamine may have some important postexercise immune supporting effects. However, there has been much speculation as to how glutamine could be of benefit to athletes wanting to gain lean muscle mass. The observation that the cells of the gastrointestinal tract require a large supply of glutamine has led to the supposition that glutamine may offer anticatabolic effects and thereby be useful in sparing muscle protein *(148)*. This position has been bolstered by evidence suggesting that glutamine is a potent agent for enhancing cellular swelling *(149)*. Such an association could have implications for muscular growth because recent evidence suggests that the state of cellular hydration (i.e., celluar swelling) is an important factor in the control of many important cell functions, including modulation of hormones, oxidative stress, and gene expression. Cell swelling also inhibits protein breakdown (i.e., anticatabolism) *(149)*.

The mechanisms proposed for improved protein turnover as mediated via glutamine-induced cell swelling are twofold. First, it may influence the function of cyclic AMP, a chemical messenger associated with many cell functions including inhibition of protein synthesis. Second, it may have a direct effect on cell stability *(149)*.

Not all research regarding the potential glutamine-associated effects on muscle building have been encouraging. For example, a recent study investigated the effect of glutamine on strength and lean tissue mass during a 6-week resistance-training program *(134)*. The study reported that glutamine did not enhance adaptations to the strength-training program. However, the scientists suggested that the lack of any beneficial effect of glutamine in the study might have been attributed to the fact that the resistance-training program may not have been stressful enough. This conclusion tends to be in line with the prevalent, and previously noted, current scientific opinions that indicate glutamine is most likely of use under conditions of stress *(124)*.

9. CONCLUSION

Recovery supplements can be classified into four overlapping categories based on their mechanism(s) of action: oral rehydration/electrolyte balance, replenishment of energy reserves, minimizing oxidative damage, and stimulation of muscle repair. Because of ongoing clinical trials and continuous advances in new ingredients, this chapter is in no way intended to be an exhaustive review of supplements that can improve recovery from exercise. In addition, as with any dietary supplement, the decision to use a specific supplement regimen is ideally made with input from a knowledgeable physician, exercise scientist, and/or dietitian (with expertise in sports nutrition) to ascertain potential contraindications.

Based on the current body of research data, it is our contention that the following supplements have the potential to safely enhance recovery from intense exercise: multivitamin-mineral blends (when dietary intake of micronutrients is suboptimal), antioxidant blends (especially in aging athletes and during multiple training sessions within the same day), sports drinks (especially during prolonged exercise in high heat/humidity and/or altitude), certain amino acids (creatine, EAAs, BCAAs, L-carnitine), and perhaps special lipids (CLA, omega-3 fatty acids). Readers are encouraged to study other chapters in this text that cover the potential applications of HMB and β-alanine for their potential benefits during training as well.

10. PRACTICAL APPLICATIONS

- Proper recovery from intense exercise begins during the preexercise period. The athlete must pay attention to macronutrient and micronutrient needs and never train in a dehydrated or glycogen-depleted state.
- With the possible exception of long-distance athletes, most athletes can benefit from daily supplementation with 3 to 5 g of creatine monohydrate.
- Sports drinks containing carbohydrates and protein (or ideally essential amino acids) are recommended immediately after training to attenuate muscle breakdown and stimulate glycogen regeneration and muscle protein synthesis. In a pinch, low-fat chocolate milk is a good substitute.
- Strong preliminary evidence suggests that supplementation with L-carnitine (1–2 g/day) may also enhance recovery from acute bouts of intense resistance exercise.
- Although speculative at this time, CLA and omega-3 fatty acids have strong theoretical bases for enhancing recovery during prolonged training.

REFERENCES

1. Montain SJ, EF Coyle. Influence of graded dehydration on hyperthermia and cardiovascular drift during exercise. J Appl Physiol 1992;73:1340–1350.
2. Nose H, T Morimoto, K Ogura. Distribution of water losses among fluid compartments of tissues under thermal dehydration in the rat. Jpn J Physiol 1983;33:1019–1029.

3. Costill DL, Sparks KE. Rapid fluid replacement after thermal dehydration. J Appl Physiol 1973;34:299–303.

4. Nose H, Mack GW, Shi X, Nader ER. Role of osmolality and plasma volume during rehydration in humans. J Appl Physiol 1988;65:325–331.

5. Shirreffs SM, Taylor AJ, Leiper JB, Maughan RJ. Post-exercise rehydration in man: effects of volume consumed and sodium content. Med Sci Sports Exerc 1996;28:1260–1271.

6. Meyer F, Bar-Or O. Fluid and electrolyte loss during exercise. Sports Med 1994;18:4–9.

7. Convertino, VA, LE Armstrong, Coyle EF, et al. American College of Sports Medicine position stand: exercise and fluid replacement. Med Sci Sports Exerc 1996;28: i–vii.

8. Casa, DJ, Armstrong LE, Hillman SK, et al. National Athletic Trainers' Association position statement: fluid replacement for athletes. J Athl Train 2000;35:212–224.

9. Maughan RJ, Owen JH, Shirrefs SM, Leiper JB. Post-exercise rehydration in man: effects of electrolyte addition to ingested fluids. Eur J Appl Physiol 1994;69:209–215.

10. Yawata T. Effect of potassium solution on rehydration in rats: comparison with sodium solution and water. Jpn J Physiol 1990;40:369–381.

11. Maughan RJ. Restoration of water and electrolyte balance after exercise. Int J Sports Med 1998;(Suppl 2):s136–s138.

12. Shirreffs SM, Maughan RJ. Rehydration and recovery of fluid balance after exercise. Exerc Sport Sci Rev 2000;28:27.

13. Maughan RJ, JB Leiper. Post-exercise rehydration in man: effects of voluntary intake of four different beverages. Med Sci Sports Exerc 1993;25(Suppl):S2.

14. Vist GE, RJ Maughan. Gastric emptying of ingested solutions in man: effect of beverage glucose concentration. Med Sci Sports Exerc 1994;26:1269–1273.

15. Scheett TP, Webster MJ, Wagoner KD. Effectiveness of glycerol as a rehydrating agent. Int J Sport Nutr Exerc Metab 2001;11:63–71.

16. Montner P, Stark DM, Riedesel ML, et al. Pre-exercise glycerol hydration improves cycling endurance time. Int J Sports Med 1996;17:27–33.

17. Magal M, Webster MJ, Sistrunk LE, Whitehead MT, Evans RK, Boyd JC. Comparison of glycerol and water hydration regimens on tennis-related performance. Med Sci Sports Exerc 2003;35:150–156.

18. Kavouras SA, Armstrong LE, Maresh CM, et al. Rehydration with glycerol: endocrine, cardiovascular, and thermoregulatory responses during exercise in the heat. J Appl Physiol 2006;100:442–450.

19. Wagner DR. Hyperhydrating with glycerol: implications for athletic performance. J Am Diet Assoc 1999;99:207–212.

20. Godek SF, Bartolozzi AR, Godek JJ. Sweat rate and fluid turnover in American football players compared with runners in a hot and humid environment. Br J Sports Med 2005;39:205–211.

21. Shirreffs SM, Maughan RJ. Whole body sweat collection in humans: an improved method with preliminary data on electrolyte content. J Appl Physiol 1997;82:336–341.

22. Stofan JR, Zachwiega JJ, Horswill CA, et al. Sweat and sodium losses during practice in professional football players: field studies. Med Sci Sports Exerc 2002;34:S113.

23. Allan JR, Wilson CG. Influence of heat acclimatization on sweat sodium concentration. J Appl Physiol 1971;30:708–712.

24. Costill DL, Cote R, Miller E, Miller T, Wynder S. Water and electrolyte replacement during repeated days of work in the heat. Aviat Space Environ Med 1975;46:795–800.

25. Stofan JR, Zachwieja JJ, Horswill CA, Murray R, Anderson SA, Eichner ER. Sweat and sodium losses in NCAA football players: a precursor to heat cramps? Int J Sport Nutr Exerc Metab 2005;15:641–652.

26. Institute of Medicine of the National Academies. Dietary Reference Intakes for Water, Potassium, Sodium, Chloride, and Sulfate. National Academies Press, Washington, DC, 2004.

27. Godek SF, Godek JJ, Bartolozzi A. Thermal responses in football and cross-country athletes during their respective practices in a hot environment. J Athl Train 2004;39:235–240.

28. NCAA Membership Service Staff. 2005-2006 NCAA Division I Manual. The National Collegiate Athletic Association, Indianapolis, 2005.

29. Yeargin SW, Casa DJ, Armstrong LE, et al. Heat acclimatization and hydration status of American football players during initial summer workouts. J Strength Cond Res 2006;20:463–470.

30. Casa DJ, Clarkson PM, Roberts WO. American College of Sports Medicine roundtable on hydration and physical activity: consensus statements. Curr Sports Med Rep 2005;4:115–127.

31. Cheuvront SN, Carter III R, Montain SJ, Sawka MN. Fluid balance and endurance exercise performance. Curr Sports Med Rep 2003;2:202–208.

32. Sawka MN, Burke LM, Eichner ER, Maughan RJ, Montain SJ, Stachenfeld NS. American College of Sports Medicine position stand: exercise and fluid replacement. Med Sci Sports Exerc 2007;39:377–390.

33. Casa DJ, Armstrong LE, Hillman SK, et al. National Athletic Trainers' Association position statement: fluid replacment for athletes. J Ath Train 2000;35:212–224.

34. Noakes D. Fluid replacement during exercise. Exerc Sports Sci Rev 1993;21:297.

35. Noakes TD, Sharwood K, Collins M, Perkins DR. The dipsomania of great distance: water intoxication in an Ironman triathlete. Br J Sports Med 2004;38:E16.

36. Speedy DB, Noakes TD, Rogers IR, et al. Hyponatremia in ultradistance triathletes. Med Sci Sports Exerc 1999;31:809–815.

37. O'Toole M, Douglas PM, Laird RH, et al. Fluid and electrolyte status in athletes receiving medical care at an ultradistance triathalon. Clin J Sport Med 1995;5:116–122.

38. Hew-Butler T, Almond C, Ayus JC, et al. Consensus Statement of the 1st International Exercise-Associated Hyponatremia Consensus Development Conference, Cape Town, South Africa 2005. Clin J Sport Med 2005;15:208–213.

39. Zambraski EJ. The renal system. In: Tipton CM,f Tate CA, Terjung RL (eds) ACSM's Advanced Exercise Physiology (pp 521–532). Lippincott Williams & Wilkins, Philadelphia, 2005.

40. Garigan TP, Ristedt DE. Death from hyponatremia as a result of acute water intoxication in an Army basic trainee. Mil Med 1999;164:234–238.

41. Speedy DB, Rogers IR, Noakes TD. Exercise-induced hyponatremia in ultra-distance triathletes is caused by inappropriate fluid retention. Clin J Sport Med 2000;10:272–278.

42. McArdle WD, Katch FI, Katch VL. Exercise Physiology: Energy, Nutrition, and Human Performance. 6th ed. Lippincott Williams & Wilkins, Philadelphia, 2007.

43. Androgue HJ, Madias NE, Hyponatremia. N Engl J Med 2000;342:1581.

44. Gardner JW. Death by water intoxication. Mil Med 2002;5:432.

45. Almond CS, Shin AY, Fortescue EB, et al. Hyponatremia among runners in the Boston marathon. N Engl J Med 2005;352:1550–1556.

46. Noakes T. Fluid replacement during marathon running. Clin J Sport Med 2003;13:309–318.

47. Institute of Medicine of the National Academies. Dietary Reference Intakes for Water, Potassium, Sodium, Chloride, and Sulfate. National Academies Press, Washington, DC, 2004.

48. Hew-Butler TD, Sharwood K, Collins M, Speedy D, Noakes T. Sodium supplementation is not required to maintain serum sodium concentrations during an Ironman triathalon. Br J Sports Med 2006;40:255–259.

49. Twerenbold R, Knechtle B, Kakebeeke TH, et al. Effects of different sodium concentrations in replacement fluids during prolonged exercise in women. Br J Sports Med 2003;37:300–303.

50. Weschler LB. Exercise-associated hyponatremia: a mathematical review. Sports Med 2005;35:899–922.

51. Barr SI, Costill DL, Fink WJ. Fluid replacement during prolonged exercise: effects of water, saline, or no fluid. Med Sci Sports Exerc 1991;23:811–817.

52. Sawka MN, Burke LM, Eichner ER, Maughan RJ, Montain SJ, Stachenfeld NS. American College of Sports Medicine position stand: exercise and fluid replacement. Med Sci Sports Exerc 2007;39:377–390.

53. Ahlborg B, Bergstrom J, Ekelund LG, Hultman E. Muscle glycogen and muscle electrolytes during prolonged physical exercise. Acta Physiol Scand 1967;70:129–142.

54. Bergstrom J, Hultman E. Synthesis of muscle glycogen in man after glucose and fructose infusion Acta Med Scand 1967;182:93–107.

55. Costill DL, Bowers R, Branam G, Sparks K. Muscle glycogen utilization during prolonged exercise on successive days. J Appl Physiol 1971;31:834–838.

56. Costill DL, Sherman WM, Fink WJ, Maresh C, Witten M, Miller JM. The role of dietary carbohydrate in muscle glycogen resynthesis after strenuous running. Am J Clin Nutr 1981;34:183–186.

57. Maehlum S, Hostmark AT, Hermansen L. Synthesis of muscle glycogen during recovery after prolonged severe exercise in diabetic and non-diabetic subjects. Scand J Clin Lab Invest 1977;37:309–316.

58. Ivy JL, Katz AL, Cutler CL, Sherman WM, Coyle EF. Muscle glycogen synthesis after exercise: effect of time of carbohydrate ingestion. J Appl Physiol 1988;64:1480–1485.

59. Keizer HA, Kuipers H, van Kranenburg G, Geurten P. Influence of liquid and solid meals on muscle glycogen resynthesis, plasma fuel hormone response, and maximal physical working capacity. Int J Sports Med 1987;8:99–104.

60. Reed MJ, Brozinick JT Jr, Lee MC, Ivy JL. Muscle glycogen storage post-exercise: effect of mode of carbohydrate administration. J Appl Physiol 1989;66:720–726.

61. Maehlum S, Hermansen L. Muscle glycogen concentration during recovery after prolonged severe exercise in fasting subjects. Scand J Clin Lab Invest 1978;38:557–560.

62. Blom PC, Hostmark AT, Vaage O, Kardel KR, Maehlum S. Effect of different post-exercise sugar diets on the rate of muscle glycogen synthesis. Med Sci Sports Exerc 1987;19:491–496.

63. Bergstrom J, Hermansen L, Hultman E, Saltin B. 1968 Diet, muscle glycogen and physical performance. Acta Physiol Scand 1968;71140–71150.

64. Villar-Palasi C, Larner J. Uridinediphosphate glucose pyrophosphorylase from skeletal muscle. Arch Biochem Biophys 1960;86:61–66.

65. Richter EA, Garetto LP, Goodman MN, Ruderman NB. Enhanced muscle glucose metabolism after exercise: modulation by local factors. Am J Physiol 1984;246:E476–E482.

66. Garetto LP, Richter EA, Goodman MN, Ruderman NB. Enhanced muscle glucose metabolism after exercise: the two phases. Am J Physiol 1984;246:E471–E475.

67. Nuttal FQ, Mooradian MC, Gannon C, Billington C, Krezowski P. Effect of protein ingestion on the glucose and insulin response to a standardized oral glucose load. Diabetes Care 1984;7:465–470.

68. Bergstrom J, Hultman E. Muscle glycogen synthesis after exercise: an enhancing factor localized to the muscle cells in man. Nature 1967;210:309–310.

69. Sherman WM, Doyle JA, Lamb DR, Strauss RH. Dietary carbohydrate, muscle glycogen, and exercise performance during 7 d of training. Am J Clin Nutr 1993;57:27–31.

70. Bussau VA, Fairchild TJ, Rao A, Steele P, Fournier PA. Carbohydrate loading in human muscle: an improved 1 day protocol. Eur J Appl Physiol 2002;87:290–295.

71. Burke LM, Collier GR, Hargreaves M. Muscle glycogen storage after prolonged exercise: effect of glycemic index of carbohydrate feedings. J Appl Physiol 1993;75:1019–1023.

72. Williams MB, Raven PB, Fogt DL, Ivy JL. Effects of recovery beverages on glycogen restoration and endurance exercise performance. J Strength Cond Res 2003;17:12–19.

73. Zawadzki KM, Yaspelkis BB 3rd, Ivy JL. Carbohydrate-protein complex increases the rate of muscle glycogen storage after exercise. J Appl Physiol 1992;72:1854–1859.

74. Wanke CA, Pleskow D, Degirolami PC, et al. A medium chain triglyceride-based diet in patients with HIV and chronic diarrhea reduces diarrhea and malabsorption: a prospective, controlled trial. Nutrition 1996;12:766–771.

75. Craig GB, Darnell BE, Weinsier RL, et al. Decreased fat and nitrogen losses in patients with AIDS receiving medium-chain-triglyceride-enriched formula vs. those receiving long-chain-triglyceride-containing formula. J Am Diet Assoc 1997;97:605–611.

76. Fan, ST. Review: nutritional support for patients with cirrhosis. Gastroenterol Hepatol 1997;12:282–286.

77. Jiang ZM, Zhang SY, Wang XR, et al. A comparison of medium-chain and long-chain triglycerides in surgical patients. Ann Surg 1993;217:175–184.

78. Bach AC, Ingenbleek Y, Frey A. The usefulness of dietary medium-chain triglycerides in body weight control: fact or fancy? J Lipid Res 1996;37:708–726.

79. Hawley JA, Brouns F, Jeukendrup A. Strategies to enhance fat utilization during exercise. Sports Med 1998;25:241–257.

80. Jeukendrup AE, Thielen JJ, Wagenmakers AJ, et al. Effect of medium-chain triacylglycerol and carbohydrate ingestion during exercise on substrate utilization and subsequent cycling performance. Am J Clin Nutr 1998;67:397–404.

81. West DB, Delany JP, Camet PM, et al. Effects of conjugated linoleic acid on body fat and energy metabolism in the mouse. Am J Physiol 1998;275 (Pt 2):R667–R672.

82. Park Y, Albright KJ, Liu W, et al. Effect of conjugated linoleic acid on body composition in mice. Lipids. 1997;32:853–858.

83. Ostrowski E, Muralitharan M, Cross RF. Dietary conjugated linoleic acids increase lean tissue and decrease fat deposition in growing pigs. J Nutr 1999;129:2037–2042.

84. Song HJ, Grant I, Rotondo D, Mohede I, Sattar N, Heys SD, Wahle KW. Effect of CLA supplementation on immune function in young healthy volunteers. Eur J Clin Nutr. 2005;59:508–517

85. Gavino VC, Gavino G, Leblanc MJ, Tuchweber B. An isomeric mixture of conjugated linoleic acids but not pure cis-9, trans-11-octadecadienoic acid affects body weight gain and plasma lipids in hamsters. J Nutr 2000;130:27–29.

86. Pariza MW, Park Y, Cook ME. Mechanisms of action of conjugated linoleic acid: evidence and speculation. Proc Soc Exp Biol Med. 2000;223:8–13.

87. Riserus U, Arner P, Brismar K, Vessby B. Treatment with dietary trans10-cis12 conjugated linoleic acid causes isomer-specific insulin resistance in obese men with the metabolic syndrome. Diabetes Care 2002;25:1516–1521.

88. Branch JD. Effect of creatine supplementation on body composition and performance: a meta-analysis. Int J Sport Nutr Exerc Metab 2003;13:198–226.

89. Kreider RB. Effects of creatine supplementation on performance and training adaptations. Mol Cell Biochem 2003;244:89–94.

90. Rawson ES, Volek JS. Effects of creatine supplementation and resistance training on muscle strength and weightlifting performance. J Strength Cond Res 2003;17:822–831.

91. Wyss M, Schulze A. Health implications of creatine: can oral creatine supplementation protect against neurological and atherosclerotic disease? Neuroscience 2002;112:243–260.

92. Kley RA, Vorgerd M, Tarnopolsky MA. Creatine for treating muscle disorders. Exp Physiol 2007;92:323–331.

93. Priml W, von Arnim T, Stablein A, et al. Effects of ribose on exercise-induced ischaemia in stable coronary artery disease. Lancet 1992;340:507–510.

94. Gross M, Dormann B, Zollner N. Ribose administration during exercise: effects on substrates and products of energy metabolism in healthy subjects and a patient with myoadenylate deaminase deficiency. Klin Wochenschr 1991;69:151–155.

95. Salerno C, D'Eufermia P, Finocchiaro R, et al. Effect of D-ribose on purine synthesis and neurological symptoms in a patient with adenylsuccinase deficiency. Biochim Biophys Acta 1999;1453:135–140.

96. Steele IC, Patterson VH, Nicholls DP. A double-blind, placebo-controlled, crossover trial of D-ribose in McArdle's disease. J Neurol Sci 1996;136:174–177.

97. Berardi JM, Ziegenfuss TN. Effects of ribose supplementation on repeated sprint performance in men. J Strength Cond Res 2003;17:47–52.

98. Close GL, Ashton T, McArdle A, Maclaren DP. The emerging role of free radicals in delayed onset muscle soreness and contraction-induced muscle injury. Comp Biochem Physiol A Mol Integr Physiol 2005;42:257–266.

99. Alessio HM. Exercise-induced oxidative stress. Med Sci Sports Exerc 1993;25:218–224.

100. Davies KJ, Quintaniha AT, Brooks GA, Packer L. Free radicals and tissue damage produced by exercise. Biochem Biophys Res Commun 1982;107:1198–1205.

101. Jackson MJ, Edwards RH, Symons MC. Electron spin resonance studies of intact mammalian skeletal muscle. Biochim Biophys Acta 1985;847:185–190.

102. Novelli GP, Bracciotti G, Falsini S. Spin-trappers and vitamin E prolong endurance to muscle fatigue in mice. Free Radic Biol Med 1990;8:9–13.

103. Quintanilha AT. Effects of physical exercise and/or vitamin E on tissue oxidative metabolism. Biochem Soc Trans 1984;12:403–404.

104. Ji LL. Exercise, oxidative stress, and antioxidants. Am J Sports Med 1996;24(Suppl):S20–S24.

105. Fischer CP, Hiscock NJ, Penkowa M, et al. Supplementation with vitamins C and E inhibits the release of interleukin-6 from contracting human skeletal muscle. J Physiol 2004;558:633–645.

106. Bloomer RJ, Goldfarb AH, McKenzie MJ. Oxidative stress response to aerobic exercise: comparison of antioxidant supplements. Med Sci Sports Exerc 2006;38:1098–1105.

107. Goldfarb AH, Bloomer RJ, McKenzie MJ. Combined antioxidant treatment effects on blood oxidative stress after eccentric exercise. Med Sci Sports Exerc 2005;37:234–239.

108. Sumida S, Tanaka K, Kitao H, Nakadomo F. Exercise-induced lipid peroxidation and leakage of enzymes before and after vitamin E supplementation. Int J Biochem 1989;21:835–838.

109. Evans WJ, Cannon JG. The metabolic effects of exercise-induced muscle damage. Exec Sport Sci Rev 1991;19:99–125.

110. Thompson D, Williams C, Garcia-Roves P, McGregor SJ, McArdle F, Jackson MJ. Post-exercise vitamin C supplementation and recovery from demanding exercise. Eur J Appl Physiol 2003;89:393–400.

111. Close GL, Ashton T, Cable T, et al. Ascorbic acid supplementation does not attenuate post-exercise muscle soreness following muscle-damaging exercise but may delay the recovery process. Br J Nutr 2006;95:976–981.

112. Childs A, Jacobs C, Kaminski T, Halliwell B, Leeuwenburgh C. Supplementation with vitamin C and N-acetyl-cysteine increases oxidative stress in humans after an acute muscle injury induced by eccentric exercise. Free Radic Biol Med 2001;15:745–753.

113. Bryer SC, Goldfarb AH. Effect of high dose vitamin C supplementation on muscle soreness, damage, function, and oxidative stress to eccentric exercise. Int J Sport Nutr Exerc Metab 2006;16:270–280.

114. Thompson D, Williams C, McGregor SJ, et al. Prolonged vitamin C supplementation and recovery from demanding exercise. Int J Sport Nutr Exerc Metab 2001;11:466–481.

115. Goldfarb AH, Patrick SW, Bryer S, You T. Vitamin C supplementation affects oxidative-stress blood markers in response to a 30-min run at 75% VO_2max. Int J Sport Nutr Exerc Metab 2005;15:279–290.

116. Hathcock JN, Azzi A, Blumberg J, et al. Vitamins E and C are safe across a broad range of intakes. Am J Clin Nutr 2005;81:736–745.

117. Consumers Union. Is E for you? Consumer Rep Health 1996;8:121–124.

118. Kraemer WJ, Volek JS, French DN, et al. The effects of l-carnitine l-tartrate supplementation on hormonal responses to resistance exercise and recovery. J Strength Cond Res 2003;7:455–462.

119. Volek JS, Kraemer WJ, Rubin MR, Gomez AL, Ratamess NA, Gaynor P. l-Carnitine l-tartrate supplementation favorably affects markers of recovery from exercise stress. Am J Physiol Endocrinol Metab 2002;282:E474–E482.

120. Spiering BA, Kraemer WJ, Vingren JL, et al. Responses of criterion variables to different supplemental doses of l-carnitine l-tartrate. J Strength Cond Res 2007;21:259–264.

121. Brass EP, Hiatt WR. The role of carnitine and carnitine supplementation during exercise in man and in individuals with special needs. J Am Coll Nutr 1998;17:207–215.

122. Kraemer WJ, Volek JS. L-Carnitine supplementation for the athlete: a new perspective. Ann Nutr Metab 2000;44:88–89.

123. Rubin MR, Volek JS, Gomez AL, et al. Safety measures of L-carnitine L-tartrate supplementation in healthy men. J Strength Cond Res 2001;15:486–490.

124. Antonio J, Street C. Glutamine: a potentially useful supplement for athletes. Can J Appl Physiol 1999;24:1–14.

125. Newsholme EA, Castell LM. Amino acids, fatigue and immunosuppression in exercise. In: Maughan RJ (ed) Nutrition in Sport, IOC Encyclopedia of Sport (p 153). Blackwell Science, Oxford, 2000.

126. Ardawi MSM, Newsholme EA. Metabolism in lymphocytes and its importance in the immune response. Essays Biochem 1985;21:1–44.

127. Elia M, Wood S, Khan K, Pullicino E. Ketone body metabolism in lean male adults during short-term starvation, with particular reference to forearm muscle metabolism. Clin Sci 1990;78:579–584.

128. Castell LM, Poortmans JR, Newsholme EA. Does glutamine have a role in reducing infections in athletes? Eur J Appl Physiol Occup Physiol 1996;73:488–490.

129. Kargotich S, Goodman C, Dawson B, Morton AR, Keast D, Joske DJ. Plasma glutamine responses to high-intensity exercise before and after endurance training. Res Sports Med 2005;13:287–300.

130. Rennie MJ, Edwards RHT, Krywawych S, et al. Effect of protein turnover in man. Clin Sci (Lond) 1981;61:627–639.

131. Castell LM. Glutamine supplementation in vitro and in vivo, in exercise and in immunodepression. Sports Med 2003;16:323–345.

132. Castell LM, Newsholme EA. The relation between glutamine and the immunodepression observed in exercise. Amino Acids 2001;20:49–61.

133. Juretic A, Spagnoli GC, Horig H, et al. Glutamine requirements in the generation of lymphokine activated killer cells. Clin Nutr 1994;13:42–49.

134. Candow DG, Chilibeck PD, Burke DG, Davison KS, Smith-Palmer T. Effect of glutamine supplementation combined with resistance training in young adults. Eur J Appl Physiol 2001;86:142–149.

135. Bassit RA, Sawada LA, Bacurau RFP, Navarro F, Costa Rosa LF. The effect of BCAA supplementation upon the immune response of triathletes. Med Sci Sports Exerc 2000;32:1214–1219.

136. Fujita S, Rasmussen BB, Cadenas JG, Grady JJ, Volpi E. Effect of insulin on human skeletal muscle protein synthesis is modulated by insulin-induced changes in muscle blood flow and amino acid availability. Am J Physiol Endocrinol Metab 2006;291:E745–E754.

137. Biolo G, Fleming RYD, Wolfe RR. Physiologic hyperinsulinemia stimulates protein systhesis and enhances transport of selected amino acids in human skeletal muscle. J Clin Invest 1995;95:811–819.

138. Claessens M, Calame W, Siemensma AD, Van Baak MA, Saris WH. The effect of different protein hydrolysate/carbohydrate mixtures on postprandial

glucagons and insulin responses in healthy subjects. Eur J Clin Nutr 2007;Sept 12 [Epub ahead of print]

139. Manders RJ, Wagenmakers AJ, Koopman R, et al. Co-ingestion of a protein hydrolysate and amino acid mixture with carbohydrate improves plasma glucose disposal in patients with type 2 diabetes. Am J Clin Nutr 2005;82:72–83.

140. Kreider RB, Miriel V, Bertun E. Amino acid supplementation and exercise performance: analysis of the proposed ergogenic value. Sports Med 1993;16:190–209.

141. Levenhagen DK, Carr C, Carlson MG, Maron DJ, Borel MJ, Flakoll PJ. Postexercise protein intake enhances whole-body and leg protein accretion in humans. Med Sci Sports Exerc 2002;34:828837.

142. Levenhagen DK, Gresham JD, Carlson MG, Maron DJ, Borel MJ, Flakoll PJ. Postexercise nutrient intake timing in humans is critical to recovery of leg glucose and protein homeostasis. Am J Physiol Endocrinol Metab 2001;280:E982–E993.

143. Rasmussen BB, Tipton KD, Miller SL, Wolf SE, Wolfe RR. An oral essential acid-carbohydrate supplement enhances muscle protein anabolism after resistance exercise. J Appl Physiol 2000;88:386–392.

144. Koopman R, Pannemans DL, Jeukendrup AE, et al. Combined ingestion of protein and carbohydrate improves protein balance during ultra-endurance exercise. Am J Physiol Endocrinol Metab 2004;287:E712–E720.

145. Koopman R, Wagenmakers AJ, Manders RJ, et al. Combined ingestion of protein and free leucine with carbohydrate increases postexercise muscle protein synthesis in vivo in male subjects. Am J Physiol Endocrinol Metab 2005;288:E645–E653.

146. Norton LE, Layman DK. Leucine regulates translation of protein synthesis in skeletal muscle after exercise. J Nutr 2006;136:533S–537S.

147. Bird SP, Tarpenning KM, Marino FE. Independent and combined effects of liquid carbohydrate/essential amino acid ingestion on hormonal and muscular adaptations following resistance training in untrained men. Eur J Appl Physiol 2006;97:225–238.

148. Watford M, Wu G. Glutamine metabolism in uricotelic species: variations in skeletal muscle glutamine synthesis, glutaminase, glutamine levels and rates of protein synthesis. Comp Biochem Physical B Biochem Mol Biol 2005;140:607–614.

149. Jayakumar AR, Rao KV, Murthy CR, Norenberg MD. Glutamine in the mechanisms of ammonia-induced astrocyte swelling. Neurochem Int 2006;48:623–628.

13 Nutritional Supplementation and Meal Timing

Jim Farris

Abstract

For the competitive athlete and the serious recreational athlete, nutritional supplementation can have a positive effect on training and on performance. There are many fad supplements on the market, and many that have come and gone. However, two nutrients have withstood the test of time and many tests in research laboratories around the world, and they continue to have positive training- and performance-enhancing effects. Carbohydrates are commonly supplemented to improve energy availability and to replace valuable muscle and liver glycogen stores. Protein supplementation usually is associated with building muscle tissue.

A relatively new line of thought in relation to these two nutrients involves the combined ingestion of carbohydrate and protein during time periods that allow optimal metabolic conditions related to energy storage and protein synthesis. Athletes along the spectrum from endurance-trained to strength-trained may benefit during activity and recovery from exercise by incorporating the timing of nutrient ingestion into their overall nutritional regimen. The literature provides strong evidence for the optimization of carbohydrate availability and positive net protein synthesis when individuals incorporate timing protocols for nutrient supplementation into their total regimen.

The evidence that these optimal metabolic conditions translate into enhanced performance is still debatable, and documentation requires further work. There are some results, however, suggesting that ideal timing of nutrient intake can enhance both aerobic and strength performance. The reader—whether coach, trainer, athlete, or student—should make a critical analysis of the information presented. As much as possible, the information presented here includes the pros and the cons related to the timing of supplemental nutrient ingestion. Because there does not appear to be any adverse effects on performance by utilizing the timing methods presented below, athletes are encouraged to incorporate these methods into their training regimen and monitor performance changes prior to any competitive event.

From: *Nutritional Supplements in Sports and Exercise*
Edited by: M. Greenwood, D. Kalman, J. Antonio,
DOI: 10.1007/978-1-59745-231-1_13, © Humana Press Inc., Totowa, NJ

Key words

Carbohydrate · Protein · Fat · Amino acid · Timing · Nutrient · Performance · Enhancement · Replacement · Supplementation · Energy · Metabolism

1. INTRODUCTION

An abundance of research articles address the nutritional needs of athletes and active individuals. As discussed in detail in other chapters of this book, it is known that persons engaged in athletic events (athlete and athletic event meaning a sports or activity-related event where an individual is training to improve performance for individual gains or training for competition on any level) need to include carbohydrates and protein in the prescribed proportions to meet the athlete's specific goals. Carbohydrates continue to be the most important energy-producing compound for athletic activity and performance. Protein's role has been better defined as important in maintaining muscle mass and providing essential amino acids for cellular protein synthesis. The use of supplemental forms of carbohydrates and protein is often a convenient way to obtain specific nutrients and easily incorporate them into an overall nutritional plan. Fat supplementation or attempting to adapt to a high fat diet does not enhance performance (1). Although it is an essential macronutrient, the use of fat in the athlete's diet for anything other than the delivery of nutrients and calories is not supported in the literature.

So what is the athlete who has the proper amounts of carbohydrate, protein, and fat in their diet supposed to do? An answer to that question should include the timing of the intake of these nutrients before, during, and after training sessions or competition. When the timing of nutrient intake is given serious consideration and utilized during training, the athlete can create an optimal metabolic response that may lead to enhanced performance.

Over the last two decades, a line of scientific investigation has lead not only to the proper nutrients but to the timing of the ingestion of these nutrients to optimize the body's internal milieu (i.e., everything from fluid, nutrients, hormones, neurotransmitters, and waste products, among others, inside and outside the cells in the interstitial fluid and circulation) to prepare the body for exercise, enhance work output during performance, and then assist in maximizing recovery from strenuous activity. The timing of nutrient intake helps athletes fine-tune their nutritional programs so they are able to use their given abilities at the maximum level. The available literature seems to indicate that—regardless of whether the athlete's focus is endurance, strength, power, or a combination of the three—timing the ingestion of nutrients creates conditions that allow optimal performance and adaptations to the demands placed on the body.

What is the role of nutritional supplements and the timing of nutrient delivery for the athlete? Most importantly, nutritional supplements that are used in an overall nutritional plan provide a relatively easily obtainable form of key nutrients (especially carbohydrates and protein) in nutrient-dense volumes and amounts. These nutritional supplements, whether they are commercially prepared or homemade, can also provide the nutrients in optimal forms, concentrations, and ratios for the intended effect. In other words, all athletes should begin with a fundamentally sound, scientifically based foundation of healthy human nutrition (such as the U.S. Department of Agriculture's MyPyramid food guidelines at www.mypyramid.gov) and can then incorporate specific nutritional supplements to meet their performance goals.

It is similar in concept to exercise training. Each athlete needs foundational abilities in strength, speed, power, and agility. Then, according to the athlete's specific event and goals, specialized exercises are added to the foundation of general athletic ability. An analogy for timing the intake of nutrients is the multiple stages of a rocket's liftoff and journey into space. Each stage has a specifically designed fuel that is utilized at specific times throughout its ascent into space. The timing of the use of these fuels is designed to maintain the propulsive thrust that helps the rocket maintain the speeds necessary to escape the earth's gravitational pull while adapting to changes in the availability of oxygen in the various levels of the atmosphere. The timing of nutrient ingestion can help the body maintain an optimum rate of replacement of energy stores in muscle and maintain the optimal rate of muscle growth while limiting the rate of muscle breakdown. Although these processes occur at a much slower rate in the human body than in the delivery of a rocket into the heavens (thank goodness or we would self-vaporize), the principle of optimizing metabolic rates to achieve maximum replacement of energy providing nutrients and synthesis of muscle protein in response to an exercise session holds in the analogy.

Humans are not machines like rockets, but the fuel an athlete ingests does have an effect on performance. Athletes who do not have a sound nutritional regimen and assume that nutritional supplements will help them perform better may be sadly mistaken. This is similar to athletes who do not build a strong physical training foundation and then decide they are going to train a few days before an event so they can do well. Neither athlete will perform to the best of their ability. The intent of this chapter is to help the athlete understand the use of nutritional supplements at the right time to optimize the effect of nutrition on overall performance. The supplements are to be considered only as part of an overall nutritional program based on sound nutritional practices and a well designed physical training program.

A training session or competitive event essentially results in the breakdown of muscle tissue and reduction of energy stores in the active muscles. During the recovery and adaptation processes that follow exercise, the body rebuilds this tissue and restores the energy providing compounds to the cells. There is a good amount of evidence in the literature to support that a carbohydrate and protein supplement ingested immediately after exercise and at regular intervals during the first few hours of recovery results in enhanced recovery of energy stores and enhanced building of muscle tissue *(2)*. This evidence is supported for both endurance and strength trained athletes. The purposes of timing nutrient intake are to 1) reduce the breakdown of tissue and reduction of energy stores during exercise, and then 2) enhance the recovery and adaptation processes that occur during the time between exercise sessions. The proper timing of nutrient ingestion can optimize training stimuli, and there is some early evidence that it may enhance functional performance in certain strength-trained populations. Although the evidence is not yet definitive, the timing of nutrient supplementation may also be beneficial to individuals recovering from inactivity-induced muscle loss due to injury or surgery *(3)* and after fatiguing ambulation in spinal cord-injured patients *(4)*. Although many exquisite animal models have led to deeper understanding of human metabolism, the response of the human athlete to nutritional strategies is the focus of this chapter.

2. PROLONGED EXERCISE

Carbohydrates, more specifically muscle glycogen followed by blood glucose availability, are the most important energy substrates for prolonged moderate-to-intense exercise performance. That statement would get majority, if not unanimous, agreement in a room full of nutrition and exercise scientists. Strategies to enhance muscle glycogen stores through exercise, high carbohydrate diets, and carbohydrate supplementation have been described very well by a number of other authors *(5–7)*. There is also little doubt that carbohydrate supplementation during prolonged exercise, along with sufficient water intake and electrolytes, enhances performance compared to water intake alone *(8–11)*. There is a current debate, however, as to whether the addition of protein to a carbohydrate-electrolyte beverage enhances endurance performance if the beverage is ingested during exercise *(12–15)* (with recent presentations at major conferences and articles in press at the time this chapter was written).

Volumes have already been written on glycogen loading, pre-event carbohydrate meals, carbohydrate ingestion during exercise, and glycogen replenishment after fatiguing exercise *(5–9,16)*. In brief, guidelines for the timing of supplemental CHO ingestion to optimize muscle glycogen stores and enhance performance during prolonged exercise are:

- At 7 days before the event perform a glycogen-depleting exercise followed by;
- 6 days of a high CHO diet with only 20 minutes of mild exercise ($< 65\%$ VO_2max per day) *(7)*. **[Key Point: Some authors recommend 7 days of exercise tapering without the glycogen-depleting bout of exercise *(6)*. The athlete is encouraged to try both methods and determine which one best fits his or her needs.]**
- Ingestion of a high-CHO meal (200–350 g CHO) 3 to 4 hours prior to exercise *(5,9)*;
- Ingestion of 50–200 g CHO 30 to 60 minutes before exercise;
- Ingestion of approximately 60 g CHO per hour during exercise in addition to water and electrolytes. The CHO should be in the form of a 6% CHO solution with electrolytes *(11)* or a similar CHO-electrolyte solution with the addition of no more than 2% protein in solution *(14, 15)*;

- Ingestion of 1.6 g CHO/kg body weight per hour taken at a rate of 0.4 g CHO/kg every 15 minutes for 4 hours after exercise followed by (17,18);
- Moderate-CHO meals throughout the remainder of the day, or high-CHO meals if there will be repeated days of prolonged exercise.

Athletes and coaches should keep in mind that glycogen loading has not been shown to enhance performance for events less than 60 to 90 minutes in duration (19). Also, so long as the athlete is following a well-balanced moderate carbohydrate diet, glycogen loading may not enhance performance above that of using only the preexercise meals and supplementation of CHO during exercise (20).

Because the value of protein supplementation is often overlooked in discussions of supplementation for endurance exercise, the focus of this section is on the use of carbohydrate-protein (CHO-PRO) supplements during and after prolonged exercise. More than three decades ago, Felig and Wahren (21) demonstrated that the amino acid alanine was released from exercising muscle during prolonged bicycle ergometer exercise and that there was an uptake of blood glucose by the exercising muscle that became of more importance as exercise duration progressed (22). Not long after that Haralambie and Berg (23) demonstrated protein utilization during prolonged exercise increased as the duration of the activity increased. The use of muscle protein was dependent on the initial availability of carbohydrate to the exercising muscle (24,25). For more than two decades it has been known that long-distance runners experience loss of muscle protein and muscle damage after training and competition (26). Not only do the valuable energy stores of carbohydrate need to be maintained and then replenished, but adequate protein must be available for maintenance, repair, and adaptation during the resting period between prolonged endurance exercise sessions.

2.1. During Exercise

2.1.1. CHO Supplementation

As stated earlier, the performance-enhancing effect of CHO supplementation during prolonged exercise lasting more than 60 to 90 minutes is well established. When CHO is provided during exercise in addition to fluid, multiple measures of performance and energy metabolism are enhanced. The maximum rate of CHO oxidation during prolonged exercise is approximately 1.0 g of CHO per minute. Therefore, the timing of CHO supplementation during exercise need not exceed this rate of utilization. In a review article covering more than 25 years of research related to CHO ingestion during exercise and performance, Jeukendrup *(27)* noted that the accepted maximum rate of oxidation of exogenous CHO oxidation during exercise is approximately 1.0 g of CHO per minute even with larger amounts of CHO intake.

A few examples of the performance-enhancing effects of CHO supplementation are found in the literature. Well trained cyclists exercised longer before becoming fatigued when given either liquid or solid CHO supplementation during interval exercise of varied intensities *(28)*. Also in this study, after 190 minutes of exercise, muscle glycogen was greater when CHO was ingested than in the placebo exercise trial. Additionally, both insulin and blood glucose levels were also maintained at higher levels when CHO was ingested in these cyclists. In another study, subjects were fed 43 g of a solid sucrose supplement and 400 ml of water (or artificially flavored placebo) at the beginning of exercise and again at 1, 2, and 3 hours of cycling exercise *(29)*. Subjects performed 4 hours of cycling followed by a sprint ride (100% VO_2max) to fatigue. Compared to the placebo beverage, the solid CHO supplemental feedings, with water, maintained blood glucose levels during exercise, slowed the rate of muscle glycogen depletion, and prolonged sprint time to fatigue at the end of 4 hours of cycling.

> **Key Point.** Carbohydrate provided at a rate of 1.0 g CHO per minute (60 g of CHO per hour) in liquid or solid + liquid form during prolonged exercise enhances performance. Not much debate here—it just works.

2.1.2. CHO-PRO SUPPLEMENTATION

It is with CHO-PRO supplementation that the newer research gets interesting. There is some evidence that the addition of protein (usually whey protein in solution at a 2% concentration) to a CHO-electrolyte beverage may enhance performance during exercise above that of the CHO beverage alone. The debate regarding the effectiveness of the addition of protein to the CHO beverage appears to be coming from investigators at different laboratories, with each group receiving funding from different companies competing in the same sports supplement-related market. As a nutritionist, coach, or athlete, it is always best to read the publications from both sides of the debate and then make your decision. Remember that implementing a new nutritional strategy should be done during training, not during a competitive event. Some of these published reports are presented below.

When trained cyclists exercised for 3 hours at variable intensities (similar to interval-type training) followed by a performance trial at 85% VO$_2$max to fatigue, a CHO-PRO supplement solution enhanced time-to-fatigue compared to CHO supplementation alone (15). Both the CHO and CHO-PRO solutions improved time-to-fatigue compared to the placebo. In a somewhat different type of exercise task, cyclists rode at 75% VO$_2$max until exhaustion and then came back 12 to 15 hours later and performed another ride to exhaustion at 85% of VO$_2$max (14). Subjects in this study rode 29% longer during the first ride and 40% longer during the second ride when they received the CHO-PRO supplement compared to the CHO-only supplement. The authors of this study also reported reductions in markers of muscle damage during exercise with the CHO-PRO beverage compared to the CHO-only beverage. In both of the studies just described, the content of CHO in the beverages was the same and the beverages were provided every 15 to 20 minutes. Because CHO content was the same, the CHO-PRO beverages had a greater caloric content. When isocaloric CHO and CHO-PRO beverages were compared, no performance advantages were found between the two conditions (13). Subjects in this study performed a first ride-to-exhaustion at 70% of VO$_2$max followed by a second ride at 80% of VO$_2$max 24 hours later. Because the beverages were isocaloric, there was a lower total CHO content in the CHO-PRO beverage. Even though the CHO content was lower in the CHO-PRO beverage, performance was similar to the CHO-only treatment. Again, markers of muscle damage were reduced with the CHO-PRO treatment.

The addition of 2% protein to a 6% CHO beverage may not enhance performance during cycling that replicates a time trial (i.e., completing a prescribed distance in the shortest amount of time possible) (12). Subjects in this study were asked to complete an 80-km laboratory-simulated time trial using three beverages (placebo, CHO-only, CHO-PRO). The beverages were ingested at the rate of 250 ml every 15 minutes (60 g of CHO per hour). Both the CHO and the CHO-PRO beverages improved time-to-completion of the time trial compared to the placebo. There was no difference in time-to-completion of the time trial between the CHO and CHO-PRO beverage conditions.

Whether the addition of protein to a carbohydrate sports beverage improves performance appears to depend on the type of prolonged exercise being performed and the total content of carbohydrate available. The timing of ingestion of the supplemental beverages at 15- to 20-minute intervals appears to be tolerable by athletes and is related to improved performance. The addition of protein to the carbohydrate sports beverage reduced markers of muscle damage and did not cause any detrimental effect on performance. Whether the addition of protein to an optimally formulated carbohydrate-electrolyte beverage improves performance above that of the carbohydrate-electrolyte beverage alone has not been established.

> **Key Point.** From a metabolic point of view, the use of additional protein in a carbohydrate sports beverage enhances the metabolic state but currently has questionable effects on the enhancement of competitive performance during prolonged exercise. Although the performance-enhancing effects of additional protein are questionable, ingestion of a supplemental beverage every 15 minutes during prolonged exercise does enhance performance compared to that with a placebo.

2.2. After Exercise (Replacement, Recovery, Regeneration, Adaptation)

The importance of the postexercise period and its role in the enhancement of subsequent performances is not always a top priority for athletes and coaches. This part of an overall training program can lead to more rapid and possibly more complete recovery before a subsequent exercise session. If the timing of nutrient intake is also given special consideration, the enhanced metabolic recovery and adaptation may lead to increased performance during subsequent bouts of exercise.

2.2.1. IMMEDIATELY AFTER EXERCISE

The body appears to be most receptive to nutrient utilization immediately after exercise and for a relatively short duration thereafter. Glycogen replacement and protein utilization appear to be optimized when these nutrients are readily available to the muscles that were active as soon as activity ceases. When a 25% CHO solution (2 g CHO/kg) is given immediately after glycogen-depleting exercise, the rate of muscle glycogen storage is nearly twice that when ingestion of the CHO solution is delayed 2 hours after the exercise session *(30)*. This early study helped establish the importance of ingesting nutrients as soon as possible after exercise optimally to replace the energy stores used during exercise. The rapid replacement of nutrients is one of the key points behind the concept of utilizing time as a factor for optimizing nutrient intake.

When a CHO supplement (1 g/kg) or a CHO-PRO supplement (CHO 0.75 g/kg + PRO 0.1 g/kg + fat 0.02 g/kg) was given to men and women immediately after and at 1 hour after exercise, glycogen resynthesis was three- to fourfold higher than when the subjects were given a placebo *(31)*. Although glycogen synthesis was slightly higher with the CHO supplement than the CHO-PRO supplement, the differences were not significant even though the amount of CHO given to each subject in the CHO-PRO supplement was less than in the CHO supplement. This suggests that there may be a synergistic effect of the CHO-PRO combination in a supplement.

In one of the initial studies demonstrating a beneficial effect of leucine on postexercise muscle protein synthesis (in rats), leucine supplementation given immediately after prolonged treadmill exercise enhanced muscle protein synthesis *(32)*. This effect of leucine supplementation was observed even though there was not a concomitant increase in plasma glucose or insulin concentrations. In this same study in rats, the combination of carbohydrate and leucine supplementation enhanced both muscle glycogen replenishment and postexercise muscle protein synthesis, whereas carbohydrate supplementation alone was reported only to enhance muscle glycogen replacement. Animal studies can provide an initial look into possible metabolic similarities in humans.

Van Loon et al. reported that co-ingestion of leucine, phenylalanine, and protein with carbohydrate results in an increased plasma insulin response and muscle glycogen resynthesis during recovery from endurance exercise *(33)*. From the same laboratory, these authors used repeated bouts of high intensity cycle ergometer exercise in interval-type sessions until a defined level of fatigue *(34)*. The exercise was designed to deplete muscle glycogen stores. Although glycogen resynthesis and plasma insulin concentrations were increased with CHO-only supplementation (0.8 g CHO/kg/hr) at 30-minute intervals after exercise, the addition of protein/amino acid (PRO-AA) supplementation to this level of CHO supplementation further increased insulin levels and glycogen resynthesis rates. In a subsequent study, Jentjens et al. *(35)* used a similar glycogen-depleting cycle ergometer exercise protocol and timing of postexercise supplementation and found that when CHO was given at 1.2 g/kg/hr muscle glycogen synthesis was not improved over the same CHO dose with additional PRO-AA supplementation (0.4 g PRO/kg/hr). Muscle glycogen was sampled immediately after exercise, at 1 hour, and again at 3 hours after exercise. Plasma insulin levels were significantly higher in the CHO-PRO trial than in the CHO trial, but muscle glycogen resynthesis was not different.

Ivy et al. had subjects ingest either a high carbohydrate supplement (CHO) or an isocaloric lower carbohydrate supplement combined with protein (CHO-PRO) immediately after 2.5 hours of intense cycling and at 2 hours after exercise during recovery *(36)*. Muscle glycogen stores were measured using nuclear magnetic resonance (NMR) spectroscopy before exercise and periodically during 4 hours of recovery. The cycling exercise significantly depleted muscle glycogen to similar levels in the treatment conditions. At 4 hours of recovery the group that received the CHO-PRO supplement had greater muscle glycogen stores than the CHO group. Plasma insulin concentrations were not different between the groups, although plasma glucose was lower during the CHO-PRO treatment than the CHO trial. This study demonstrated that the addition of protein to a carbohydrate supplement can significantly enhance the replenishment of muscle glycogen stores after prolonged cycling exercises compared to the supplementation of an isocaloric amount of carbohydrate alone. Endurance athletes, like strength athletes, can also benefit from optimizing nutrition to maintain muscle mass and muscle protein content. Although not reported in this study, it is highly possible that the CHO-PRO subjects also had a more optimal metabolic profile for the maintenance or replacement of amino acids released from the exercising muscles during this type of glycogen depleting exercise.

In an effort to address the uptake of protein and amino acids after exercise, investigators had five men and five women cycle at 60% of VO_2max for 60 minutes and then observed the cyclists for 3 hours after exercise on three occasions *(37)*. Three supplements were given immediately after exercise in a randomized order, and amino acid uptake in the legs and whole body was measured after each supplemental trial. The three supplements were a noncaloric placebo, a carbohydrate supplement (CHO) with a small amount of fat in it (8 g CHO + 3 g lipid), and a CHO supplement with the addition of 10 g of protein (CHO-PRO). When the subjects were given the placebo or the CHO supplement, there was a net loss of leg and whole-body protein that included a net release of essential amino acids from the legs. Not only was this response reversed with the CHO-PRO supplement, there was a 33% increase in plasma essential amino acids along with an increase of the following in the legs: amino acid extraction, glucose uptake, and protein synthesis. There was also an increase in whole-body protein synthesis with the CHO-PRO supplement. This demonstrates that even during moderate to mild prolonged exercise sessions endurance athletes can lose valuable muscle protein if the timing and the composition of the postexercise supplement is not designed for optimal metabolic replacement of energy and amino acids essential to the maintenance of muscle mass.

In another study (38), glucose uptake and protein synthesis by active muscle was increased threefold when a CHO-PRO supplement was given immediately after 60 minutes of leg exercise compared to that at 3 hours after exercise. In a study including performance-related results, subjects had greater performance on a second exercise trial when they were given a CHO-PRO supplement during recovery from a previous 2-hour glycogen-depleting cycle ergometer session at 65% to 75% VO$_2$max compared to a standard commercial sports beverage (39). Both supplements were given immediately and at 2 hours after the glycogen-depleting exercise session. The CHO-PRO supplement averaged 0.8 g CHO/kg and 0.2 g PRO/kg. The sports beverage was approximately 0.3 g CHO/kg. Ingestion of the CHO-PRO supplement resulted in a 17% greater plasma glucose response, 92% greater insulin response, and 128% greater muscle glycogen storage compared to that seen with the commercial sports beverage. This study indicates that the sports beverage lacked the provision of carbohydrate needed to optimize glycogen replacement, and it was related to a poorer performance in a subsequent exercise bout. The higher carbohydrate supplement would have likely resulted in a greater rate of muscle glycogen replacement, even without the addition of protein to the beverage. Keep in mind that sports beverages are usually designed to optimize fluid replacement. This is important for endurance performance and should not be minimized. However, a supplement that is designed for recovery may be more optimal at energy and protein replacement than the typical sports beverage.

Based on the studies discussed in this section, it is likely that the CHO-PRO supplement not only resulted in a greater rate of glycogen replacement but enhanced muscle protein synthesis in these subjects during recovery from prolonged exercise. In subjects who performed 2 hours of glycogen-depleting exercise, a CHO-PRO supplement provided a higher rate of muscle glycogen replacement than did a CHO supplement or a PRO supplement under the same conditions *(18)*. In each trial, the subjects were given the supplement immediately after exercise and at 2 hours after exercise. These studies support the use of a CHO-PRO supplement immediately after and every hour for at least the first 2 hours after prolonged exercise to optimize glycogen replacement and muscle protein synthesis.

Cyclists performing two 60-minute time trials separated by 6 hours had higher glycogen resynthesis rates when given carbohydrate-protein supplements (4.8 kcal/kg; CHO 0.8 g/kg, PRO 0.4 g/kg) compared to an isocaloric CHO-only supplement (1.2 g/kg). The CHO-PRO supplement given at 1 and 2 hours after the first exercise bout enhanced glycogen resynthesis compared to the CHO-only supplement given at the same time period or the control condition. However, performance during the second time trial was not different between the three experimental nutrition groups *(40)*. The difference in supplement contents in this study may explain the discrepancy in relation to the glycogen replacement reported by Jentjens et al. *(35)*. Regardless, the effect of adding a PRO-AA supplement to CHO supplements appears to have the potential to optimize glycogen replenishment after endurance exercise without a detrimental effect on performance.

> **Key Point.** From these series of studies, it appears that the provision of a CHO-PRO or CHO-AA supplement immediately and every 30 minutes after exercise for 3 to 5 hours maximizes the replenishment of muscle glycogen stores. It is also fairly evident that the addition of protein or amino acid supplements to the postexercise carbohydrate supplement minimizes muscle protein breakdown, and the rate of muscle protein synthesis is optimized. The effect on subsequent endurance exercise performance of the CHO-PRO recovery supplement, compared to the

CHO-only recovery supplement, regardless of timing is still questionable. This pattern of nutritional supplementation should create an ideal internal environment for recovery and adaptation for the endurance athlete. This information may be useful for the athlete who is participating in daily training sessions or who is competing in multiday endurance events.

2.3. Dehydration: Fluid, the Forgotten Supplement

Sauna-induced dehydration (2.95% decrease in body weight) decreased muscular endurance at 25% of maximum voluntary contraction compared to the euhydrated state. Interestingly, there was a trend, but without significance, toward a decrease in isometric maximum voluntary contraction when the subjects were in the dehydrated state compared to the euhydrated state *(41)*. Two hours of sauna-induced dehydration (1.5% loss of body mass) resulted in a significant decrease in one-repetition maximum bench press in weight-trained males *(42)*. This decrease in strength was reversed by rehydration with water over the course of a 2-hour recovery period after dehydration.

When triathletes who competed in an Olympic distance triathlon hydrated early in the race compared to later in the race, their overall times were faster. The main improvement was the reduction in the run time for the early-hydrating athletes compared to the later-hydrating athletes. The timing between hydration points for the two groups was similar, with the only difference being that one group began later in the race than the other *(43)*. The saying, "drink early and drink often" is reinforced by the results of this report. Acute dehydration-related weight loss in rowers, who dehydrate to make their weight category, may best be treated by the replacement of lost fluid and may not need the supplementation of CHO *(44)*. Fluid alone and a combination of fluid + CHO was better than CHO alone. Although performance was slightly better in the fluid + CHO trial than the fluid-only trial, it was not significantly better. This indicates that although the fluid + CHO beverage was helpful to performance after recovering from acute dehydration, fluid intake alone had a larger impact on subsequent performance than did CHO alone.

The examples above are presented to underscore the importance of hydration to performance. Recommendations for adequate hydration include the following guidelines.

- Ingest approximately 1 g of CHO per kilogram body weight per hour by supplementing a beverage with sugars such as sucrose, glucose, or maltodextrins. Refrain from beverages that are high in fructose.
- Ingest approximately 450 mg of sodium per hour (minimum) during prolonged exercise to help maintain plasma sodium concentration and to maintain the desire to drink fluids.
- Drink enough fluids to minimize dehydration during prolonged exercise. The range for fluid volume is approximately 400 mL to more than 1.5 liters per hour. The volume required depends on a person's sweating rate in the given environmental conditions (45).

3. STRENGTH/RESISTANCE TRAINING

There has traditionally been a focus on protein as a major dietary component for individuals involved in strength training and muscle building exercises. Some of the classic studies have debated the amount of protein needed to maintain a positive nitrogen balance and not just achieve nitrogen balance. After all, nitrogen balance involves the replacement of what is lost and not necessarily the addition or increase of muscle mass and other tissues. Individuals involved with regular resistance training do benefit from about 1.5 to 2.0 times the recommended daily allowance (RDA) for protein ($\sim$1.4–1.8 g/kg/day); more than this amount does not appear to be of any additional benefit (46,47) assuming the athlete's calorie intake and food variety are sufficient to meet the demands of exercise.

The old debates related to the amount of protein needed for gaining muscle mass have evolved into new and exciting investigations into the timing and the type of protein that produces optimal training responses. The ingestion of small amounts of protein, or even some select amino acids, before and immediately after resistance training sessions appears to produce greater gains in muscle mass and strength than training alone (48). As with endurance performance where the value of protein as a nutrient for optimum performance is often overlooked, in resistance training the value of carbohydrate is often obscured in the orations of the protein pundits. Not only is supplemental carbohydrate valuable for maintaining and restoring muscle glycogen stores during resistance exercise, it appears to enhance the postexercise anabolic effect and reduce the postexercise catabolic activity when ingested along with proteins or amino acids before and after resistance exercise *(49)*. There is limited evidence that carbohydrate supplementation enhances resistance exercise performance, but there is ample evidence pointing to the relation of carbohydrate availability and the rate of muscle glycogen resynthesis (and to a lesser extent net protein synthesis) after performing resistance exercise. Therefore, the focus of this section will be on the timing of ingesting supplemental carbohydrates (CHO), proteins (PRO), and amino acids (AA) with the goal of optimizing the body's response to the demands of resistance training. The literature can be separated into the category of responses related to acute exercise bouts and the category of prospective, or prolonged, resistance training for clarity of presentation.

3.1. Acute Exercise Bout Studies—Strength/Resistance

In response to an acute bout of resistance exercise, net muscle protein synthesis is increased for at least 48 hours after exercise *(50)*. The increased net synthesis appears to be present when either eccentric or concentric exercises are performed. To optimize the increased protein synthesis after exercise, it seems logical to provide the body with the proper nutrients so protein synthesis can be enhanced. The amount and type of protein required by strength and resistance training athletes is covered in Chapters 7 and 10 of this text. It appears that protein intake of less than 2.8 g/kg body weight does not have a negative effect on kidney function in healthy males who are engaged in resistance training *(51)*. However, this high protein intake is not necessary for resistance-trained individuals to maintain nitrogen balance. In fact, it is quite possible that endurance athletes require more protein to maintain nitrogen balance than do bodybuilders *(52)*.

3.1.1. SUPPLEMENTATION BEFORE, DURING, AND IMMEDIATELY AFTER RESISTANCE EXERCISE

It is well established that muscle glycogen depletion occurs during resistance training exercise and that the rate of muscle glycogen depletion is related to the intensity of the training regimen *(53,54)*. In subjects performing eight sets of 10 repetitions at approximately

85% of their one-rep maximum, a CHO supplement (1 g/kg) given immediately after exercise and then 1 hour later resulted in a decrease in muscle protein breakdown *(55)*. Although this study did not show an increase in protein synthesis (mainly because a protein or amino acid supplement was not given in addition to the CHO supplement), it did show that CHO supplementation did result in a more positive protein balance compared to a placebo.

Key Point. CHO supplement immediately after exercise provides a more positive protein balance than placebo.

When subjects ingested 1.0 g CHO/kg or placebo prior to an isotonic exercise session and an additional 0.5 g CHO/kg every 10 minutes during the exercise session, muscle glycogen remained at higher levels than when these same subjects were given placebo *(56)*. The authors used an isokinetic leg exercise test to determine performance. There was no difference between the placebo or CHO groups for performance in this acute exercise protocol. In another study *(57)*, these authors demonstrated that CHO ingestion during an isometric-resistance leg exercise protocol improved total work and average work compared to the placebo condition. The group ingesting the CHO supplement also had a higher plasma glucose level after completing the exercise protocol than did the placebo group.

Key Point. CHO supplementation *during* acute bouts of resistance exercise reduces muscle glycogen depletion, enhances blood glucose concentration, and may contribute to improved performance. At minimum, performance is not hindered, and the body's energy stores can be maintained at higher levels.

3.1.2. SUPPLEMENTATION IMMEDIATELY AFTER RESISTANCE EXERCISE

Multiple research reports in which the investigators were trying to maximize potential treatment outcomes begin their supplementation protocols immediately after the exercise bout. The reasoning behind this method of supplementing immediately after exercise is that exercise elevates blood flow in active muscles, and increased insulin levels in the blood appear to enhance blood flow to peripheral musculature. For example, muscle blood flow is increased fivefold above resting during exercise in muscles *(58)*, and a systemic increase in insulin concentration can result in increased blood flow to the peripheral limbs *(59,60)*. When CHO or CHO-PRO drinks are given immediately and at 2 hours after resistance exercise, insulin levels were elevated compared to a PRO-only or a placebo drink *(61)*.

Using an intravenous amino acid infusion (alanine, phenylalanine, leucine, lysine) immediately after a leg resistance exercise routine, investigators found that the stimulatory effect of supplemental amino acids is enhanced by a prior bout of exercise and suggested that the increased blood flow to the muscles that was measured after exercise played a significant role in the enhanced muscle protein synthesis *(62)*. Along with the increased blood flow in exercising muscles it appears that increased insulin levels in the blood result in greater muscle blood flow. Increased blood flow results in increased delivery of nutrients and removal of waste products. The enhanced muscular blood flow does not remain indefinitely; and the sooner nutrients can be delivered to active muscles, the sooner the recovery and adaptation processes can begin (i.e., recovery and adaptation can be optimized).

Key Point. This is a nice example that underscores the importance of *combining exercise and nutrition* for optimal stimulation of growth and development processes in the human body.

You might be thinking that if exercise increases blood flow, and delivery of nutrients is related to blood flow, why not have the nutrients "on board" before the exercise session? Well, here is an answer: Ingesting a CHO-AA supplement immediately after exercise may not be as beneficial as ingesting the same supplement right before a strength training session. Investigators used the same strength training protocol as discussed below for the legs and reported that during the total time period of exercise and during 2 hours of recovery protein synthesis in the leg muscles was significantly greater for the same subjects when the CHO-AA solution was ingested immediately before exercise than when the same solution was ingested immediately after exercise *(63)*. When only the 2 hours of recovery after exercise were compared between the pre- and postexercise supplemented conditions, there was a trend, although not significantly different, for the preexercise supplemented condition to have greater protein synthesis than the postexercise supplemented condition. In both experimental conditions (preexercise supplement only vs. postexercise supplement only) leg blood flow and insulin levels were elevated after ingestion and during the first hour after exercise but returned to resting levels by the second hour of recovery. As the authors pointed out, the elevation of insulin and blood flow to the exercising muscles in the preexercise supplemented condition may be the reason for enhanced delivery of amino acids in the overall time duration of the exercise and recovery period measured in this study. Muscle protein synthesis is related to amino acid delivery to the leg *(62,64,65)*. Insulin is necessary for protein synthesis to occur *(66–68)*; however, increased insulin does not stimulate muscle protein synthesis but it does reduce muscle protein breakdown *(69)*.

Key Point. The main point from this study is that ingestion of a CHO-AA supplement immediately before resistance exercise appears to have better overall efficiency of amino acid delivery to the muscle and more positive amino acid balance (inferring better protein synthesis) than when the CHO-AA is ingested only right after the exercise session. This does not mean that a postexercise supplement is not needed. The availability of nutrients after exercise may have been optimized in the group that was given the supplement before the exercise session. This is a supplement timing issue.

In a subsequent investigation from the same university it was determined that the addition of protein (whey, AAs) to a postexercise CHO supplement prolonged the anabolic effect compared to an isocaloric CHO-only postexercise supplement *(70)*. Interestingly, this group of investigators replicated their previous experiment using only a protein supplement ingested either before or after exercise *(71)*. The protein supplement, taken either before or after exercise, did not enhance markers of muscle protein synthesis like the CHO-AA supplement used in their previous study. Again, this supports the need for a CHO-PRO supplement after exercise.

At a different university laboratory, both a CHO-PRO supplement and a CHO supplement significantly increased nonoxidative leucine disposal (marker of protein synthesis) when given immediately after exercise and again 1 hour after exercise compared to a placebo *(72)*. Although the values for protein synthesis in the CHO-PRO supplement trial were not significantly greater than in the CHO supplement in this study, the CHO-PRO supplement had average values that were approximately 6% greater than those with the CHO supplement.

Yet another group of investigators attempted to address the individual or combined effects of CHO or AA supplementation on the hormonal response to a single strength training exercise session. Bird et al. reported that both CHO and CHO+AA enhance insulin and glucose concentrations and reduce cortisol concentrations during and after exercise compared to placebo *(73,74)*. These authors also reported a synergistic effect of CHO+AA in reducing markers of muscle catabolism when supplement drinks are ingested during strength training exercise sessions *(73)*.

> **Key Point.** Combined with the above point, the evidence from these three independent groups of investigators supports the need for carbohydrate in nutritional supplements as a facilitator of muscle amino acid uptake and subsequent muscle protein synthesis when supplements are taken before and/ or after exercise.

3.1.3. Supplementation 1–4 hours After Strength/Resistance Training

A study that has one of the highest reported rates of glycogen replacement after an acute bout of resistance exercise used frequent timing of supplemental carbohydrate to achieve glycogen replacement rates of approximately 10 mmol/kg/hr *(17)*. The subjects in this study performed 10 sets of 10 repetitions of eccentric, or concentric, leg extension exercises after they had just finished 75 minutes of cycling exercise that was designed to facilitate glycogen depletion. The subjects were then given 0.4 g CHO/kg every 15 minutes for 4 hours (1.6 g CHO/ kg/hr). Whether the high rate of glycogen replacement was due to the degree of glycogen depletion resulting from the protocol used, the high amounts of CHO given, or the frequency of ingestion cannot be determined from this study. However, the frequent ingestion of a relatively high amount of carbohydrate every 15 minutes for 4 hours after the exercise performed in this study resulted in one of the highest rates of glycogen replenishment that has been reported. The investigators did not include a protein or amino acid supplement in this study; but as already noted and as is repeated later, there appears to be a connection between carbohydrate and protein supplementation after exercise, indicating that a combination of the two is more optimal for recovery than either substrate alone. Endurance athletes can also benefit from the results of this study and should utilize frequent supplemental feedings for the first 4 hours after prolonged exercise to optimize replenishment of glycogen stores.

Rassmussen et al. used a repeated-measures crossover design and had subjects perform 10 sets of eight repetitions of leg press and eight sets of eight repetitions of leg extension exercises on two occasions *(75)*. At 1 hour and again at 3 hours after the resistance exercise bouts, the subjects were given either a placebo or a CHO-AA solution

(6 g AA/35 g CHO in 500 ml H_2O). For example, after one exercise session a single subject was given the placebo 1 hour after exercise and then the CHO-AA drink 3 hours after exercise. After the next exercise session, the subject was given the CHO-AA drink 1 hour after exercise and the placebo 3 hours after exercise. This allowed for some interesting timing comparisons as well as comparison of the CHO-AA drink at 1 and at 3 hours after exercise. The authors reported that insulin concentrations and glucose uptake peaked 20 to 30 minutes after each ingestion of the CHO-AA drink at 1 and 3 hours after exercise but not after ingesting the placebo drink. Also, muscle protein synthesis was increased after each CHO-AA supplementation but not after the placebo. However, there was a similar rate of muscle breakdown between the placebo and the CHO-AA drink conditions, which indicated to the authors that the ingestion of CHO-AA drinks at 1 and 3 hours after strength training exercises promoted muscle building (anabolism) but did not inhibit muscle breakdown. The effect of the CHO-AA beverage appeared to be as strong at 3 hours after exercise as it was at 1 hour after exercise. Others have reported that when amino acids are combined with carbohydrate in a recovery-type beverage and ingested at 1 hour and at 2 hours after leg extension resistance exercise, muscle protein synthesis is greater than when either carbohydrate or amino acid is given alone *(76)*. Interestingly, the metabolic responses to the ingestion of the recovery drink at the second hour were similar in magnitude to the prior responses when the drink was ingested 1 hour after exercise. These investigators did not compare postexercise metabolic responses to the immediately postexercise supplementation of the CHO-AA, although it would have been an interesting addition to these studies. These reports also demonstrate that there is a metabolic advantage to ingesting supplements at repeated time intervals for up to 3 hours after exercise.

When essential amino acids (EAAs) are ingested at 1 and 2 hours after resistance exercise, there is a dose-dependent effect of the EAA on muscle protein synthesis *(77)*. The authors reported that ingestion of 6 g of EAAs doubled protein synthesis after resistance exercise compared to the ingestion of 3 g of EAAs plus 3 g of nonessential amino acids (NEAAs). Also mentioned in this report was that NEAAs are apparently not needed for muscle protein synthesis.

Key Point. Combining the above information with studies looking at ingestion of a CHO-PRO/AA supplement at the beginning of an exercise session and during exercise, and those reporting on supplementation immediately after exercise, leads to the recommendation that CHO-PRO/AA supplementation occur immediately before exercise, during exercise, immediately after exercise, and at least hourly (if not more frequently) for the first 4 hours after the exercise session to optimize the synthesis of muscle protein, minimize the breakdown of muscle protein, and optimize the replenishment of energy storage in muscle.

3.1.4. 4+ HOURS THROUGHOUT THE REST OF THE DAY—STRENGTH/ RESISTANCE

The ingestion of whey protein isolates, although thought to have high bioavailability, may not be appropriate for maintaining optimal anabolic conditions after the initial postexercise recovery periods *(78,79)*. Proteins that are digested more slowly (e.g., casein, milk proteins) and other complete protein sources appear to promote the amino acid turnover profiles related to optimal protein availability and an enhanced protein building effect for the later recovery period after exercise.

3.1.5. PROSPECTIVE RESISTANCE TRAINING STUDIES AND SUPPLEMENTATION TIMING

Prospective training studies are important because they take the metabolic information gained from the acute exercise bout studies and apply it to training regimens. Such investigations provide evidence that reinforces, weakens, refutes, or justifies further investigation into the possible performance enhancing effects proposed by the acute-exercise bout studies. In other words, do the metabolic changes observed with nutritional supplementation after a single exercise session result in improved performance above that of exercise training alone? This is the question that every athlete should apply to all nutritional supplementation theories. A brief review of these types of study follows.

The ingestion of a PRO-CHO-fat beverage immediately after strength training exercise sessions for 12 weeks in elderly men (mean age 74 years) produced significant increases strength and muscle size compared to a matched group that received the same supplementation 2 hours after the strength training sessions

(90). In this study, nutrient intake immediately after exercise demonstrated enhanced gains over delayed intake of the same supplement.

Using the method of supplementing before and immediately after exercise, Andersen and colleagues *(91)* took subjects through 14 weeks of resistance training and gave them either a protein supplement or an isocaloric carbohydrate supplement throughout the training program. Subjects receiving the protein supplement had a significant change above baseline measures regarding hypertrophy of type I and type II fibers, whereas the carbohydrate supplement group did not demonstrate significant changes above baseline. At the end of the training sessions the protein group demonstrated improvement over the carbohydrate group in only one of four measures of performance. In this case, protein supplementation given immediately before and after each training session resulted in a greater increase in muscle mass than a carbohydrate-only supplement, but the performance advantage of protein supplementation alone was questionable. This study could have provided more information with the addition of a CHO-PRO supplement group, which would have addressed the combination of PRO + CHO for optimization of response. It is interesting that the performance of the CHO-only group was similar to that of the PRO-only group in three of four of the performance measures.

During a 10-week strength training regimen, a CHO-PRO supplement taken after each training session did not produce significant improvements in body composition or performance above that of ingestion of an isocaloric CHO-only supplemental beverage *(92)*. The authors indicated that there was a nonsignificant trend for greater gains in the CHO-PRO group and suggested that a longer duration of training may have changed their conclusions. Even though functional improvements were not observed in this study with the CHO-PRO supplement, a decrease in function was not observed either.

During a 12-week strength training program, previously untrained subjects who ingested a CHO-AA solution during training sessions had significantly greater increases in muscle fiber cross-sectional area than those receiving a placebo solution *(93)*. Although not significantly different, the CHO-AA solution appeared to have greater cross-sectional gains than did either the CHO solution or the AA solution alone. The authors states that the synergistic effect of combining the CHO and AA supplements optimizes the anabolic effect of strength training sessions compared to supplementing with either one alone. The authors did not report any performance results, however, so inferring actual strength or performance gains from this study would not be valid.

In a prospective study that followed strength-trained males over a 10-week period of training, ingestion of a PRO-CHO-creatine supplement right before and immediately after workout sessions had greater gains in strength (in two of three measures), lean body mass, cross-sectional area, and contractile protein content of type II fibers and higher creatine and muscle glycogen levels than another group of matched subjects who were given the same feedings except in the morning and the evening (94). This study provides some evidence that the timing of nutritional supplements during strength training can result in enhanced bio-logical and functional performance gains.

In a subsequent study by Cribb et al. (95), resistance-trained males were followed during 11 weeks of a structured and supervised resistance exercise program under four nutritional supplement conditions: carbohydrate-creatine (CHO-Cr), creatine-whey protein (Cr-WP), whey protein only (WP), or carbohydrate only (CHO). The supplements were ingested in the morning, immediately after the exercise session, and again in the evening. Strength increases in the CHO-Cr, Cr-WP, and WP groups were greater than in the CHO group for all three performance measures (1RM for bench press, pull-down, and squat). The supplement groups that included crea-tine also demonstrated greater increases in muscle size than the CHO-only group with the WP group, demonstrating intermediate changes that were not different from any of the other groups. Because of the timing of supplementation used in this study, it is difficult to speak directly to the benefit of a nutrient timing regimen except that the subjects did ingest a supplement immediately after exercise. Also, adding creatine to the CHO supplement produced greater performance gains than the CHO supplement alone. No significant differences between creatine and whey protein supple-ments could be determined in this study.

There is not a lot of strong evidence in direct favor of enhanced performance by applying nutrient timing as part of an extended training regimen. There is yet to be a published study that compares the physical and functional changes of a nutrient timing group and a standard supplement group over the course of a supervised resistance exercise regimen. In a study of this type, the nutrient timing group should receive supplemental feedings of CHO-PRO/AA immediately before exercise, during exercise, immediately after exercise, and at frequent intervals for 4 hours after exercise. This supplementation schedule would be in addition to regular daily meals. The standard supplement group would receive the same amount of the supplement except only at regular daily meals. Ideally in the standard group, the supplements would be taken outside the time-frame that is believed to be optimal for nutrient utilization, and the calories and nutrient content of the diets in the two groups would be similar. When this type of study is published and replicated by at least two independent laboratories, there will be some solid evidence to write about and discuss.

For now, there are a few hints that nutrient timing may be related to enhanced physical and functional gains over the course of a resistance training regimen. It is important that interested individuals understand that the effectiveness of the timing of nutritional supplements over the course of a prolonged training regimen is still somewhat questionable according to the available published literature. This is disappointing in light of the significant metabolic changes that are observed after acute bouts of resistance exercise when timing of supplement intake is utilized. At the very least, there does not appear to be any detrimental effect of incorporating this type of nutritional timing regimen into an overall training program. There is still work to be done in this promising area of study.

4. IMMUNE FUNCTION AND OXIDATIVE STRESS

Strenuous exercise can have a negative effect on the body's immune system. There is some limited evidence that supplemental vitamin intake and supplemental carbohydrate intake can be effective in maintaining immune function and help to keep the athlete in an optimal state of health. Avoiding illness during training and competition may not necessarily lead to a season personal record, but not becoming sick can definitely keep life on the positive side of physical and mental performance. Also, the fact that strenuous exercise places stressful demands on the human immune system is a strong reminder that proper rest and recovery are an extremely important part of any training regimen. The information presented below is a concise overview of supplementation as it relates to immune function and the reduction of oxidative stress related to exercise.

4.1. Between Exercise Sessions

In untrained males performing 60 minutes of cycling at 70% of VO_2max, a high-CHO diet limited the rise in cortisol, decrease in plasma glutamine concentration, and amount of neutrophilia during the postexercise period compared to a low-CHO diet *(80)*. These subjects followed the experimental diets for 3 days before each exercise session. This indicates that nutritional regimens before events can effect immune-related responses to strenuous endurance exercise in untrained subjects. Again, an overall nutritional plan has an important role in optimizing metabolism and performance.

4.2. During Exercise

In an earlier investigation, runners training for and competing in the Los Angeles marathon were found to be at a greater risk for an infectious illness if their training distances were more than 97 km per week compared to those runners training at 32 km per week or less. The high-distance runners were also faster in their marathon times than the low-distance trained runners. The greater distances and intensities used by the high-distance group appears to have been detrimental to their immune systems and resulted in an increased incidence of episodes of infectious illnesses *(81)*. When prolonged exercise is performed to fatigue, carbohydrate supplementation (5% w/v beverage) may not have a protective effect on the immune system *(82)*. Samples taken at fatigue from these cyclists did not indicate that carbohydrate supplementation was helpful to immune function when compared to placebo. Still, these subjects exercised for 30% longer when carbohydrate was ingested during exercise. Branched-chain amino acid (BCAA) supplementation during one-legged knee extensor exercise for 60 minutes at approximately 70% of maximal working capacity resulted in a decrease in markers of muscle protein breakdown but did not increase performance above that of the control condition *(83)*. When supplemental carbohydrate beverage is given during 2.5 hours of running at about 76% VO$_2$max, the subjects' immune system and inflammatory response markers were significantly decreased immediately after exercise when compared to the same exercise when the subject was given a placebo beverage *(84)*. Interleukin-6 (IL-6), IL-1-receptor antagonist (IL-1ra), and plasma cortisol levels were all decreased at the end of the exercise session when supplemental carbohydrate was provided during exercise. The decreased markers of inflammatory response immediately after exercise may indicate a more favorable immune system profile when carbohydrate supplements are given during prolonged exercise. In competitive ultra-marathon runners, the supplementation of 600 mg of vitamin C daily appeared to reduce the incidence of upper respiratory tract (URT) infections over a 2-week period after an ultra-marathon race *(85)*. Nearly 70% of the runners who received a placebo reported symptoms of URT infection compared to only one-third of the racers who received supplemental vitamin C. The reported severity and duration of the URT symptoms were also less in the vitamin C-supplemented group than in the placebo group.

The effect of vitamin C on improved immune function and oxidative stress markers after prolonged exercise does not appear to be greater than the effect of carbohydrate supplementation during the exercise bout *(86)*. Subjects were provided either supplementation of carbohydrate plus vitamin C or supplementation with only carbohydrate during an ultra-marathon running race. The addition of vitamin C to the carbohydrate supplementation did not provide any notable benefit to the athletes above that of the carbohydrate supplement only. A fairly comprehensive systematic review of the literature in 2004 concluded that there does appear to be a protective effect of vitamin C supplementation that reduces the incidence of the common cold in persons exposed to episodes of intense exercise *(87)*. However, the supplementation must be given regularly and daily before the exercise sessions because there does not appear to be any beneficial effect of vitamin C after the onset of the cold. There is a need for prospective investigations—studies that follow athletes for prolonged periods of time such as throughout a competitive season or over the course of a year—that attempt to determine if there is, in fact, a decreased incidence of infectious illnesses due to the use of nutritional supplements.

A number of studies have indicated that carbohydrate supplementation during exercise contributes to a more favorable immune system profile after strenuous exercise. Individuals who are untrained and are beginning a resistance training program may exhibit decreased immune function compared to trained persons (88). Supplementation therefore, may be beneficial to those beginning a resistance training program. The adaptations of the body in response to the training period may be more important than the effect of antioxidant supplementation (89). Given the level of evidence in the literature, there appears to a mild protective effect of antioxidant vitamin supplementation on the immune system and oxidative stress markers in persons exposed to prolonged endurance exercise or untrained persons. These effects may lead to an overall cumulative effect that may be beneficial to active individuals. There does not appear to any negative effect on performance when these compounds are taken in moderation.

Athletes, trainers, and coaches should note that there is currently no evidence supporting any performance enhancement by supplementation with antioxidant vitamins. Also, there does not appear to be any benefit of taking large, or mega-dose (more than 10 times the RDA), amounts of these vitamins. The effect of these antioxidant supplements does not appear to be greater than, or enhance, the immune or antiinflammatory benefits observed by the supplementation of carbohydrate during the exercise sessions of endurance or resistance athletes. The athlete should be sensible if choosing to use these supplements as part of an overall nutritional plan. If the choice is made to supplement with these compounds, doses no greater than the RDA should be taken on a regular daily basis before exposure to intense or novel exercise sessions and throughout training and competitive seasons to optimize any potential beneficial effect.

5. CONCLUSION

This chapter began with questions related to how often an athlete should eat and whether athletes could become bigger, stronger, faster, or perform more work before fatigue if they utilized the concept of timing their intake of supplemental nutrition. The supplements of main focus in this chapter were not the fad or "chic" supplements but, instead, nutrients necessary for energy and tissue building: carbohydrates and complete proteins or amino acids. The value of carbohydrate supplementation for enhancing prolonged exercise is well established. There are multiple review articles on muscle glycogen loading, carbohydrate supplementation during exercise, and replenishment of muscle glycogen stores immediately after exercise. These articles provide clear evidence that carbohydrate supplementation before, during, and after prolonged exercise results in increased performance and enhanced replenishment of the muscular energy stores of carbohydrates.

The importance of adequate fluid intake during prolonged exercise to minimize dehydration and maintain performance capacities must also be remembered. Supplementation of water alone or carbohydrate alone is not as effective during exercise as the combination of carbohydrates and electrolytes in a fluid replacement beverage.

The use of protein supplementation for athletes engaged in prolonged exercise has been overshadowed by the focus on carbohydrate utilization. Endurance athletes utilize significant amounts of muscular protein and amino acids during prolonged exercise. These nutrients need to be replaced to maintain muscle mass and nitrogen balance. The addition of protein and certain amino acids to the carbohydrate supplementation regimen of endurance athletes appears to optimize the uptake and utilization of protein by exercising muscles, especially after the exercise session. There is scarce, if any, unequivocal evidence that the addition of protein to carbohydrate supplementation improves performance in endurance athletes during a single bout of exercise or throughout the course of an extended training regimen. In a similar but reverse fashion, the importance of carbohydrate for optimal responses to resistance training has been overshadowed by the focus on protein utilization. The supplementation of carbohydrate with protein appears to have a synergistic effect on anabolic factors related to muscle growth compared to either of the two supplements taken alone.

In addition to the synergistic metabolic effects of a combined carbohydrate-protein (CHO-PRO) supplement, it is evident that the timing of the ingestion of these nutrients can provide additional metabolic advantages for athletes involved in resistance training.

These metabolic advantages have been most strongly demonstrated with supplementation immediately before, during, immediately after, and up to 4 hours after a single bout of resistance exercise. When the CHO-PRO supplement is ingested during this window of optimal metabolic utilization of nutrients, there is an enhanced uptake of amino acids, protein, and carbohydrates by active muscle compared to when nutrient intake is delayed or does not occur. What is not clear for both prolonged exercise and resistance exercise is whether the incorporation of a nutrient timing protocol into an extended training regimen effectively enhances performance compared to regular dietary patterns of equivalent nutrient contents and caloric levels. Hints are beginning to appear in the literature that the effects of resistance training may be enhanced by utilizing nutrient timing and CHO-PRO for both physical and functional performance gains at the end of a prescribed training regimen. These performance-enhancing effects have yet to be convincingly demonstrated over the course of an extended endurance training regimen.

6. PRACTICAL APPLICATIONS

The practical applications can be based within two lines of thinking. First, timing the ingestion of nutrients (nutrient timing) provides performance advantages for endurance athletes and possibly for resistance training athletes. Second, nutrient timing leads to enhanced provision of nutrients for postexercise replacement and for an enhanced anabolic metabolic profile during and after exercise. The details for endurance athletes do not differ greatly from the specifics for the resistance training athlete.

For the endurance athlete, a daily nutritional plan that has more than 50% of its calories from carbohydrates and 1.2 to 1.8 g of protein per kilogram body weight per day provides the dietary foundation. Some authors recommend up to 60% of calories from carbohydrates, which requires supplementation of carbohydrates, usually in solution. To optimize glycogen stores and carbohydrate availability, a high-carbohydrate pre-event meal ingested 4 hours before exercise followed by a pre-event high-carbohydrate snack eaten within 45 minutes of exercise is recommended. During exercise the supplementation of 60 g of carbohydrate per hour provided in a water and electrolyte beverage provides energy and needed fluid. The beverage should not contain more than 6% carbohydrate for optimal fluid replacement, which is the equivalent of 1 liter of a 6% carbohydrate-electrolyte solution per hour. After exercise, frequent feedings beginning immediately and then at least every 30 minutes thereafter, a carbohydrate-protein replacement supplement for the first 4 hours after exercise optimizes glycogen replenishment and protein replacement in active muscles. Whey protein appears to be more rapidly absorbed and more readily utilized by recovering muscles during this initial postexercise time period. The same is true for the high glycemic index sugars (e.g., glucose, sucrose) during the preexercise period, during exercise, and during the initial postexercise time-frame. Outside this initial postexercise window, continuing a regular high-carbohydrate diet focusing on complex carbohydrates and complete proteins that are more slowly absorbed (e.g., casein) in regular meals and snacks is recommended.

A similar nutritional plan and nutrient timing can be used for athletes engaged in resistance training. The dietary foundation does not need to be as high in carbohydrates and the protein content does not need to exceed 2 g/kg/day. To gain muscle mass, calorie intake must exceed expenditure until the desired lean tissue mass is obtained. The other main difference for resistance trained athletes is that protein should be combined with carbohydrate at every supplemental time point in the nutrient timing window (immediately before, during, immediately after, and frequently for the first 4 hours after exercise—considered the nutrient timing window). This form of supplementation is related to enhanced physical gains and possibly to enhanced functional gains during regulated resistance training regimens.

REFERENCES

1. Burke LM, Keins B. "Fat adaptation" for athletic performance: the nail in the coffin. J Appl Physiol 2006;100:7–8.
2. Fielding RA, Parkington J. What are the dietary requirements of physically active individuals? New evidence on the effects of exercise on protein utilization during post-exercise recovery. Nutr Clin Care 2002;5:191–196.
3. Holm L, Esmarck B, Mizuno M, et al. The effect of protein and carbohydrate supplementation on strength training outcome of rehabilitation in ACL patients. J Orthop Res 2006;24:2114–2123.
4. Nash MS, Meltzer NM, Martins SC, Burns PA, Lindley SD, Field-Fote EC. Nutrient supplementation post ambulation in persons with incomplete spinal

cord injuries: a randomized, double-blinded, placebo controlled case series. Arch Phys Med Rehabil 2007;88:228–233.

5. Costill DL, Hargreaves M. Carbohydrate nutrition and fatigue. Sports Med 1992;13:86–92.

6. Sherman WM, Costill DL, Fink WJ, Hagerman FC, Armstrong LE, Murray TF. Effect of a 42.2-km footrace and subsequent rest or exercise on muscle glycogen and enzymes. J Appl Physiol 1983;55:1219–1224.

7. Goforth HW, Laurent D, Prusaczyk WK, Schneider KE, Petersen KF, Shulman GI. Effects of depletion exercise and light training on muscle glycogen supercompensation in men. Am J Physiol Endocrinol Metab 2003;285:E1304–E1311.

8. Coggan AR, Coyle EF. Carbohydrate ingestion during prolonged exercise: effects on metabolism and performance. Exerc Sport Sci Rev 1991;19:1–40.

9. Coggan AR, Swanson SC. Nutritional manipulations before and during endurance exercise: effects on performance. Med Sci Sports Exerc 1992;24: S331–S335.

10. Convertino VA, Armstrong LE, Coyle EF, et al. American College of Sports Medicine position stand: exercise and fluid replacement. Med Sci Sports Exerc 1996;28:i–vii.

11. Murray B. The role of salt and glucose replacement drinks in the marathon. Sports Med 2007;37:358–360.

12. Van Essen M, Gibala MJ. Failure of protein to improve time trial performance when added to a sports drink. Med Sci Sports Exerc 2006;38:1476–1483.

13. Romano-Ely BC, Todd MK, Saunders MJ, Laurent TS. Effect of an isocaloric carbohydrate-protein-antioxidant drink on cycling performance. Med Sci Sports Exerc 2006;38:1608–1616.

14. Saunders MJ, Kane MD, Todd MK. Effects of a carbohydrate-protein beverage on cycling endurance and muscle damage. Med Sci Sports Exerc 2004;36:1233–1238.

15. Ivy JL, Res PT, Sprague RC, Widzer MO. Effect of a carbohydrate-protein supplement on endurance performance during exercise of varying intensity. Int J Sports Nutr Exerc Metab 2003;13:382–395.

16. Jeukendrup AE, Jentjens RL, Moseley L. Nutritional considerations in triathalon. Sports Med 2005;35:163–181.

17. Doyle JA, Sherman WM, Strauss JL. Effects of eccentric and concentric exercise on muscle glycogen replenishment. J Appl Physiol 1993;74:1848–1855.

18. Zawadzki KM Yasselkis BB 3rd, Ivy JL. Carbohydrate-protein complex increases the rate of muscle glycogen storage after exercise. J Appl Physiol 1992;72:1854–1859.

19. Hawley JA, Shabort EJ, Noakes, TD, Dennis SC. Carbohydrate-loading and exercise performance: an update. Sports Med 1997;24:73–81.

20. Burke LM, Hawley JA, Schabort EJ, St Clair-Gibson A, Mujika I, Noakes TD. Carbohydrate loading failed to improve 100-km cycling performance in a placebo-controlled trial. J Appl Physiol 2000;88:1284–1290.

21. Felig P, Wahren J. Amino acid metabolism in exercising man. J Clin Invest 1971;50:2703–2714.

22. Wahren J, Felig P, Ahlborg G, Jorfeldt L. Glucose metabolism during leg exercise in man. J Clin Invest 1971;50:2715–2725.

23. Haralambie G, Berg A. Serum urea and amino nitrogen changes with exercise duration. Eur J Appl Physiol Occup Physiol 1976;36:39–48.

24. Lemon PW, Mullin JP. Effect of initial muscle glycogen levels on protein catabolism during exercise. J Appl Physiol 1980;48:624–629.

25. Wagenmakers AJ, Beckers EJ, Brouns F, et al. Carbohydrate supplementation, glycogen depletion, and amino acid metabolism during exercise. Am J Physiol 1991;260:E883–E890.

26. Apple FS, Rogers MA, Sherman WM, Ivy JL. Comparison of serum creatine kinase and creatine kinase MB activities post marathon race versus post myocardial infarction. Clin Chim Acta 1984;138:111–118.

27. Jeukendrup AE. Carbohydrate intake during exercise and performance. Nutrition 2004;20:669–677.

28. Yaspelkis BB 3rd, Patterson JG, Anderia PA, Ding Z, Ivy JL. Carbohydrate supplementation spares muscle glycogen during variable-intensity exercise. J Appl Physiol 1993;75:1477–1485.

29. Hargreaves M, Costill DL, Coggan A, Fink WJ, Nishibata I. Effect of carbohydrate feedings on muscle glycogen utilization and exercise performance. Med Sci Sports Exerc 1984;16:219–222.

30. Ivy JL, Katz AL, Cutler CL, Sherman WM, Coyle EF. Muscle glycogen synthesis after exercise: effect of time of carbohydrate ingestion. J Appl Physiol 1988;64:1480–1485.

31. Tarnopolsky MA, Bosman M, Macdonald JR, Vendeputte D, Martin J, Roy BD. Postexercise protein-carbohydrate and carbohydrate supplements increase muscle glycogen in men and women. J Appl Physiol 1997;83:1877–1883.

32. Anthony JC, Anthony TG, Layman DK. Leucine supplementation enhances skeletal muscle recovery in rats following exercise. J Nutr 1999;129:1102–1106.

33. Van Loon LJ, Kruijshoop M, Verhagen H, Saris WH, Wagenmakers AJ. Ingestion of protein hydrolysate and amino acid-carbohydrate mixtures increases postexercise plasma insulin responses in men. J Nutr 2000;130:2508–2513.

34. Van Loon LJ, Saris WH, Kruijshoop M, Wagenmakers AJ. Maximizing postexercise muscle glycogen synthesis: carbohydrate supplementation and the application of amino acid or protein hydrolysate mixtures. Am J Clin Nutr 2000;72:106–111.

35. Jentjens RL, van Loon LJ, Mann CH, Wagenmakers AJ, Jeukendrup AE. Addition of protein and amino acids to carbohydrates does not enhance postexercise muscle glycogen synthesis. J Appl Physiol 2001;91:839–846.

36. Ivy JL, Goforth HW Jr, Damon BM, McCauley TR, Parsons EC, Price TB. Early postexercise muscle glycogen recovery is enhanced with a carbohydrate-protein supplement. J Appl Physiol 2002;93:1337–1344.

37. Levenhagen DK, Carr C, Carlson MG, Maron DJ, Borel MJ, Flakoll PJ. Postexercise protein intake enhances whole-body and leg protein accretion in humans. Med Sci Sports Exerc 2002;34:828–837.

38. Levenhagen DK, Gresham JD, Carlson MG, Maron DJ, Borel MJ, Flakoll PJ. Postexercise nutrient intake timing in humans is critical to recovery of

leg glucose and protein homeostasis. Am J Physiol Endocrinol Metab 2001;280:E982–E993.

39. Williams MB, Raven PB, Fogt DL, Ivy JL. Effects of recovery beverages on glycogen restoration and endurance exercise performance. J Strength Cond Res 2003;17:12–19.

40. Berardi JM, Price TB, Noreen EE, Lemon PW. Postexercise muscle glycogen recovery enhanced with a carbohydrate-protein supplement. Med Sci Sports Exerc 2006;38:1106–1113.

41. Bigard AX, Sanchez H, Claveyrolas G, Martin S, Thimonier B, Arnaud MJ. Effects of dehydration and rehydration on EMG changes during fatiguing contractions. Med Sci Sports Exerc 2001;33:1694–1700.

42. Schoffstall JE, Branch JD, Leutholtz BC, Swain DE. Effects of dehydration and rehydration on the one-repetition maximum bench press of weight-trained males. J Strength Cond Res 2001;15:102–108.

43. McMurray RG, Williams DK, Battaglini CL. The timing of fluid intake during an Olympic distance triathlon. Int J Sport Nutr Exerc Metab 2006;16:611–619.

44. Slater GJ, Rice AJ, Sharpe K, Jenkins D, Hahn AG. Influence of nutrient intake after weigh-in on lightweight rowing performance. Med Sci Sports Exerc 2007;39:184–191.

45. Murray B. The role of salt and glucose replacement drinks in the marathon. Sports Med 2007;37:358–360.

46. Lemon PW. Protein and amino acid needs of the strength athlete. Int J Sport Nutr 1991;1:127–145.

47. Evans WJ. Protein nutrition and resistance exercise. Can J Appl Physiol 2001;26:S141–S152.

48. Lemon PW, Berardi JM, Noreen EE. The role of protein and amino acid supplements in the athlete's diet: does type or timing of ingestion matter? Curr Sports Med Rep 2002;1:214–221.

49. Tipton KD, Elliot TA, Cree MG, Aarsland AA, Sanford AP, Wolfe RR. Stimulation of net muscle protein synthesis by whey protein ingestion before and after exercise. Am J Phys Endocrinol Metab 2007;292:E71–E76.

50. Phillips SM, Tipton KD, Aarsland A, Wolf SE, Wolfe RR. Mixed muscle protein synthesis and breakdown after resistance exercise in humans. Am J Physiol 1997;273:E99–E107.

51. Poortmans JR, Dellalieux O. Do regular high protein diets have potential health risks on kidney function in athletes? Int J Sport Nutr Exerc Metab 2000;10:28–38.

52. Tarnopolsky MA, MacDougall JD, Atkinson SA. Influence of protein and training status on nitrogen balance and lean body mass. J Appl Physiol 1988;64:187–193.

53. Robergs RA, Pearson DR, Costill DL, et al. Muscle glycogenolysis during differing intensities of weight-resistance exercise. J Appl Physiol 1991;70:1700–1706.

54. Tesch PA, Colliander EB, Kaiser P. Muscle metabolism during intense, heavy resistance exercise. Eur J Appl Physiol Occup Physiol 1986;55:362–366.

55. Roy BD, Tarnopolsky MA, MacDougall JD, Fowles J, Yarasheski KE. Effect of glucose supplement on protein metabolism after resistance training. J Appl Physiol 1997;82:1882–1888.

56. Haff GG, Koch AJ, Potteiger JA, et al. Carbohydrate supplementation attenuates muscle glycogen loss during acute bouts of resistance training. Int J Sports Nutr Exerc Metab 2000;10:326–339.

57. Haff GG, Schroeder CA, Koch AJ, Kuphal KE, Comeau MJ, Potteiger JA. The effects of supplemental carbohydrate ingestion on intermittent isokinetic leg exercise. J Sports Med Phys Fit 2001;41:216–222.

58. Kalliokoski KK, Kemppainen J, Larmola K, et al. Muscle blood flow and flow heterogeneity during exercise studied with positron emission tomography in humans. Eur J Appl Phys 2000;83:395–401.

59. Lasko M, Edelman SV, Brechtel G, Baron AD. Decreased effect of insulin to stimulate skeletal muscle blood flow in obese man: a novel mechanism for insulin resistance. J Clin Invest 1990;85:1844–1852.

60. Cardillo C, Kilcoyne CM, Nambi SS, Cannon RO 3rd, Quon MJ, Panza JA. Vasodilator responses to systemic but not to local hyperinsulinemia in the human forearm. Hypertension 1998;32:740–745.

61. Chandler RM, Byrne HK, Patterson JG, Ivy JL. Dietary supplements affect the anabolic hormones after weight-training exercise. J Appl Physiol 1994;76:839–845.

62. Biolo G, Tipton KD, Klein S, Wolfe RR. An abundant supply of amino acids enhances the metabolic effect of exercise on muscle protein. Am J Physiol Endocrinol Metab 1997; 273:E122–E129.

63. Tipton KD, Rasmussen BB, Miller SL, et al. Timing of amino acid-carbohydrate ingestion alters anabolic responses of muscle to resistance exercise. Am J Physiol Endocrinol Metab 2001;281:E197.

64. Biolo G, Maggi SP, Williams BD, Tipton KD, Wolfe RR. Increased rates of muscle protein turnover and amino acid transport after resistance exercise in humans. Am J Physiol Endocrinol Metab 1995;268:E514–E20.

65. Tipton KD, Ferrando AA, Phillips SM, Doyle D Jr, Wolfe RR. Postexercise net protein synthesis in human muscle from orally administered amino acids. Am J Physiol Endocrinol Metab 1999;276:E628–E634.

66. Farrell PA, Fedele MJ, Vary TC, Kimball SR, Lang CH, Jefferson LS. Regulation of protein synthesis after acute resistance exercise in diabetic rats. Am J Physiol Endocrinol Metab 1999;276:E721–E727.

67. Fedele MJ, Hernandez JM, Lang CH, et al. Severe diabetes prohibits elevations in muscle protein synthesis after acute resistance exercise in rats. J Appl Physiol 2000;88:102–108.

68. Fluckey JD, Vary TC, Jefferson LS, Farrell PA. Augmented insulin action on rates of protein synthesis after resistance exercise in rats. Am J Physiol Endocrinol Metab 1996;270:E313–E319.

69. Biolo G, Williams BD, Fleming RY, Wolfe RR. Insulin action on muscle protein kinetics and amino acid transport during recovery after resistance exercise. Diabetes 1999;48:949–957.

70. Borsheim E, Aarsland A, Wolfe RR. Effect of an amino acid, protein, and carbohydrate mixture on net muscle protein balance after resistance exercise. Int J Sport Nutr Exerc Metab 2004;14:255–271.

71. Tipton KD, Elliot TA, Cree MG, Aarsland AA, Sanford AP, Wolfe RR. Stimulation of net muscle protein synthesis by whey ingestion before and after exercise. Am J Phys Endocrinol Metab 2007;292:E71–E76.

72. Roy BD, Fowles JR, Hill R, Tarnopolsky MA. Macronutrient intake and whole body protein metabolism following resistance exercise. Med Sci Sports Exerc 2000;32:1412–1418.

73. Bird SP, Tarpenning KM, Marino FE. Effects of liquid carbohydrate/essential amino acid ingestion on acute hormonal response during a single bout of resistance exercise in untrained men. Nutrition 2006;22:367–375.

74. Bird SP, Tarpenning KM, Marino FE. Liquid carbohydrate/essential amino acid ingestion during a short-term bout of resistance exercise suppresses myofibrillar protein degradation. Metabolism 2006;55:570–577.

75. Rassmussen BB, Tipton, KD, Miller SL, Wolf SE, Wolfe RR. An oral essential amino acid-carbohydrate supplement enhances muscle protein anabolism after resistance exercise. J Appl Physiol 2000;88:386–392.

76. Miller SL, Tipton KD, Chinkes DL, Wolf SE, Wolfe RR. Independent and combined effects of amino acids and glucose after resistance exercise. Med Sci Sports Exerc 2003;35:449–455.

77. Borsheim E, Tipton KD, Wolf SE, Wolfe RR. Essential amino acids and muscle protein recovery from resistance exercise. Am J Phys Endocrinol Metab 2002;283:E648–E657.

78. Dangin M, Boirie Y, Garcia-Rodenas C, et al. The digestion rate of protein is an independent regulating factor of postprandial protein retention. Am J Physiol Endocrinol Metab 2001;280:E340–E348.

79. LaCroix M, Bos C, Leonil J, et al. Compared with casein or total milk protein, digestion of milk soluble proteins is too rapid to sustain the anabolic postprandial amino acid requirement. Am J Clin Nutr 2006;84:1070–1079.

80. Gleeson M, Blannin AK, Walsh NP, Bishop NC, Clark AM. Effect of low- and high-carbohydrate diets on the plasma glutamine and circulating leukocyte responses to exercise. In J Sport Nutr 1998;8:49–59.

81. Nieman DC, Johanssen LM, Lee JW, Arabatzis K. Infectious episodes in runners before and after the Los Angeles marathon. J Sports Med Phys Fitness 1990;30:316–328.

82. Bishop NC, Blannin AK, Walsh NP, Gleeson M. Carbohydrate beverage ingestion and neutrophil degranulation responses following cycling to fatigue at 75% VO_2max. Int J Sports Med 2001;22:226–231.

83. Maclean DA, Graham TE, Saltin B. Branched-chain amino acids augment ammonia metabolism while attenuating protein breakdown during exercise. Am J Physiol 1994;267:E1010–E1022.

84. Nehlson-Cannarella SL, Fagoaga OR, Nieman DC, et al. Carbohydrate and the cytokine response to 2.5 h of running. J Appl Physiol 1997;82:1662–1667.

85. Peters EM, Goetzsche JM, Grobbelaar B, Noakes TD. Vitamin C supplementation reduces the incidence of postrace symptoms of upper-respiratory-tract infection in ultramarathon runners. Am J Clin Nutr 1993;57:170–174.

86. Palmer FM, Nieman DC, Henson DA, et al. Influence of vitamin C supplementation on oxidative and salivary IgA changes following an ultramarathon. Eur J Appl Physiol 2003;89:100–107.

87. Douglas RM, Hemila H, D'Souza R, Chalker EB, Treacy B. Vitamin C for preventing and treating the common cold. Cochrane Database Syst Rev 2004;18:CD000980.

88. Potteiger JA, Chan MA, Haff GG, et al. Training status influences T-cell responses in women following acute resistance exercise. J Strength Cond Res 2001;15:185–191.

89. Tauler P, Aquilo A, Gimeno I, Fuentespina E, Tur JA, Pons A. Response of blood cell antioxidant enzyme defenses to antioxidant diet supplementation and to intense exercise. Eur J Nutr 2006;45:187–195.

90. Esmark B, Andersen JL, Olsen S, Richter EA, Mizuno M, Kiaer M. Timing of postexercise protein intake in important for muscle hypertrophy with resistance training in elderly humans. J Physiol 2001;535(Pt 1):301–311.

91. Andersen LL, Tufekovic G, Zebis MK, et al. The effect of resistance training combined with timed ingestion of protein on muscle fiber size and muscle strength. Metabolism 2005;54:151–156.

92. Chromiak JA, Smedley B, Carpenter W, et al. Effect of a 10-week strength training program and recovery drink on body composition, muscular strength and endurance, and anaerobic power and capacity. Nutrition 2004;20:420–427.

93. Bird SP, Tarpenning KM, Marino FE. Independent and combined effects of liquid carbohydrate/essential amino acid ingestion on hormonal and muscular adaptations following resistance training in untrained men. Eur J Appl Physiol 2006;97:225–238.

94. Cribb PJ, Hayes A. Effects of supplement timing and resistance exercise on skeletal muscle hypertrophy. Med Sci Sports Exerc 2006;38:1918–1925.

95. Cribb PJ, Williams AD, Stathis CG, Carey MF, Hayes A. Effects of whey isolate, creatine, and resistance training on muscle hypertrophy. Med Sci Sports Exerc 2007;39:298–307.

Part IV
Present and Future Directions
of Nutritional Supplements

14 Future Trends
Nutritional Supplements in Sports and Exercise

Marie Spano and Jose Antonio

Abstract

The field of sports nutrition is defined not only by dietary recommendations for various athletes, research and new supplements that are on store shelves but also by the direction of the industry itself. Consumer spending, media coverage, professional athlete endorsement of various supplements, lawsuits, regulations in governing bodies and clinical research all have an impact on the direction and growth of the sports nutrition industry. To date, no supplement has affected sports nutrition as much as creatine and the company that both funded most of the research supporting the ergogenic benefits of creatine and capitalized on such research. There is no current leader in the sports nutrition market. Instead, companies are vying among steady competition for space on store shelves and overall product sales.

Key words

Industry · Trends · Market · Supplements · Low carbohydrate · Sales · Growth · Research

1. INTRODUCTION

Over the past several years, the sports nutrition industry has been growing at a steady rate; and according to forecasts from market research firms and the *Nutrition Business Journal* (NBJ), it is expected to continue this growth in future years. As sports nutrition research uncovers new supplements that have the potential to make us run faster, lift more, and perform better, consumer interest grows and the demand for knowledge increases. Individual athletes as well as teams are increasingly seeing the benefits of sports nutrition

From: *Nutritional Supplements in Sports and Exercise*
Edited by: M. Greenwood, D. Kalman, J. Antonio,
DOI: 10.1007/978-1-59745-231-1_14, © Humana Press Inc., Totowa, NJ

counseling as well. When races are won by mere fractions of a second and games may be lost in overtime due to fatigue or cramping, coaches and athletes alike cannot afford to overlook any potential ergogenic aid.

In addition to new research, marketing, trends (e.g., the low carbohydrate era), media coverage, lawsuits, and ever-changing regulations in sport all define the sports nutrition industry and the direction in which it moves (1).

2. HOW IS THIS INDUSTRY DEFINED?

The sports nutrition industry is a conglomerate of several types of products including supplements, sports and energy beverages, nutrition bars, protein powders, and even foods that are marketed specifically to athletes. However, when analyzing this industry and examining sales data, the sports nutrition industry is often lumped with the weight loss industry (making it the sports nutrition and weight loss, or SNWL, industry) because attributing products solely to one category (without overlapping sales from dual buyers in both categories) or another proves to be an arduous task. In addition, the SNWL industry includes sales from companies that carry only sports nutrition products marketed solely to athletes along with several crossover products such as nutrition bars that may be marketed to both the athlete and general person with a busy lifestyle. Teasing out information on who is buying various products and why they are buying them (to pursue an athletic goal versus just making a lifestyle choice or for aesthetic reasons), is nearly impossible. Therefore, sales of products that have multiple uses associated with them, such as nutrition bars and energy beverages, are included in the SNWL industry (1).

3. WHAT IS SELLING?

The SNWL industry amassed a total $16.8 billion in sales in 2005, a 7% increase over the previous year (Table 1). This category was led by a strong showing from sports and energy beverages, which grew 27% over the previous year, bringing in $7.1 billion (sports drinks represented 4.6 billion). The sports drinks subcategory (which includes all beverages marketed as a sports beverage or those

Table 1
United States Sport Nutrition and Weight Loss Category Market:
2003–2005

SNWL category	2003	2004	2005	2005% Growth
Weight-loss pills	1796	1704	1626	−5
Weight-loss LMRs	2295	2096	2047	−2
Sports supplements*	1980	2100	2220	6
Low-CHO foods**	889	2237	1871	−16
Nutrition bars	2069	2208	1947	−4
Sports/energy drinks	4750	5600	7100	27

From Sports nutrition and weight loss markets VI. Nutr Bus J
2006;11(9):1–12, with permission. *Includes powders, pills and hardcore
bodybuilding RTDs. **Not including bars and meal replacements.
LMR, liquid meal replacement; CHO, carbohydrate.
Data are millions of dollars.

marketed to enhance performance, such as Powerade and All Sport)
has been led by Gatorade since its inception (Gatorade accounts
for approximately 80% of sales) (1). New entrants during the
past few years have started to differentiate themselves by adding
protein to their electrolyte-replacement, glucose-containing solutions
as a result of research indicating that carbohydrate-protein beverages
may delay fatigue and attenuate muscle damage during endurance
exercise (2).

In addition to what we typically think of as sports beverages,
energy drinks (Red Bull, Redline, Monster Energy, Vault, and similar
drinks that contain stimulants) have also represented this category
well over the past several years. Although these are included in the
sports drinks and energy beverages category, they have gained more
notoriety for being mixed with various types of liquor at bars and
parties as well as keeping college students and busy professionals
from falling asleep during the day. These drinks have done so well
that this entire category gained significant notoriety when stock of
Hansen's Natural (Monster Energy) climbed dramatically over a
5-year period. In 2005, the sports supplements category grew by 6%
for a total of $2.2 billion (1).

As of this writing, weight loss supplements have yet to regain ground after the U.S. Food and Drug Administration (FDA) ruled that ephedra was too dangerous to be kept on store shelves. From 2003 to 2005, weight loss pills experienced a $0.5 billion decrease in sales from $2.1 billion to $1.6 billion. With the continual introduction of additional prescription weight loss medicines, weight loss supplements have the potential to slide even further unless stellar introductions to this category stand out from the crowd. Other weight loss products have also experienced a hit due to a decrease in demand for low carbohydrate products. Low carbohydrate foods declined 31% from 2004 to 2005. Formerly "low carbohydrate" companies have responded to this drop by repositioning themselves as "low glycemic" or "carbohydrate controlled," thereby trying to capture those looking for a healthy lifestyle versus solely being associated with dieting. Nutrition bars ($1.9 billion in sales in 2005) continued a slow, steady decline in 2005. Despite a decrease in overall sales in the bar category, several new entrants made their way to store shelves in 2005. Even cereal bar and candy companies have tried to gain a part of this market by creating bars marketed as nutrition or energy bars. According to NBJ, bars that stand out have differentiated themselves from the crowd. These include raw, organic, and natural ingredient bars *(1)*.

4. TRENDS

Low carbohydrate products have continued on a downward trend over the last few years. The new marketing strategy of low glycemic and carbohydrate controlled may do well depending on developing research in this area and how well this research is disseminated. Currently, trans fats have stolen the spotlight as the nutritional bad guy, making carbohydrate content all but an afterthought. Nutrition bars have also been steadily declining in total sales over the past few years. This has been blamed on lack of innovation and saturation in this category. Regardless, expect to see more organic, gluten-free, raw, and other specialized bars hit the market. Finally, as mentioned previously, weight loss supplements are on the decline with their future dependent on prescription weight loss medicine sales, media spin, legal action, and innovation *(1)*.

Although ephedra scared many consumers from taking thermogenic aids, new products are hitting the shelves that contain no central nervous system (CNS) stimulants and have clinical data forthcoming (the brown seaweed containing Fucothin from Garden of Life for example). If these products do well, it is possible that other companies will create nonstimulatory weight loss products as well, giving this category a much needed boost.

Mass appeal? Many companies have opted to bring their sports nutrition brand to the mass market in an effort to reach more consumers, thereby generating greater total sales. For instance, a product may start out in the sports nutrition market, advertising in fitness publications, and then change its marketing strategy to gain broader appeal in the mass market. In fact, mass market sales accounted for a whooping two-thirds of all SNWL sales in 2005, so companies often cannot resist this crossover. By making this switch, most companies know that price is a key driver, and so they may reformulate products using less expensive ingredients in an effort to bring the price down. By doing this and placing products on grocery, convenience, and club store shelves, companies are faced with the potential to lose core customers once the brand goes "public" *(1)*.

4.1. A Breakthrough in Carnitine Research

L-Carnitine was popularized in the 1980s as a fat-burning supplement. The premise behind this concept is sound; carnitine is a "carrier" that transports fatty acids across the mitochondrial membrane for β-oxidation to occur. This is a fancy way of saying that carnitine plays an essential role in fat oxidation *(3)*. However, previous research indicates that supplemental L-carnitine does not increase skeletal muscle carnitine content, making it all but useless in terms of direct enhancement of endurance or fat loss *(4–6)*.

Recently, scientists have discovered a way to make carnitine supplementation more effective. Raising serum insulin levels at the same time carnitine is taken appears to increase muscle carnitine content. In a study from the University of Nottingham, scientists used a 6-hour euglycemic hyperinsulinemic insulin clamp to infuse insulin at a steady rate in seven healthy men. One hour after the start of the insulin clamp, subjects were infused with 60 mM carnitine or an equivalent volume of saline over the remaining 5 hours of the insulin infusion period. Total carnitine content in skeletal muscle increased significantly, by 15%, in subjects infused with carnitine, whereas those infused with saline experienced no change. In addition, muscle glycogen content increased significantly in both groups during the insulin infusion. At 24 hours after the study, muscle glycogen content increased again, above study values in the carnitine-supplemented group. Muscle pyruvate dehydrogenase complex activity, which plays an important role in fatty acid synthesis and whole-body carbohydrate regulation, was decreased by 30% in the carnitine-supplemented group. These results suggest that the increase in muscle carnitine content led to decreases in carbohydrate oxidation, increased glycogen storage, and increased fatty acid oxidation *(7)*. Earlier research also supports the notion that increased insulin levels support muscle carnitine content during carnitine supplementation.

The same group of researchers examined a more practical way to raise insulin levels—by ingesting simple carbohydrates. In this study, 94 g of glucose (in a beverage) was used to facilitate increases in insulin during carnitine supplementation (3 g/day). In comparison to placebo and carnitine 3 g/day, the group supplemented with carbohydrate showed a significant decrease in urinary carnitine excretion, indicating increased carnitine retention associated with concurrent carbohydrate intake *(6)*.

As we know, carbohydrates vary tremendously in their ability to help us rapidly synthesize glycogen after exercise. The recommendation has always been to ingest simple sugars that rapidly spike insulin levels, thereby replenishing glycogen stores quickly as well. However, even simple carbohydrates differ; higher-molecular-weight carbohydrates are superior in their ability to help stimulate glycogen synthesis *(8)*.

5. DRIVING SALES AND GROWTH

Consumer demand drives sales, but how does a company increase such demand? Effective marketing is essential, as is product taste. If it does not taste good, chances are it will not continue to sell. Product cost also affects sales. Prices are determined based on raw ingredient costs, production costs, and of course consumer demand. As commodity prices increase baseline product cost goes up and therefore selling price increases. Whereas some consumers are willing to pay a premium for certain products, others are not; and cheaper versions of such products may jump in to steal sales.

To unlock the key to massive growth, a company should emulate what EAS, a prominent sports nutrition company, did during the 1990s—combine sound scientific research with effective marketing. Stated simply, "products that do the best have the most evidence behind them," according to Daniel Fabricant, PhD, Vice President of Scientific and Regulatory Affairs for the Natural Products Association. After doing this, a company must continue to be innovative regarding their marketing materials and products.

Although there is a great deal of interest in scientific research on sports nutrition products and novel ingredients, funding a sound clinical research trial is pricey; therefore, many companies skimp on product testing and clinical studies and, instead, rely on extrapolated research from other companies to make product claims. In other instances, manufacturers have taken basic or clinical research studies on ingredients or another company's product, combined with the knowledge of advanced physiology, and woven them together to entice consumers. Indeed, there is a big discrepancy: Some supplements are adequately researched, whereas research is sorely lacking on others.

Other factors that affect sales and growth include positive or negative media coverage; product innovation; market saturation (e.g., nutrition bars); exaggerated marketing claims leading to unrealistic consumer expectations and subsequent disappointment or, worse, legal action; lack of consumer education; and inaccurately labeled supplements.

Both sales and growth in this industry depend, in part, on the economy as well. Supplements, bars, and protein powders are not necessities for most but, instead, luxuries and therefore items that are dropped from the shopping cart by those who are not competing in a bodybuilding show or aiming to win their next Ironman triathlon.

Product considerations include the following.

- What demographic is being targeted? Are there multiple target audiences?
- Supplement form (powder, pill, capsule, drink)

- Method/methods of sale (internet, multilevel marketing company, through health food and vitamin stores, through large membership-only stores)
- Function (energy, fat loss, recovery, muscle building)
- Scientific merit (Are there clinical trials behind the product? If not, is extrapolated science being used to help promote the product? Or is science not even mentioned?)
- Environmental factors (cost, taste, variety, availability/access, circle of influence)
- Quality (What ingredients are used, how are they processed and how much of each?) *(9)*

6. FUTURE MARKET PREDICTIONS

Expect to see further growth and differentiation in the sports drinks category with other companies jumping on board to create drinks with added protein and various other ingredients. Although the weight loss category has dropped sales, our nation is growing in size; and the demand for anything "weight loss" is likely to grow with it.

Given the research on both essential amino acids (EAAs) and branched-chain amino acids (BCAAs), it is surprising that few EAA and BCAA products currently exist. As additional published research continues to extol the benefits of EAAs and BCAAs, expect to see companies developing more "amino shooter" (Champion Nutrition) type products for preworkout as well as sports drinks and recovery products enhanced with EAAs or BCAAs.

Creatine and protein are still staples and bring in an estimated $220 million and $1.4 billion, respectively, in sales according to the NBJ *(1)*. Therefore, companies are still trying to capitalize on their appeal by slightly tweaking them to stand out on store shelves and gain part of this market share.

Significant research interest in the following supplements will ensure an increase in product offerings that contain the following: β-alanine, EAAs, BCAAs, leucine, protein (soy, whey, casein), L-carnitine, phosphatidylserine, epigallocatechin (EGCG), and green tea extract. In addition, recovery is a hot research topic, and therefore a breakout product that can mitigate muscle damage, soreness, and inflammation may lead the pack.

7. RESEARCH DIRECTIONS

Years after its introduction to the market, creatine still stands out as the supplement with the most research backing its efficacy. However, sports scientists are uncovering the potential of new supplements and ingredients and changes to existing ones. EAAs, BCAAs, β-alanine, and nutrient timing are currently paving the research path.

β-alanine, an intracellular buffer, is a rate-limiting substrate in carnosine production, working largely as a buffer to offset acid production in the muscle. Delaying acid production may delay muscular fatigue and hasten recovery. Why not just take carnosine? When carnosine enters the digestive system, it is hydrolyzed into histidine and β-alanine, which are then taken up by skeletal muscle and synthesized into carnosine. Owing to immediate hydrolysis, carnosine cannot be taken up into the muscle intact.

What has the research thus far told us? In one study, 13 male subjects (age 25.4 ± 2.1 years) supplemented with 4 to 6 g of β-alanine every day for 10 weeks increased muscle carnosine 58.8% and 80.0% after 4 and 10 weeks, respectively. An exhaustive cycling test at 110% maximum power was then completed, which resulted in a 13% increase in total work after 4 weeks of supplementation and a further increase of 3.2% after 10 weeks, with no changes in the placebo group (10). Two other studies examined the effects of a combination of β-alanine and creatine on various aspects of athletic performance. In a double-blind, placebo-controlled, randomized study, researchers assigned 55 males (24.5 ± 5.3 years) to one of four groups for 4 weeks of treatment: placebo, creatine 5.25 g/day, β-alanine 1.6 g/day, or β-alanine 1.6 g/day + creatine 5.25 g/day. The participants then completed exhaustion tests on a cycle ergometer. No between-group differences were noted pretest to posttest. The group supplemented with β-alanine + creatine demonstrated significant improvement from pretest to posttest in five of the eight indices of cardiorespiratory endurance measured. More specifically, this group improved maximum oxygen uptake and power output at metabolic thresholds as well as the percentage of maximum oxygen consumption, which was maintained at ventilation threshold (the point at which ventilation deviates from a steady linear increase and instead increases exponentially). The creatine-only group showed pretest to posttest improvement in total time to exhaustion and power output at ventilation threshold. The β-alanine-only group showed an increase in power output at metabolic thresholds from pretest to posttest. The results of this study suggest that supplementation with creatine + β-alanine may enhance submaximum endurance performance (11).

In a second study, 33 male college football players were randomized into one of three groups over the course of a 10-week resistance training program: β-alanine 3.2 g/day + creatine 10.5 g/day, creatine 10.5 g/day only, or placebo (dextrose 10.5 g/day). Significantly greater improvements in strength, muscle mass, and percent body fat were noted in the β-alanine + creatine group compared with either the creatine-only group and the placebo group (12). Although intense training increases muscle levels of creatine and carnosine, supplementation even over brief periods can raise the levels even higher. Therefore, loading with β-alanine 4 to 6 g/day may lead to prolonged exercise bouts by delaying acid production. This, in turn, might enable athletes to recover from strenuous training more rapidly. Additionally, the combination of β-alanine + creatine may increase strength and muscle mass more than creatine alone.

Over the past few years quite a bit of research has focused on the timing of EAA and protein intake and muscle protein synthesis (MPS). The exact dosage of EAA per kilogram body weight necessary to stimulate MPS is unknown, but research has determined that there is a dose-response relation to EAA consumption and MPS, although as little as 6 g EAA effectively stimulates MPS (13) and nonessential amino acids play no role in MPS (13,14). MPS responds differently to the timing of EAA intake versus protein (pre- or post-exercise), and amino acid delivery to the muscle is greater with ingestion of EAA than intact protein prior to exercise (15). Current research-based recommendations include taking EAA before exercise and a combination of whey and casein within 2 hours after exercise.

Expect to see future research delving deeper into more precise nutrient ratios and timing to yield increased MPS while minimizing fat gain (15,16). In addition, future research will most likely continue to explore an athlete's protein needs during exercise. Does a carbohydrate-protein beverage increase time-to-fatigue and/or work performance in endurance athletes in comparison to an iso-caloric carbohydrate-only beverage? Moreover, do additional studies support published research indicating that the addition of protein during an endurance bout may indeed decrease muscle damage *(17)*?

7.1. Legal Matters: An Interview with Rick Collins

Question: What laws affect the sports nutrition industry?

Answer: Congress amended the Federal Food, Drug and Cosmetic Act by passing the Dietary Supplement Health and Education Act of 1994 (DSHEA). DSHEA defines the term "dietary supplement" and establishes a regulatory framework for dietary supplements. It protects consumer access to dietary supplements while giving the FDA regulatory authority over the industry. For example, the FDA can remove from the market products that pose a "significant or unreasonable" risk to consumers—as they did with ephedra supplements in 1994—or that are otherwise deemed misbranded, contaminated, or adulterated. DSHEA requires that supplement labeling be accurate. The Food, Drug and Cosmetic Act (FDCA) sets forth the guidelines for dietary supplement claims and the FDA works in conjunction with the Federal Trade Commission (FTC) to regulate supplement claims and advertising.

When consumers are harmed by products, they may sue the responsible parties for their injuries. These lawsuits may seek individual monetary damages. If a group of consumers are harmed, they may collectively file a class action lawsuit seeking redress for all similarly situated persons. Also, there have been cases where elite competitive athletes, including Olympic athletes and football players, have blamed positive doping tests on contaminated sports nutrition products. In some cases, these athletes have recovered monetary damages from the manufacturers.

Criminal laws also may affect the industry. For example, federal legislation passed in 2004 classified a wide variety of steroid precursor products as anabolic steroids. Formerly marketed to sports nutrition consumers as "prohormones" or "prosteroids," the new classification of these products makes them illegal to possess, manufacture, or sell under the Controlled Substances Act.

Question: Is any future legislation on the horizon that may affect the sports nutrition industry?

Answer: There are three significant pieces of new legislation affecting the Sports Nutrition industry today. In 2006, Congress passed the Dietary Supplement and Nonprescription Drug Consumer Act (S.3546), a bill that requires dietary supplement manufacturers to report to the FDA serious adverse events associated with product use. This legislation, taking effect on December 22, 2007, may have a significant impact on the dietary supplement industry, particularly small businesses.

The federal legislation criminalizing numerous steroid precursor supplements explicitly exempted the popular supplement DHEA (dehydroepiandrosterone). Senate Bill 762 seeks to reclassify DHEA as a controlled substance, effectively removing it from health food store shelves. The bill is identical to legislation proposed in 2005 that died in committee. There are those who believe that DSHEA does not do enough to regulate the supplement market and who would like to impose a stringent regulatory framework similar to the prescription drug model. However, many consumers are concerned that such heightened regulation would deprive them of the range of supplement choices they currently enjoy. Consequently, when Senate Bill 1082 was proposed, it caused quite a stir throughout the industry. The bill is entitled, "An act to amend the Federal Food, Drug, and Cosmetic Act and the Public Health Service Act to reauthorize drug and device user fees and ensure the safety of medical products, and for other purposes." However, the proposed legislation contains a provision establishing the Reagan-Udall Foundation for the FDA. Ambiguity with regard to the purpose of the Foundation caused some to suggest that it threatened the future of DSHEA. In response, Senators Enzi and Kennedy have clarified their intention for the legislation, indicating that, "the purpose of the nonprofit Foundation is to lead collaborations among the FDA, academic research institutions and industry designed to bolster research and development productivity, provide new tools for improving safety in regulated product evaluation, and in the long-term make the development of those products more predictable and manageable."

8. REGULATIONS IN SPORT

Through the years various governing bodies have stepped up their regulations in regard to dietary supplement use among their athletes. Events such as the BALCO scandal only make these governing bodies more aware of substances that have the potential to give an athlete an unfair advantage or subject an unknowing athlete to a positive drug test. From the U.S. Olympic Committee (USOC) to major league baseball, each sports association has a different list of substances it bans. This list only gets longer as scientists create various new designer steroids that are often undetectable by current testing programs.

The National Collegiate Athletic Association (NCAA) too bans several substances classified into the following categories: stimulants, anabolic agents, diuretics, substances banned for specific sports, street drugs, peptide hormones, and analogues. In addition to the typical banned substances, an NCAA bylaw allows for Division 1 institutions to provide only non-muscle-building nutritional supplements for the purpose of additional calories and electrolytes. Amino acids and nutrition supplements with more than 30% of their calories from protein are considered "muscle building" and are therefore not allowed—a regulation that has been surrounded by much controversy. However, this regulation has opened the doors for various companies that target their sales to large universities to create "NCAA-compliant" products including protein powders and bars that are just under 30% protein. Vitamins and minerals, energy bars (that are under 30% protein), calorie replacement drinks, and noncaffeinated sports drinks are permitted by the NCAA; therefore, a company stands to gain significant sales from a university if it can provide several of these products along with NCAA-compliant meal replacement powders containing the allowed limit of protein.

Alas, regulations in sports have paved the way for companies to make new products that help athletes meet their nutritional needs without testing positive. In addition, such regulations have opened the opportunity for companies to create programs to evaluate and certify supplements that meet the criteria for being "banned substance free." Third party testing provides skeptical athletes and coaches with a greater level of assurance regarding the product they are consuming.

8.1. Creating a Category

8.1.1. JOSE ANTONIO

What follows is not a scientific treatise but an analysis of how marketing and positioning affects products and brands. Certainly, science plays a role in positioning a product or brand. Nonetheless, in the sports nutrition market, or any market for that matter, the fundamental marketing/sales goal is the creation of a new and successful category. In the automobile industry, perhaps one the most successful "new" categories is the SUV. Despite the fact that many so-called SUVs look like your mom and dad's old-fashioned station wagon, the SUV category is robust and here to stay. In the exercise equipment world, we see new "categories" of products as well. What about the sports nutrition category?

In recent sports nutrition history, perhaps the biggest "new" category creation is creatine, particularly creatine monohydrate. Since the landmark work of Roger Harris and Eric Hultman, creatine sales have reached hundreds of millions. In fact, one could argue that the advent of creatine "legitimized" the sports supplement industry. Here was a supplement against which no one could reasonably argue. The data are plentiful regarding the safety and efficacy of creatine (18–27). Interestingly, in the hopes of finding the next blockbuster product, there have been "subcategories" of creatine that have been launched but with limited success. We had creatine "jubes" (creatine packaged in a Tootsie Roll textured consumable), a subcategory that was not a success. Then, there was liquid creatine in quality packaging. However, at the time of writing this chapter, there is no compelling evidence that a true liquid creatine exists in the marketplace (i.e., liquid creatine that contains predominantly creatine)—meaning that there are no double-blind, placebo-controlled studies showing this to be true. Finally, one of the newer subcategories is CEE, or creatine ethyl ester. Certainly, there are no data to show that CEE is better or worse than creatine monohydrate. Again, double-blind, placebo-controlled studies that compare the two products head to head are needed.

Interestingly, once a category (e.g., creatine) has become enormously successful, the mere omnipresence of the product often pushes it to becoming a commodity. In that case, the commodity (i.e., creatine monohydrate) becomes another undifferentiated product (or service) that is marketed, sold, and purchased based mainly on price rather than any distinguishing features or quality. Hence the reason for supplement companies seeking to differentiate themselves from the crowd by launching a CEE, a creatine jube, a liquid creatine, or some other such product. Regardless of the efficacy (or lack thereof), the goal is to position one's brand or product as superior, better, sexier, and premium compared to the competitors. Differentiate or die—or so the saying goes for companies in the sports nutrition/supplement market trying to stay alive.

Other categories, which range from those that are failures to others with amazing success stories, include myostatin inhibitors (failure), nitric oxide supplements (successful), β-alanine (a new ingredient category), energy drinks (a billion dollar business with Red Bull as the premier brand), vitamin enhanced waters (another billion dollar category), fortified coffee (a new and growing category, the value of which remains to be determined), sports drinks (Gatorade), and nutrient timing among others.

Several sample categories are listed here.

- Myostatin inhibitors—The quick launch and utter failure of the myostatin inhibitor category could be ascribed, in my opinion, solely to the fact that the products did not work. Furthermore, the notion that supplements could block myostatin or inhibit it seemed far-fetched. Does this category exist today? No.
- Nitric oxide (NO)—This category is interesting in that the primary ingredient in these products is arginine. Is arginine new? No. What else is typically added to NO products? The answer is creatine. However, if you look at the original NO product—one based on the presence AAKG, or arginine α-ketoglutarate—you can see that there are few data showing a positive effect on body composition (19).
- β-Alanine—This is a "new" ingredient that has a modicum of supportive data (10,11,20–22). The sports nutrition industry is hoping that this ingredient becomes a "creatine-like" category. If that occurs, it will spawn subcategories (similar to how creatine spawned CEE and liquid creatine, among others).
- Energy drinks—Red Bull is perhaps the best known brand in this category. Combining taurine, caffeine, glucuronolactone, and sugar into a carbonated drink started this category. It resulted in other Red Bull imitators or knockoffs, including other brands such as Monster, AMP, and Daredevil. Interestingly, there are supportive data for Red Bull having an ergogenic effect (23,24).
- Vitamin Water—Coca-Cola purchased Vitamin Water maker Glaceau for $4.2 billion in cash. Again, some of the simplest ideas are often the

most successful. Putting vitamins (a commodity) into water (which is free) resulted ultimately in a product/brand worth billions of dollars.

- Fortified coffee—Here is a concept as simple as Vitamin Water. Take coffee, the second most frequently consumed beverage in the word, and add dietary supplements to it. The first to appear on the market was Javalution, makers of JavaFit coffee; however, there are other brands now such as Boaters Coffee, and the brand sold in 7-Eleven stores, Fusion Energy Coffee. Will the fortified coffee category be as successful as vitamin water?
- Sports drinks—In the sports drink category, Coca-Cola's POWERade is a distant second to Gatorade, which still commands 80% market share of the estimated $1.5 billion sports drink category. The question is why companies keep trying to enter the "-ade" war (e.g., POWERade, Accelerade) when they should be thinking of creating a new category.
- Nutrient timing—This category, interestingly, is not a product but a concept, which has been addressed by a huge volume of scientific research *(18,25)*. Capitalizing on this category is difficult in that it requires consumer education. Nevertheless, there are credible nutrient timing products (most of them protein- or essential amino acid-based) on the market today. Pacific Health Laboratories has expanded on this category via education (i.e., *Nutrient Timing: The Future of Sports Nutrition* by John Ivy and Robert Portman) and a $30 million to $40 million dollar advertising campaign in 2007.

Summary: Is there a secret to product success?

Answer: It is multifactorial. It includes the positioning of the brand, clever marketing, and being the first to market, among other factors. Certainly, however, the real excitement for any business person or entrepreneur is the creation of a new category.

9. CONCLUSION

The sports nutrition industry is shaped by a variety of factors. The direction of future growth is dependent on the products produced, research backing such products, supplement company scandals, lawsuits, and media coverage, among other things. NBJ indicates that this industry is expected to continue to grow at a steady rate. In fact, the potential for upward growth is unlimited for any company that produces sound products based on scientific research or based on a whole new category. In addition, the fizzled low carbohydrate era combined with the ephedra debacle has left a gap in weight loss products and supplements that is yearning to be filled. The sports world is simply a microcosm for the rest of the society; and therefore the SNWL industry mimics societal and economic trends with the future sure to unfold in an exciting fashion.

REFERENCES

1. Sports nutrition and weight loss markets. VI. Nutr Bus Jl 2006;11(9).
2. Saunders MJ, Kane MD, Todd M, et al. Effects of a carbohydrate-protein beverage on cycling endurance and muscle damage. Med Sci Sports Exerc 2004;36:1233–1238.
3. Champe PC, Harvey RA. Metabolism of dietary lipids biochemistry. In: Champe PC, Harvey RA (eds) Lippincott's Illustrated Review: Biochemistry. 2nd ed. Lippincott, Philadelphia, 1994, pp 181–182.
4. Brass EP. Supplemental carnitine and exercise. Am J Clin Nutr 2000;72:618S–623S.
5. Wächter S, Vogt M, Kreis R, et al. Long-term administration of L-carnitine to humans: effects on skeletal muscle carnitine content and physical performance. Clin Chim Acta 2002;318:51–61.
6. Stephens FB, Constantin-Teodosiu D, Laithwaite D, et al. Insulin stimulates l-carnitine accumulation in human skeletal muscle. FASEB J 2006;20:377–379.
7. Stephens FB, Constantin-Teodosiu D, Laithwaite D, et al. An acute increase in skeletal muscle carnitine content alters fuel metabolism in resting human skeletal muscle. J Clin Endocrinol Metab 2007;91:5013–5018.
8. Piehl Aulin K, Soderlund K, Hultman E. Muscle glycogen resynthesis rate in humans after supplementation of drinks containing carbohydrates with low and high molecular masses. Eur J Appl Physiol 2000;81:346–351.
9. Sports Nutrition: Assessing Market Trends and the Mind of the Athlete. Nutraceuticals World webinar, May 2007. http://www.nutraceuticalsworld.com/webinar.php
10. Hill CA, Harris RC, Kim HJ, et al. Influence of β-alanine supplementation on skeletal muscle carnosine concentrations and high intensity cycling capacity. Amino Acids 2007;32:225–233.
11. Zoeller RF, Stout JR, O'Kroy JA, Torok DJ, Mielke M. Effects of 28 days of beta-alanine and creatine monohydrate supplementation on aerobic power, ventilatory and lactate thresholds, and time to exhaustion. Amino Acids 2007;33:505–510.
12. Hoffman J, Ratamess N, Kang J, et al. Effect of creatine and beta-alanine supplementation on performance and endocrine responses in strength/power athletes. Int J Sport Nutr Exerc Metab 2006;16:430–446.
13. Borsheim E, Tipton KD, Wolf SE, et al. Essential amino acids and muscle protein recovery from resistance exercise. Am J Physiol Endocrinol Metab 2002;283:E648–E657.

14. Tipton KD, Ferrando AA, Phillips SM, Doyle D Jr, Wolfe RR. Postexercise net protein synthesis in human muscle from orally administered amino acids. Am J Endocrinol Metab 1999;276:E628–E634.

15. Tipton KD, Rasmussen BB, Miller SL, et al. Timing of amino acid-carbohydrate ingestion alters anabolic response of muscle to resistance exercise. Am J Physiol Endocrinol Metab 2001; 281:E197–E206.

16. Biolo G, Tipton KD, Klein S, et al. An abundant supply of amino acids enhances the metabolic effect of exercise on muscle protein. Am J Physiol 1997;273(Pt 1):E122–E129.

17. Romano-Ely BC, Todd MK, Saunders MJ, et al. Effect of an isocaloric carbohydrate-protein-antioxidant drink on cycling performance. Med Sci Sports Exerc 2006;38:1608–1616.

18. Ziegenfuss TN, Berardi JM, Lowery LM. Effects of prohormone supplementation in humans: a review. Can J Appl Physiol 2002;27:628–646.

19. Campbell B, Roberts M, Kerksick C, et al. Pharmacokinetics, safety, and effects on exercise performance of l-arginine alpha-ketoglutarate in trained adult men. Nutrition 2006;22:872–881.

20. Hill CA, Harris RC, Kim HJ, et al. Influence of beta-alanine supplementation on skeletal muscle carnosine concentrations and high intensity cycling capacity. Amino Acids 2007;32:225–233.

21. Hoffman J, Ratamess N, Kang J, et al. Effect of creatine and beta-alanine supplementation on performance and endocrine responses in strength/power athletes. Int J Sport Nutr Exerc Metab 2006;16:430–446.

22. Harris RC, Tallon MJ, Dunnett M, et al. The absorption of orally supplied beta-alanine and its effect on muscle carnosine synthesis in human vastus lateralis. Amino Acids 2006;30:279–289.

23. Alford C, Cox H, Wescott R. The effects of red bull energy drink on human performance and mood. Amino Acids 2001;21:139–150.

24. Horne JA, Reyner LA. Beneficial effects of an "energy drink" given to sleepy drivers. Amino Acids 2001;20:83–89.

25. Willoughby DS, Stout JR, Wilborn CD. Effects of resistance training and protein plus amino acid supplementation on muscle anabolism, mass, and strength. Amino Acids 2007;32:467–477.

Index

AA pool, 326
Absorbitol, 241
Achievement goal theory
    task-oriented *vs.* ego-oriented
        individuals, 47
Acute exercise bout studies—strength/
        resistance
    4 + hours throughout rest of day—
        strength/resistance, 472
    supplementation before, during, and
        immediately after resistance
        exercise, 466–467
        CHO provides positive protein
            balance, 466–467
    supplementation 1–4 hours after
        strength/resistance training,
        470–472
        CHO-PRO/AA, 470–472
    supplementation immediately after
        resistance exercise, 467–470
        CHO-AA and CHO-PRO, 469–470
Adonis complex, 49
    incidence of muscular dissatisfaction, 50
    muscle building supplements, 49–50
Adulterated products
    penalties for conviction of introducing, 5
Adverse event
    defined, 19
    serious, 19
Adverse Event Reporting (AER) legislation, 19
β-alanine, 330, 499
Amino acid, 354, 385, 413
Anabolic androgenic steroids (AASs)
    abuse, 50–51
    positive psychological outcomes
        associated with AAS
        administration, 51
Anabolic hormones, 205
Anabolic Steroid Control Act
    of 1990, 15–16
        unlawful steroid distribution or
            possession subject to punishment, 16
    of 2004, 16–18
        prohormone supplements, effects of,
            16–17
    performance-enhancing substances in
        sports, use of, 14–15
Anabolic steroids
    analysis of new compounds, 17
    defined, 17
    educational programs for children
        regarding dangers of, 17
Anabolic window, 197
Androstenedione
    ban on, 10
    exemption criteria, 11–12
Anticatabolic supplements, 211–214
Antioxidants, 342–344, 357, 433
    functions, 343
    supplementation of vitamin C reduced
        muscle damage, 344
    supplementation of vitamin E
        decreases creatine kinase
        leakage, 343
Arginine
    roles, 331
Arginine/α-ketogluturate, 331–333
    growth hormone and influence on body
        composition and muscle function,
        effects on, 333
    pharmacokinetics, safety, and efficacy
        of, 332
Aromatase, 334
Aromatase
    inhibitors, 333–336
        body composition and changes in
            serum hormone levels, effect
            on, 335
        increase testosterone
            levels, 335
Aromatization, 210
Ascorbic acid, *see* Water-soluble vitamins,
        vitamin C
Attitudes, 45

Balance theory, 46
Banned drugs
    androstenedione, 10–12
    ephedra supplements, 8–10
BCAAs, *see* Branched-chain amino acids
        (BCAAs)
Biotin, *see* Vitamin H; Water-soluble
        vitamins, vitamin B₇
Bitter orange
    mechanism of action, 230–236
    pharmokinetics and cardiovascular
        effects, examining, 232–233
    weight loss, effects on, 234–235
    Xenadrine-EFX on metabolism and
        substrate utilization, acute effects
        of, 235–236
Blue-green algae (case study)
    adverse effects, 21–22
    producing hepatotoxins (microcystins),
        20–21
Body image
    and eating disorders, 47–49
    problems related to, 48
    social physique anxiety, 48
    supplement use to lose weight, 48
Branched-chain amino acids (BCAAs),
        100, 281
Brindelberry, 241

Caffeine, 51–53, 283, 349–351,
        386–387
    on aerobic and anaerobic exercise, 350
    beneficial effect of, 51–52
    enhance performance, 387
    improvement in time-to-exhaustion and
        work output, 350–351
    Parkinson's symptoms, effect on, 53
    as psychological ergogenic, 51
    reversal of withdrawal effects,
        controlling, 52
    state of arousal, linked to, 52–53
Caffeine or caffeine and aspirin
        (ECA stack)
    efficacy for treatment of obesity, 228
    See also Herbal ephedrine
Calciferol, *see* Fat-soluble vitamins,
        vitamin D
Calcium, 247–249
    adiposity, role in, 247
    body weight, effect on, 248
    carbonate supplements, 149
    citrate supplements, 149
    mechanism of action, 247–248

Caloric expenditure, 107
Caloric intake, 79, 80, 106
Calories (or joules), 76
Carbohydrate, 96–100, 347–349
    classes of, 97
    definition, 97
    Glycemic Index of common
        foods, 98–99
    and protein intake, 439
    simple, 97
Carbohydrate–protein supplements
    casein-whey proteins and effects on
        muscle anabolism, 201
    increasing protein synthesis and
        degradation, 200
    liquid form, 201
Cardiovascular drift, 371
L-Carnitine, 434, 495
Carnosine, 329
Carnosine/β-alanine, 329–331
    beneficial effects of supplementing, 331
Carotenoids, 135
Casein protein
    combined with whey increase in lean
        muscle mass, 195
Casein-whey proteins
    promoting net muscle protein, 201
Chitosan
    body weight, effects on, 240–241
Chloride, 422
Cholesterol
    good, 102
    high density lipoproteins (HDLs), 102
    low-density lipoproteins (LDLs), 102
CHO-PRO supplementation, 457–459
    effectiveness of, 457–458
    ingestion of supplemental beverage every
        15 minutes enhance performance,
        458–459
CHO supplementation, 456–457
    performance-enhancing effect
        of, 456
    rate of CHO oxidation, 457
    timing of, 457
Chromium, 239
Chromium picolinate (CrP)
    body weight and body composition,
        effect on, 240
    body weight reduction, effect
        on, 239–240
Citrus aurantium, 230–236
    See also Bitter Orange
Colostrum, 394

Conduction, 371
Conjugated linoleic acid (CLA),
        244–247, 428
    body weight and body composition,
        effect of, 245–246
    positive effects on body composition, 245
Conjugated linoleic acids (CLAs)
    health and performance-enhancing
        benefits, 346
    immune system, plasma lipids and
        glucose, effect on, 346
Convection, 371
CortiSlim™, 232
Creatine, 201–205, 323–326, 429
    cognitive benefits in humans, 55
    CrM and CrP supplementation, effect
        of, 325
    dosage, 202
    increases lean body mass, 202, 203
    loading on anaerobic working capacity
        (AWC), 325
    longer-term supplementation,
        202–203
    neuroprotective effects, 54
    short-term supplementation, 202
    side effects, 202
    supplementation with CrM on muscular
        performance, 325
    supplements
        as bodybuilding and strength-
            boosting, 390–391
        creatine content in select
            foods, 391
        role in cellular energy production, 391
        supplementation on endurance
            performance, 392
    in treatment of TBI, Huntington's
        disease and cognitive functioning,
        54, 55
Creatine monohydrate (CrM)
    benefits of, 284
Creatine-phosphocreatine (Cr-PCr) system
        in muscle fiber
    functions of, 261
Cyamposis tetragonolobus (Indian cluster
        bean), 250
Cyanocobalamin, see Water-soluble
        vitamins, vitamin B_{12}

Dehydration, 172, 371
    affecting temperature regulation and
        physiological functioning, 175
    impair exercise performance, 176

recommendations for hydration
        guidelines, 465
Dehydroepiandrosterone (DHEA), 8, 11
Dehydrogenation, 103
Descriptive norm, 45
Dietary fiber, 251
Dietary guidelines, 86
Dietary reference intakes (DRIs), 122
    adequate intake (AI), 122
    estimated average requirement
        (EAR), 122
    recommended dietary allowance
        (RDA), 122
    tolerable upper intake level (UL), 122
Dietary supplements, 3–4, 251–252
    supplements containing banned
        substances/drugs, 36
Dietary Supplement and Nonprescription
        Drug Consumer Protection
        Act, 19
Dietary Supplement Health and Education
        Act (DSHEA), 3, 34
Dihydroxyacetone (DHA), 236

ECA stack, 228
Electrolyte balance, 422
Electrolytes, 422
Elephant yam, 250
Elsinore pill, 227
Endurance athlete, 370
    nutritional supplements for, 370
    proper training and nutrition, 369
    supplementation improving
        performance, 370
Endurance enhancement
    carbohydrate and protein needs, 82–83
    fats, role of, 83
Energy expenditure, 105–106
Ephedra, 227–230
    adverse effects with preexisting
        cardiovascular or cerebrovascular
        conditions, 229–230
    efficacy and safety of, 228
    risks and benefits of, 227
    See also Ma huang
Ephedra supplements
    ban on, 8–10
Ephedrine, 283
    role in enhancing endurance exercise
        performance, 388
Ephedrine plus caffeine (EC)
    risks and benefits associated with, 229
    safety and efficacy of, 228

Ergogenic aids, 50, 260
Ergogenic aids, 374
Essential amino acids, 100, 101, 193, 194, 278
Essential and nonessential micronutrients
        and sport
    minerals
        macrominerals, 149–153
        microminerals/trace elements,
            154–155
        vegetarian athlete, 156–159
    practical application, 160
    vitamins
        fat-soluble vitamins, 135–140
        water-soluble vitamins, 127–135
Essential nutrients, 345
Ethics of sport, 399
Evaporation, 371
Evolution of sports nutrition, effect of
        government regulation on
    Adverse Events Regulation &
        Legislation, 18–20
    advertising, 23–24
        research substantiating advertising
            claims, 24
    Anabolic Steroid Control Act
        of 1990, 15–16
        of 2004, 16–18
    contamination or adulteration, 20–22
        blue-green algae, examining, 20–21
    cost of nontruth, 26–27
        substantiation need for claims, 27
        unsubstantiated marketing claims, 26
    dietary ingredients, 5–7
        application process, 7–8
    Dietary Supplement Health and
        Education Act of 1994, 3–4
        FDA in regulation of dietary
            supplements, 3–4
    doping—contaminated supplements and
        banned ingredients, 12–14
    FDA regulatory action
        androstenedione, 10–12
        ephedra supplements, 8–10
    government protections from dietary
        supplement hazards and
        risks, 4–5
        FDA authorized to protect consumers
            from dietary supplements, 4–5
    legal precedence, 24–26
        factors evaluating substantiation, 26
        FDA guidance regarding claims and
            compliance guidelines for dietary
            supplements, 24
        substantiation from perspective of
            research, 22–23
        need for lemon law, 23
Exercise, immediately after, 459–464
    CHO-PRO/CHO-AA supplementation
        maximizes replenishment
        of muscle glycogen
        stores, 460–464
Exercise-associated hyponatremia
        (EAH), 419
Exercise performance, 283
Exercise recovery, 409–410
    classification of supplements, 411,
        412–413
        immune function, 434–436
        minimizing oxidative damage,
            431–434
        muscle fuel energy, 423–431
        rehydration, 411–422
        repair of muscle damage, 436–440
    phases of recovery, 410–411
        fast (partial), 410
        slow (complete), 410
Extracellular fluid, (ECF), 168

Fast (partial), exercise recovery, 410
    from catabolic to anabolism, cellular
        transitions, 410–411
Fat, 101–103
    functions, 102
    and lipids, 101–102
    triglycerides, 102
Fat-soluble vitamins, 136–137
    adverse effects with
        oversupplementation, 135
    vitamin A, 135–138
        adverse effects associated with
            overdose of, 138
        funtions, 135
    vitamin D (calciferol), 138–139
        adverse effects, 138
        forms, 138
        source of, 138
    vitamin E, 139–140
        deficiency, effects of, 139
        functions, 139
        oxidative stress, effects
            on, 139–140
    vitamin K, 140
        deficiency, effects of, 140
        forms, 140
        functions, 140
    See also Phylloquinones

See also Vitamin supplementation to
    improved athletic performance,
    benefits of
Fatty acids
    essential and trans, 102–103
FDA, *see* Food and Drug Administration
Federal Trade Commission (FTC), 4, 25
Fiber
    soluble and insoluble, 97
Fluid regulation for life and human
    performance
    carbohydrate drinks and exercise
        performance, 180–181
        benefits of carbohydrate intake, 181
        carbohydrate concentration, 181
    electrolytes replacement, 177–179
        maintaining fluid balance and
            exercise, 178
        rehydration and recovery of fluid
            balance, 178
        restoring fluid balance, 178
        water replenishment, 177–178
    exercise and hydration requirements,
        172–176
        dehydration and exercise
            performance, effects of, 172–174
        fluid intake on core temperature and
            running time, effect of, 173
        weight gain in wrestlers, 174
        weight loss, effect of, 175
    function of water in human body, 168
        extracellular fluid, (ECF), 168
        intracellular fluid, (ICF), 168
    glycerol, 179–180
        facilitates hyperhydrating, 180
        facilitating intestinal water
            absorption and extracellular fluid
            retention, 180
    hyponatremia
        preventing development of, 179
        symptoms, 179
    nature of water in human body, 167
    practical applications, 182–183
        recommendations concerning fluid
            intake before, during, and after
            exercise, 182–183
    water balance, 169–172
        dietary reference intake values for
            total water, 169
        factors affecting, 169
        fluid loss and replenishment, 170
        water intake sources, 169–170
        water loss, reasons for, 170–171

water balance and exercise
    after exercise, 177
    before exercise, 176–177
    during exercise, 177
    replacement of water, 176
    risks associated with dehydration,
        preventing, 176
Fluid replacement, 178, 180
Food, Drug, and Cosmetic Act (FDCA), 3
Food and Drug Administration (FDA), 3
    ban on androstenedione, 10–12
    defined, 4
    prohibiting ephedra supplements, 8–10
    risk-to-benefit analysis on ephedra
        supplements, 9–10
Free radical, 431
FTC, *see* Federal Trade Commission (FTC)
Future trends
    driving sales and growth, 496–498
        product considerations, 497–498
    future market predictions, 498
    industry defined, 492
    regulations in sport, 503–506
        creating category, 504–506
    research directions, 499–502
        legal matters, 501–502
    sales growth, 492–494
        United States sport nutrition
            and weight loss category
            market, 493
    trends, 494–496
        breakthrough in carnitine research,
            495–496

Garcinia cambogia, 241–244
    comparing effects of three HCA-
        containing products, 244
    HCA on body weight, effect
        of, 242–243
    HCA on fat oxidation and metabolic rate,
        effect of, 243–244
    HCA in combination with CrP and
        herbal caffeine, 244
    See also Brindelberry
"Genetic sugar," 430
Ginkgo biloba
    as therapeutic aid, 56
Ginko biloba extract (EGb)
    beneficial effects on Alzheimer's disease
        and dementia, 58
    for medical and research purposes, 57
    positive effects of, 56
GI value, 98

Glucomannan *(Amorphophallus konjac)*
   effect of, 250
   See also elephant yam
Glutamine, 339–340, 434–436, 439–440
   beneficial effects of supplementation, 396
   cellular hydration, 440
   functions, 339, 395
   role in enhancement of immune
      system, 339
   supplementation and training, effect of,
      339–340
Glutathione reductase, 433
Glycemic index (GI), 376
   BCAA supplements on exercise
      performance, 386
   caffeine and ephedrine, effects of, 389
   caffeine's ergogenic effect, 386–387
   carbohydrate (CHO) concentration,
      effect of, 378
   carbohydrate-protein
      combination in stimulating protein
         synthesis, 382
      on cycling endurance and muscle
         damage, effects of, 379
      on reducing muscle protein
         degradation and speeding
         postexercise recovery, effects
         of, 379
   creatine content in select foods, 391
   ideal nutrient composition for sports
      drink during exercise, 380
   of carbohydrate, effect on glucose and
      insulin responses, 377
   protein and carbohydrate alone and in
      combination on protein synthesis,
      effect of, 383
   typical amino acid composition of whey,
      casein, and soy isolates, 385
   typical caffeine content in common
      beverages, pills, and other
      products, 388
Glycerin, 389
Glycerol, 389–390, 415–416
   in postexercise rehydration
      supplements, 416
   See also Glycerin
Glycogen, 97
   resynthesis, 274–275, 303
Glycogen storage immediately after exercise
   postexercise synthesis of glycogen, 424
Glycogen supercompensation
   differences between simple
      carbohydrates, 426

Growth hormone, 206–208
   arginine aspartate, effects of, 207–208
   arginine stimulates, 207
   improving body composition, 206
Growth of sports nutrition industry, 496–498
Growth of supplements, 34
Guarana, 228, 229, 232, 234, 235, 252
Guar gum
   adverse effects, 250
   effectiveness for reducing body weight,
      determining, 250

Herbal caffeine, 227–230
Herbal ephedrine, 230
HMβ, 267, 288
Huntington's disease, 55
   transgenic mouse model of, 55
Hydration, 178, 180
Hydration level
   methods to determine, 177
β-Hydroxy-β-methylbutyric acid (HMB),
      212–213, 340–342
   increasing lean body mass, effect on, 213
   inhibits protein degradation, 341
   role in regulation of protein
      breakdown, 340
   strength and body composition, effects
      on, 341
   supplementation on muscle damage, 341
Hypertrophy, 190
Hyponatremia
   factors leading to, 420
Hyponatremia, 179

IGF-1, 206
   See also Mechano growth factor
Immune function, exercise recovery,
      434–436
   glutamine, 434–436
Injunctive norm, 45
Insensible perspiration, 170
Insulin
   net protein balance, role in
      improving, 199
Intracellular fluid, (ICF), 168
Isoflavones (antioxidants)
   decreasing risk of developing
      cardiovascular disease and cancer,
      195–196

7-Keto® (3-acetyl-7-oxo
      dehydroepiandrosterone), 8
A-Ketoisocaproic acid (KIC), 212

Lactate, 372
Lactic acid, 372
Lean body mass, increasing, 192
Lean System 7 (weight loss product), 235
"Lemon laws," 23
Lipid peroxidation, 431
Lipids
  conjugated linoleic acid, 428–429
Low carbohydrate, 494

Macrominerals
  body weight and sweat losses, effects
    on, 149–150
  calcium, 149–150
    compounds, 149
  magnesium, 151–152
    deficiency, effects of, 151
    functions, 151
    physical performance, effect on,
      151–152
  phosphorus, 150–151
    deficiency, effects of, 150
    oversupplementation, 150–151
  physical performance, role in, 143–148
  potassium, 152–153
    exercise performance, effects on,
      152–153
    functions, 152
  sodium and chloride, 153
    fluid replacement guidelines for
      exercise, 154
    recommended intake, 153
  sulfur, 152
    functions, 152
    sources of, 152
Macronutrient intake
  determining, 106–109
    balanced diet, 108
    based on body weight and activity
      recommendations, 109
    daily recommendations, 108
Macronutrient intake for physical activity
  determining intake needs
    caloric needs, 105–106
    case study, 110
    macronutrient intake, 106–109
    metabolic expenditure of given
      activities, 107
  macronutrients
    carbohydrates, 96–100
    fat, 101–103
    metabolic usage, 103–104
    protein, 100–101

maintaining optimal health during
    training and competition, 112–113
  factors contributing to incidence of
    infection in athletes, 112
  macronutrients, role of, 112
nutrient timing, 110–112
nutritional considerations for vegetarian
    athlete
  carbohydrate, 115
  creatine supplementation, 116
  energy, 114
  fat, 115–116
  protein, 114–115
Macronutrient profiles, 82–88
Magnesium
  role in cellular reactions, 337
Ma huang, 227
Maintenance diet for male athlete training
    ≥5 days per week
  final meal plan for maintenance diet, 90
  food group categories: one day's
    menu, 91–92
Mechano growth factor, 206
Medium-chain triglycerides (MCTs)
  role of, 428
Metabolife-356®, 229
Metabolism, 170, 297
Microminerals/trace elements
  iron, 154–155
    deficiency, effects of, 155
    functions, 154–155
    performance impairment, 155
Micronutrients, role of, 121
Minerals
  functions, 142
Monosaccharides, 97
Motivational climate, 43
Motivational theories
  resulting in reward contingencies, 41
Muscle damage, repair of, 436–440
Muscle dysmorphia, 49, 50
  effects of, 50
  See also Reverse anorexia
Muscle fuel energy, exercise recovery, 423
  carbohydrate, 423–424
    glycogen restoration, 423
    glycogen storage immediately after
      exercise, 424
  creatine, 429–430
  D-ribose, 430–431
  glycogen supercompensation, 425–426
    differences between simple
      carbohydrates, 426

Muscle fuel energy (*cont.*)
   insulin regulates synthesis of
      glycogen, 425
   lipids
      conjugated linoleic acid, 428–429
   protein, 427–428
      general guidelines for postexercise
         carbohydrate and protein
         intake, 427
      glycogen synthesis pathway
         mechanisms, 427
   sustained post-exercise glycogen
      storage, 425
   postexercise rate of glycogen
      storage, 425
Muscle glycogen, 423
Muscle hypertrophy, 436
Muscle mass and weight gain nutritional
      supplements
   anabolic hormone enhancers
      growth hormone, 206–208
      insulin-like growth factor-1, 206
      testosterone, 209–210
   anticatabolic supplements
      β-Hydroxy-β-methylbutyrate, 212–213
      glutamine, 214
      α-Ketoisocaproic acid, 212
   aromatase inhibitors, 210–211
      to increase testosterone levels, 211
   carbohydrate–protein combinations
      insulin, amino acids, and protein
         synthesis, 198–199
      resistance training in absence of
         nutritional intake, 197–198
      timing of ingestion and importance of,
         200–201
   creatine
      lean body mass, effects on, 202–203
      physiological mechanisms for
         increasing lean body mass,
         203–204
      satellite cell activity, 204–205
   importance of net protein balance,
      190–191
   nitric oxide boosters, 214–216
   protein supplements
      requirements, 192–193
      types of, 193–196
   skeletal muscle hypertrophy, role of genes
      in, 191–192
Muscle repair, 413
Muscle substrates
   creatine, 429–430

dietary sources of, 429
   D-Ribose, 430–431
Myogenic regulatory factors, 204
Myonuclear domain theory, 204, 205
Myosin, 190

National Collegiate Athletic Association
      (NCAA), 503
Neo-Synephrine, 231
Net protein balance, positive,
      190, 197
New dietary ingredient (NDI)
   application process
      with example, 8
      submitting for review of safety to
         FDA, 7–8
   criteria for adulteration, 6
   defined, 6
   reasonable expectation of safety, 6–7
Niacin, *see* Water-soluble vitamins,
      vitamin B_3
Nitric oxide, 214–216
   affecting metabolic control during
      exercise via multiple mechanisms,
      214–215
   physiological functions, 214
Nonvitamain, nonmineral supplements
      (NVNM), 35
Nutrient database software (for meal
      planning), 87–88
   USDA Nutrient Database for Standard
      Reference, 88
Nutrient dense diet, 75, 77–78
   foundational food group categories, 78
   variety among food groups and from
      within each group, 77–78
Nutrient-dense diet, designing
   establishing viable energy requirements
      combination of nutritional
         supplementation to a quality
         nutritional dietary profile for
         athletes, 81
      evaluating amount of caloric needs for
         athletes, 80–81
      optimal training/performance
         outcomes, 80–82
Nutrient density, 92
   defined, 76–77, 92
Nutrient intake, timing of
   purposes of, 454
Nutrients, 345
Nutrient timing, 84, 110–112
   importance of, 196

maximize and replenish glycogen
  stores, 111
Nutrition, 3
"Nutritional design", 96
Nutritional ergogenic aids, 374
Nutritional sports ergogenic, 383
Nutritional supplements, 373
  defined, 78
Nutritional supplementation, 80
  benefits for athletes, 79
  performance enhancement, role
    in, 79–80
Nutritional supplementation and meal
    timing
  immune function and oxidative stress
    between exercise sessions, 473
    during exercise, 473–475
    prospective resistance training studies
      and supplementation timing,
      475–478
    vitamin C, effect of, 474
  practical applications, 480–481
  prolonged exercise
    after exercise (replacement, recovery,
      regeneration, adaptation),
      459–464
    carbohydrate-protein (CHO-PRO)
      supplements, 456
    dehydration: fluid, 464–465
    during exercise, 456–459
    guidelines for timing of supplemental
      CHO ingestion, 455–456
  strength/resistance training
    acute exercise bout studies, 466–472
Nutritional supplement combinations,
    effective
  addition of protein to carbohydrate
    supplement for increased
    efficiency of muscle glycogen
    storage, 303–304
  chronic adaptations: supplement
    combinations promote muscle
    hypertrophy and strength,
    285–292
  combinations enhance aerobic/ anaerobic
    performance, 283–285
  combinations shown to enhance
    anaerobic/ aerobic exercise
    performance, 292–295
  creatine + protein or creatine +
    carbohydrate for better muscle
    hypertrophy, 299–302
  change in contractile protein, 302

change in lean body mass, 301
PCr-Cr system in exercise metabolism,
    multifaceted role of muscle,
    297–299
supplement combinations
  enhance muscle anabolism, 278–282
  enhance muscle glycogen, 267–277
  enhance phosphagen system, 260–267
supplement timing double gains in muscle
    mass, 304–306
  body composition changes after 10
    weeks of training, 305
Nutritional supplements complementing
    nutrient-dense diets, role of
designing nutrient-dense diet
  establishing viable energy
    requirements, 80–82
determining macronutrient profiles
  endurance enhancement, 82–83
  weight gain and muscle growth, 83
establishing adequate dietary
    foundations, 75–76
  details of nutrient density, 76
example diet plans
  maintenance diet for male athlete
    training ≥5 days per week, 89–92
nutrient-dense diet, 77–78
  foundational food group
    categories, 78
nutrient-dense diet, role of supplements
    in, 79–80
  creatine and protein supplement, 79
nutrient density defined, 76–77
  amount of energy available in
    athlete's diet, 76
  calorie intake, 76
  macronutrient and micronutrient
    contents of select fluids, 77
  nutritionally dense restoration timing
    consideration, 83–85
  estimation of daily energy needs of
    men and women based on activity
    intensity, 85
  guidelines for nutritional timing
    for athletic and exercise
    populations, 84
nutritional supplements defined, 78
nutritional aids category, 78
translating nutrients into food, 86–88
  Food Guide Pyramid (dietary
    guidelines), 86
  nutrient database software (for meal
    planning), 87–88

Nutritional supplements complementing
      nutrient-dense diets (*cont.*)
   nutrients and calories per serving of
      food from each food group and
      from teaspoons of added sugar, 87
Nutritional supplements for endurance athletes
   additional concerns, 398–399
   factors limiting endurance athletic
      performance
      lactate, 372–373
   immune system and endurance
      erformance, 392–398
      exercise affecting thyroid hormones
         and testosterone levels, 398
      importance of nutrient timing,
         393–394
      signs and symptoms, 392
   practical applications, 400–401
   supplements for endurance athletes
      glycemic index, 376–392
      nutritional ergogenic aids, 374
      partial list of glycemic index of
         foods using glucose as standard,
         375–376
      preexercise nutrition, 374
Nutritional supplements for strength power
      athletes
   practical applications and conclusion,
      351–360
   supplements for strength and power
      athletes, 353–359
   supplements enhance immune function
      conjugated linoleic acids, 346–347
      vitamins and minerals, 344–346
   supplements enhance strength, power, or
      hypertrophy directly
      arginine/α-ketogluturate, 331–333
      aromatase inhibitors, 333–336
      carnosine/β-alanine, 329–331
      creatine, 323–326
      protein and amino acids, 326–329
      zinc magnesium aspartate, 336–338
   supplements for strength and power
      athletes, 353–359
   supplements promote recovery
      antioxidants, 342–344
      β-hydroxy-β-methylbutyric acid,
         340–342
      glutamine, 339–340
   supplements provide energy and
      enhanced workouts
      caffeine, 349–351
      carbohydrate, 347–349

Nutritional supplements to enhance
      recovery
   classification of recovery supplements,
      411, 412–413
   exercise recovery, 409–410
   immune function
      glutamine, 434–436
   minimizing oxidative damage
      L-carnitine, 434
      oxidation and reduction reactions,
         432–433
      vitamins C and E, 433–434
   muscle fuel energy
      carbohydrate, 423–424
      glycogen supercompensation,
         425–426
      lipids, 428–429
      other muscle substrates,
         429–431
      protein, 427–428
      sustained post-exercise glycogen
         storage, 425
   phases of recovery, 410–411
      characteristics, 410–411
      fast recovery, 410
      slow recovery, 410
   practical applications, 441
   rehydration
      carbohydrate-electrolyte solutions, 415
      electrolyte balance, 422
      exercise-associated hyponatremia,
         419–422
      fluid balance in American football
         players, 416–419
      glycerol, 415–416
      water and sodium, 413–415
   repair of muscle damage
      anabolic hormones and putative and
         established secretagogues, 437
      carbohydrate and protein intake, 439
      glutamine, 439–440
      leucine and muscle protein synthesis,
         438–439
      protein and amino acid intake,
         437–438
"Nutritional timing system", 85
Nutrition Business Journal
      (NBJ), 491
NV^TM (weight loss beauty pill), 47

Obesity, 226
   incidences of overweight and obese in
      U.S., 226

treatment for
    ma huang and guarana, effects of, 229, 230
    mixtures of ephedrine and caffeine, 229
Oligosaccharides, 97
Operant conditioning, 41–44
    punishment, 44
    reinforcers of behavior, 44
Over-the-counter (OTC) drugs, 226
    Hydroxycut™, 228
Overtraining
    syndrome, 392
Overweight, 226–227
Oxidation, 431, 432
    See also Reduction
Oxidation-reduction/redox reactions, 432
Oxidative damage, minimizing, 431–434
    L-carnitine, 434
    oxidation and reduction reactions, 432–433
    Vitamins C and E, 433–434
Oxidative stress, 342, 431

Pantothenic acid, *see* Water-soluble vitamins, vitamin B_5
Pellagra, 132
Perceived behavioral control, 45
Persuasion and conformity
    balance theory, 46
    muscle-building supplements, 44, 46
    normative influences on behavior, 45
    Theory of Planned Behavior (TPB), 44
        applied to sport/exercise supplement, 45
Pharmacological sports ergogenics, 386
Phosphagen system
    CrM on performance and strength training, effects of, 261
    PCr system role in muscle energy metabolism, 260
Phosphocreatine (PC), 323
Phylloquinones, 140
Physical performance, 135
Physiological mechanisms
    for increasing lean body mass, 203–204
        creatine and factors involved in gene expression, effects of, 203
        creatine on myogenic regulatory factor gene expression, effects of, 204
Physiological sports ergogenics, 389
Plantago ovata or *Plantago psyllium,* 251

D-Pinotol, 267, 296
Polysaccharides, 97
Potassium, 422
Preexercise hydration, 176
Prohormones/prosteroids, 16
Protein, 100–101, 289
    and amino acids, 326–329
        beneficial effects of postexercise protein supplementation, 438
        enhances muscle protein synthesis, 328
        intake, 437–438
        postexercise net protein synthesis, effects on, 327
        protein synthesis, increases, 328, 352
        role in development of skeletal muscle, 327
    balance, 278, 279, 282, 289
    complete, 101
    essential and nonessential amino acids, 101
    incomplete, 101
    role in maintaining muscle mass and providing essential amino acids for cellular protein synthesis, 462
    simple, 100
Protein hydrolysate, 438
Protein supplements, 192–196
    classification of amino acids, 194
    complete, 194
    to increase lean body mass, 192
    types of
        casein protein, 195
        classification of amino acids, 194
        egg protein, 196
        soy protein, 195–196
        whey protein, 194–195
Proteolysis, 340
P/S ratio, 102
Psyllium, 251
Punishment, 44
Pyridoxine, *see* Water-soluble vitamins, vitamin B_6
Pyruvate (PYR), 236–239
    exercise, effects of, 238
    PYR during training on body composition, effect of, 238
    weight loss, effects on, 236–237
    See also Dihydroxyacetone (DHA)

Radiation, 371
Reactive oxygen species (ROS), 431

Recovery from exercise, 409
    See also Exercise recovery
Reducing agent/reductant, 432
Reduction, 432
Rehydration, exercise recovery, 411–413
    carbohydrate-electrolyte solutions, 415
    electrolyte balance, 422
    exercise-associated hyponatremia,
        419–422
    fluid balance in American football
        players, 416–419
    glycerol, 415–416
    recommendations for marathon
        runners, 421
    water and sodium, 413–415
Reinforcers, 44
Relax-Aid, 14
Replenishment of energy reserves, 412
Resistance training, 264–265, 282
Resistance training in absence of nutritional
        intake, 197–198
    improved net protein balance, 197
    rates of protein synthesis and breakdown,
        197–198
Reverse anorexia, 49
Riboflavin, *see* Water-soluble vitamins,
        vitamin B_2
D-Ribose, 430
    See also "Genetic sugar"

St. John's wort
    negative interactions with
        drugs, 59
    reduced depression, 59
    risk–benefit of, 58–60
Satellite cell activity, 204–205
    influence of creatine and protein
        supplementation on, 205
Saturation, 102
Scurvy, 135
Skeletal muscle hypertrophy, 190
    factors regulating
        gene transcription, 191, 206
        protein translation, 190, 206
        satellite cell activity, 204–205
    positive net protein balance for, 197
    protein supplements for, 192
    recommendations for protein intake, 193
    role of genes in, 191–192
        sports supplements increasing protein
            synthesis, 191
        synthesis and breakdown, 191
    See also Muscle hypertrophy

Skinny Water®, 242
Social physique anxiety, 48, 50
Sodium, 422
Sodium bicarbonate, 284
Sodium/water solution for rehydration, 414
Soy protein
    decrease or prevent exercise-induced
        damage to muscle, 196
Sport and exercise, psychology of
        supplementation in
    biobehavioral effects of selected
        supplements for performance,
        fitness, and health
        caffeine, 51–53
        creatine, 54–56
        Ginkgo biloba, 56–58
        St. John's wort, 58–60
    use rates and motivation
        definition of dietary supplement,
            35–37
        motivational theories applied to
            supplement use, 40–50
        prevalence of supplement use, 37–40
Sports
    drinks, 76, 412
    nutrition, 3, 196
Sports nutrition and weight loss, or SNWL
        industry, 492
Sports nutrition industry, 492
Sport-specific eating, 96
Starch, 97
Strength athletes
    supplement needs of, 322–323
"Substantiation doctrine," 25
Supplementation, 263, 273, 277, 282
Supplement combinations
    enhance aerobic/ anaerobic performance
        benefits of CrM, 284
        caffeine–ephedrine, 283–284
        CrM + sodium bicarbonate, 284–285
    enhance anaerobic/ aerobic exercise
        performance
        BCAA + arginine + glutamine, 293
        CrM + β-alanine, 293
    enhance muscle anabolism
        AA-CHO, 281–282
        PRO-CHO, 278–279
    enhance muscle Cr accumulation
        ALA + CrM + CHO, 266–267
        CrM + CHO *vs.* CrM, 263
        CrM + D-pinitol, 266
        CrM + PRO, 263–264
        CrM + PRO + CHO, 263–265

high-CHO diet with CrM *vs.* high-
        CHO diet without CrM, 263
enhance muscle glycogen
    CHO/AA, 276
    high-GI CHOs *vs.* low-GI
        CHOs, 273
    PRO-CHO, 274–275
promote muscle hypertrophy and
    strength
    CrM and β–alanine, 289
    CrM-CHO *vs.* CrM-PRO, 286
    CrM + PRO + CHO, 285–286, 292
    CrM to PRO-CHO, 287
    CrM with magnesium or HMβ, 288
    supplement timing with CrM-PRO-
        CHO, effects of, 291
    whey + BCAA + glutamine, 290
Supplements
    biobehavioral effects of, 34
Supplement use
    motivational theories applied to
        achievement goal theory, 47
        adonis complex, 49–50
        body image and eating disorders,
            47–49
        operant conditioning, 41–44
        persuasion and conformity, 44–47
    practical applications, 62–63
    prevalence of
        commonly used vitamins/minerals in
            United States, 38–39
        reasons for using herbal/
            supplements, 37
        sport-specific use, 40
        usuage rates, 37
    problems associated with, 36–37
    in sports, 36

Testosterone
    anabolic effect, 334
    free/bioavailable, 209
    Tribulus Terrestris, 210
    ZMA, 209–210
        testosterone levels, effects
            on, 209–210
Tetrahydrogestrinone (THG), 17
Theory of Planned Behavior (TPB), 44
    applied to sport/exercise supplement
        use, 45
Thiamine, *see* Water-soluble vitamins,
    vitamin B₁
Timing nutrient intake
    purposes of, 454

Total energy expenditure (TEE), 105
    exercise energy expenditure and activities
        of daily living, 106
    resting metabolic rate (RMR),
        determining, 105
    thermogenesis, 106
Tryptophan, 384

Ultimate Thermogenic Fuel™, 232
Unreasonable risk of illness or injury
    supplements containing ephedra/
        ephedrine alkaloids, 5
Upper respiratory tract infections (URTIs),
    112–113
USDA Food Guide Pyramid, 86, 89

Vegetarian athlete
    boron, 158–159
        bone mineral density, effects on, 159
        functions, 158
    chromium, 157–158
        chromium supplementation and
            exercise, effects of, 158
        forms, 157
        side effects associated with
            deficiency, 158
    food sources of specific micronutrients, 157
    lacto-, 156
    lacto/ovo-, 156
    other minerals, 159
    ovo-, 156
    vegans, 155–156
    zinc, 156–157
        deficiency, effects of, 156
        supplementation, 156–157
Vitamin A, 135–138
Vitamin B$_1$ (thiamine), 127–131
Vitamin B$_2$ (riboflavin), 131
Vitamin B$_3$ (niacin), 132
Vitamin B$_5$ (pantothenic acid), 132
Vitamin B$_6$ (pyridoxine and related
    compounds), 133
Vitamin B$_7$ (biotin), 133
Vitamin B$_{12}$ (cyanocobalamin), 134
Vitamin C (ascorbic acid), 134–135
Vitamin D (calciferol), 138–139
Vitamin E, 139–140
Vitamin H, 133
Vitamin K, 140
Vitamins, 122–127
    Dietary Reference Intakes (DRIs)
        recommended intakes of elements,
            125–126

Vitamins (*cont.*)
  recommended intakes of vitamins, 123–124
  fat-soluble, 135–140
  water-soluble, 127–135
Vitamin supplementation to improved athletic performance, benefits of, 128–130, 136–137

Water
  factors affecting range of water content, 167
  as sport rehydration beverage, 413–414
Water intoxication, 179, 420
  See also Hyponatremia
Water-soluble vitamins, 128–130
  as antioxidant, 127
  folic acid, 133–134
    deficiency, 133
    exercise performance with folate supplementation, effects on, 134
    functions, 133
  vitamin B_7 (biotin), 133
    functions, 133
  vitamin B_{12} (cyanocobalamin), 134
    deficiency, effects of, 134
    functions, 134
  vitamin B_3 (niacin), 132
    deficiency, effects of, 132
    functions, 132
  vitamin B_5 (pantothenic acid), 132
    functions, 132
  vitamin B_6 (pyridoxine and related compounds), 133
    exercise performance with pyridoxine supplementation, effects on, 133
  vitamin B_2 (riboflavin), 131
    supplementation, effect on, 131
    symptoms of deficiency, 131
  vitamin B_1 (thiamine), 127–131
    deficiency, 127
    exercise performance with thiamine supplementation, effects on, 131

  forms, 127
  functions, 127
  vitamin C (ascorbic acid), 134–135
    exercise performance, effects on, 135
    functions, 134–135
  See also Vitamin supplementation to improved athletic performance, benefits of
Weight gain and muscle growth nutritional requirements, 83
Weight loss nutritional supplements
  nutritional supplements increase energy expenditure
    bitter orange *(Citrus aurantium)*, 230–236
    ephedra alkaloids and herbal caffeine, 227–230
    pyruvate, 236–239
  other dietary supplements, 251–252
  supplements increase satiety
    glucommannan, guar gum, and psyllium, 250–251
  supplements modify carbohydrate and fat metabolism
    calcium, 247–249
    chitosan, 240–241
    chromium picolinate, 239–240
    conjugated linoleic acid, 244–247
    Garcinia cambogia (hydroxycitric acid), 241–244
Whey protein
  faster acting, 194
  optimize protein synthesis, 194–195

Zinc, 336
Zinc magnesium aspartate
  benefits to resistance training athletes, 337
  deficiency, effects of, 336–337
  ZMA on anabolic hormones and muscle function, effects of, 338

Printed in the United States of America